CORD ALGEBRA 1

Mathematics in Context

Developed by the
Center for Occupational Research and Development
Waco, Texas

JOIN US ON THE INTERNET
WWW: http://www.thomson.com
EMAIL: findit@kiosk.thomson.com   A service of I(T)P®

South-Western Educational Publishing
an International Thomson Publishing company I(T)P®

Cincinnati • Albany, NY • Belmont, CA • Bonn • Boston • Detroit • Johannesburg • London • Madrid
Melbourne • Mexico City • New York • Paris • Singapore • Tokyo • Toronto • Washington

CORD Staff
Director: John Souders, Jr., Ph.D.
Mathematics Specialist: Michael Crawford, Ph.D.
Educational Specialist: Lewis Westbrook
Chief Scientist: Leno Pedrotti, Ph.D.
Applications Specialist: John Chamberlain

South-Western Educational Publishing Staff
Publisher/Team Leader: Thomas Emrick
Project Manager: Suzanne Knapic
Marketing Manager: Colleen Skola
Consulting/Developmental Editor: Richard Monnard
Manufacturing Coordinator: Jennifer Carles

Consultants

Wes Evans
Mathematics Teacher
Midway High School
Waco, Texas

Donna Long
Mathematics Coordinator
MSD Wayne Township
Indianapolis, Indiana

Editorial Services: Monotype Editorial Services
Production Services: Monotype Composition Company, Inc.
Design: Zender + Associates, Inc.

Copyright © 1998
by the Center for Occupational Research and Development
Waco, Texas

Published and distributed
by South-Western Educational Publishing
5101 Madison Road
Cincinnati, OH 45227
1-800-354-9706 Fax: 1-800-487-8488

ISBN: 0-538-67121-1

1 2 3 4 5 6 7 8 WS 01 00 99 98 97

Printed in the United States of America

I(T)P®

International Thomson Publishing Company
South-Western Educational Publishing is an ITP Company. The ITP trademark is used under license.

CHAPTER 1

Integers and Vectors

CHAPTER 2

Scientific Notation

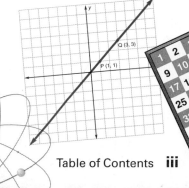

CHAPTER 3

Using Formulas

CHAPTER 4

Solving Linear Equations

CHAPTER 5

Graphing Linear Functions

CHAPTER 6

Nonlinear Functions

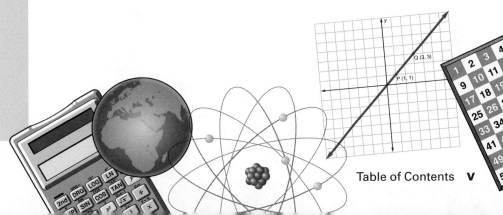

CHAPTER 7

Statistics and Probability

CHAPTER 8

Systems of Equations

CHAPTER 9

Inequalities

CHAPTER 10

Polynomials and Factors

CHAPTER 11

Quadratic Functions

CHAPTER 12

Right Triangle Relationships

Also

ALGEBRA AT WORK

"Why do I have to take an algebra course?" This is a great question and deserves an answer. Algebra is one of the cornerstones of today's technology boom. Without algebra there would not exist a single TV, radio, telephone, microwave oven, or other gadget that makes modern life so comfortable and interesting. Once you learn to use algebra, you will have the opportunity to help design and create the next generation of technology. You will secure yourself a place in tomorrow's workforce. But since all of this may seem mind boggling and too far in the future, here are just three of the ways algebra is used at work today.

USING ALGEBRA

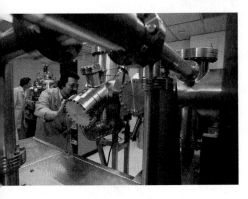

Chen works for an aerospace company designing solar panels for satellites. His company is upgrading its environmental monitoring satellites by adding several new cameras to each. This upgrade is expensive. To reduce costs, the company wants to continue using the same type of solar panel that is already on the satellites. Chen is asked to find if the solar panels can handle the additional power requirements of the upgraded satellite. The power output of a solar panel depends on its area. The solar panels are rectangular in shape and are 3 feet wide by 5 feet high. To find the area of a solar panel, Chen does the following calculation:

$$\text{Solar panel area} = 3 \text{ ft} \times 5 \text{ ft} = 15 \text{ square feet}$$

Chen uses this area to find the total power output of the solar panels. After completing his calculations, Chen reports that the solar panels will power the upgraded satellite.

John is in charge of environmental safety at a large oil refinery. At 10 AM one morning, he received notice that one of the storage tanks on an oil tanker had begun leaking. John immediately deployed a cleanup crew. The cleanup crew reported back at 10:20 AM. The oil spill had already spread 200 feet to the north of the tanker. John knew he had a responsibility to report both the direction of the spill and its speed to the proper authorities. To find the speed of the oil spill, John does the following calculation:

$$\text{Oil spill speed} = \frac{200 \text{ ft}}{20 \text{ min}} = 10 \ \frac{\text{ft}}{\text{min}}$$

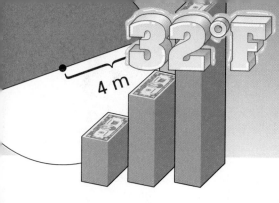

ALGEBRA AT WORK

John reported that the spill was advancing at a rate of $10 \frac{\text{ft}}{\text{min}}$ in a northerly direction.

Tracy works for a data processing company. She is responsible for maintaining the company's computer system. The operating manual states that the room containing the computer system must be kept at a temperature of less than 20°C. The thermostat in the computer room reads 74°F. To determine if she needs to take any action, Tracy calculates the following:

$$\text{Temperature °C} = (74 - 32) \times \frac{5}{9} = 23.3°C$$

Tracy takes immediate action to lower the computer room temperature.

Because Chen, John, and Tracy understand how to use algebra, they are able to solve problems that affect their jobs. Each computation they do is based on a mathematical expression called a **formula**. Chen, John, and Tracy's job performance is directly related to how skillfully they use these formulas to solve problems.

To succeed in the workplace, you must be able to use problem solving skills. *CORD Algebra 1* is designed to provide you with these skills. Each chapter contains a number of exercises that relate directly to workplace problems. As you learn algebra, you will use it to solve problems involving agriculture, business and marketing, family and consumer science, health, and industrial technology. You will have a chance to sharpen your problem solving skills further by completing laboratory activities. These activities will test your knowledge of algebra and introduce you to important workplace skills, such as measuring, data collection, and data analysis. *CORD Algebra 1* will help you learn mathematics by using both your mind and your hands.

What is Algebra?

Algebra is a branch of mathematics that enables you to work with symbols instead of numbers. Take another look at the three

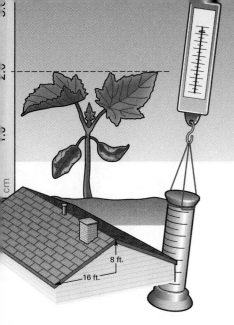

ALGEBRA AT WORK

situations involving Chen, John, and Tracy. They all used numbers. How did Chen, John, and Tracy know how to associate the numbers in the problems—such as length, height, time, speed, and temperature—with each other? In Chen's case, he used the formula for the area of a rectangle. John used a rate formula, and Tracy used a temperature conversion formula. Algebra provides tools for relating numbers, symbols, and formulas. Here is how it works.

Before Chen could begin multiplying numbers, he had to know how to calculate the area of a rectangle. The formula for the area of a rectangle is

$$A = lw$$

In this formula, the symbol A represents the area of a rectangle with length l and width (or height) w. The symbols A, l, and w are called **variables**. A, l, and w are variable because they can represent many different values of area, length, or width. A mathematical expression that contains a variable is called an **algebraic expression**. Examine Chen's calculations:

$$\text{Solar panel area} = 3 \text{ ft} \times 5 \text{ ft} = 15 \text{ square feet}$$

$$A = l \times w$$

As you can see, Chen's calculation is based on the area formula. He simply replaced the variables l and w with numerical values that represent the solar panel's dimensions.

John used a rate formula to find the speed of the oil spill.

$$v = \frac{d}{t}$$

In this formula, t represents the time it takes for the oil to spread a certain distance, d represents the distance the oil moves, and v is the speed of the oil spill.

Complete the activities that follow. Then you will see how solving problems in the real world depends on understanding how algebra works.

ALGEBRA AT WORK

ACTIVITY 1   The Rate Formula

In John's calculation, identify the numbers that stand for the variables t, d, and v.

$$\text{Oil spill speed} = \frac{200\ \text{ft}}{20\ \text{min}} = \frac{10\ \text{ft}}{\text{min}}$$

Complete the following table using the rate formula.

v	d	t
$200\ \dfrac{\text{miles}}{\text{hours}}$	400 miles	2 hours
?	186,000 miles	1 second
?	12 meters	2 seconds
?	650 feet	50 seconds

Which speed in the table is closest to the speed of light?

Tracy used a temperature conversion formula that converts Fahrenheit temperature to Celsius temperature.

$$T_C = (T_F - 32) \times \frac{5}{9}$$

In this formula, the variable T_C represents a Celsius temperature and T_F represents a Fahrenheit temperature.

ACTIVITY 2   The Celsius Formula

Use Tracy's temperature conversion formula to convert 212°F to a Celsius temperature.

As you have just seen, algebra and its use of symbols gives us the means to write formulas. In turn, these formulas give us directions on how to calculate important quantities such as area, speed, and temperature. Algebra also provides a way for rearranging formulas to solve problems.

ALGEBRA AT WORK

Chen's company discovered a problem with the new satellite. The weight and location of the added cameras make the satellite hard to control and difficult to place in the proper orbit. One solution to this problem is to reduce the length of the solar panels to 3 feet. Chen knows the area of the solar panels has to stay at 15 square feet; otherwise the satellite will not have enough power to operate. This means the width of the solar panel has to increase. To find this new width, Chen rearranges the formula.

$$w = \frac{A}{l}$$

This new formula can be called the *width formula*.

ACTIVITY 3   The Width Formula

Calculate the new width of the satellite solar panel.

Once John reports the oil spill to the harbor authorities, he is faced with the challenge of containing the spill. John has no containment equipment at his refinery. The closest storage facility is 2 hours away. John knows it will take at least 4 hours to receive the equipment after he makes initial contact with the storage facility. To order the correct amount of equipment, John needs to know the approximate distance the oil spill will spread in 4 hours. John uses algebra to help him rewrite the rate formula in the following form:

$$d = vt$$

Now, the formula meets the need for calculating the distance, d, that the oil spill will spread after time, t.

ACTIVITY 4   The Distance Formula

How far will the oil spread in 4 hours?

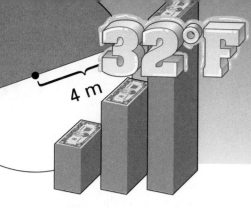

ALGEBRA AT WORK

Tracy has just received a new piece of equipment to install in her company's computer system. The instructions for operating this equipment state that it must be operated at a room temperature of less than 15°C. To what Fahrenheit temperature should Tracy set the room thermostat to meet the new equipment requirement? To answer this question, Tracy uses algebra to rewrite the temperature conversion formula.

$$T_F = \frac{9}{5} T_C + 32$$

ACTIVITY 5   The Fahrenheit Formula

Calculate the new temperature setting for the computer room thermostat. 59°F

Functions

You have seen that algebra can be used to write formulas. You have also found that algebra allows you to rewrite formulas. But the real power of algebra lies in how it is used to create formulas. Recall the formula for the area of a rectangle.

$$A = lw$$

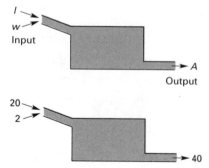

You can think of this formula as an input-output machine. If you input a value for l and w, the output is the area of a rectangle.

This formula also shows us that the output of the machine is dependent on the input. In the case of the area formula, the area of a rectangle depends on its length and width. Does this make sense? To answer this question, complete the following tables:

Table 1	l	w	A
	20 feet	2 feet	? 40 square feet
	20 feet	6 feet	? 120 square feet
	20 feet	10 feet	? 200 square feet

ALGEBRA AT WORK

Table 2	l	w	A
	20 feet	2 feet	? 40 square feet
	30 feet	2 feet	? 60 square feet
	40 feet	2 feet	? 80 square feet

In Table 1, the length l remains fixed, and the width w is varied. As w varies, so does the area of the rectangle, A. Thus, the area of a rectangle does depend on its width. Table 2 demonstrates that the area of a rectangle also depends on its length. Because of the dependence of A on both l and w, the variable A is called a **dependent variable**. The variables l and w are called **independent variables**. Dependent variables are the outputs of our input-output machine, and independent variables are the inputs.

ACTIVITY 6   The Rate Formula

Complete the following tables for the rate formula. Then identify the dependent and independent variables.

Table 3	d	t	v
	30 feet	2 seconds	? 15 $\frac{\text{feet}}{\text{second}}$
	30 feet	10 seconds	? 3 $\frac{\text{feet}}{\text{second}}$
	30 feet	15 seconds	? 2 $\frac{\text{feet}}{\text{second}}$

Table 4	d	t	v
	30 feet	2 seconds	? 15 $\frac{\text{feet}}{\text{second}}$
	60 feet	2 seconds	? 30 $\frac{\text{feet}}{\text{second}}$
	90 feet	2 seconds	? 45 $\frac{\text{feet}}{\text{second}}$

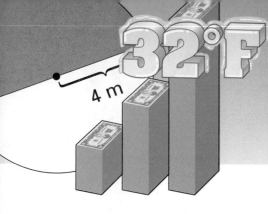

ALGEBRA AT WORK

The relationship between the area of a rectangle and its length and width is called a **function**. A function is a mathematical way of describing certain relationships between quantities. The relationship between speed, distance, and time in the rate formula also is a function. The relationship between Celsius and Fahrenheit temperatures in a temperature conversion formula is a function as well. Starting in Chapter 5 and continuing for the rest of your work in algebra, functions will become a central part of your study. Here are just two examples of how functions have changed our technology today.

Think about diving into a lake. What happens to your ears as you go deeper into the water? You feel more pressure. Thus, there is a relationship between the pressure water exerts on your ears and the height of water above you. This relationship is a function. Scientists discovered this relationship back in the 1700s. From this relationship, scientists have found formulas that help design today's fastest and most maneuverable jet airplanes.

Turn on a radio. Stand a certain distance from the radio and then slowly begin to move away from it. What happens to the loudness of the sound? It decreases. The loudness of the sound is related to the distance from the source of the sound. This relationship is a function that scientists can express as a formula. This same formula is used extensively in designing today's high resolution stereo speakers.

Algebra and Your Future

The future is full of exciting technological challenges—faster planes and more powerful stereo systems. If you want to be a part of this future, you must be able to use the concepts of algebra. As the previous examples have shown, algebra is the language of technology. Becoming fluent in this language will help secure your future success. So why delay the future? Now is the time to begin your study of algebra.

CORD ALGEBRA 1

Mathematics in Context

PART A 1

CHAPTER 1

WHY SHOULD I LEARN THIS?

Many of the tools of modern technology—computer chips, lasers, and high resolution TV—were invented by people skilled in the use of numbers and vectors. As technology becomes a bigger part of our everyday lives, the need for these skills is growing. Build the skills for your future now by exploring the concepts in this chapter.

INTEGERS AND VECTORS

OBJECTIVES

1. Identify integers.
2. Find the absolute value of an integer.
3. Add, subtract, multiply, and divide integers.
4. Find the magnitude and direction of a vector.
5. Solve problems involving integers and vectors.

Numbers such as 3 are used to model many everyday situations.

   1. A city is 3 feet below sea level. 2. A stock index drops 3 points.

   3. You walk north 3 miles.   4. A friend walks south 3 miles.

   5. A store takes in 3 dollars more than it pays out.

To show that you walked north 3 miles, you can use the positive number, $+3$. To show that your friend walked 3 miles south, you can use the negative number, -3. The number 3 shows how far each of you walked. The positive and negative signs show which direction.

Directed line segments or *vectors* represented by arrows also model many situations. In each of the vector drawings, a specific direction and magnitude is represented.

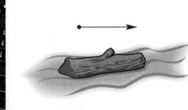

River Current 3 mph, east   Force on Rope 3 lb, 45°N of E

Notice each vector represents some actual physical quantity such as a river current or a pulling force.

This chapter will show you how to use positive and negative numbers and how to draw and use vectors to solve problems. As you view the video for this chapter, notice how people on the job use integers and vectors.

LESSON 1.1 IDENTIFYING INTEGERS

A temperature rise of 5 degrees can be represented by the symbol +5 and read, "positive five." A temperature drop of 3 degrees can be represented by the symbol −3 and read "negative three."

Positive and Negative Numbers

A number preceded by a positive sign (+) is a positive number. A number preceded by a minus sign (−) is a negative number. If a number is not preceded by either sign, it is considered positive.

Critical Thinking   How is placing a minus sign in front of a number different than placing it between two numbers?

A **number line** is used to picture positive and negative numbers. Follow the steps in the following activities to draw a number line.

ACTIVITY 1   Constructing a Number Line

1 In the center of a sheet of paper, draw a point and label it 0. This zero point is the center of your number line and is called the **origin.**

●
0

2 Start at the origin and draw a horizontal line to the right. Show that this part of the number line continues forever by drawing an arrow pointing to the right.

3 Mark a point on the line about a thumb's width to the right of the origin. Label this point 1. The distance from 0 to 1 on the number line is called the **unit** for the number line. Once you choose the unit length, it must remain the same for the entire line.

4 Now start at 1. Use a compass or piece of string to mark off one unit to the right of 1. Label this mark 2.

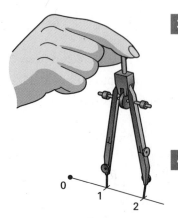

5 Now continue to mark units until you reach the arrowhead. Keep the distance between marks equal to 1 unit. Remember to label each mark.

The marks you labeled to the right of zero represent **positive integers.** One way to describe the set of positive integers is to use three dots.

$$1, 2, 3 \ldots$$

The three dots indicate that the numbers continue in the same way. Notice that there is no largest positive integer.

Now continue your number line to represent the negative integers.

ACTIVITY 2   Negative Numbers on the Number Line

1 Start at the origin and extend the line to the left. Draw an arrow pointing to the left to show that it goes forever to the left of 0.

2 Mark equal units to the left of 0. What point is one unit to the left of 0? Complete your number line. These numbers are called **negative integers.**

3 How many negative integers do you think there are? Is there a smallest negative integer?

Your number line should look similar to this one. The marks on the number line represent the set of all integers.

The Set of Integers
The set of integers includes all the positive integers, all the negative integers, and zero.

At what two times did the radar equipment come closest to automatic shutdown? If you were adjusting the air conditioning, when would you want it to operate most effectively?

From the Desk of:
Michael Chen

To: Facility Engineering

Bob,

As you know we have just installed some new radar receiving equipment in building A on the Mangano Test Range. This equipment is very temperature sensitive. It will automatically shutdown, if building A's temperature changes by more than 5°C during any one-hour period.

During the last shift, the environmental control system in building A recorded the temperatures shown on the attached printout. Please have an air conditioning technician check the HVAC system for the building. We must avoid an equipment shutdown since it takes a couple of hours for restarting. My current test schedule and budget will not tolerate this kind of delay.

Thanks,
Mike

	ENVIRONMENTAL CONTROL DATA FOR		
Time	Temperature, C	Temperature Change, C	Relati Hum
8am	22		
9am	26	4	
10am	25	−1	.
11am	26	1	
12	28	2	
1pm	28	0	
2pm	27	−1	
3pm	28	1	
4pm	25	−3	
5pm	24	−1	

LESSON ASSESSMENT

Think and Discuss

1 What is an integer?

2 Explain how a minus sign can be used in two ways.

3 Give two examples where negative integers model workplace situations.

4 How is a meter stick like a number line?

5 Explain how a number line is used to model the countdown of a space shuttle launch.

Practice and Problem Solving

Write an integer to model each situation.

6. An elevator goes down 15 floors.

7. A construction detour causes a 25 minute longer drive to work after school.

8. The City Floral Company lost $18,500 the first year of business.

9. To land at the local airport, a plane must lose 2340 feet of altitude.

10. The stock market increased 22 points in the first hour of trading.

11. The deepest point in the Atlantic Ocean is 8648 meters below sea level.

12. A hot air balloon rose 515 feet.

13. Tim deposited $36 into his savings account.

14. Reynaldo gained 4 pounds during the first week of weight training.

15. The Smiths enlarged their patio by 400 square feet.

16. After a chemical reaction, the temperature of the metal dropped 52 degrees.

17. A mountain climber descends 1000 feet in one hour.

Mixed Review

Solve each problem.

You must decide whether to repair your car or replace it. However, you only have $2000 to spend. After checking several repair shops, you find it will cost $885 to repair the engine and $640 to repair the outside of the car. A good used car will cost $4800. You can get $1200 for trading in your old car.

18. What is the total cost to repair your old car?

19. After the trade-in, how much will you pay to buy a used car?

20. Given the information you now have, which choice would you make? Explain why.

LESSON 1.2 ABSOLUTE VALUE

The Number Line

In Lesson 1.1, you used a number line to represent the set of integers. Numbers to the right of zero are positive. Numbers to the left of zero are negative. Zero is neither positive nor negative. A thermometer held horizontally is a good model for the integers. Most thermometers show two number lines with different scales. The scale is the size of one unit—the distance between 0 and 1.

°F -60 -50 -40 -30 -20 -10 0 10 20 30 40 50 60 70 80 90 100 110 120

°C -50 -40 -30 -20 -10 0 10 20 30 40 50

ACTIVITY 1   Using an Integer Model

Use either the °C or °F scale to answer questions 1–7 on your paper.

1 On which side of zero are the positive values?

2 On which side of zero are the negative values?

3 Which temperature is farther to the right: 20° or 30°?

4 Which temperature is higher (hotter): 20° or 30°?

5 Which temperature is farther to the right: −20° or −30°?

6 Which temperature is higher (hotter): −20° or −30°?

7 Which temperature is farther to the right: 20° or −30°?

Critical Thinking If two numbers are on a number line, how can you determine which one is larger? Do any of the answers change if you use a different scale?

Think about the comparison of −20° and −30°. On a number line, −20 is to the right of −30. Thus, −20 is greater than −30. It is also true that −20° is a higher temperature than −30°. As you move to the *right* on the number line, the numbers *increase*.

Use a number line to determine which number is greater.

a. 4 or -6 b. -3 or -7 c. -4 or -2

The sentence *-20 is greater than -30* can be written

$$-20 > -30$$

The symbol $>$ is read "is greater than."

The sentence *-30 is less than -20* can be written

$$-30 < -20$$

The symbol $<$ is read "is less than."

Notice that the point of the less than or greater than symbol always points to the number of lesser value.

Absolute Value

Notice that -20 is 20 units from zero on the number line. The distance from zero to a number is called the **absolute value** of the number. The absolute value of -20 is 20. The symbol for absolute value is $|\ |$. Thus,

$$|-20| = 20$$

To remember the meaning of the absolute value symbol, read the first vertical line as "the distance of" and the second vertical line as "from zero."

Absolute Value
The absolute value of a number is its distance from zero.

Since -30 is 30 units from zero,

$$|-30| = 30$$

How far from zero is $+30$? Notice that $+30$ is exactly the same distance from zero as -30. Both numbers are 30 units from zero. Therefore, $+30$ and -30 have the same absolute value.

$$|+30| = |-30| = 30$$

Equal Absolute Values

If two numbers are the same distance from zero on the number line, the numbers have the same absolute value.

Ongoing Assessment

Use $<$ or $>$ to compare the absolute values.

a. $|-10|$ and $|-4|$ b. $|-3|$ and $|-5|$ c. $|-9|$ and $|9|$

Here is one way to think of the relationship between -20 and -30.

-20 degrees Fahrenheit is hotter than -30 degrees Fahrenheit.

$$-20°\text{F} > -30°\text{F}$$

But 30 feet underwater (-30) is a greater depth than 20 feet underwater (-20).

$$|-30| > |-20|$$

In other words, -30 is farther from zero (the water surface) than -20.

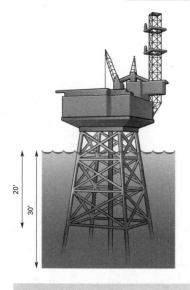

LESSON ASSESSMENT

Think and Discuss

1 How can you use a thermometer to model the integers?

2 Compare the distance from the zero point on a thermometer for two temperatures that have the same absolute value.

3 Describe the relative location on a thermometer for one temperature that is greater than another.

4 What is the distance between two temperatures on a thermometer if the absolute values of the two temperatures are equal?

5 Why is $|-10| > |-5|$?

Compare the integers. Use < or >.

6. −3; 4   **7.** −6; −8   **8.** 5; −5   **9.** −4; −5

10. −1; 0   **11.** −20; −15   **12.** 12; 1   **13.** 0; −9

14. A time line is an example of a number line. Draw a time line and show where each event should be placed.

   A. Buddha lived in China around 500 BCE.

   B. The Old Egyptian Empire was just beginning around 2700 BCE.

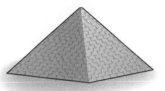

   C. The great Epics of India were being written around 1200 BCE.

   D. Hammurabi was the ruler of the Tigris-Euphrates Valley around 1800 BCE.

   E. The Hebrews left Egypt around 1500 BCE.

   F. The Phoenicians traded throughout the Mediterranean around 1000 BCE.

Evaluate each expression.

15. $|-7|$ **16.** $|10|$

17. $|0|$ **18.** $|-23|$

19. $|-6|$ **20.** $|-18|$

21. $-|-25|$ **22.** $|35-17|$

Each employee production team at the Captain Tire Company has a goal of producing 278 tires per day. Each day, your team's production is recorded as a deviation from the 278 tires. If you produce 281 tires, the deviation is +3. If you produce 277 tires, the deviation is −1. On Monday, you received this part of an e-mail message:

> Daily bonuses will be paid to production teams only when your team's product deviation is greater than −5

23. Will you make a bonus if your team produces 270 tires for the day?

24. How many tires does your team have to produce to make a bonus?

Mixed Review

25. Rashid's take-home pay is $2045 each month. Rashid has completed the budget shown below. Find the amount Rashid will have left after he pays his monthly expenses.

Rent	$475	Utilities	$125
Phone	$ 46	Cable	$ 52
Car payment	$279	Car expense	$ 85
Food	$418	Charge cards	$ 78

LESSON 1.3 ADDING INTEGERS

You can use a number line to model integer addition. Adding a positive number results in a move to the right. Adding a negative number results in a move to the left.

Adding Integers with the Same Sign

The temperature is 12°F at noon. Suddenly the temperature increases by 3°. What is the current temperature?

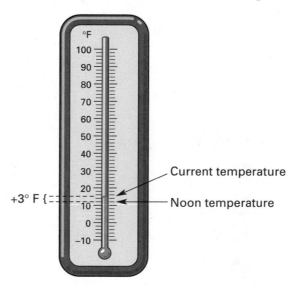

+3° F {

Current temperature

Noon temperature

The thermometer shows the new temperature is 15°. This is an example of adding two integers with the same sign. In this case, both integers are positive. Thus,

$$+12 + (+3) = +15 \text{ or } 12 + 3 = 15$$

Adding two positive integers is just like adding two numbers in arithmetic. What happens when you add two negative integers?

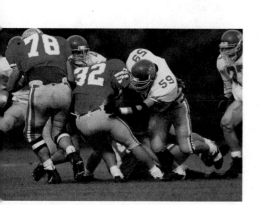

If a football team loses four yards on the first down, the loss is represented by -4. If they lose another three yards (-3) on the next down, they have lost seven yards (-7) in all. A mathematical sentence is used to represent this situation.

$$-4 + (-3) = -7$$

In this sentence, the $+$ sign is used as an arithmetic operation. This sign tells you to add two negative numbers. The second negative number is enclosed in parentheses to avoid confusing the + symbol for addition with the symbol that shows the sign of the number.

To model the operation $-4 + (-3) = -7$ on a number line, begin at zero. Draw an arrow called a **vector** four units to the left to represent -4. From (-4), draw another vector 3 units to the left to represent -3. Notice that the vectors are defined both by magnitude (the length of the arrow) and direction. At what point is the tip of the second arrow?

This model shows that

$$-4 + (-3) = -7$$

ACTIVITY 1   Adding Negative Integers

1 Choose 5 pairs of negative numbers.

2 Use a number line to add each pair.

3 Write a mathematical sentence to model each addition.

4 What is the sign of each sum?

When two or more negative numbers are added together, the result is always a negative number. This is exactly the same rule that you use with positive numbers: When two or more positive numbers are added together, the result is always a positive number.

Adding Integers with the Same Sign
When you add two or more integers with the same sign,

    1. add the absolute values,

and

    2. write the total with the sign of the integers.

A calculator can be used to add negative integers. First, add -328 and -49 on your calculator. Then compare your method to the one shown on the opposite page.

Enter the number 328

Press the ⊞⊟ key. (Remember, the ⊞⊟ key changes the sign of the number in the display.)

Press ⊞

Enter the number 49

Press the ⊞⊟ key

Press ⊜

Did you get −377?

Critical Thinking   You can also add the absolute values. First press the keys for 328 + 49. Then, write the total with a negative sign in front. Explain why this works.

Adding Integers with Different Signs

How do you add two numbers that have different signs? Recall the football example. This time, let the first down result in a loss of 7 yards and the next down result in a gain of 4 yards. The final result of the two plays is a net loss of 3 yards. A mathematical sentence can be used to represent this situation.

$$-7 + 4 = -3$$

You can also model $-7 + 4 = -3$ on a number line. Draw a vector from zero to −7 to represent a loss of seven yards. Since the next play is a gain of 4 yards, draw a vector from −7 four units to the right. Where do you finish on the number line?

ACTIVITY 2   Adding Integers Using Absolute Value

1   Explain why the following sketch shows the sum of −7 + 4.

```
            +4
      ←──────────→
                   −7
      ←──────────────
  ←──┼──┼──┼──┼──┼──┼──┼──┼──┼──┼──┼──┼──┼──→
    −8 −7 −6 −5 −4 −3 −2 −1  0  1  2  3  4
```

2   Notice that the vector representing −7 is longer than the vector representing 4. How does this help you determine the sign of the answer?

3   What is the difference between the absolute value of −7 and the absolute value of 4? What is the sign of the sum?

4   Explain how to use absolute value to find the sum of 3 and −8.

Did you notice that it is a lot easier to add two numbers with different signs than it is to explain it?

> ### Adding Integers with Different Signs
> When you add two integers with different signs
>
> 1. subtract the absolute values,
>
> and
>
> 2. write the difference with the sign of the integer having the larger absolute value.

To add $-7 + 4$, first subtract the absolute values.

$$|-7| - |4| = 7 - 4 = 3$$

Since $|-7| > |4|$, write the difference as -3.

$$-7 + 4 = -3$$

Ongoing Assessment

Find each sum.

a. $-6 + -7$ b. $-2 + 5$ c. $9 + (-14)$

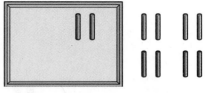

CULTURAL CONNECTION

For a very long time, people have used the idea of negative numbers. Among the first to work with both positive and negative numbers were the Chinese. They did so by using colored rods. Any two colors will work, as long as they are easily contrasted. Here is how a Chinese merchant might have calculated his finances by using red rods to represent negative integers and blue rods to represent positive integers.

Negative 6

Positive 4

The merchant might show a loss of 6 followed by a gain of 4 by placing 6 red rods and 4 blue rods in a container.

The merchant would then remove 4 red rods and 4 blue rods. Why?

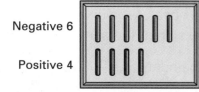

Since there are 2 red rods left in the container, the net result is a loss of 2.

Use the colored rod model to explain the rule for adding integers with different signs. Find the difference of the number of rods. Use the sign associated with the greater number of rods.

LESSON ASSESSMENT

Think and Discuss

1 Explain how to use a number line to add two positive integers.

2 Explain how to use a number line to add two negative integers.

3 Explain how to use a number line to add a positive and a negative integer.

4 How is absolute value used to add two integers with the same sign?

5 How is absolute value used to add two integers with different signs?

Practice and Problem Solving

Add.

6. $-3 + 5$ **7.** $-8 + 8$ **8.** $6 + (-4)$

9. $-9 + (-6)$ **10.** $-23 + 12$ **11.** $-14 + (-11)$

12. $13 + (-7)$ **13.** $-16 + (-18)$ **14.** $-21 + 15$

15. $-23 + (-9)$ **16.** $-9 + (-23)$ **17.** $27 + (-27)$

18. $45 + (-26)$ **19.** $-63 + (-37)$ **20.** $-59 + (-67)$

21. During one week of testing a new generator, electrical output dropped 328 kilowatts. During the next three weeks, output increased 143 kilowatts, decreased 37 kilowatts, and increased 219 kilowatts. Use integers to describe the net decrease or increase.

A proton has a charge of positive one. An electron has a charge of negative one. Find the total charge of an ion with

22. 12 protons and 15 electrons.

23. 21 protons and 19 electrons.

24. Mark the following amounts on a number line.

a. gain of $48   **b.** loss of $36   **c.** loss of $12
d. loss of $18   **e.** gain of $27   **f.** loss of $27

The chart shows several altitudes above or below sea level.

Caspian Sea	−28 meters
Death Valley	−86 meters
Mount Everest	+8848 meters
Mount McAuthur	+4344 meters
Pacific Ocean	−10,912 meters
Sea Level	0 meters

Compare the altitudes of each pair by writing an inequality.

25. Death Valley; Caspian Sea

26. Mount Everest; Mount McAuthur

27. Sea Level; Death Valley

28. Mount Everest; Pacific Ocean

29. Mount McAuthur; Sea Level

30. List all six altitudes in order from least to greatest.

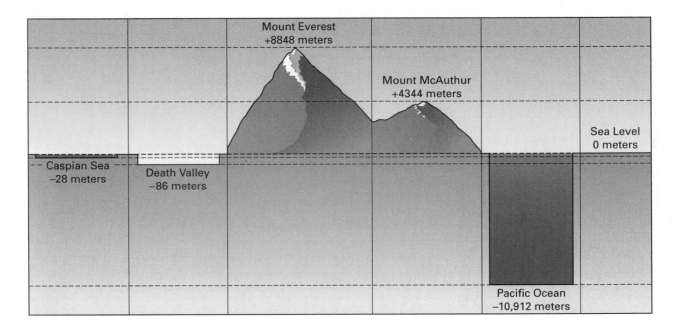

Mount Everest
+8848 meters

Mount McAuthur
+4344 meters

Sea Level
0 meters

Caspian Sea
−28 meters

Death Valley
−86 meters

Pacific Ocean
−10,912 meters

LESSON 1.4 SUBTRACTING INTEGERS

The number line is also a good model for integer subtraction. To model $8 - (-4)$, think of -4 in another way.

Opposites

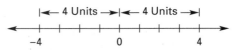

Notice the position of -4 and 4 on the number line. Although they are the same distance from zero, they are on **opposite** sides of zero. From the number line, you can see that

-4 is "*opposite of* 4" and 4 is "*opposite of* -4."

> **The Opposite of an Integer**
> Two integers are opposites if they are
>
> 1. on opposite sides of zero,
>
> and
>
> 2. the same distance from zero.

Thus,

$$\text{the opposite of } (-4) = 4$$

$$\text{and the opposite of } 4 = -4$$

A minus sign $(-)$ is used to indicate "the opposite of" a number. The opposite of -4 can also be written as $-(-4)$. Thus, "the opposite of negative four is positive four" can be written as follows:

$$\text{opposite sign} \rightarrow -(-4) = 4$$
$$\uparrow$$
$$\text{negative sign}$$

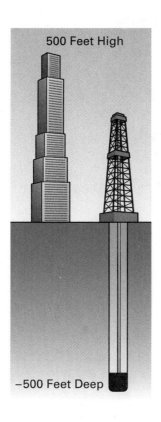

500 Feet High

−500 Feet Deep

Some calculators have a ⊞⊟ key. If your calculator has one, experiment with it. How is this key used to find the opposite of an integer? How is the ⊞⊟ key used to find the absolute value of an integer?

Adding Opposites

ACTIVITY 1   The Sum of Two Opposites

1 Choose five pairs of opposite numbers.

2 Find the sum of each pair.

3 What is the sum of each pair of opposites?

> **Addition of Opposites**
> The sum of two opposites is always zero.

Opposites are used to find the difference between two integers.

ACTIVITY 2   Use a Calculator to Subtract Integers

1 Complete each set of directions using a scientific calculator.

a. $-10 -(-6)$

Enter 10.
Press the ⊞⊟ key.
Press the ⊟ key.
Enter 6.
Press the ⊞⊟ key.
Press the ⊜ key.

b. $-10 + 6$

Enter 10.
Press the ⊞⊟ key.
Press the ⊞ key.
Enter 6.
Press the ⊜ key.

2 What is the result in each case? Why are the results equal?

3 Which method do you prefer? Explain why.

4 How are the expressions $-10 -(-6)$ and $-10 + 6$ the same? How are they different?

5 Explain how you can use addition to solve a subtraction problem.

The calculator problem is an example of the rule for subtracting a negative integer. However, the rule also works when you subtract a positive integer.

Subtracting Integers
You can subtract an integer by adding its opposite.

Both adding and subtracting occur often in the workplace. For instance, the wings of an airplane are attached to the airplane fuselage using rivets. After the rivets are in place, they are cured by heating them to 125°C and holding them at this temperature for a specified period of time. This curing process insures that the joint between the wings and the fuselage has maximum strength.

During the curing process, a technician monitors the temperature and records hourly readings.

Time	Deviation from 125°C
8 AM	+1
9 AM	+5
10 AM	−4
11 AM	+2
12 AM	0

Since the temperature is supposed to be 125°C, +1 indicates a temperature of 126°C and −3 indicates a temperature of 122°C.

Every four hours, the technician is required to report the hourly changes in the temperature to the engineer in charge of the plane's assembly.

To find the hourly change between 8 AM and 9 AM, the technician subtracts a positive number.

$$(+5) - (+1) = 5 - 1 = 4$$

Thus, during this hour, the temperature increased 4° from 126°C to 130°C. How can you find the temperature change between 9 AM and 10 AM?

$$(-4) - (+5) = ?$$

Since you can subtract an integer by adding its opposite, change +5 to −5 and add −5 to −4. Mathematically, this statement is modeled as

$$(-4) + (-5) = -9$$

From 9 AM to 10 AM the temperature decreased by 9° from 130°C to 121°C.

The temperature change from 10 AM to 11 AM is

$$(+2) - (-4) = (+2) + (+4) = +6$$

From 10 AM to 11 AM, the temperature increased by 6° from 121°C to 127°C. What was the temperature change from 11 AM to 12 AM?

Ongoing Assessment

Solve each problem without paper and pencil.

1. $9 + 6$ 2. $9 + (-6)$ 3. $9 - 6$ 4. $9 - (-6)$

5. $-9 + 6$ 6. $-9 + (-6)$ 7. $-9 - 6$ 8. $-9 - (-6)$

LESSON ASSESSMENT

Think and Discuss

1 Explain how to subtract an integer by using its opposite.

2 You buy a $15 CD at a music store. Use a number line to show that adding $15 to the account balance of the music store is the opposite of subtracting $15 from your checking account balance.

Practice and Problem Solving

Subtract.

3. $9 - (-5)$ **4.** $8 - 11$

5. $-7 - (-6)$ **6.** $-4 - 3$

7. $-12 - (-15)$ **8.** $-14 - 16$

9. $18 - (-14)$ **10.** $21 - (-19)$

Perform the indicated operations.

11. $-21 + (-15) - (-16)$ **12.** $38 - 45 + 26$

13. $-29 + 53 - 42$ **14.** $-31 - (-48) - (-16)$

15. $-56 + (56) - 18$ **16.** $75 - 96 + 43$

Tara made two deposits and two withdrawals from her bank account. She started with $781.08 in her account. Find the amount Tara has in her account after each transaction.

17. September 15; + $116.58   **18.** September 20; − $216.25

19. October 1; − $201.88   **20.** October 10; + $196.39

Mixed Review

The average rainfall for February is 4.21 inches. Following is the deviation from the average for five different years. Find the February rainfall for each year.

21. +1.85   **22.** −0.68   **23.** −1.36   **24.** −1.5   **25.** +0.7

The height of the Empire State Building is 1250 feet. Use integers to compare each of the following buildings with the Empire State Building.

26. Sears Tower; 1454 feet   **27.** John Hancock; 1127 feet

28. Amoco; 1136 feet   **29.** World Trade North; 1368 feet

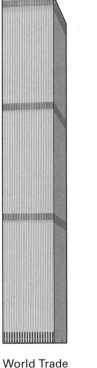

World Trade
Center North,
1368 feet

Empire State
Building,
1250 feet

LESSON 1.5 MULTIPLYING AND DIVIDING INTEGERS

You can use a multiplication table to find the product of two positive integers such as 2×3.

×	0	1	2	3
3	0	3	6	9
2	0	2	4	6
1	0	1	2	3
0	0	0	0	0

This multiplication table has four rows and four columns. The row labels are at the left of the table. The column labels are at the top of the table. The numbers in the table are entries. The entry in row 2 and column 3 is the product of 2 and 3. Since $2 \times 3 = 6$, the entry is 6.

Critical Thinking   Why are there so many zeros in the table?

Multiplying Negative Numbers

ACTIVITY 1   Extending the Multiplication Table

1 Copy this partial table.

×	−3	−2	−1	0	1	2	3
3				0	3	6	9
2				0	2	4	6
1				0	1	2	3
0	0	0	0	0	0	0	0
−1				0			
−2				0			
−3				0			

2 The table is made up of four parts, called **quadrants.** What are the signs of the entries in the top right quadrant? Explain how this table shows that the product of two positive integers is a positive integer.

3 Start at 9. Move across the first row of entries from right to left. Do the entries increase or decrease? By how much? Describe the pattern in each row as you move from right to left.

4 Use the pattern from step 3 to complete the entries in the top left quadrant of the table.

×	−3	−2	−1	0	1	2	3
3	?	?	?	0	3	6	9
2	?	?	?	0	2	4	6
1	?	?	?	0	1	2	3
0	0	0	0	0	0	0	0

5 Does your table look like the one below?

×	−3	−2	−1	0	1	2	3
3	−9	−6	−3	0	3	6	9
2	−6	−4	−2	0	2	4	6
1	−3	−2	−1	0	1	2	3
0	0	0	0	0	0	0	0

Explain how you can use the table to show that the product of a negative integer and a positive integer is a negative integer.

6 Extend the table downward so that it has rows numbered −1, −2, and −3. Use patterns to complete the lower half of the table. Describe the patterns you see in the table.

7 Does your table look like the one below?

×	−3	−2	−1	0	1	2	3
3	−9	−6	−3	0	3	6	9
2	−6	−4	−2	0	2	4	6
1	−3	−2	−1	0	1	2	3
0	0	0	0	0	0	0	0
−1	3	2	1	0	−1	−2	−3
−2	6	4	2	0	−2	−4	−6
−3	9	6	3	0	−3	−6	−9

Explain how you can use the table to show that the product of two negative integers is a positive integer.

Multiplying Integers

The product of two integers with the same sign is positive.

The product of two integers with different signs is negative.

Here is a short form of the rule for multiplying signed numbers:

Same signs → **Positive product**
Different signs → **Negative product**

When using a calculator to find a product, first decide on the sign of the product. For example, when multiplying $128 \times (-78)$, look at the signs of the integers. In this case, you are multiplying two integers with different signs. The product will be negative.

Now multiply the absolute values of the numbers and write the product with the correct sign. Thus, $128 \times (-78) = -9984$.

You can also find the product $128 \times (-78)$ by using the ⊞ key on a scientific calculator.

$128 \times (-78) = ?$

Enter 128
Press ⊠
Enter 78
Press ⊞
Press ⊟

Does the window display -9984?

Dividing Integers

Multiplying and dividing are **inverse operations**; that is, they undo each other. The rule for finding the sign of the quotient of two integers is exactly the same as for multiplying integers.

Dividing Integers

The quotient of two integers with the same sign is positive.

The quotient of two integers with different signs is negative.

Find each quotient mentally.

a. $12 \div 4$ b. $12 \div (-4)$ c. $-12 \div 4$ d. $-12 \div (-4)$

Now find $-13{,}650 \div 26$ using your calculator.

It's *still* not full! Bring more zeros!

Be sure to press the ⊕⊖ key after entering 13650. Did you get -525? Now try

$$-13{,}650 \div (-26.1)$$

The display should show 522.9885057. Rounded to the nearest whole number, the answer is 523.

Critical Thinking   Use your calculator to divide 5 by zero. What happens? Why is it impossible to divide by zero?

LESSON ASSESSMENT

Think and Discuss

1 How do you know which sign to use when you multiply two integers?

2 What are the steps for multiplying -24.5 by -6 on a calculator? What is the product?

3 What are the steps for dividing -7.5 by -0.5 on a calculator? What is the quotient?

4 Explain why 1.13 is not a reasonable quotient for -3.4 divided by 3.

5 When more than two integers are multiplied, how is the sign of the product chosen?

Practice and Problem Solving

Perform the indicated operations.

6. -3×-8 **7.** $-6 \div -3$ **8.** -4×7 **9.** $-18 \div 2$

10. 8×-7 **11.** $63 \div -9$ **12.** -12×-12 **13.** -24×3

14. 15×-6 **15.** $-25 \div -3$ **16.** $-8 \div -6$ **17.** $38 \div -8$

To find the average of 5 numbers, find the sum of the numbers and then divide that sum by 5.

18. Find the average of 8, 3, 7, 9, and 3.

19. Subtract each number in Exercise 18 from the average of the numbers. Write the differences using negative and positive integers.

20. Find the average of −16, 11, 12, −20, and 18.

21. Find the difference between each number in Exercise 20 and the average of the numbers. Write the differences using negative and positive integers.

Mixed Review

23. How much greater than −26 is −15? Explain your reasoning.

Use the table for Exercises 24−26.

COPIES	
1–99	$0.10
100–499	$0.09
500–999	$0.08
1000+	$0.07

24. How much does it cost to make 499 copies?

25. How much does it cost to make 500 copies?

26. How much does it cost to make 445 copies?

27. A customer said that if he had between 445 and 500 sheets to copy, he would rather pay for 500 sheets. Explain why.

LESSON 1.6 IDENTIFYING VECTORS

In a previous lesson, you used **vectors** to add and subtract integers. A vector is modeled by an arrow that shows both magnitude and direction. The length of the arrow indicates the magnitude of the vector. The arrow indicates the direction of the vector. For example, the ship's velocity vector has magnitude 3 and direction to the right. This vector indicates that the ship's velocity is 3 miles per hour to the east.

3 miles/hr, east

Ongoing Assessment

Use a number line and draw a vector to show a move from 1 to 4. What is the length of the vector? What is the direction of the vector?

The arrow used to show the speed and direction of the ship could represent any vector with a magnitude of 3 and the same direction. For example, it could represent a force of three pounds pulling to the right. The vector could also represent a current flowing at three miles per hour downstream.

Force on Rope 3 lb, 45°N of E

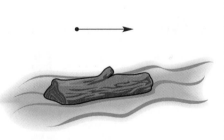

River Current 3 mph, east

The way the arrow points tells you the direction of the vector. The length of the arrow tells you the magnitude or size. To determine magnitude, you must have a scale such as a number line or a number written next to the arrow.

Vectors can show the position of an object after two or more different moves.

Suppose a ship begins at point *A* and sails 10 miles north to point *B*. It then turns and sails 10 miles east to point *C*. Vectors *AB* and *BC* describe the two parts of the ship's journey. Vector *AB* has a magnitude of 10 miles and a direction of north. Vector *BC* also has a magnitude of 10 miles, but the direction is east. Why are vectors *AB* and *BC* *not* equal?

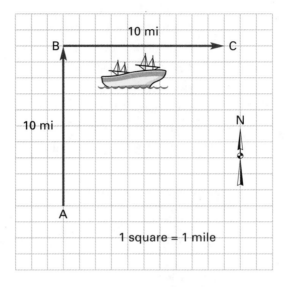

Vectors can model the addition and subtraction of several integers. Remember that subtracting an integer is the same as adding its opposite. The **opposite of a vector** is a vector that is the same size, but drawn in the opposite direction.

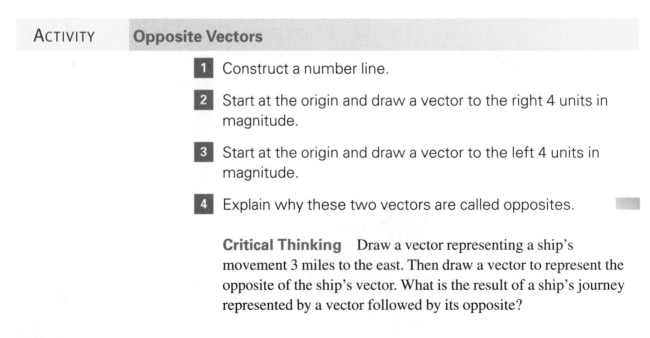

ACTIVITY   **Opposite Vectors**

1 Construct a number line.

2 Start at the origin and draw a vector to the right 4 units in magnitude.

3 Start at the origin and draw a vector to the left 4 units in magnitude.

4 Explain why these two vectors are called opposites.

Critical Thinking   Draw a vector representing a ship's movement 3 miles to the east. Then draw a vector to represent the opposite of the ship's vector. What is the result of a ship's journey represented by a vector followed by its opposite?

Use vectors to find $6 + (-4) - (-3)$.

SOLUTION

Start at 0 and draw a 6 vector.

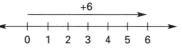

Start at 6 and draw a -4 vector.

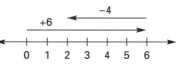

Start at 2 and draw the vector that is
the opposite of a -3 vector.

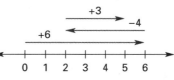

The total is 5.

Ongoing Assessment

Use vectors to find $-4 - 3 + (-5)$. Then explain how to find the
total without using vectors.

WORKPLACE COMMUNICATION

What is the "obvious" error in the
display of velocity vectors during
the test?

Modern Workplace, Inc.
Interoffice Memo

To:
From: Vera Modawell, Software Quality Control Manager
Subject: Jeffery Given, Programmer
         Correction of Air Traffic Control Software

We have identified the problem encountered last week in testing the new software being
developed for the Federal Aviation Administration. The purpose of the new software is to use
data obtained from Global Positioning Satellites in order to display aircraft position and velocity
vectors on Air Traffic Controller computer monitors.

The sequence of aircraft velocity vectors that should have appeared on the monitors is shown
below on the left. The erroneous sequence, that the new software displayed during the test, is on
the right.

Sequence of velocity vectors
that should have been displayed.

Sequence of velocity vectors
that were displayed during the test.

As can be seen above, the error is obvious and can be easily corrected in the software. We will be
ready for another test tomorrow.

LESSON ASSESSMENT

Think and Discuss

1 What two quantities are displayed by all vectors?

2 Draw a diagram showing a sailboat's movement if it sails 6 miles west and then turns south and sails 8 miles.

3 Draw the opposite of the vector that represents a journey 6 miles to the north.

4 Show how to use vectors to find $-7 + 4 - (-3)$. What is the total?

Practice and Problem Solving

Find the total using vectors.

5. $6 + (-5) + (-3)$

6. $-3 + 5 - (-2)$

7. $-6 - (-8) + 9$

8. $4 + 3 - (6)$

9. $-2 - (-6) - (-3)$

10. $0 - (-5) - 2$

Find the total without using vectors.

11. $-15 + 36 - 23$

12. $32 - (-35) + 18$

13. $-37 - (-12) - 28$

14. $43 - (-27) + (56)$

15. $51 + (-48) - (-52)$

16. $-23 - (-62) - (-36)$

17. $39 - (85) - (39)$

18. $-68 + (52) - (-84)$

19. $-16 - 37 + 15 - 28$

20. $48 - (-36) + (-72) - 68$

21. Mark opened a savings account for $75. He withdrew $30 on Wednesday and $20 on Friday. On the following Monday, Mark deposited $15. Represent each transaction with an integer. How much does Mark have in his savings account after the final deposit?

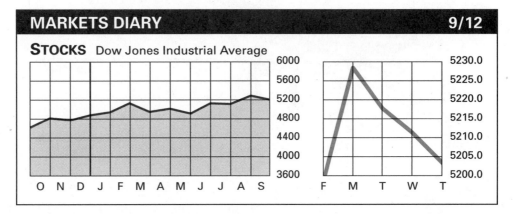

The chart shows the change in the stock market during four days' time. If the market started at 5220.25, give the value of the market at the close of each day.

22. Mon.; +8.5

23. Tue.; −11.125

24. Wed.; −6.375

25. Thu.; −7.75

26. Compare the changes in the stock market by drawing a number line and placing the changes for each day in the correct order. Which day has the biggest decline?

Combining Vectors

You can use vectors to help solve many problems.

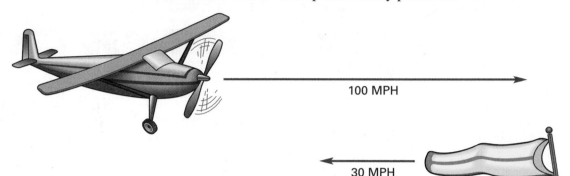

100 MPH

30 MPH

An airplane is flying with an airspeed of 100 miles per hour. The airplane is also flying directly into a headwind of 30 miles per hour. The plane does not actually travel 100 mph over the ground. To find the speed of the plane over the ground, the velocity vectors shown in the diagram are *combined*.

When you combine vectors, you need to think about the **absolute value** of each vector. The absolute value of a vector is its length or magnitude. For example, the length of the lower arrow represents the magnitude of the wind speed.

Absolute Value of a Vector
The absolute value or magnitude of a vector is the measure of its length.

Ongoing Assessment

What is the magnitude of the vector representing

a. the airplane speed?　　　b. the wind speed?

Now you can combine the velocity vectors for the airplane and the wind to find the velocity vector of the airplane over the ground.

1 On a piece of graph paper, make a scale drawing for the airplane and the wind velocity vectors. Let one square represent 10 miles per hour. Label your beginning point—the tail of the vector—as point *A*.

2 Draw a line segment 10 units long heading directly east from point *A*. Label the end point—the head of this vector—as point *B*. This vector represents the plane's velocity.

3 From point *B*—and a little below vector *AB* so that the lines do not overlap—draw another line segment 3 squares long heading due west. Label the head of this vector point *C*. This second vector represents the wind speed.

4 The result of adding these two vectors is the vector from point *A*, the beginning point, to point *C*, the ending point. How long is the vector from *A* to *C*? What is its direction? Compare your picture to the one below.

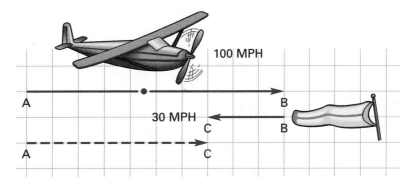

5 The direction of the vector from *A* to *C* tells you the direction of the plane. The length of the vector tells you the ground speed; that is, how fast the plane is actually passing over the ground. What is the airplane's velocity with respect to the ground?

Vectors at Right Angles

The velocity vectors in the airplane example point in opposite directions. Sometimes vectors are directed at right angles to each other. The drawing on the next page is similar to the one drawn for the ship example in Lesson 1.6. In this drawing, the vector *AC* is included with vectors *AB* and *BC*.

Use this drawing to complete Activity 2.

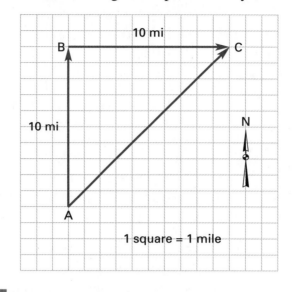

1 square = 1 mile

1 What is the scale for vectors *AB* and *BC*?

2 What is the direction of vector *AB*? What is its magnitude (in miles)?

3 What is the direction of vector *BC*? What is its magnitude (in miles)?

4 What is the direction of vector *AC*? What is its approximate magnitude (in miles)?

5 What does vector *AC* represent?

Combining Two Vectors

Justine and Jorge use ropes to pull a heavy cart. Jorge pulls to the left with a force of 50 pounds. Justine pulls at a 45-degree angle with a force of 30 pounds.

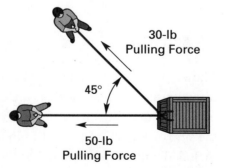

You can draw a picture of what happens by drawing vectors of magnitude 50 pounds and 30 pounds in the directions of the two forces (along each rope). These two vectors are combined to find the resultant force on the cart.

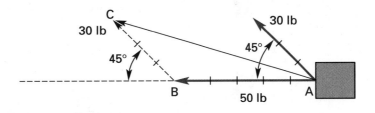

The method for combining the vectors is as follows:

First redraw the 30-pound vector with its tail at the head of the 50-pound vector. Notice that the 30 pounds is still directed at 45 degrees. Then draw the **resultant vector** from point *A* to point *C*, which shows the combined effect of the two separate forces.

ACTIVITY 3   Combining Forces

Use graph paper to make a scale drawing of the separate vectors representing Justine and Jorge pulling forces. Let each square on the graph paper represent 10 pounds of force.

1. Label your starting point *A*. Draw a vector 5 units long going directly to the left of point *A*. Label the arrowhead of this vector point *B*.

2. From point *B*, draw another vector 3 units long upward at a 45-degree angle from your first vector. Label the head of this vector point *C*. The result of adding these two vectors is the resultant vector *AC*.

3. Use a protractor to find the angle between the resultant vector and the vector *AB*.

4. Complete this statement.

Combining these two forces is the same as using one force of approximately _?_ pounds pulling at about _?_ degrees north of west.

LESSON ASSESSMENT

1 Research the meaning of each of the following terms. Explain how a vector can be used to represent the quantity.

a. acceleration   **b.** displacement   **c.** velocity

2 How are two vectors combined to find the resultant vector?

Describe a situation where vectors are used to represent

3 two forces acting in the same direction.

4 two forces acting in opposite directions.

5 two forces acting at a 90-degree angle to each other.

Practice and Problem Solving

Use graph paper to sketch vector diagrams for each situation.

6. A canoe floats due east on a river at a speed of 10 miles per hour across the water. The river current flows east at a speed of 3 miles per hour.

7. A canoe floats due south on a river at a speed of 10 miles per hour across the water. The river current flows to the south at a speed of 3 miles per hour.

A jet flew from Dallas to Atlanta with an airspeed of 400 miles per hour. The wind has a speed of 100 miles per hour as shown below.

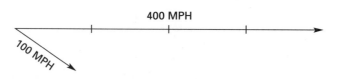

8. Combine the two velocity vectors and sketch the resultant vector over the ground.

9. Is the jet's ground speed greater than or less than its air speed? Why?

On the return flight from Atlanta to Dallas, the velocity vector of the wind is the same.

10. Combine the two velocity vectors and sketch the resultant velocity vector of the jet over the ground.

11. Is the jet's ground speed greater than or less than its air speed? Why?

Mixed Review

The EZ-Duzzit Company budgeted a total of $5580 for Project Number 2468, to be completed over a span of six months. The management projected that the money would be equally spent during each of the six months. During the first four months, the following total amounts were reported spent: March $570, April $1071, May $1247, and June $928.

12. What is the budgeted monthly amount for each of the six months of Project Number 2468?

13. Make a table of monthly spending. For each month, compare the actual spending to the budgeted amount, and compute the amount "over/under budget." Represent each amount using an integer.

14. What is the total amount "over/under budget" for the first four months of the project?

The best selling card at City Card Shop is the Top Flight Card. During one 8-day period, the owner recorded the following daily sales of the Top Flight Card: 30, 41, 44, 50, 24, 32, 40, 43

Examining the data, the owner guessed the average daily sale for the Top Flight Card might be 40.

15. Subtract each actual daily sale from the owner's guessed average. Write the differences using integers.

16. Find the average of the integers you found in Exercise 15.

17. Add the integer you found in Exercise 16 to the guessed average.

18. Find the average of the daily sales by adding the sales and dividing by 8. How does this average compare to the number you found in Exercise 17?

19. Use the guessing method to find the average for the following set of test scores. 91, 94, 79, 80, 79, 75

MATH LAB

Equipment   Timer
                Calculator

Problem Statement

You will compare your pulse rate to the average pulse rate of your class following three different states of physical activity.

Procedure

a   Locate your pulse. The two pulses you can find most easily are the radial pulse and the carotid pulse. The radial pulse is located on your wrist near the base of your thumb. The carotid pulse is on the side of your throat beside your jaw.

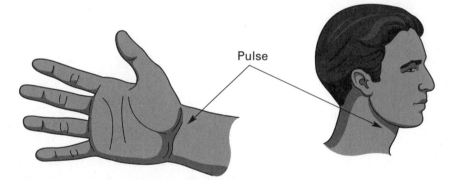

Pulse

Choose the location where your pulse is easiest for you to find.

b   Find your pulse and count the number of beats in a 10-second period. Multiply this number by 6 to get the number of beats per minute. Write the number of beats per minute on a sheet of data paper as your "resting pulse rate."

c   Run in place for one minute. Immediately count your pulse beats for a 10-second period. Multiply this number by 6 to get the beats per minute. Record the number of beats as your "exercising pulse rate."

d Rest for five minutes. Then count beats for a 10-second period. Multiply this number by 6 to get the beats per minute. Record this number as your "recovery pulse rate."

e On the chalkboard, write the headings for "resting pulse rate," "exercising pulse rate," and "recovery pulse rate." Record all the data for each class member on the chalkboard. Copy all the data to your record.

f Calculate the average pulse rate of the class for each state of activity. Use integers to report the difference between your pulse rate and the class average for each kind of rate.

g Use integers to report the increase from "resting pulse rate" to "exercising pulse rate" and the decrease from "exercising pulse rate" to "recovery pulse rate." Report these changes for your own pulse rates and for the average pulse rates for the class.

Activity 2: Drawing Displacement Vectors

Equipment   Directional compass
Calculator
Protractor
Tape measure
Two stakes

Problem Statement

A displacement vector is a distance traveled (magnitude) in a given direction. A distance of 50 feet at a compass heading of 45° (northeast) is an example of a displacement vector. You will move from a beginning point to an ending point by following directions given to you in terms of displacement vectors.

Procedure

a Select a location for one of the stakes and call it point *A*. Starting at point *A*, measure 20 feet along a compass heading of 55°. This locates point *B*. From point *B*, measure 15 feet along a compass heading of 145°. This locates point *C*. From point *C*, measure 10 feet along a compass heading of 265°. This locates point *D*. At point *D*, locate the other stake.

b Find the distance from point A to point D. Use your compass to determine the direction from A to D.

c Make a scale drawing using arrows as the displacement vectors in Step **a.** Draw a vector from point A to point B. From point B, draw a second vector to point C. From point C, draw a third vector to point D. Draw a final vector from A to D with the tail of the vector at A and the head at D. This vector is the resultant of the three displacement vectors.

d Based on the scale you used to make your drawing, determine the magnitude of the resultant vector.

e With a protractor, measure the heading (angle) of the resultant vector. Compare these results to the measured direction and distance from the beginning stake to the end stake. How do the two sets of measurements compare?

Activity 3: Combining Force Vectors

Equipment   Two spring scales with 5000-gram capacities
Three golf tees
Calculator
Masking tape
Plumb line
Key ring
String (2 feet long)
500-gram hook weight
Protractor
Pegboard

Problem Statement

You can represent weight as a vector. The direction of this vector is usually toward the center of the Earth. You will "weigh" an object with two scales that are pulling at angles other than vertical. These two scales represent vectors. When the vectors are added together, they are equivalent to the weight of the object.

Procedure

a Tie the key ring 3 feet from the bob on the plumb line.

b Position three golf tees on the pegboard as shown in the drawing. Slip the key ring over the center tee. Tie a loop in each end of a 2-foot piece of string. Slip one loop over the hook at the end of one of the scales. Slip the other loop over the hook on the other scale. Hang the two scale rings on the two outer tees. Hook the 500-gram weight over the string at a point $\frac{1}{3}$ of the way from one end of the 2-foot string, then tape the hook neatly to the string with masking tape. Adjust the angle of the pegboard so that the plumb line crosses the 2-foot string at the position where the weight is hooked. The completed setup is shown in the drawing. Be sure that the plane of the pegboard is near-vertical and perpendicular to the floor. Do not let the scales "drag" against the pegboard.

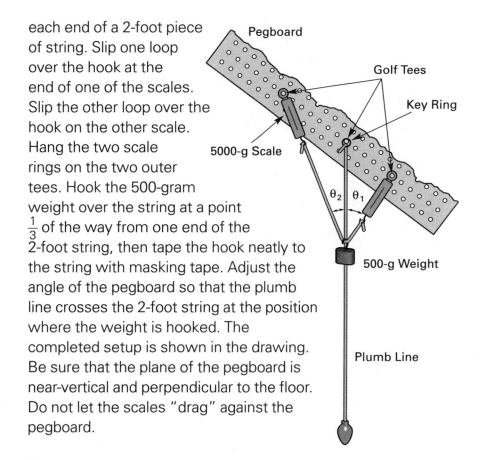

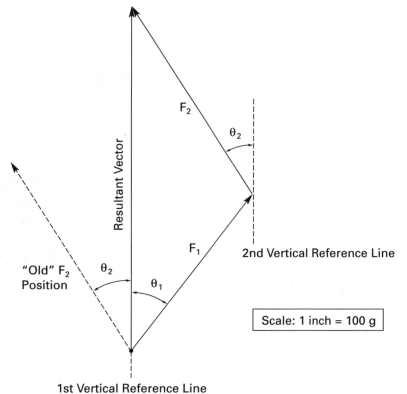

Scale: 1 inch = 100 g

c Read the weight shown on each of the scales. On a sheet of data paper, record the weight from the scale on the right as F_1. Record the weight from the scale on the left as F_2.

d Measure the angle between the plumb line and the scale on the right (force F_1) and write this value on a sheet of data paper as θ_1. Measure the angle between the plumb line and the scale on the left (force F_2) and write the value on a sheet of paper as θ_2.

e Move the 500-gram weight to the halfway point of the 2-foot string and repeat Steps **b, c,** and **d.**

f Move the weight to a point $\frac{1}{4}$ of the way along the 2-foot string and repeat Steps **b, c,** and **d.** This gives you three different pairs of weight vectors. Each pair of weight vectors is equivalent to the suspended 500-gram weight.

g Draw the first vector pair. Begin by drawing a vertical reference line. Measure the angle θ_1. The drawing will help you see how to do this. Use a scale of 1 inch to 100 grams and draw the force F_1. Put an arrowhead on the end of the vector (away from the reference line) to show the direction of the vector. Now draw another vertical reference line through the arrowhead of F_1. Measure the angle θ_2 from the second vertical reference line and draw the vector for the force F_2 beginning at the head of F_1. Finally, draw a vector from the tail of F_1 to the head of F_2. This is the resultant vector. It represents the combined effects of F_1 and F_2 and should be equal, in scaled units, to the 500-gram weight. The resultant vector should be 5 inches long and drawn along the first vertical reference line.

h Draw vector diagrams for the other two vector pairs. Compare the three vector diagrams. For each resultant vector drawn, the length should be 5 inches, and the vector should point along the first vertical reference line. Do your three drawings show this to be true?

MATH APPLICATIONS

The applications that follow are like the ones you will encounter in many workplaces. Use the mathematics you have learned in this chapter to solve the problems. Wherever possible, use your calculator to solve the problems that require numerical answers.

The altitude of Owens Telescope Peak in southern California is 3367 meters above sea level. Just a few kilometers away, the floor of Death Valley is 86 meters below sea level.

1 Let sea level be represented by 0. Express the altitude of each location as an integer.

2 What is the gain of altitude for a helicopter pilot who flies from the floor of Death Valley to the top of Owens Telescope Peak?

Indicate whether the value associated with each of the items below is considered a positive or a negative value.

3 A tax increase

4 Price reduction during a sale

5 The change in your checking account balance after you write a check

6 Temperature change when a cold front "passes through"

7 Stock market index rising 3 points

8 Difference in price paid for an item, after manufacturer's rebate

9 Change in pulse rate from a resting position to jogging

10 Tide at 2 feet above normal

11 Discounted airfares

12 A plunge in the wholesale price index

Suppose your travel alarm clock runs about one minute slow each day. On the back of the clock you see an adjustment lever in a slot.

(slow down) — • • • • • • • • • • + (speed up)

13 Which direction should you move the lever to adjust the speed?

14 If each dot above the slot represents a change of $\frac{1}{2}$ minute per day, by "how many dots" should you adjust your clock? Represent this adjustment as an integer.

A small airplane flies cross-country under various wind conditions. Through all the conditions, the plane maintains an air speed of 115 miles per hour and a compass heading due east. For each of the conditions, sketch the vectors representing the plane's air velocity (speed and direction), the wind velocity (speed and direction), and the ground velocity (speed and direction).

15 Wind from the west with a speed of 25 mph.

16 Wind from the east with a speed of 20 mph.

17 Wind from the south with a speed of 30 mph.

To run a cable across Murky Creek, you must determine the distance across the creek. Points A and D locate utility poles.

Start at point A. Walk due south for 21 meters. Then turn due east and walk 22.5 meters across the bridge. Finally, walk due north 9 meters to reach the desired point D.

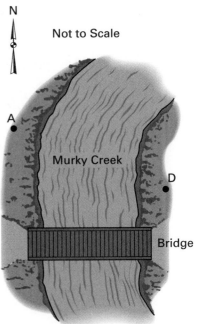

18 Make a scale drawing of the distances and directions traveled. The final point should be labeled D.

19 The resultant vector joining point A to point D on your scale drawing is equivalent to the sum of the vectors that represent the distances traveled on each leg of the trip. Measure the magnitude of vector AD on your scale drawing. How many meters is it across Murky Creek from point A to point D?

A business firm buys a second building across the street from its current location. Most of the office space is located on the second floor of each building. A raised walkway is proposed to reduce the walking time from one office area to the other. An overhead view of both the proposed raised walkway and the current route is shown in the drawing.

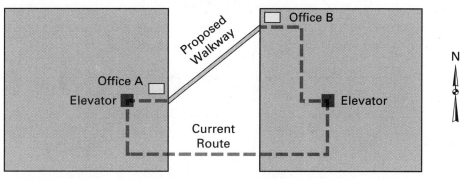

20 How many vectors are required to describe the path taken from office A to office B through hallways and doorways along the current route? (Remember the elevator rides between the second and first floors!)

21 What single vector can be used to represent the sum of these individual vectors?

22 Assume the distance from the beginning of one "dash" to the next is 10 feet. Make a scale drawing of the addition of the vectors representing the current route. Omit the elevator rides. About how long is the proposed walkway?

23 If the current route requires 10 minutes (assume the two elevator rides last 2 minutes combined), estimate the time for the trip across the walkway.

A hardware supply business keeps parts in small bins along one wall. You are looking for a particular part. A clerk who sees you looking in bin A says, " . . . no problem! From the bin you're now looking in, go down three rows and over five bins."

A	B	C	D	E	F
K	L	M	N	O	P
U	V	W	X	Y	Z
e	f	g	h	i	j
o	p	q	r	s	t

24 Make a quick sketch of the bins. On your sketch, draw the two vectors described by the clerk. What is the letter of the box that holds your part?

25 Write another combination of vectors to describe the location of the bin containing your part.

26 Draw the vector representing the sum of the two vectors in Exercise 24 and Exercise 25. How can you describe this resultant vector? Of the two methods to locate your part (the two vectors described by the clerk, or the one vector sum), which is the better way to describe the location of the bin you wanted?

AGRICULTURE & AGRIBUSINESS

An experimental process for increasing the growth rate of wheat is being tested and compared to the results predicted by many months of computer simulation. The predicted and actual values of growth rate increase for several trials of this process have been recorded in a table.

Test Results for Wheat Growth Rate Increases		
	Predicted Percent Increase	Actual Percent Increase
Trial A	1.103	1.007
Trial B	4.198	4.225
Trial C	14.611	14.803
Trial D	8.014	7.891

27 To evaluate the results of the test, you must compare the actual values to the predicted values. Copy the table. Add a column to show the difference "Actual − Predicted." Use a positive or negative number to show each difference.

28 You need to report the average difference for the test. Add the differences and divide by the number of trials.

29 Find the average using the absolute values of the differences. How does this approach differ from Exercise 28?

The average rainfall for one county is 0.8 inches per week during this time of year. The rainfall for the past five weeks is recorded in a table.

Weekly Rainfall	
Week 1	0.0"
Week 2	1.0"
Week 3	0.5"
Week 4	0.8"
Week 5	0.1"

30 Write the positive and negative difference of each week's rainfall from the average for this time of year. This difference is called a **deviation**.

31 What is the meaning of a positive deviation? A negative deviation?

32 What is the total deviation for this period of five weeks? Is the rainfall above normal or below normal?

When you apply insecticide with a sprayer, you must calibrate your applicator. If the application rate is higher or lower than the recommended amount, you must make an adjustment. A common method of testing the application rate is to hang small bags to collect the insecticide at a set distance and speed. The amounts from four trials at different application rates are recorded in a table. According to your calculations, each sample should have 9 ounces.

Calibration Samples

Trial	Sample amount
1	7 oz
2	12 oz
3	10 oz
4	9 oz

33 For each sample, compute the difference between the actual amount and the target amount of 9 ounces. Which trial was farthest from the target application rate?

BUSINESS & MARKETING

A newspaper reported the performance of various investments since the beginning of the year.

Results of Various Investments This Year			
Investment	Dec 31	Today	Percent change
Japanese yen	121.00	130.95	+8.2%
Long-term T-bonds	2658.27	2816.47	+6.0%
Dow Industrials	1938.83	1983.26	+2.3%
Silver (ounce)	6.68	6.27	−4.6%
Gold (ounce)	486.20	445.30	−8.4%

34 Which investments have increased in value?

35 Which investments have decreased in value?

36 Which investment had the greatest increase?

37 Which investment had the greatest loss?

38 A monthly government report indicates a change from last month in the gross national product (GNP) of −1.8%. Does this change reflect a growth or a shrinkage of the GNP as compared to last month?

The net income/loss for Rocky Road Corporation is shown with a bar graph.

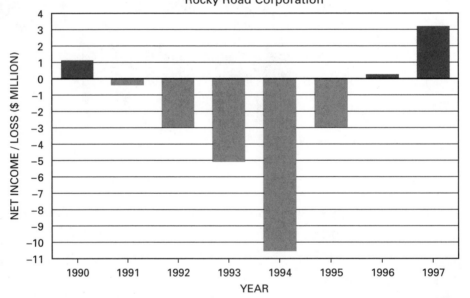

NET INCOME / LOSS
Rocky Road Corporation

39 Were most of the years shown in the graph losses or profits for Rocky Road Corporation?

40 Which year had the greatest income?

41 Which year had the greatest loss?

42 Do the recent years' net income/loss (the last 3–4 years) show a trend that is improving or worsening?

The total cost of merchandise sold in a clothing store is computed by adding or subtracting the following amounts.

Total cost of merchandise	=	+ cost of beginning inventory
		+ billed cost of net purchases
		+ inbound transportation cost
		− cost of ending inventory
		− cash discounts
		+ workroom/alteration costs

43 A clothing store showed the following figures for a six-month period: beginning inventory, $13,651; billed cost of net purchases, $87,055; transportation charges $4,879; cash discounts earned on purchases, $7,510; cost of ending inventory $12,402; workroom costs, $748. Compute the total cost of the merchandise sold.

The Smooth-Zipper Corporation publishes a balance sheet that compares this year's performance with last year's. Here is a portion of the balance sheet.

Smooth-Zipper Corp. Balance Sheet December 31			
Assets	**Last year**	**This year**	**Increase/Decrease**
Cash	$12,500	$13,800	
Inventory	42,400	47,500	
Accounts receivable	24,700	22,000	
Land	$12,000	$12,500	
Buildings	35,000	37,200	
Equipment	8,600	8,000	
Total assets	$135,200	$141,000	

44 Complete the last column of the statement to report the increase or decrease for each asset.

45 What is the difference in value between the assets for this year and last year? Compare the sum of the third column of integers with the difference in total assets.

During the week, the warehouse for Comfy Footwear posted these adjustments to the inventory.

Inventory Adjustments for Week Ending 2/29	
Date	**Adjustment**
2/24	+16
2/25	−28
2/25	−2
2/26	+1
2/26	−3
2/27	+12
2/28	+8
2/29	−14

46 What is the total of the adjustments made for this week?

47 The inventory at the beginning of the week is 18,384 units. By the end of the day on 2/29, 720 units have been shipped from the warehouse. Allowing for the adjustments, what is the inventory at the end of this week?

An office manager is preparing a report that compares performance to schedule for the past quarter. A list of projects and their completion dates are recorded in a table.

	Project Schedule Summary	
Project ID	**Scheduled completion date**	**Actual completion date**
1010	3/30	3/29
4242	3/10	3/10
5142	2/17	2/15
6231	3/25	3/28
6323	2/15	2/12
7652	1/29	1/28
7774	2/8	2/10

48 Find the difference (number of days) between the scheduled completion date and the actual completion date. Early project completion is represented by a negative integer. Record each difference as an integer.

49 How many days were lost or gained by all the projects for the quarter?

You have purchased 144 pairs of earrings to sell in your store. You are going to sell each pair at a retail price of $4.98. During the first 30 days, you sold 60 pairs. Your total expenses for selling the 60 pairs totaled $280.87. You sold the remainder during the next 30 days, when the expenses for these pairs were $339.24.

50 Compute the total retail value of the earrings sold during the first 30 days and during the second 30 days.

51 Subtract the expenses from the total retail profit for each of the two periods to find the net profit (or loss).

52 Find the total net profit or loss by adding the net profit from each 30-day period.

A market research study for River City reported the median income for various age groups. You decide to compare this report with a similar report done five years ago. During the past five years, wages should have increased about 22% to keep up with inflation. The table shows the results of the two studies.

Comparison of Market Research Studies for River City		
Median Annual Incomes		
Age group	**Previous study**	**Current study**
12–18	$6,000	$7,000
19–25	$17,500	$21,500
26–45	$47,500	$58,500
46–65	$49,000	$59,000
66 and above	$33,500	$41,500

Copy the table. Add a third column to your table with the heading, "Projected Income."

53 For each age group, calculate the projected income necessary to keep pace with 22% inflation over the five years. Record each value.

54 Compare the projected incomes with the results of the current study. Add a fourth column to your table showing the difference between the current and the projected incomes. Record each difference as an integer.

55 Which sentence describes your use of a negative integer?

a. Income has not kept pace with inflation.
b. Income has exceeded the value projected by the inflation rate?

56 Which age group(s) in River City is within $500 of its projected income?

You work as a public health official who interviews clients at a woman's shelter to see if they need nutritional assistance. Your county has established the following guidelines for three categories of individuals:

Required Nutrition (servings per day)			
Food group	Children	Pregnant women	Other women
Fruits and vegetables	4	4	4
Breads and grains	4	4	4
Meats and other protein sources	2	2	2
Dairy products	3	4	2

Each client's eating habits are recorded. If a total of three or more deficiencies is discovered, the client is eligible for assistance. For example, one missed serving in a food group is one deficiency, two missed servings is two deficiencies, and so on. Excess servings in a food group have no affect on eligibility.

You are interviewing a pregnant woman who has one small child. You have recorded the eating habits of the mother and child.

Servings per Day		
Food group	Child	Pregnant mother
Fruits and vegetables	2	3
Breads and grains	5	3
Meats and other protein sources	1	2
Dairy products	3	2

57 Compute the difference between the actual and recommended number of servings per day for both the mother and the child. Record the differences as integers.

58 Record the total number of deficiencies for both the mother and the child.

59 Use the guidelines to determine if either the mother or the child is eligible for assistance.

When a patient is recovering from an illness, it is important to monitor the patient's fluid intake and output. A fluid imbalance must be remedied as soon as possible. A record of a patient's fluids for several 8-hour periods during a hospital stay are recorded in a table.

Intake-Output Record for J. Smith			
Date	**Shift**	**Intake**	**Output***
3/10	A	1200 cc	900 cc
3/10	B	1000 cc	800 cc
3/10	C	1050 cc	950 cc
3/11	A	1300 cc	1250 cc
3/11	B	1500 cc	1600 cc
3/11	C	1200 cc	1300 cc
3/12	A	1100 cc	1250 cc

*Includes estimated perspiration, etc.

60 A fluid balance is indicated by a difference between the intake and output of nearly zero. Make a table of the differences between the intake and output for each shift shown in the table. Use integers to record your differences.

61 Which of your differences indicates an imbalance due to insufficient output of fluids?

62 Are the absolute values of these differences useful in spotting a condition of fluid imbalance? What information have you lost if you show only the absolute values of the differences on your chart?

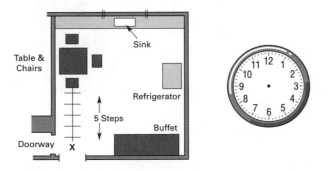

To help a person who has recently lost eyesight orient himself in a new room, you may describe the location of some of the important objects in that room. One method is to use clock positions. 12 o'clock is straight ahead, 3 o'clock is to the right etc. Consider the room shown in the illustration. Notice the location of the labeled objects. The vision-impaired person is standing in the doorway at the point labeled X, facing the table and chairs.

63 Trace the basic room layout. Notice the scale provided, in terms of "steps." Draw the vectors that show the steps and direction from *X* needed to reach each of the labeled objects or locations:

a. Nearest chair
b. Sink
c. Refrigerator
d. Nearest edge of buffet
e. Other doorway (assume person enters from doorway at bottom of drawing)

64 Write the words that describe each of these vectors. Use the clock position to communicate the direction of each vector.

Eyeglass prescriptions define the amount of curvature required in a lens to correct a patient's vision. The amount of curvature or sphere power is expressed in units of diopters (D). To correct a near-sighted condition requires a negative value for sphere power, while to correct a far-sighted condition requires a positive value. A value close to zero indicates very little correction. Examine the values of sphere power listed below.

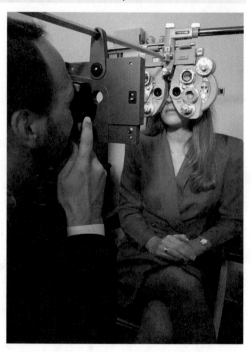

Values of Sphere Power Prescribed	
Patient	**Sphere Power (D)**
A	−2.00
B	+0.50
C	+1.00
D	−1.25
E	−3.75
F	+2.00

65 Draw a number line. For each patient, label the position on your number line representing the sphere power prescribed. Label the range of corrections for the near-sighted patients, and the range of correction for far-sighted patients.

66 Which patient requires the most correction for near-sightedness? For far-sightedness?

Some tax forms from the Internal Revenue Service have specific instructions for filling them out. One form has the following instruction for line 43.

Subtract line 42 from line 39.
Enter the result (but not less than zero).

For each set of values for line 42 and line 39 shown below, determine the correct value for line 43.

67 Line 39: $12,600
Line 42: $4267

68 Line 39: $2328
Line 42: $2230

69 Line 39: $3125
Line 42: $3419

You are balancing your checkbook with your calculator. Your starting balance is $264.88. You have written checks for $38.76, $144.69, $51.25, and $30.03. Your last bank statement showed that $5.16 interest was credited to your account and $2.50 was deducted for a service charge.

70 Make a list of the transactions posted to your account. Indicate subtractions from your account balance as negative numbers. Indicate additions to your account as positive numbers.

71 What is the final balance of your account after these transactions? What does the sign of the number tell you?

You are a flight attendant who travels worldwide with an international airline. As you fly, you experience many changes in local time zones. You have discovered that the local time increases when you travel east (+) and decreases when you travel west (−). The table shows some of the cities your airline travels to and the difference in local time compared to the time in London.

Location	Hours Changed
Sydney, Australia	+10
Tokyo, Japan	+ 9
Hong Kong, China	+ 8
Tehran, Iran	+ 4
Moscow, Russia	+ 3
Helsinki, Finland	+ 2
Rome, Italy	+ 1
London, England	0
Rio de Janeiro, Brazil	− 3
Halifax, Nova Scotia	− 4
New York, New York, USA	− 5
Chicago, Illinois, USA	− 6
Denver, Colorado, USA	− 7
Los Angeles, California, USA	− 8
Honolulu, Hawaii, USA	−10

72 Suppose you leave London and travel to Helsinki. How should you adjust your watch when you arrive in Helsinki?

73 The next day you fly from Helsinki to Moscow and then on to Rome. Identify the hour changes for these cities. What is the total difference in these values from Helsinki to Rome? How should you adjust your watch when you arrive in Rome?

74 Soon after this, you leave Rome, fly west to New York and then on to Denver. How should you adjust your watch when you arrive in Denver?

You have received a monthly credit card statement with a summary of your transactions.

+	Previous Balance	$294.39
−	Payments	200.00
−	Credits	17.81
+	Purchases and cash advances	128.33
+	Debit adjustments	0.00
+	Finance charge	1.87
	* New Balance	$206.78

* An amount followed by a minus sign is a credit or a credit balance, unless otherwise indicated.

75 How much did you pay during the last month to reduce your credit card debt? Why is it listed as a negative number?

76 What is the meaning of a positive value in the summary?

77 What circumstances would create a condition referred to by the footnote ("* An amount followed . . . ")?

INDUSTRIAL TECHNOLOGY

An automobile's ammeter shows whether the battery is charging (+) or discharging (−), and the current in the circuit. Examine the ammeter displays.

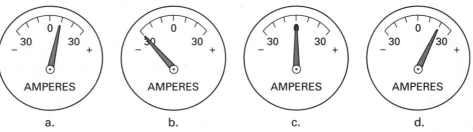

a. b. c. d.

78 Write the reading from each display. Use the sign to indicate a charging or a discharging condition.

79 If a fuse fails, it will do so at the maximum current. Write the absolute value of the current readings from Exercise 78. Which ammeter reading shows the current with the greatest absolute value?

A furniture manufacturing line accepts hardwood lumber that has a specification (spec) weight of 20.7 pounds. When the lumber pieces enter the plant, they proceed down a conveyor belt and pass over a weigh scale set to the spec weight. You view the window of the weigh scale as each piece passes and record the deviations from "spec."

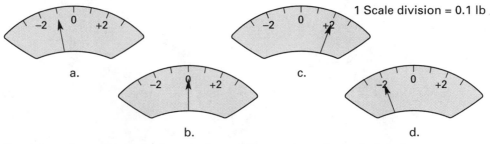

80 Examine the weigh scale windows. Record each deviation from spec.

81 Find the average deviation from spec weight for your observations.

82 Use the average deviation to determine the average weight of the four pieces you observed.

Suppose a tank is filled at the rate of 40 gallons of water per minute. Meanwhile, a valve is opened to drain 32 gallons per minute into a pipeline.

83 Express each rate as an integer.

84 Use these integers to compute the number of gallons of water flowing into the tank in 10 minutes and the number of gallons drained from the tank in 10 minutes.

85 If the tank starts out empty, what is the total amount of water in the tank after 10 minutes?

The ignition timing of an automobile is referenced to "top dead center," more commonly known as TDC. The specification for a certain car calls for a timing of 10° before TDC. An inspection of a customer's poorly running car reveals the ignition timing is actually firing at 2° *after* TDC.

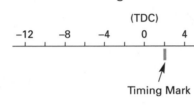

(TDC)

−12   −8   −4   0   4

Timing Mark

86 Express the specification timing and the actual timing as integers.

87 How much does the actual timing differ from the specification timing?

You and a coworker carry a tool chest that weighs 86.6 pounds. Using the handles on each end of the chest, the two of you carry the chest as shown in the drawing.

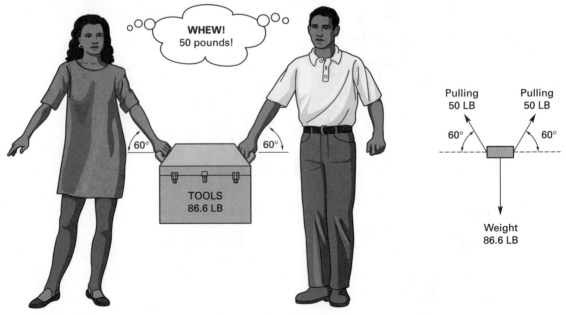

WHEW! 50 pounds!

60°   60°

TOOLS 86.6 LB

Pulling 50 LB   Pulling 50 LB

60°   60°

Weight 86.6 LB

88 Suppose that each of you pulls with a force of 50 pounds at an angle of 60°, as shown in the drawing. What is the resultant (vector sum) of these two pulling forces? What is the effect of the resultant force on the tool chest?

89 What would the resultant vector be if you and your coworker each pulled with a force of 50 pounds at an angle of 45°? What is the effect of this resultant force on the tool chest?

90 If each of you lift straight up—at 90°—what is your lifting force on the tool chest? Show your answer with a vector diagram.

Examine the simplified side view of a metal framework requiring an interior weld. Note the initial position of the welding tip. You have been assigned the task of writing a program for the robotic welder. This progam will move the welding tip from the location shown to the weld location.

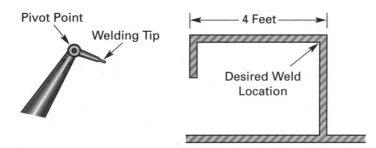

91 Sketch the metal framework and initial position of the welding tip. Draw a set of vectors your program could contain that would move the welding tip to the desired location without striking the frame. (Consider only the positioning of the tip. For this exercise, do not worry about the rest of the robot arm.)

92 How many different answers are there to this problem?

93 Draw the resultant vector that is the sum of your vector instructions. How many different answers are there to this question?

CHAPTER 1 ASSESSMENT

Skills

Find the absolute value.

1. $|-8|$ **2.** $|25|$ **3.** $|-16|$

Which integer is larger?

4. -6 or -3 **5.** -8 or -12 **6.** 0 or -3

Add, subtract, multiply, or divide.

7. 9×-6 **8.** $-23 - 51$ **9.** $-9 + (-5)$

10. $18 - (-14)$ **11.** $-36 \div -9$ **12.** $56 \div -7$

13. $-28 + 35$ **14.** $24 - (-17)$ **15.** -13×-6

16. $-21 \div 3$ **17.** $-16 - (-16)$ **18.** $-84 \div -4$

Applications

19. The stock market lost 35 points over 5 days. Write the average daily loss as an integer.

20. The record high temperature for a city is 107°F. The record low is -25°F. Find the difference between the high and low record temperatures.

The Bay Country Store keeps a record of daily sales. Use the table to complete the Exercises.

Day	Daily Sales Previous week's sales	This week's sales
Monday	$848	$812
Tuesday	$915	$975
Wednesday	$639	$612
Thursday	$986	$846
Friday	$743	$850

21. Write each daily sales difference using an integer.

22. What is the average difference in the daily sales?

Draw a vector diagram on graph paper for problems 23–26. Find the magnitude and direction of each resultant vector in problems 24–26.

23. One person pulls on a cart with a force of 65 pounds due east. Another person pulls with a force of 75 pounds in a direction 35° south of east.

24. An airplane flying east with an air speed of 435 miles per hour.

25. An airplane flying east with an air speed of 415 miles per hour with a *tail wind* of 15 miles per hour.

26. An airplane flying east with an air speed of 425 miles per hour into a *head wind* of 15 miles per hour.

Math Lab

27. Use integers to report the change in pulse rate from exercising at 155 beats per minute (bpm) to complete recovery at 68 bpm.

28. Use graph paper to find the resultant displacement vector when the following displacement vectors are added in the order shown.
    30 feet north
    50 feet west
    15 feet south

29. Two spring scales are attached to a weight as shown in the diagram.

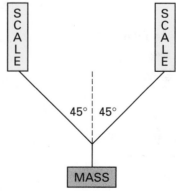

Each scale reads 353.5 grams. What is the weight of the mass?

CHAPTER 2

WHY SHOULD I LEARN THIS?

Scientific notation is a language used in today's high tech workplace. This chapter explores how these mathematical concepts are used in professions ranging from aeronautics to health care. Find out how scientific notation can be part of your future career.

SCIENTIFIC NOTATION

OBJECTIVES

1. Write large and small numbers in powers-of-ten notation.
2. Read and write numbers in scientific notation.
3. Use a scientific calculator to compute numbers displayed in scientific notation.
4. Use numbers written in scientific notation to solve problems.

Many people must use *very large* and *very small* numbers every day on the job. People who work in the space industry, in national defense, or at the World Bank often use large numbers. For example, astronomers look through telescopes to study stars and planets that are great distances from the Earth. Scientists and technicians working in a biology lab or a center for disease control use very small numbers when working with viruses and human cells.

People working in science related jobs use very large and very small number facts. For example,

An ounce of gold contains approximately
86,700,000,000,000,000,000,000 atoms.

One atom of gold has a mass of
0.00000000000000000000327 grams.

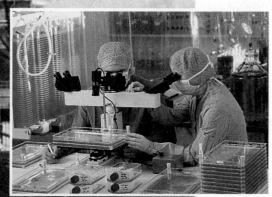

The average volume of an atom of gold is
0.0000000000000000000001695 cubic centimeters.

The mass of the earth is about
6,000,000,000,000,000,000,000 metric tons.

The diameter of a single human red blood cell is about
0.000007 meters.

In this chapter you will use a shortcut way to write very large and very small numbers. This shortcut is called *scientific notation*. As you view the video, watch for the use of very large and very small numbers. Notice how people use them at work.

LESSON 2.1 POWERS OF TEN

Exponents and Powers

You already know how to abbreviate the expression

$$2 + 2 + 2 + 2 + 2 = 10$$

by writing it as $5 \times 2 = 10$. There is also a way to abbreviate repeated multiplication. In the sentence

$$2 \times 2 \times 2 \times 2 \times 2 = 32$$

2 is repeated as a factor 5 times, and the product is 32. You can abbreviate this sentence by writing the number 5 up and to the right of the 2.

$$2^5 = 32$$

In this form, 2 is called the base, and 5 is called the exponent. This sentence is read "2 to the 5th power is 32."

You can raise any integer to a positive power.

Positive Exponent

A positive exponent tells you how many times to use the base as a factor in a repeated multiplication.

For example,

$$(-4)^3 = (-4) \times (-4) \times (-4) = -64$$

When you raise a negative number to a power, use parentheses to show the integer is used as a factor. Thus $(-4)^2 = 16$, but $-4^2 = -16$.

Some powers have special names. The third power is the cube of a number and the second power is the square of a number. The first power of a number is the number itself.

Ongoing Assessment

What is the 4th power of 2? What is -3 squared? What is 5 cubed?

Critical Thinking Suppose you raise a negative integer to a positive power. How can you tell if the resulting value is positive or negative?

There are usually several ways to raise a number to a power on a calculator. On most scientific calculators, there is an x-square key, x^2. There is also a power key, y^x, that raises any number to any power. Experiment with your calculator. Use the power key to check the answers to the assessment questions.

Most programmable calculators use the caret key, ^, to find powers. When you enter $-2\verb|^|4$, the calculator displays 16.

ACTIVITY 1 Powers of 2

2:00 p.m.

2:15 p.m.

1 Make a table of the first 5 positive powers of 2.

2 Explain why the positive powers of two are sometimes referred to as "doubling powers."

3 Suppose the number of bacteria in a culture doubles each minute. If there are 2 bacteria in a culture at 2 PM, how many will be in the culture at 2:15 PM?

Positive Powers of Ten

The numbers 1000, 10,000, or 100,000 are all powers of ten. When you write powers of ten, you write the base 10 with an exponent. For example, the second power of 10, or 10 squared, is written 10^2.

Other examples

1000	$=$	10^3	base 10, exponent 3
10,000	$=$	10^4	base 10, exponent 4
100,000	$=$	10^5	base 10, exponent 5

In each case, the positive exponent tells you how many times to write down the factor 10 and multiply to get the number. You can see a pattern for writing powers of ten by examining the following table

$$10^1 = 10.$$
$$10^2 = 10 \times 10 = 100.$$
$$10^3 = 10 \times 10 \times 10 = 1000.$$
$$10^4 = 10 \times 10 \times 10 \times 10 = 10,000.$$

The decimal point moves one place to the right each time you multiply by another 10.

> **Multiplying by 10**
> Multiplying by 10 is the same as moving the decimal point one place to the right.

ACTIVITY 2 Powers of 10

Use the powers of ten pattern to complete each statement.

1 To write 10^4 as an integer, place __?__ zeros after the 1.

2 To write 10^3 as an integer, place __?__ zeros after the 1.

3 To write 10^2 as an integer, place __?__ zeros after the 1.

4 What is the relationship between the exponent and the number of zeros after the 1?

5 Extend and complete the table for 10^5 and 10^6.

Now do the problem "backward." First write down the integer and then find the power of ten it equals. The first line in the following table is done to get you started. Finish the other two lines.

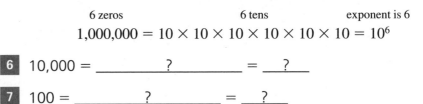

$$\overset{\text{6 zeros}}{} \qquad \overset{\text{6 tens}}{} \qquad \overset{\text{exponent is 6}}{}$$
$$1{,}000{,}000 = 10 \times 10 \times 10 \times 10 \times 10 \times 10 = 10^6$$

6 $10{,}000 = $ _____?_____ = __?__

7 $100 = $ _____?_____ = __?__

8 Explain the relationship between the number of zeros after the 1 and the exponent of 10.

LESSON ASSESSMENT

Think and Discuss

1 What is the meaning of an exponent?

2 How do you find the third power of a positive integer?

3 How do you know how many zeros follow the one when you find any power of ten?

4 How do you write ten million as a power of ten?

Practice and Problem Solving

Write each expression using an exponent.

5. 3×3 **6.** $2 \times 2 \times 2 \times 2 \times 2$

7. -5×-5 **8.** $10 \times 10 \times 10$

9. $-3 \times -3 \times -3 \times -3 \times -3$ **10.** $1 \times 1 \times 1$

Write each expression as an integer.

11. 3^3 **12.** 4^2 **13.** $(-1)^3$

14. 10^6 **15.** 2^7 **16.** $(-2)^3$

17. An electrical generation plant produces over 10^6 watts of power. Write this power level as an integer.

18. The distance between the Earth and the sun is about 10^8 kilometers. Write this distance as an integer.

19. Write 5×10^4 as an integer.

20. Write 7×10^6 as an integer.

Which is greater? Explain why.

21. 3^4 or 4^3 **22.** $(-2)^3$ or $(-1)^2$ **23.** 5^4 or 4^5

Mixed Review

Suppose you are keeping the weight loss statistics in pounds for a fitness club. The table shows the weekly gains and losses for five of your customers.

Person	Week 1	Week 2	Week 3	Week 4
Jose	−2	−1	−2	+1
Tim	−3	−2	−4	+1
Cathy	−4	−1	−2	−5
Maria	−2	+2	−2	+2
Pat	0	+2	0	+2

24. What is the total weight gain or loss for each customer?

+2 −3

25. Draw a number line and place each person in order according to weight gain or loss.

26. What is the average weight gain or loss per week for each customer?

27. What is the average weight gain or loss per week for the group?

28. Compare the average gain or loss per week for each customer to the weekly average for the entire group. By how much more or less than the weekly group average did each of the customers gain or lose? Express your answer as an integer.

The depth gauge of an unmanned submarine is tested at −500 (500 feet below the ocean's surface). If the depth gauge is incorrectly calibrated and is off by a factor of +0.1, the depth gauge error will be

$$(-500) \times +0.1 = -50 \text{ feet}$$

This means the depth gauge will record a depth of

$$-500 - (-50) = -450 \text{ feet}$$

What will the depth gauge record for each of the following calibration error factors?

29. −0.01　　**30.** − 1.3　　**31.** +0.05　　**32.** −0.5

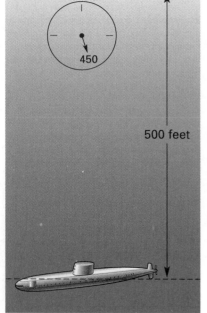

450

500 feet

LESSON 2.2 OTHER POWERS OF TEN

Negative Exponents

In Lesson 2.1, you found a pattern for writing powers of ten with positive exponents. Is there a similar pattern when the exponents are negative? A negative exponent used with base ten tells you how many times to write the factor $\frac{1}{10}$ and multiply to get the answer. That is, 10^{-4} is the same as $(\frac{1}{10})^4$.

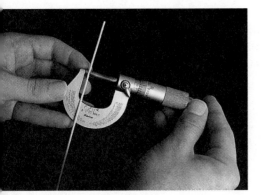

Copy the following table and try to find a pattern.

$$10^{-4} = \frac{1}{10} \times \frac{1}{10} \times \frac{1}{10} \times \frac{1}{10} = \frac{1}{10000} = 0.0001$$

$$10^{-3} = \frac{1}{10} \times \frac{1}{10} \times \frac{1}{10} = \frac{1}{1000} = 0.001$$

$$10^{-2} = \frac{1}{10} \times \frac{1}{10} = \frac{1}{100} = 0.01$$

$$10^{-1} = \frac{1}{10} = 0.1$$

How does the number of decimal places compare with the exponent?

Again, there is a pattern. The absolute value of the exponent tells you the number of times to write the factor $\frac{1}{10}$ and multiply to get the answer, as well as the number of decimal places in the answer.

Check your understanding of the table and the pattern for negative powers of ten by working through Activity 1.

ACTIVITY 1   **Raising 10 to a Negative Power**

1 Extend and complete the table above for 10^{-5} and 10^{-6}. Then, write a relationship giving the exponent, the number of times $\frac{1}{10}$ is used as a factor, and the number of decimal places in the answer.

2 Now do the problem "backward." First write the decimal number. Then write the power of ten it equals. The first line is done to get you started. Complete the others.

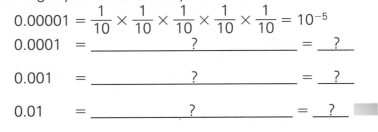

$$0.00001 = \frac{1}{10} \times \frac{1}{10} \times \frac{1}{10} \times \frac{1}{10} \times \frac{1}{10} = 10^{-5}$$

$$0.0001 \quad = \underline{\hspace{4cm}?\hspace{4cm}} = \underline{\ ?\ }$$

$$0.001 \quad = \underline{\hspace{4cm}?\hspace{4cm}} = \underline{\ ?\ }$$

$$0.01 \quad = \underline{\hspace{4cm}?\hspace{4cm}} = \underline{\ ?\ }$$

CULTURAL CONNECTION

Early civilizations used symbols for fractions. However, adding fractions using a common denominator was unknown. The early Egyptians used a method that involved rewriting each fraction as a **unit fraction sum.** Unit fractions are fractions with a 1 in the numerator.

For example, a scribe would write the fraction $\frac{2}{7}$ as $\frac{1}{4} + \frac{1}{28}$. It was the job of the scribes to

make tables of as many unit fractions as the accountants and surveyors needed. But making the tables was difficult because the Egyptians wrote their numbers and letters in hieroglyphics on papyrus. Thus, the fraction $\frac{1}{4}$ was written with an ellipse as the numerator and a dot as the denominator.

How would you write $\frac{5}{6}$ using unit fractions?

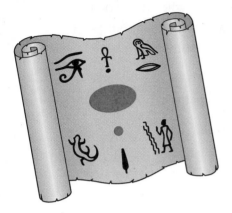

$$\frac{2}{7} = \frac{1}{4} + \frac{1}{28}$$

Zero Exponents

You have written the numbers for powers of ten, such as 10^6, 10^5, 10^4 ..., down to 10^{-5} and 10^{-6}. But you passed over one particular power of ten that belongs in this sequence. What does 10^0 (ten to the zero power) equal? Our previous way of writing 10 as a factor and multiplying does not work here. The statement "*write ten zero times and multiply*" does not make sense. Complete the following Activity and find a pattern.

ACTIVITY 2   **10 to the Zero Power**

1 Copy the powers of ten table.

$$10^3 \ = 1000.$$
$$10^2 \ = 100.$$
$$10^1 \ = 10.$$
$$10^0 \ = ????$$
$$10^{-1} = 0.1$$
$$10^{-2} = 0.01$$
$$10^{-3} = 0.001$$

2 Start at the first line. As the exponent decreases, what is happening to the number of zeros? Notice that whenever you divide a line by 10, the result is on the line below it.

3 To maintain this pattern, what is the value of 10^0?

> **Zero power of 10**
> 10^0 is another name for 1.

Critical Thinking Use a pattern to show that 2^0 is also equal to 1.

LESSON ASSESSMENT

Think and Discuss

1 Explain what each of the numbers in 3^{-2} means.

2 Explain why 10^0 and 2^0 are equal to 1.

3 How do you write the decimal representation for 10^{-3}?

4 How do you choose the exponent when writing 0.000001 as a power of ten?

5 Explain why a positive number with a negative exponent is not negative.

Practice and Problem Solving

Write each expression as an integer or a ratio without exponents.

6. 10^{-4} **7.** 3^{-5} **8.** 4^0 **9.** 10^{-1}

10. 2^{-3} **11.** 5^{-2} **12.** 10^{-7} **13.** 1^{-7}

14. 10^3 **15.** 10^{-3} **16.** $10^3 \times 10^{-3}$

Write each expression with a negative exponent.

17. $\dfrac{1}{3} \times \dfrac{1}{3} \times \dfrac{1}{3} \times \dfrac{1}{3}$ **18.** $\dfrac{1}{5} \times \dfrac{1}{5} \times \dfrac{1}{5}$ **19.** $\dfrac{1}{10} \times \dfrac{1}{10}$

20. The probability of rolling a 4 on a six-sided number cube is $\frac{1}{6}$. the probability of rolling a 4 twice in a row is $\frac{1}{6} \times \frac{1}{6}$. The probability of rolling a 4 three times in a row is $\frac{1}{6} \times \frac{1}{6} \times \frac{1}{6}$. What is the probability of rolling a 4 four times in a row? Express your answer as a product of fractions and as a number with a negative exponent.

21. A computer can read a number stored on its disk drive in about one nanosecond, or 10^{-9} seconds. Write this time as a ratio and as a decimal without an exponent.

Mixed Review

The lengths (in miles) of some of the principal rivers in the world are given in the table. Use $<$ or $>$ to compare the lengths of each pair of rivers.

Amazon	4000
Danube	1776
Ganges	1560
Mississippi	2340
Nile	4160
Ohio	1310
Rio Grande	1900
St. Lawrence	800
Volga	2194

22. Amazon; St. Lawrence

23. Volga; Mississippi

24. Danube; Ganges

25. Use an integer to show how much longer or shorter than the Rio Grande each of the rest of the rivers is.

LESSON 2.3 POWER OF TEN NOTATION

Using a Shortcut

Recall the very large number of atoms in an ounce of gold

86,700,000,000,000,000,000,000

Notice that the number 867 is followed by 20 zeros. In Lesson 2.1 you learned that 1 followed by twenty zeros is the same as writing 10^{20}. Thus, you can write

$$86,700,000,000,000,000,000,000 = 867 \times 10^{20}$$

which is now in **power of ten notation.**

A single human red blood cell is about 0.000007 meters in diameter. How can you rewrite this diameter in power of ten notation?

First write the number 7 as 7.0. Recall that $7.0 \div 10$ or $7 \times \frac{1}{10}$ is another name for 0.7. See if you can find a pattern in the following table.

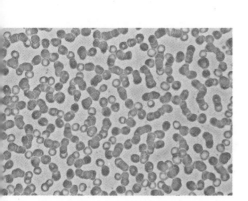

0.000007 m

$$7.0 \times \frac{1}{10} = 0.7$$

$$0.7 \times \frac{1}{10} = 0.07$$

$$0.07 \times \frac{1}{10} = 0.007$$

$$0.007 \times \frac{1}{10} = 0.0007$$

$$0.0007 \times \frac{1}{10} = 0.00007$$

Notice that multiplying a number by $\frac{1}{10}$ is the same as moving the decimal point one place to the left.

> **Multiplying by $\frac{1}{10}$ or dividing by 10**
> Multiplying a number by $\frac{1}{10}$ or dividing the number by 10 moves the decimal point one place to the left.

Think again about the number for the diameter of the red blood cell, 0.000007 meters. The previous discussion shows that you can write 0.000007 as 7 multiplied by six factors of $\frac{1}{10}$.

$$0.000007 = 7 \times \frac{1}{10} \times \frac{1}{10} \times \frac{1}{10} \times \frac{1}{10} \times \frac{1}{10} \times \frac{1}{10}$$

But you know that writing the factor $\frac{1}{10}$ six times is the same as writing 10^{-6}. So,

$$7 \times \frac{1}{10} \times \frac{1}{10} \times \frac{1}{10} \times \frac{1}{10} \times \frac{1}{10} \times \frac{1}{10} = 7 \times 10^{-6}$$

Thus, you can write the diameter of a human red blood cell as 0.000007 meters or as 7×10^{-6} meters.

The Decimal System

Since you use exactly *ten* digits (0–9) to write numbers, our system is called the **decimal system**. Each digit in a number is associated with exactly one power of ten. Thus, you can convert numbers written in decimal notation to **exponential notation**.

Decimal Notation	Expanded Notation	Exponential Notation
345	$3 \times 100 + 4 \times 10 + 5 \times 1$	$3 \times 10^2 + 4 \times 10^1 + 5 \times 10^0$
7.28	$7 \times 1 + 2 \times \frac{1}{10} + 8 \times \frac{1}{100}$	$7 \times 10^0 + 2 \times 10^{-1} + 8 \times 10^{-2}$
90.6	$9 \times 10 + 0 \times 1 + 6 \times \frac{1}{10}$	$9 \times 10^1 + 0 \times 10^0 + 6 \times 10^{-1}$

Ongoing Assessment

Write 34.06 in exponential notation.

LESSON ASSESSMENT

Think and Discuss

1 The distance from the sun to Mars is about 228 million miles. Explain how to write this distance in power of ten notation.

2 When 10 is raised to a negative power, how do you know how many places to move the decimal point? Do you move it to the left or right?

3 What happens to the decimal point when a number is multiplied by $\frac{1}{10}$?

4 Explain how to write a number in exponential notation.

Write each amount in power of ten notation.

5. The Earth is about 93,000,000 miles from the sun.

6. A basketball player makes $25,000,000 on a 5-year contract.

7. There are approximately 5,500,000,000 people in the world.

8. The land area of the Earth is about 52,000,000 square miles.

9. The average diameter of a polio virus is 0.000025 millimeters.

10. It takes light about 0.000000001 seconds to travel 30 centimeters.

11. The national debt is over $4,000,000,000,000.

12. The nearest star other than the sun is about 40,000,000,000,000 kilometers away.

Write each number in exponential notation.

13. 23,683   **14.** 45.39   **15.** 4.189   **16.** 726.05

Mixed Review

Simplify.

17. $-49 \div 7$ **18.** -6×-8 **19.** $-6 + (-3)$

20. $15 - (-7) - 4$ **21.** $9 - 16 + (-12)$ **22.** $-3 \times -9 \times -2$

23. One week Roberta works 35 hours at $6.45 per hour. The same week Ricardo works 38 hours at $5.95 per hour. What is the difference in their pay for the week?

24. Business financial statements show a loss recorded in parentheses. The profit and loss from Donita's plumbing business for the first year is shown in her quarterly report. What is the total profit or loss for the year?

Q1	Q2	Q3	Q4	Total
($1246.59)	$4759.35	$8427.89	($851.17)	?

LESSON 2.4 SCIENTIFIC NOTATION

You can write the estimated world population as a power of ten.

$$5.5 \times 10^9$$

This number is also written in **scientific notation.**

> **Scientific Notation**
> A number is in **scientific notation** when it is a number greater than or equal to 1, but less than 10, multiplied by a power of ten.

Large Numbers and Scientific Notation

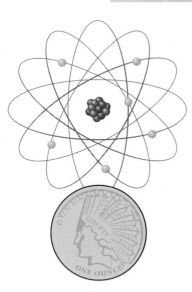

Look again at the number of atoms in an ounce of gold. In power of ten notation, you write 867×10^{20}. The number 867 is multiplied by a power of ten, but 867 is greater than 10. Thus, 867×10^{20} is not in scientific notation.

To rewrite 867×10^{20} in scientific notation, you must begin with a number that is between one and ten. Thus you must replace 867 with 8.67. But, to retain the value 867, you must multiply 8.67 by 10^2.

The result is

$$867 = 8.67 \times 10^2$$

Now 86,700,000,000,000,000,000,000 can be written in scientific notation.

$$86,700,000,000,000,000,000,000 = 867 \times 10^{20}$$
$$= 8.67 \times 10^2 \times 10^{20}$$

Multiplying by 10 twice, and then by 10 twenty times, is the same as using 10 as a factor 22 times. This means that

$$10^2 \times 10^{20} = 10^{22}$$

Thus,

$$86,700,000,000,000,000,000,000 = 8.67 \times 10^{22}$$

which is now in scientific notation.

Critical Thinking Explain why 5.06×10^{15} is in scientific notation.

EXAMPLE 1 Writing a Number in Scientific Notation

Write 270,000,000 molecules (the number of hemoglobin molecules in a single human red blood cell) in scientific notation.

SOLUTION

Place a pointer to the *right of the first nonzero digit*.

$$270,000,000. \text{ molecules}$$
$$\uparrow$$

Count the spaces as you move from the decimal point to the pointer ($\uparrow$). (Remember, there is a decimal point to the right of the last zero.) As you move from the decimal point to the pointer ($\uparrow$), count 8 spaces.

$$270,000,000. \text{ molecules}$$
$$\uparrow \quad \text{8 spaces}$$

Move the decimal point to the pointer and multiply the number by 10 with an exponent equal to the number of places you moved—in this case, 8.

The result is 2.7×10^8 molecules.

Thus, 2.7×10^8 is in scientific notation because 2.7 is between 1 and 10 and it is multiplied by a power of 10.

Small Numbers and Scientific Notation

The same process works when you need to write a very small number in scientific notation.

EXAMPLE 2 Small Numbers in Scientific Notation

Write 0.000007 meters in scientific notation.

SOLUTION

Place a pointer to the right of the first digit that is not zero.

$$0.000007 \text{ meters}$$
$$\uparrow$$

Count the decimal places as you move from the decimal point to the pointer. As you move right count 6 spaces.

$$0.\underbrace{000007}_{6 \text{ spaces}} \text{ meters}$$

Move the decimal point to the pointer and multiply by $\frac{1}{10}$ with an exponent equal to the number of places you moved (in this case 6). Remember that $(\frac{1}{10})^6$ is the same as 10^{-6}.

The result is 7×10^{-6} meters.

7×10^{-6} is in scientific notation because 7 is between 1 and 10 and is multiplied by a power of ten.

Ongoing Assessment

Write each number in scientific notation.

a. 93,000,000   b. 0.000025

Converting Scientific Notation to Decimal Notation

Sometimes it is useful to rewrite numbers in the opposite direction; that is, to convert from scientific notation to decimal notation.

EXAMPLE 3 Writing a Number in Decimal Notation

Write 1.6022×10^4 in decimal notation.

SOLUTION

Place a pointer at the decimal point.

$$1.6022 \times 10^4$$
$$\uparrow$$

Count the spaces as you move left or right from the pointer (left if the exponent is negative, right if the exponent is positive). Move the same number of spaces as the absolute value of the exponent (add zeros if needed). In this case, move right 4 spaces.

Put a decimal point after the fourth place.

16022. is in decimal notation.

Write 4.023×10^7 in decimal notation.

Numbers written in scientific notation and containing negative exponents can also be written in decimal notation.

EXAMPLE 4 Negative Exponents and Scientific Notation

Write 1.6022×10^{-19} coulombs (the charge of an electron) in decimal notation.

SOLUTION

Place a pointer at the decimal point.

$$1.6022 \times 10^{-19} \text{ coulombs}$$
$$\uparrow$$

Move left from the pointer (since the exponent is negative) 19 places (19 is the absolute value of -19), adding zeros as needed.

$$00000000000000000016022 \text{ coulombs}$$
$$\uparrow$$

Move the decimal point to the left of the last zero you wrote and add a zero in front of the decimal point to protect it from being accidentally omitted.

$0.00000000000000000016022$ coulombs is in decimal notation.

LESSON ASSESSMENT

Think and Discuss

1 The average length of a bacteria is about 0.0015 centimeters. How do you write this number in scientific notation?

2 The average distance from the sun to the Earth is about 150,000,000 kilometers. How do you write this distance in scientific notation?

3 Compare power of ten notation and scientific notation. How are they alike and different?

4 When you convert a number from scientific notation to decimal notation, how do you determine how to move the decimal point?

Practice and Problem Solving

Write each amount in scientific notation.

5. 56.7 **6.** 0.0089 **7.** 245,000

8. 0.00000145 **9.** 93,673 **10.** 0.000003

11. 45,000,000 **12.** 0.00203 **13.** 135.737

14. The national debt is about 4 trillion dollars.

15. At sea level there are about 20,000,000,000,000,000,000 molecules per cubic centimeter.

16. The thickness of this page is about 0.00008 meters.

17. The wavelength of red light is about 0.0000065 meters.

Write each amount in decimal notation.

18. 4.7×10^4 **19.** 1.0×10^{-5} **20.** 7.9×10^{-7}

21. The distance light travels in one year is about 5.87×10^{12} miles.

22. The diameter of the largest moon of Jupiter is about 3.5×10^3 miles.

23. The length of a microchip is about 3×10^{-7} meters.

24. The star Alpha Centauri is about 4.1×10^{12} kilometers from Earth.

25. Alicia needs $4,675 to buy a used car. She has saved $225 for each of the last 15 months. If she continues to save at the same rate, how many more months will it take Alicia to save enough money to buy the car?

26. Julius Caesar was born in 100 BCE. He died when he was 56 years old. In what year did Caesar die?

27. The countdown clock currently shows T−85 seconds and counting. The second stage rocket will go into last check at T+20 seconds. How long after the current countdown time will the second stage check take place?

28. Your normal temperature is about 98.6°F. One morning your temperature registered 101°F. Use a positive or negative number to compare your temperature with normal.

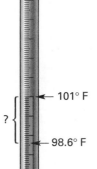

← 101° F

? {

← 98.6° F

LESSON 2.5 USING SCIENTIFIC NOTATION

In Lesson 2.4, you found that multiplying $10^2 \times 10^{20}$ results in the same product as using 10 as a factor 22 times or 10^{22}. This idea leads to a rule for multiplying two numbers that have the same base but different exponents.

Multiplying Powers of Ten

10^2 means 10×10. 10^3 means $10 \times 10 \times 10$.

$10^2 \times 10^3$ means $10 \times 10 \times 10 \times 10 \times 10$ or 10^5

What do you notice about the exponents of the factors, 10^2 and 10^3 and the exponent in the product 10^5? Written in a different way, the problem looks like this:

$$10^2 \times 10^3 = 10^{2+3} = 10^5$$

> **Multiplying Powers of Ten**
> To multiply powers of ten, add the exponents.

Critical Thinking Use the rule for multiplying powers of ten to show why 10^0 is equal to 1.

Dividing Powers of Ten

Is there a similar rule for dividing powers of ten?

$10^5 = 10 \times 10 \times 10 \times 10 \times 10$ $10^3 = 10 \times 10 \times 10$

$$\frac{10^5}{10^3} = \frac{10 \times \cancel{10} \times \cancel{10} \times \cancel{10} \times 10}{\cancel{10} \times \cancel{10} \times \cancel{10}} = 10 \times 10 = 10^2$$

Thus,

$$10^5 \div 10^3 = 10^{5-3} = 10^2$$

The same method works for $10^4 \div 10^7$.

First, rewrite the division problem as a fraction.

$$10^4 \div 10^7 = \frac{\cancel{10} \times \cancel{10} \times \cancel{10} \times \cancel{10}}{10 \times \cancel{10} \times \cancel{10} \times \cancel{10} \times \cancel{10} \times 10 \times 10} = \frac{1}{10 \times 10 \times 10}$$

$\underset{\text{dividend}}{\,} \underset{\text{divisor}}{\,}$

The result is

$$\frac{1}{1000} \text{ or } 10^{-3}$$

Thus, $10^4 \div 10^7 = 10^{4-7} = 10^{-3}$.

These two examples suggest a rule for dividing powers of ten.

Dividing Powers of Ten

To divide one power of 10 by another power of 10, subtract the exponent of the divisor from the exponent of the dividend.

Ongoing Assessment

Multiply or divide and write the result as a power of ten.

a. $10^{-5} \times 10^8$ b. $10^{-2} \times 10^{-3}$ c. $10^{-4} \div 10^{-6}$ d. $10^2 \div 10^{-5}$

Your scientific or graphics calculator uses an exponential shift key to convert, multiply, and divide numbers in scientific notation. The exponential shift key is usually designated by the EE (enter exponent) key. On some calculators you must use the 2nd key to write a number in scientific notation. Some calculators automatically convert numbers to scientific notation when they have more digits than can be displayed.

Here is how one calculator displays 256 in scientific notation.

Press the 2nd key along with the SCI key (the 5 key)
Enter 2 5 6
Press =

The display shows 2.56^{02}, which represents 2.56×10^2. Experiment with your calculator to see how it converts numbers to scientific notation.

Critical Thinking   How can you use your calculator to convert a number from scientific notation to decimal notation?

Multiplying Numbers in Scientific Notation

There are 2.7×10^8 hemoglobin molecules in a single human red blood cell. There are about 5.0×10^6 human red blood cells in one cubic millimeter (a small drop) of blood. How many hemoglobin molecules are in one cubic millimeter of blood?

To find the number of hemoglobin molecules in one cubic millimeter of blood, multiply the number of molecules by the number of red blood cells in one cubic millimeter of blood.

$$(2.7 \times 10^8) \times (5.0 \times 10^6)$$

Rewrite this problem to group the numbers between one and ten together and the powers of ten together.

$$(2.7 \times 5.0) \times (10^8 \times 10^6)$$

Use your calculator to multiply the numbers in the first parentheses. You can either multiply $10^8 \times 10^6$ mentally by adding the 8 and 6 to get 10^{14}, or you can use your calculator.

Here is one way to use a scientific calculator to find the number of molecules.

Press ②.⑦ EE ⑧
Press ✕
Press ⑤.⓪ EE ⑥
Press ＝

The calculator displays 1.35^{15}. 1.35×10^{15} is the number of molecules written in scientific notation.

Ongoing Assessment

You have calculated that there are 1.35×10^{15} molecules of hemoglobin in one cubic millimeter of blood. A donation of blood to the American Red Cross consists of 5×10^5 cubic millimeters of blood. How many hemoglobin molecules were donated?

You can work a problem with negative exponents in the same way as positive exponents.

$$(2.0 \times 10^{-3}) \times (1.4 \times 10^{-2}) = (2.0 \times 1.4) \times (10^{-3} \times 10^{-2})$$

$$= 2.8 \times 10^{-5}$$

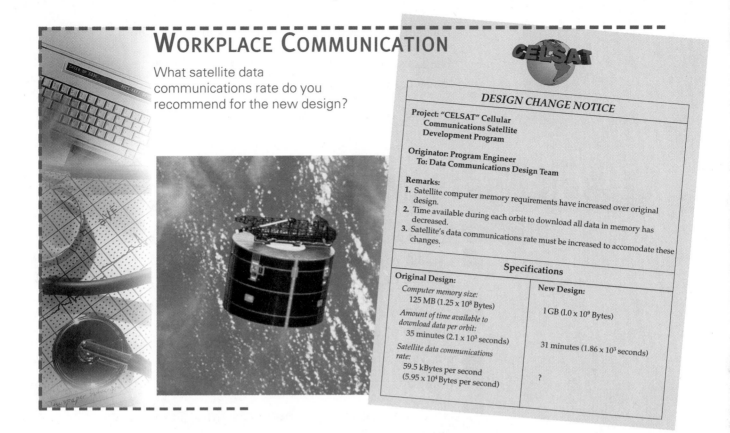

WORKPLACE COMMUNICATION

What satellite data communications rate do you recommend for the new design?

DESIGN CHANGE NOTICE

Project: "CELSAT" Cellular Communications Satellite Development Program

Originator: Program Engineer
To: Data Communications Design Team

Remarks:
1. Satellite computer memory requirements have increased over original design.
2. Time available during each orbit to download all data in memory has decreased.
3. Satellite's data communications rate must be increased to accomodate these changes.

Specifications

Original Design:	New Design:
Computer memory size: 125 MB (1.25×10^8 Bytes)	1 GB (1.0×10^9 Bytes)
Amount of time available to download data per orbit: 35 minutes (2.1×10^3 seconds)	31 minutes (1.86×10^3 seconds)
Satellite data communications rate: 59.5 kBytes per second (5.95×10^4 Bytes per second)	?

Dividing Numbers in Scientific Notation

Suppose you have a gold nugget that measures 5 millimeters on a side and has a mass of 2.41 grams. There are 7.4×10^{21} atoms of gold in the nugget. What is the mass of one atom of gold?

To find the mass of one atom, divide the mass of the nugget by the number of atoms in the nugget. First rewrite 2.41 in scientific notation as 2.41×10^0.

Now, write the division as a fraction

$$\frac{2.41 \times 10^0}{7.4 \times 10^{21}}$$

Use your calculator to simplify the fraction.

Enter 2.41
Press ⌷EE⌷ ⌷0⌷
Press ⌷÷⌷
Enter 7.4
Press ⌷EE⌷ ⌷2⌷ ⌷1⌷
Press ⌷=⌷

Thus, 3.2568×10^{-22} is the number of grams per atom.

The keystrokes used in the activity are sometimes written in the following way.

$2.41 \boxed{\text{EE}} \quad 0 \boxed{\div} 7.4 \boxed{\text{EE}} \quad 21 = 3.2568 \times 10^{-22}$

LESSON ASSESSMENT

Think and Discuss

1 Explain why you add the exponents when you multiply powers of ten.

2 Explain why you subtract the exponents when you divide powers of ten.

3 How do you multiply two numbers written in scientific notation?

4 How do you divide two numbers written in scientific notation?

5 Describe the steps used on your calculator when you multiply or divide two numbers written in scientific notation.

Practice and Problem Solving

Simplify. Write each answer as a power of ten.

6. $10^4 \times 10^{-3}$ **7.** $10^{-2} \div 10^{-4}$ **8.** $10^3 \div 10^7$

9. $10^0 \div 10^3$ **10.** $10^{-5} \times 10^9$ **11.** $10^{-7} \div 10^6$

12. $10^{-4} \times 10^4$ **13.** $10^{12} \div 10^{12}$ **14.** $10^{-3} \times 10^{-3}$

Express each product or quotient in scientific notation.

15. $(7 \times 10^5) \times (5 \times 10^3)$ **16.** $(2.6 \times 10^{-4}) \times (3.7 \times 10^6)$

17. $(5.4 \times 10^{-3}) \div (1.7 \times 10^5)$ **18.** $(7.1 \times 10^{-2}) \div (8.5 \times 10^{-7})$

19. You are setting up a laser electronics timing experiment. You need to position two mirrors so that the laser beam travels the distance between them in 6 nanoseconds (6×10^{-9} seconds). Since the speed of light is 3×10^8 meters per second, distance traveled is (3×10^8 meters per second) $\times$ (6×10^{-9} seconds).

How far apart should you position the mirrors? Express your answer in power of ten notation, scientific notation, and decimal notation.

20. In another laser experiment, you measure the speed of light in an optical fiber. The laser beam travels through 6 meters of fiber in 22 nanoseconds. You calculate

$$\text{speed of light} = (6 \text{ meters}) \div (22 \times 10^{-9} \text{ seconds})$$

Write 6 and 22×10^{-9} in scientific notation. Then compute the speed of the laser beam in the optical fiber. Write your answer in scientific notation.

Mixed Review

21. In March the normal low temperature in Denver is 25°F. The record low temperature in March is −11°F. How much greater is the normal low temperature than the record low temperature?

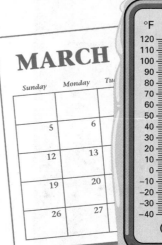

22. There are 8.64×10^4 seconds in one day. Write this number in decimal notation.

23. The surface temperature of the sun is about 6000 degrees Celsius. Write this number in scientific notation.

24. A special balance can weigh an object as light as 0.00000001 gram. Write this number in scientific notation.

Measurement is an important part of many jobs. When you measure the dimensions of an object, you are comparing the dimensions to a standard unit of measure. On most jobs, you will measure length, mass, or capacity. These measurements are usually made in feet, pounds, or gallons. But in science and industry, these measurements are more often made using the basic standards of the metric system: meters, grams, and liters.

The metric system is a decimal system. Exponents of ten and scientific notation make it easy to convert measurements within the metric system. Prefixes used in the metric system are shown in the following table.

Value	Exponent	Symbol	Prefix
1 000 000 000 000	10^{12}	T	tera
1 000 000 000	10^{9}	G	giga
1 000 000	10^{6}	M	mega
1 000	10^{3}	k	kilo
100	10^{2}	h	hecto
10	10^{1}	da	deca
0.1	10^{-1}	d	deci
0.01	10^{-2}	c	centi
0.001	10^{-3}	m	milli
0.000001	10^{-6}	μ	micro
0.000000001	10^{-9}	n	nano
0.000000000001	10^{-12}	p	pico

You can write these prefixes before any metric unit of measure. For, example, they can precede the meter, the gram, or the liter. Although the meter is the basic unit for measuring length, the centimeter and the kilometer are used to measure smaller or larger distances.

You also use metric prefixes for units of time such as seconds. The data in computers move at speeds measured in nanoseconds (10^{-9} seconds) and even picoseconds (10^{-12} seconds). The prefixes are used for many electrical units, such as ohms, amperes, volts, and watts. A megaohm is a million ohms. A microamp is one one-millionth of an ampere.

The following table helps you convert from one metric unit to another. Notice that the basic units (meter, gram, and liter) are in the middle box of the first column and in the middle box of the top row.

	TO MILLI-	TO CENTI-	TO DECI-	TO METER GRAM LITER	TO DECA-	TO HECTO-	TO KILO-
KILO-	$\times 10^6$	$\times 10^5$	$\times 10^4$	$\times 10^3$	$\times 10^2$	$\times 10^1$	
HECTO-	$\times 10^5$	$\times 10^4$	$\times 10^3$	$\times 10^2$	$\times 10^1$		$\times 10^{-1}$
DECA-	$\times 10^4$	$\times 10^3$	$\times 10^2$	$\times 10^1$		$\times 10^{-1}$	$\times 10^{-2}$
METER GRAM LITER	$\times 10^3$	$\times 10^2$	$\times 10^1$		$\times 10^{-1}$	$\times 10^{-2}$	$\times 10^{-3}$
DECI-	$\times 10^2$	$\times 10^1$		$\times 10^{-1}$	$\times 10^{-2}$	$\times 10^{-3}$	$\times 10^{-4}$
CENTI-	$\times 10^1$		$\times 10^{-1}$	$\times 10^{-2}$	$\times 10^{-3}$	$\times 10^{-4}$	$\times 10^{-5}$
MILLI-		$\times 10^{-1}$	$\times 10^{-2}$	$\times 10^{-3}$	$\times 10^{-4}$	$\times 10^{-5}$	$\times 10^{-6}$

(TO CONVERT)

Follow these steps to use the chart.

1. Find the name in the column at the left for the unit you are converting.

2. Go across this row to the column with the heading of the unit you are converting to.

3. Multiply the unit you are converting by the power of ten shown in the box.

Do the following Activity to see how to use the chart to convert metric measures.

ACTIVITY   **Converting Metric Measures**

How do you change 35 milliamps to amperes? An ampere is another basic unit like the meter, gram, or liter.

TO CONVERT	TO MILLI-	TO CENTI-	TO DECI-	TO METER GRAM LITER	TO DECA-	TO HECTO-	TO KILO-
KILO-	$\times 10^6$	$\times 10^5$	$\times 10^4$	$\times 10^3$	$\times 10^2$	$\times 10^1$	░
HECTO-	$\times 10^5$	$\times 10^4$	$\times 10^3$	$\times 10^2$	$\times 10^1$	░	$\times 10^{-1}$
DECA-	$\times 10^4$	$\times 10^3$	$\times 10^2$	$\times 10^1$	░	$\times 10^{-1}$	$\times 10^{-2}$
METER GRAM LITER	$\times 10^3$	$\times 10^2$	$\times 10^1$	░	$\times 10^{-1}$	$\times 10^{-2}$	$\times 10^{-3}$
DECI-	$\times 10^2$	$\times 10^1$	░	$\times 10^{-1}$	$\times 10^{-2}$	$\times 10^{-3}$	$\times 10^{-4}$
CENTI-	$\times 10^1$	░	$\times 10^{-1}$	$\times 10^{-2}$	$\times 10^{-3}$	$\times 10^{-4}$	$\times 10^{-5}$
MILLI-	░	$\times 10^{-1}$	$\times 10^{-2}$	$\times 10^{-3}$	$\times 10^{-4}$	$\times 10^{-5}$	$\times 10^{-6}$

1 Since you are converting milliamps, find "milli" in the column at the left.

2 Since you are converting to amperes, move across the row until you reach the column with the basic units for the heading. Since ampere is a basic metric unit, you can add ampere (or any other standard metric unit) to the top center box of the chart. The power of ten in the box is 10^{-3}.

3 Multiply 35 by 10^{-3}.

$$35 \text{ milliamps} = 35 \times 10^{-3} \text{ amperes}$$

Remember, multiplying by a negative power of ten moves the decimal point to the left as many places as the absolute value of the exponent. Thus 35×10^{-3} amperes is the same as 0.035 amperes.

Critical Thinking   Change 4.5 meters to centimeters. What happens to the decimal point when you multiply 4.5 by 10^2?

LESSON ASSESSMENT

Think and Discuss

1 Why are standard units of measure needed?

2 Explain how to use the metric conversion table to convert centimeters to meters.

3 Explain how to use the metric conversion table to convert kilograms to milligrams.

4 Why is the metric system rather than the English system preferred by scientists and engineers?

Practice and Problem Solving

Use the metric conversion table on page 29 to change

5. 34 meters to kilometers.

6. 6.8 milliliters to liters.

7. 5.8 centimeters to millimeters.

8. 17 kilowatts to watts.

9. 350 milliseconds to seconds.

10. 6 millimeters to centimeters.

11. 0.8 kilograms to milligrams.

12. 0.006 meters to millimeters.

13. 8.3 centimeters to meters.

14. 1000 grams to kilograms.

15. 142 milliamps to amps.

16. 500 liters to milliliters.

17. What power of ten is used to convert picoseconds to seconds?

18. What power of ten is used to convert gigabytes to bytes?

The windchill is determined from air temperature and wind speed.

Wind Speed in Miles Per Hour	Air Temperature in °F			
	20	10	0	–10
0				
10	3	–9	–22	–34
20	–10	–24	–39	–53
30	–18	–33	–49	–64

Use the windchill table to find each of the following.

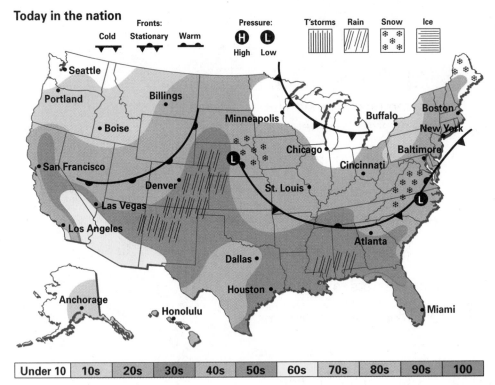

Today in the nation

19. In Boston the air temperature is 20°F and the wind is blowing 20 miles per hour. What is the windchill?

20. In Chicago the windchill is reported to be −22 with a 10 mile per hour wind. What is the air temperature?

21. The air temperature in Buffalo is 10°F. The wind suddenly increases from 10 to 20 miles per hour. Use an integer to describe the change in windchill.

22. Suppose your watch loses 1.5 minutes per day. Use integers to describe the total minutes lost after 10 days?

23. The diameter of the sun is about 1,392,000 kilometers. Write the diameter in scientific notation.

Perform each operation. Give your answer in scientific notation.

24. $(8 \times 10^4) \times (1.4 \times 10^{-6})$ **25.** $(1.458 \times 10^{-5}) \div (2.7 \times 10^9)$

Math Lab

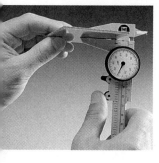

Activity 1: Measuring Average Paper Thickness

Equipment   Calculator
Vernier caliper
Micrometer caliper
Four large books of 500 or more numbered pages

Problem Statement

You will find the average paper thickness of four books. You will then compare your answer to the paper thickness measured with a micrometer caliper.

Procedure

a   Use the vernier caliper to measure the total thickness (to the nearest thousandth of an inch) of pages 1-500 for each book. Record the thickness under these headings:

Book A   Book B   Book C   Book D

b   Calculate an average paper thickness for each book by dividing the measured thickness by 250 (there are two numbered pages for each sheet of paper). Write your group's answer for the average page thickness of each book in scientific notation. Record the results for each group.

c   Calculate the class average of the paper thickness for each book. Write each class-averaged paper thickness in scientific notation.

d   How does the average paper thickness calculated by your group compare to the class-averaged paper thickness for each book? Which is more accurate?

e   Use a micrometer caliper to measure the thickness of a single sheet of paper in each book. Record your answer in scientific notation. How does the paper thickness measured directly with the micrometer caliper compare with each paper thickness your group calculated? Which is more accurate?

Activity 2: Counting Sand Grains

Equipment Calculator
Graduated cylinder
Micrometer caliper
Coarse sand such as kitty litter

Problem Statement

You will count the number of grains of sand in a small volume and use this number to estimate the number of grains in a larger volume. You will also estimate the volume of an average grain of sand.

Procedure

a Use the graduated cylinder to measure 1 milliliter of sand. Count the grains of sand in 1 milliliter.

b From your count, how many grains of sand would you estimate are in a cubic meter? (1 milliliter is equal to 1 cubic centimeter.) Write your answer in scientific notation.

c From your count, how many grains of sand would you estimate are on a strip of beach 1500 meters long by 200 meters wide down to a depth of 10 centimeters? Write your answer in scientific notation.

d From your answer to **a,** what is the average volume (in cubic centimeters) of one grain of sand? Write your answer in scientific notation.

e Isolate 4 or 5 grains of sand and measure their diameter with a micrometer caliper. From these measurements, calculate an average diameter.

Then use the formula for the volume of a sphere ($V = \frac{1}{6}\pi\, d^3$) to get the approximate volume for one grain of sand. Write your answer (in cubic centimeters) in scientific notation.

f How do the values for **d** and **e** compare?

Activity 3: Measuring Water Molecules

Equipment   Calculator
String
Spring scale with 5000-gram capacity
Graduated cylinder, 500-milliliter capacity
Water supply and drain

Problem Statement

You will measure the volume and weight of a sample of water. You will use this data to calculate the number of molecules in the sample and the average volume and weight of each molecule.

Procedure

a Tie a string around the top of the graduated cylinder (see illustration). Weigh the empty graduated cylinder and string by hooking the loop on the spring scale. Record this weight.

b Measure 500-milliliters of water in the graduated cylinder. Use the spring scale to find the combined weight of the graduated cylinder and water. Record the combined weight.

c Subtract the weight of the empty graduated cylinder and string from the combined weight of the water and graduated cylinder. This is the weight of the water sample. There are 6.02×10^{23} molecules of water in 18 grams of water. Determine the number of molecules in your 500-milliliter sample. Write your answer in scientific notation.

d Based on the data in **c,** calculate the average volume per molecule and the average weight per molecule. Write your answer in scientific notation.

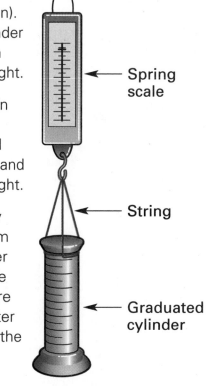

Spring scale

String

Graduated cylinder

MATH APPLICATIONS

The applications that follow are like the ones you will encounter in many workplaces. Use the mathematics you have learned in this chapter to solve the problems. Wherever possible, use your calculator to solve the problems that require numerical answers. A table of conversion factors can be found beginning on page A1.

A tropical year is 365.24220 days. A year expressed on a Gregorian calendar is 365.2425 days.

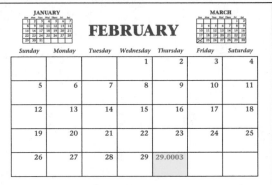

1 What is the difference between a tropical year and the Gregorian calendar year? Express your answer in scientific notation.

2 How many years will it take for this difference to amount to one day?

3 Some home heaters emit pollutants into the air. The maximum acceptable level of carbon monoxide in a room is about 9 parts per million. For nitrogen dioxide a value of about 0.05 parts per million is the suggested amount. Express the maximum acceptable concentration of each gas as a percentage written in scientific notation.

A laser printer prints characters at a density of 12 dots per millimeter on a piece of paper 216 millimeters wide and 279 millimeters long.

4 How many dots can be printed across the paper?

5 How many dots can be printed down the paper?

6 How many dots can be printed on the surface of the paper? Express your answer in scientific notation.

7 If each dot just touched an adjacent dot, what is the diameter of a dot? Express your answer in scientific notation.

The nearest star is about 4.5 light years away. A light year is the distance light can travel in one year. The speed of light is approximately 186,000 miles per second.

8 Convert the speed of light to miles per year by multiplying the speed by the number of seconds in a year. This is the distance light travels in a year.

9 How far away (in miles) is the nearest star? Express your answer in scientific notation.

A recent report of leading corn-growing states in the United States shows the following annual amounts.

State	Annual Corn Growth (Bushels)
Iowa	1,739,900,000
Illinois	1,452,540,000
Nebraska	802,700,000
Minnesota	744,700,000
Indiana	654,000,000
Wisconsin	378,000,000
Ohio	360,000,000
Michigan	273,600,000
Missouri	213,400,000
South Dakota	180,000,000

10 What is the total number of bushels of corn grown by these states? Express your answer in scientific notation.

11 A bushel of corn weighs about 56 pounds. How many pounds of corn did these states produce?

12 The total annual production was reported to be 8,201,000,000 bushels. How many bushels were produced by the states not listed? Express your answer in scientific notation.

When diesel fuel burns, it releases about 138,400 Btu of energy per gallon. A tractor uses 215 gallons of diesel fuel during 56 hours of operation.

13 How many Btu of energy are released by the 215 gallons of fuel? Express your answer in scientific notation, rounded to two decimal places.

14 Divide the total energy released by 56 hours to find the rate of energy released in Btu per hour. A rate of energy release of 2546 Btu per hour is equivalent to 1 horsepower. Find the number of horsepower released by the diesel fuel over 56 hours of operation.

15 The tractor is doing work at a rate of 70 horsepower. This rate of work is what percent of the horsepower released by the fuel? This is the "thermal efficiency" of the engine.

In one region of Hong Kong, 55,000 persons live on about 24 acres.

16 Convert the 24 acres to square feet. Express your answer in scientific notation.

17 On the average, how many people live on each square foot of space in this region?

18 Multiply the answer to Exercise 17 by the number of square miles per square foot to determine about how many people are living in each square mile in this region.

19 Repeat the above calculations using the population for your city or town, and the area of your town. How does the population per square mile for Hong Kong compare to where you live?

BUSINESS & MARKETING

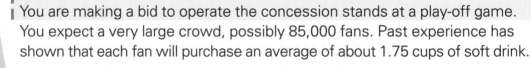

You are making a bid to operate the concession stands at a play-off game. You expect a very large crowd, possibly 85,000 fans. Past experience has shown that each fan will purchase an average of about 1.75 cups of soft drink.

20 How many cups of soft drink can you expect to sell? Express your answer in scientific notation.

21 If each cup is filled with 6 ounces of ice, how many ounces of ice can you expect to use? Express your answer in scientific notation.

A newspaper publisher uses 0.17 pounds of ink for every 1000 pages of print. The newspaper averages 80 pages per day. The distribution averages about 130,000 papers per day.

22 What is the average amount of ink on each page of newsprint? Express your answer in scientific notation.

23 What is the approximate total number of pages of newsprint produced during a week? Express your answer in scientific notation.

24 About how much ink is used during a week's printing? Express your answer in both scientific notation and decimal notation.

A report shows that annual consumer spending for professional services rose from $260,000,000,000 to $410,000,000,000 over a 10-year period.

25 Express each amount in scientific notation.

26 What is the growth in spending for services each year? Express your answer in scientific notation.

A newspaper reports that the average price of a new home in the United States is $93,000. The report estimates that 65,000 new homes were sold nationwide during a recent period.

27 Approximately how much money was spent on new homes during this period? Express your answer in scientific notation.

28 A friend states that "megabucks" are spent on new homes. Using your knowledge of prefixes and powers of ten, is your friend right to use the term "megabucks"? If so, how many "megabucks" are being spent?

29 The subcompact division of Luxury Motors has produced a total of 17,400,000 cars at its 3 assembly plants. Reports of dissatisfaction with the steering wheel design of these models have been reported from 281 owners. What percent of the owners have expressed dissatisfaction with the steering wheel? Express your answer in scientific notation.

A computer printout has been "formatted" to report its values in scientific notation.

Beginning inventory	7.125 E + 04
Additional purchases	2.450 E + 03
Production adds	1.035 E + 05

30 What is the total of the three figures reported?

31 The data were rounded off for the report. Write each of the three numbers in decimal notation and indicate where you think the rounding occurred in each number.

32 Suppose that a laser technician wanted to see more digits (less rounding). She suggests reporting the inventory in terms of "thousand units." Rewrite the data in terms of "thousand units."

Suppose the national budget is $1.8 trillion ($1,800,000,000,000).

33 Express the budget figure in scientific notation.

34 If the population of the country is 250,000,000, how much is budgeted per person in the country?

A report states that credit card users charged a total of $150 billion during a year. During the reported time period, lenders collected a total of $12.6 billion in interest on outstanding charges.

35 Express each of the dollar figures in scientific notation.

36 What percent of the charged amount is the interest amount collected by the lenders?

A computer sales brochure advertises that its new computer system has a 16-megabyte memory and can store a total of nearly a half-gigabyte of data on its three hard disk drives.

37 If a single character occupies a byte of memory or disk space, how many characters can be stored in the 16-megabyte memory? Express your answer in scientific notation and decimal notation.

38 How many characters can be stored on each of the three disk drives? Express your answer in scientific notation and decimal notation.

39 A warehouse has 7 rooms. Each room has 8 rows of 120 bays. Each bay holds 25 pallets. Each pallet contains 40 tires. Each tire averages 35 pounds. Approximately how many pounds of tires can the warehouse hold? Express your answer in scientific notation.

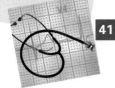
A synthetic thyroid preparation that is available in 25-microgram amounts for each tablet is prescribed to a patient. The bottle of medication contains 30 tablets.

40 How many grams of medication are in each tablet? Express your answer in scientific notation.

41 How many grams of medication are in each bottle of medication? Express your answer in scientific notation.

A group of researchers conducted independent studies of a bacterium. The researchers recorded the average length of the bacteria.

Researcher	Average Reported Length
Choy	0.001633 micrometers
Robert	1.542 nanometers
Salsido	0.001491 micrometers
Dana	1584 picometers

42 Express each measurement in scientific notation in units of meters.

43 Determine the average of the four researchers' results. Express your answer in scientific notation.

44 What is the difference between the largest and smallest reported result? (This is called the "range" of the reported results.)

45 Suppose a particular bacterium can divide into two bacteria, once each hour. Thus, in one hour, one bacterium becomes two. After two hours, two bacteria become four (2×2). In three hours, they become eight ($2 \times 2 \times 2$), and so on. If this process continues for a total of 48 hours, how many bacteria will there be? Write this number in scientific notation and in decimal notation.

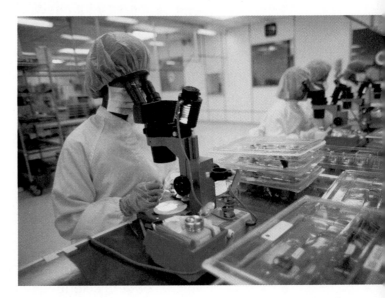

Three samples of ocean salt are weighed on an electronic balance. The samples weigh 0.052476 grams, 0.052482 grams, and 0.052490 grams.

46 Express each weight in scientific notation.

47 What is the difference in weight between the largest and smallest samples? Express your answer in scientific notation.

48 What is the prefix that denotes the smallest increment of weight reported using this balance (the last digit)?

INDUSTRIAL TECHNOLOGY

The spark plugs in most automotive engines must provide a spark for every two revolutions of the engine. When a car is driven at normal speeds, the engine turns at about 2400 revolutions per minute (rpm).

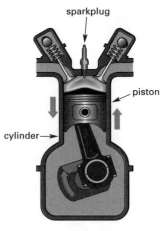

sparkplug

piston

cylinder

49 If you assume an average speed of 1 mile per minute and an annual mileage of 10,000 miles, approximately how many minutes is the car driven during the year?

50 At 2400 revolutions per minute, or 1200 sparks per minute, during the year's driving, about how many times does the spark plug fire during the year? Express your answer in scientific notation.

51 Cosmic rays entering the earth's atmosphere from outer space typically have an energy of about 2 GeV, or 2 giga-electron volts. Express the energy of these rays in electron volts (eV) using scientific notation.

A thermocouple made with lead and gold wires is designed to measure temperature. The thermocouple produces an electric potential of 2.90 microvolts for each Celsius degree above 0°C.

52 Express the thermocouple output as volts per Celsius degree in scientific notation.

53 What voltage would you expect from a thermocouple at 15°C? Express your answer in scientific notation.

A certain isotope of radioactive uranium has a half-life of 4.5×10^9 years. This means that in 4.5×10^9 years, a sample of this isotope will have one-half the level of radioactivity it has today.

54 What is the half-life of this isotope written in decimal notation?

55 If a generation is considered to be 40 years, how many generations will it take for this isotope's radioactivity to be reduced by half?

Radio and television frequencies are given in hertz, or cycles per second. Listed below are frequencies for other common forms of electromagnetism. Use the prefixes to convert each frequency to scientific notation in units of Hz.

Type of Broadcast	Frequency
56 Household electricity	60 Hz
57 AM radio	1080 kHz (kilohertz)
58 Shortwave radio	10 MHz (megahertz)
59 FM radio	102 MHz (megahertz)
60 Radar	8 GHz (gigahertz)
61 Microwave communication	12 GHz (gigahertz)
62 Visible light	400 THz (terahertz)

A quartz crystal is commonly used in a digital watch to provide a stable time base. The crystal oscillates with a frequency of 5.0 megahertz. One hertz is equivalent to one vibration or cycle per second.

63 Express the quartz frequency of vibration in scientific notation, with units of hertz.

64 With this frequency, very small time divisions are possible. The smallest time division is the time for one cycle of the crystal vibration. This cycle period is computed by dividing 1 by the frequency. What is the smallest time division for this quartz crystal in seconds? Express your answer in scientific notation.

A commonly measured value of chemical solutions is pH. This is a measure of hydrogen-ion activity. The hydrogen-ion activity of pure water at 25°C is 0.0000001 mole per liter. A highly acidic solution has 1.0 mole per liter of hydrogen-ion activity, while a highly basic solution has an activity of 0.00000000000001 mole per liter.

65 Express each of the three hydrogen-ion activities in scientific notation.

66 The pH value of a solution is simply the absolute value of the exponent of ten in the measure of its hydrogen-ion activity. What is the pH value of the highly acidic solution above, of pure water, and of the highly basic solution?

A computer is advertised as having a processing speed of "11 mips," or 11 million instructions per second.

67 Express this speed in scientific notation.

68 How long does it take to process each instruction at such a speed?

69 How many "nanoseconds" is this?

Steel can be stretched when a stress is applied. The "modulus of elasticity" for steel is 30×10^6 pounds per square inch (psi). The elongation (or lengthening because of a stress) of a steel beam can be calculated by multiplying the length of the beam by the stress (in psi) and dividing by the modulus of elasticity.

70 Convert the length of a 12-foot steel beam to inches.

71 Suppose a stress of 5000 psi is applied to the beam. Compute the elongation of the beam. Express your answer in scientific notation.

A furnace used to process aluminum consumes approximately 25,000,000 watt-hours of energy per ton (2000 lb) of aluminum processed.

72 Express the energy consumed in scientific notation.

73 If a furnace processes 3200 tons of aluminum during a given period, about how much energy is used? Express your answer in scientific notation.

74 Energy is usually reported in kilowatt-hours (kWh) rather than watt-hours. Convert the answer to Exercise 73 to kWh.

An elevator with a mass of 1200 kg is lifted upward at a constant speed of 2.0 meters per second.

75 Multiply the mass of the elevator by 9.8 newtons per kg to find the force needed to lift the elevator. Express your answer in scientific notation.

76 Multiply the force by the speed to find the power (in watts) needed to lift the elevator.

77 Power is commonly reported in kilowatts (kW). How many kilowatts of power are needed to lift this elevator?

Electricity travels close to the speed of light, 3.0×10^8 meters per second. A surge suppressor used to protect computer equipment advertises a 1-ns clamping time (that is, it can suppress a voltage pulse in 1 nanosecond).

78 Express the advertised clamping time in units of seconds using scientific notation.

79 Multiply the speed of the electricity in a wire by the clamping time to estimate how far along the wire (in meters) the leading edge of a voltage pulse will travel before the remainder of the pulse is suppressed.

Security coding of automatic garage door openers is done by setting a bank of switches. For example, with three such switches that can be set to ON or OFF, the number of possible settings is $2 \times 2 \times 2$, or 8 settings. Some switches have three possible settings, indicated as $+1$, 0, or -1, yielding a total of $3 \times 3 \times 3$, or 27 possible settings for three switches.

80 How many possible settings are in a switch bank of 8 switches with three settings for each? Express your answer in scientific notation.

81 How many possible settings are in a bank of 12 switches with three settings for each? Express your answer in scientific notation.

CHAPTER 2   ASSESSMENT

Skills

Write each number as a power of ten.

1. 1000   **2.** 10,000,000   **3.** 0.00001

Write each number in decimal form.

4. 1.1×10^2 **5.** 4.5×10^{-6} **6.** 1.89×10^4

Write each number in scientific notation.

7. 0.00068   **8.** 25,000,000   **9.** 1,045,000,000

Write each number in decimal form.

10. 3^4 **11.** 5^{-3} **12.** $(-2)^5$

Express each product or quotient in scientific notation.

13. $(8.3 \times 10^{-2}) \times (1.3 \times 10^8)$ **14.** $(7.5 \times 10^5) \div (1.5 \times 10^{-3})$

Convert the metric units and write each answer in scientific notation.

15. 75 centimeters to kilometers.

16. 432 seconds to microseconds.

Applications

Express each answer in scientific notation.

17. If 6.02×10^{23} molecules of gas have a mass of 28 grams, what is the mass of one molecule of the gas?

18. If 1 cubic centimeter of water has a mass of 1 gram, what is the volume in cubic centimeters of 2538 kilograms of water?

19. One gallon of water has a mass of 3629 grams. If 18 grams of water contain 6.02×10^{23} molecules, how many molecules are in 1 gallon of water?

20. One cubic centimeter of water has a mass of 1 gram. A cubic tank is 8 meters on each edge. How many grams of water does the tank hold?

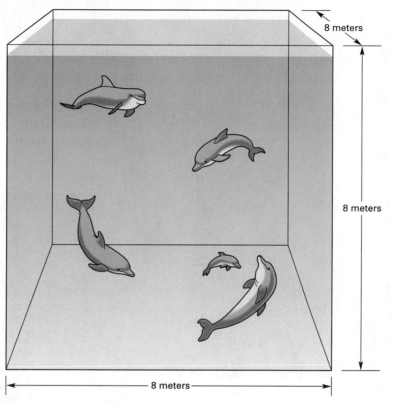

Math Lab

21. Which measuring instrument is the most accurate for finding the average page thickness in a book?

a. micrometer caliper
b. vernier caliper
c. machinist ruler

22. The average volume of a sand grain on a beach is 0.01 cubic centimeters. Estimate the number of grains of sand in a section of beach that is 1000 meters long, 100 meters wide, and 1 meter deep. Express your answer in scientific notation.

23. In the math lab "Measuring Water Molecules", why was the weight of the empty graduated cylinder and string measured?

CHAPTER 3

WHY SHOULD I LEARN THIS?

Formulas are the foundation of technology—they describe our world and how it works. Learn how to build formulas that you can use in your future career.

USING FORMULAS

OBJECTIVES

1. Substitute and evaluate values in expressions.
2. Read and write a formula.
3. Use your calculator to solve problems with formulas.

A formula is used to show how quantities are related to one another.

Although you have used many formulas before, you may not have stopped to think about what a formula does or how one is created. A formula in mathematics is a way of using symbols to write a sentence.

Sentence: The perimeter of a rectangle is the sum of twice its length plus twice its width.

Formula: $p = 2l + 2w$

Do you think it is easier to remember the sentence or the formula?

Nearly every occupation uses some kind of formula. Here are some examples.

Carpenters use formulas to check right angles.
Hospital personnel use formulas to decide dosage.
Bricklayers use formulas to mix mortar.
Dietitians use formulas to balance nutrition amounts.
Environmentalists use formulas to determine concentrations.
Photographers use formulas to mix chemicals
  Potters use formulas to create glazes.

As you watch the video for this chapter, notice how often you see people using formulas to do their jobs.

A **numerical expression** contains addition, subtraction, multiplication, or division operations. In the first two chapters, you evaluated many simple numerical expressions, such as 3^2 and 6×5. The expression 6×5 can also be written with a raised center dot or parentheses to indicate multiplication.

$$6 \cdot 5 = 30 \qquad 6(5) = 30 \qquad (6)5 = 30$$

How do you find the value of a numerical expression that contains several different operations?

Order of Operations

A flooring contractor needs to tile two rectangular areas. The dimensions of the two areas are 6 feet by 5 feet and 3 feet by 5 feet. How many square feet of tile should the contractor order?

To find the total number of square feet, the contractor must find the value of $6 \cdot 5 + 3 \cdot 5$. Which operation, addition or multiplication, is performed first?

If you add first and then multiply, the value is $6 \cdot 8 \cdot 5$ or 240.

If you multiply first and then add, the value is $30 + 15$ or 45.

To make sure everyone gets the same value when they evaluate the same numerical expression, mathematicians have developed the following **order of operations.**

The Order of Operations

First, perform all operations within parentheses.

Second, perform all operations involving exponents.

Third, multiply or divide in order from left to right.

Finally, add or subtract in order from left to right.

When the flooring contractor follows the order of operations, his result is

$$6 \cdot 5 + 3 \cdot 5 = 30 + 15$$
$$= 45$$

The contractor should order enough tile to cover 45 square feet.

If you want to perform the addition first, use parentheses.

$$6 \cdot (5 + 3) \cdot 5 = 6 \cdot 8 \cdot 5$$

$$= 240$$

Ongoing Assessment

Evaluate each expression.

a. $9 \cdot 2 - 24 \div -6$ **b.** $5 + 2(4 - 7)^2$

Scientific calculators usually follow the order of operations. Use your calculator to evaluate $6 \cdot 5 + 3 \cdot 5$. If your calculator does not display 45, you will need to use parentheses or the memory key.

Algebraic Expressions

Every time you hire an electrician, it costs $50 for the house call. The total bill, however, will depend on the number of hours the electrician is on the job. A table is useful in finding the cost for the total number of hours on the job when the hourly cost is $30.

Hours	Expression	Cost
1	$50 + 30 \cdot 1$	80
2	$50 + 30 \cdot 2$	110
3	$50 + 30 \cdot 3$	140

The table shows that the cost depends on the number of hours the electrician works on the job. How can you find the cost after any number of hours on the job? If you think of the letter h as representing the number of hours, the cost is

$$50 + 30 \cdot \text{number of hours } (h)$$

or simply, $50 + 30 \cdot h$.

In the expression $50 + 30 \cdot h$, the letter h is a **variable.** The numbers 50 and 30 are **constants.**

A **variable** is a symbol, usually a letter, used to represent a number or some other object. However, a letter that always represents the same number is a constant. For example, the Greek letter π (pi) always represents the same number (approximately 3.1416). Thus, π is a constant.

An expression containing numbers, operations, and one or more variables is called an **algebraic expression.** Each of the following expressions represents the multiplication of a and b.

$$ab \qquad a(b) \qquad (a)b \qquad (a)(b)$$

In the algebraic expression representing the electrician's cost,

$$50 + 30h$$

50 and $30h$ are the **terms** of the expression. In the term $30h$, 30 is the numerical coefficient, or just the **coefficient** of h.

Ongoing Assessment

Practice the language of algebra by answering the following questions:

a. In the expression $ab + ac$, which operation is performed first?

b. What is the value of $5n$ if $n = 7$?

c. Name the variable in $5n$.

d. Name the constant in $5n$.

e. Name the coefficient in $5n$.

To evaluate an algebraic expression, first substitute the given numbers for the variables. Then simplify the resulting numerical expression using the order of operations.

Evaluate the expression $5x - y^2$ when x is -2 and y is 3.

SOLUTION

1. Substitute the numbers for the variables. $5x - y^2$

$$5(-2) - (3)^2$$

2. Follow the order of operations. $5(-2) - 9$

$$-10 - 9$$

$$-19$$

Critical Thinking For what value of b, will $25 - b^2 = 9$?

LESSON ASSESSMENT

Think and Discuss

1 Why is it important to have an order of operations?

2 Compare the meanings of a variable, a constant, and a coefficient. Give an example of each.

3 What is the difference between a numerical expression and an algebraic expression?

4 Explain how to evaluate an algebraic expression.

Practice and Problem Solving

Evaluate each numerical expression.
5. $25 + 4 \bullet 10$ **6.** $9 \div 3 - 15 \div 5$ **7.** $4 - (9 - 12)$

8. $8 - (7 - 9)^3$ **9.** $36 \div 6 \div 3$ **10.** $36 \div (6 \div 3)$

Evaluate each algebraic expression.
11. $6a - 10$, when a is -3 **12.** $100 - 5c$, when c is 100

13. $-5(x + 6)$, when x is 14   **14.** $8y - 3y$ when y is -2

15. $2m + 2n$, when m is 6 and n is 4

16. $s^3 - t^2$, when s is -5 and t is -10

Evaluate each expression when *r* is −6, *s* is 4, and *t* is −2.

17. $rs - t$ **18.** $r \div t + st$ **19.** rst

20. $rs - rt$ **21.** $(r + s)^2$ **22.** $r^2 + s^2$

23. A stockbroker has developed the expression $P \div (G + Y)$ to help her determine if the price of a given stock is too high. Evaluate the expression when *P* is 20.25, *G* is 4.50, and *Y* is 2.25.

24. The expression $P(0.08L + l)$ helps a clothing store manager estimate projected sales for a given item. Evaluate the expression when *P* is 20, *L* is 350, and *l* is 45.

25. The manufacturing cost in dollars for a type of wooden shelf is found using the expression $br + ac$. Find the manufacturing costs when *b* is 4, *r* is 8.5, *a* is 48, and *c* is 1.5.

Mixed Review

26. The Johnson Brothers Company has lost $240 for 6 straight weeks. Use integers to write an expression to represent the total losses. Simplify the expression.

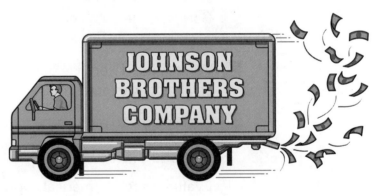

27. The estimated age of the earth is 4.7×10^9 years. Write the age of the earth as in decimal notation.

Simplify and write each answer in scientific notation.

28. $(3 \times 10^3) \bullet (6 \times 10^{-5})$ **29.** $(3.6 \times 10^2) \div (1.2 \times 10^4)$

LESSON 3.2 EQUATIONS AND FORMULAS

In Lesson 3.1, you evaluated algebraic expressions for given values of a variable. Complete Activity 1 and see if you can find a pattern.

ACTIVITY 1   Equal Expressions

Evaluate each pair of expressions. Remember to use the order of operations. Symbols such as parentheses ( ) and brackets [ ] are called **grouping symbols.** Do the operations in the innermost set of grouping symbols first.

Column 1	Column 2
1 $3 \bullet 4 + 3 \bullet 6$	$3(4 + 6)$
2 $(8 \bullet -7) + (8 \bullet -3)$	$8[-7 + (-3)]$
3 $6 \bullet 28 + (6 \bullet -8)$	$6[28 + (-8)]$

4 What do you notice about the answer to each pair of expressions?

Equations and Basic Properties

An **equation** is a sentence that states two expressions are equal. Equations are used to model the data in all kinds of situations.

ACTIVITY 2   The Distributive Property

1 A carpet installer is installing two rectangular carpets, one measuring 3 feet by 4 feet and the other measuring 3 feet by 6 feet. To find the total carpet area for this job, simplify the following expression:

$$3 \bullet 4 + 3 \bullet 6$$

2 On another job, he needs to carpet a 3 foot by 10 foot area. The area for the second job is 3 • 10 or

$$3(4 + 6)$$

3 Are the areas of the two jobs equal?

The equation

$$3 • 4 + 3 • 6 = 3(4 + 6)$$

models the carpet installer's situation. The pattern represented by this equation can be written with variables.

$$ab + ac = a(b + c)$$

In an equation, you can write either expression first. Thus, the following equation also represents the pattern.

$$a(b + c) = ab + ac$$

In this form, the equation is called the **distributive property.** There are many different forms of the distributive property. Two of the most important forms are given below.

The Distributive Property
For all numbers a, b, and c,

$$a(b + c) = ab + ac$$

and

$$a(b - c) = ab - ac$$

Certain equations in algebra are called **basic properties.** A basic property is true for every value you substitute. Of course, it is impossible to try all the values to make sure. However, if the equation is false for even a single substituted value, then it is not a basic property. To show that $a - b = b - a$ is not a basic property, substitute 2 for a and 1 for b. This substitution is called a **counterexample.**

Critical Thinking   Which of the following statements are basic properties? Explain your answer.

a. $a + b = b + a$ **b.** $(x - y) - z = x - (y - z)$

Formulas

A **formula** is a general rule or principle representing a real situation or application. A formula is written in the form of an equation. For example, the formula stating the rule that the diameter of a circle is twice its radius is written as follows:

$$d = 2r$$

A formula is just *one* possible way to state a rule. The following diagram gives three additional ways to give the same rule.

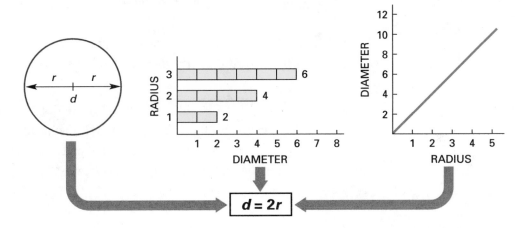

A formula such as $d = 2r$ is often the shortest, easiest way to write the relationship between two or more quantities. You have already worked with formulas such as these:

$$V = e^3 \qquad p = 2l + 2w$$

Formulas are shortcut ways to write mathematical rules or relationships. Where do formulas come from? For each formula, some person discovered the relationship that is expressed by the formula.

For example, people have known most of the formulas for finding areas since ancient times. Here are three formulas involving the area (*A*), the length (*l*), and the width (*w*) of a rectangle.

Find the area if you know the length and width: $A = lw$

Find the length if you know the area and width: $l = A \div w$

Find the width if you know the area and length: $w = A \div l$

ACTIVITY 3   Using the Area Formula

1 Use the area formula to find the missing dimensions.

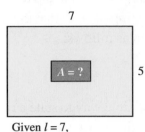
Given $l = 7$,
and $w = 5$, find A.

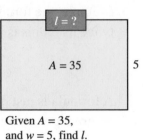
Given $A = 35$,
and $w = 5$, find l.

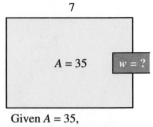
Given $A = 35$,
and $l = 7$, find w.

2 Write three different problems related to a construction project that use the area formula in each of its forms.

Ongoing Assessment

The formula for the area of a rectangle can be used to find the area of a square. A square is a rectangle with equal sides. Thus, the formula for finding the area of a square with side, s, is

$$A = s^2$$

What is the area of a square with sides of 4 centimeters?

ACTIVITY 4   Writing a Formula

1 Complete the following chart. The measurements are in inches.

Measurement	Clothing Size				
	2	3	4	5	6
a. Chest	21	22	23	?	?
b. Waist	20	$20\frac{1}{2}$	21	?	?
c. Height	35	38	41	?	?

2 Write formulas to describe each pattern in the chart.

How do you know when to use a formula? And more importantly, how do you know which formula to use?

You can write the formula converting temperatures between the Celsius (C) and Fahrenheit (F) scales in two ways.

$$F = \frac{9}{5}C + 32 \qquad C = \frac{5}{9}(F - 32)$$

A temperature of 36°F converts to what equivalent temperature on the Celsius scale?

First, choose the version of the formula that has the variable you need on the left side of the equal sign. Since you are looking for a temperature in °C, choose the second version of the formula.

$$C = \frac{5}{9}(F - 32)$$

Substitute the numerical value in the problem for the variable it matches in the formula. Since the Fahrenheit temperature is 36°,

$$C = \frac{5}{9}(36° - 32°)$$

Do you recognize the procedure here? In an earlier lesson, Tracy used this formula to find the temperature in a computer room.

Use your calculator to evaluate the formula. If you round your answer to two decimal places, you should get C = 2.22. How did you do? Did you use the parentheses keys? Can you find the value for C without using the parentheses keys? Do not forget the units in the answer. 36°F is about 2.22°C.

WORKPLACE COMMUNICATION

Modern Workplace has 915 employees. Which plan should Mr. Morgan choose for the first month's paper supply? What else must he consider?

Discount Paper Warehouse and Recycling Center
Shiela Price, Director of Marketing and Sales
987-6543

Dale Morgan, Purchasing Agent
Modern Workplace, Inc.

Dear Mr. Morgan:

Thank you for choosing *Discount Paper Warehouse* for your computer and copier paper needs for next year. We offer three ordering plans for paper that is usable in computer printers and copiers, with prices dependent on the size of your order:

Plan A: Order fewer than 10 boxes for $25.00 per box

Plan B: Order between 10 and 20 boxes for $22.50 per box

Plan C: Order more than 20 boxes for $20.00 per box

To assist you in planning for this year's paper needs, I recommend you use the following formula developed by our sales staff:

Number of Boxes = $0.035d(0.015N + 5)$

where N is the number of employees and d is the number of days the supply will last.

Please call me at your earliest convenience to place your order.

Thanks Again,

Shiela

PS: We offer an additional 5% discount if you allow *Discount Paper Warehouse and Recycling Center* to recycle your clean white paper.

LESSON ASSESSMENT

Think and Discuss

1 What are the similarities and differences between an expression and an equation?

2 How can you tell whether or not an equation is a basic property?

3 How does a formula differ from an equation or a basic property?

4 Explain how to find the area of a square if you know the length of one side.

Practice and Problem Solving

Tell whether or not each equation is a basic property.

5. $a + (b + c) = (a + b) + c$

6. $a + 0 = 0$

7. $ab = ba$

8. $a \div b = b \div a$

9. $(a + b)^2 = a^2 + b^2$

10. $(a + b)^2 = a^2 + 2ab + b^2$

11. $-(a + b) = -a + (-b)$

12. $a - (b - c) = (a - b) - c$

13. $a(b - c) = ab - ac$

14. $a - b = a + (-b)$

Use a formula to find the area, length, or width of a rectangle with the given dimensions.

	Area	Length	Width
15.	?	48 mm	8 mm
16.	36 m²	9 m	?
17.	108 mi²	?	9 mi

18. Wallpaper comes in rolls that are 50 feet long and 18 inches wide. How many rolls of wallpaper will it take to cover 450 square feet of wall space?

19. A living room is 15 feet by 15 feet. A rug 12 feet by 12 feet is placed in the center of the room. How much of the floor is not covered by the rug?

20. The Deck Company is bidding on staining a rectangular deck that is 18 feet by 30 feet. If it bids on the job at $5 per square yard, what is The Deck Company's bid?

21. The weather report for the 4th of July is 35 degrees Celsius. What is the Fahrenheit temperature?

22. Water boils at 212 degrees Fahrenheit. What is the boiling point of water in degrees Celsius?

23. Which is hotter: 104°F or 40°C?

24. How can you convince someone that 0°C is the same as 32°F?

If *D* represents the distance you travel, *R* represents the rate you are moving, and *T* represents the time you travel, write a formula for

25. Distance.   **26.** Rate.   **27.** Time.

Use the formulas you wrote to solve these problems.

28. If it takes 4 hours to travel 240 miles, how fast are you driving?

29. If you run 3 hours at 5 miles per hour, how far will you run?

30. If a plane is flying at 240 miles per hour, how long will it take to fly 720 miles?

Mixed Review

Richard is using this population chart for statistical records.

Austin	465,632
Charlotte	395,934
Cleveland	505,616
Long Beach	429,321
New Orleans	496,938
Seattle	516,259

31. Rank the cities in order from least population to greatest population.

32. Use integers to write the difference between the population of each city and a benchmark population of 500,000.

33. The diameter of a molecule of water is approximately 0.0000000276 centimeters. Write the diameter in scientific notation.

34. Compute your age in seconds as of today at noon. Write your age in scientific notation.

35. The thickness of a page in a novel is about 8×10^{-5} meters. How thick is a 1200-page novel?

LESSON 3.3 CIRCLES

You can see circles in many of the designs found in architecture and in nature.

Circumference

The distance across a circle through the center is the **diameter** of the circle. The **circumference** is the distance around the circle. Complete the Activity to see what all circles have in common. You will need some string and a ruler for this Activity.

ACTIVITY	The Ratio of Diameter to Circumference

1 Find several circular objects.

2 Use your string and a ruler to find the circumference and the diameter of each object.

3 For each circle, divide the circumference by the diameter. Round the answer to the nearest hundredth.

4 Check with your classmates. What constant number can be used as the approximate quotient of your divisions?

5 This constant number is called π. To the nearest hundredth, π is equal to 3.14.

The Activity leads directly to the formula for finding the circumference (C) of a circle when you know its diameter (d).

Circumference of a Circle
$$C = \pi d$$

Your calculator should have a π key. Press it to find what value it displays. Suppose a pipe has a diameter of 4.5 centimeters. To find the circumference of the pipe, follow these steps.

Enter 4.5

Press $\times$

Press π

Press $=$

Round the product to 14.14. The circumference of the pipe is about 14.14 centimeters. The symbol $\approx$ is used when the answer is not exact. Thus, you can write

$$C \approx 14.14 \text{ cm}$$

This is read "C is approximately equal to 14.14 centimeters."

Ongoing Assessment

You are insulating a replacement section of air conditioning ductwork. The ductwork is circular and has a diameter of 14 inches. What length of insulation is needed to wrap around the duct?

Critical Thinking Remember that the diameter is twice the length of the radius. What is the formula for the circumference of a circle with radius r?

Area

The **area** of a circle is a measure of the space inside the circle. The area (A) of the circle is determined by its radius (r).

Area of a Circle
$A = \pi r^2$

Area is measured in square units such as square feet (ft^2) and square meters (m^2). The area of a circle with a radius of 1 meter is 3.14 square meters. The circumference of the same circle is 6.28 meters.

A circular swimming pool is 32 feet in diameter. A 5-foot wide sidewalk around the pool requires a nonskid surface. What area must the nonskid surface cover?

CULTURAL CONNECTION

Many ancient civilizations discovered how to find the area of a circle. Some Oriental cultures used 3 for the value of π. The Egyptians recorded a value close to 3.1604 in the Rhind papyrus written about 1600 BCE.

One of the closest approximations was used by the Greek mathematician, Archimedes, about 287 BCE. He determined that π must be

between $3\frac{1}{7}$ and $3\frac{10}{71}$. The Egyptian scientist Ptolemy used 3.1416 in about 150 AD. But it was not until the 18th century that it was finally proven that π cannot be written as a fraction or rational number. For this reason, π is called an **irrational number.** Today, computers have found π to enough places to fill a book. Find the value of π to nine decimal places.

LESSON ASSESSMENT

Think and Discuss

1 Explain the meaning of radius, diameter, and circumference of a circle. Draw a circle and label each term.

2 Explain how to find the circumference of a circle if you know the radius.

3 Explain how to find the area of a circle if you know the radius.

4 Which fraction, $\frac{22}{7}$ or $\frac{223}{71}$, is a better approximation for π?

Practice and Problem Solving

Find the circumference and area to the nearest hundredth of each circle having the given radius.

5. 2 meters   **6.** 4 feet   **7.** 8 centimeters

8. 2.7 yards   **9.** 6.3 millimeters   **10.** 1.25 kilometers

Find the area of each shaded region.

11.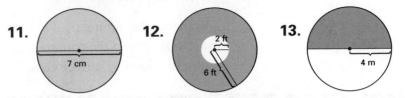
7 cm

12.
2 ft
6 ft

13.
4 m

14. The diameter of a truck wheel is 120 centimeters. What is the circumference of the wheel?

15. A patio is designed in the shape of a semicircle. If the radius of the patio is 10 feet, what is the area of the patio?

16. A pipe has a diameter of 3 inches. A welder must connect a wire clamp around the outside of the pipe. Find the length of the wire needed to wrap around the pipe.

Mixed Review

The average depth of the Pacific Ocean is 12,925 feet below sea level. The average depth of the Atlantic Ocean is 11,730 feet below sea level.

17. Write each depth as an integer.

18. What is the difference in the average depths of the two oceans?

19. The diameter of some white blood cells is about 4.03×10^{-4} inches. Suppose 1 million of these cells are aligned. Write the length of the line in scientific notation.

20. The force (F) on an object is found from the product of its mass (m) and acceleration (a), or $F = ma$. Find the mass of an object, in kilograms, when a force of 29.4 Newtons accelerates the object at a rate of 3 meters per second per second.

LESSON 3.4 VOLUME

Boxes

The dimensions of a rectangular box are length (l), width (w), and height (h). The **volume** (V) of a box is the amount of space contained in the box. The volume is also the capacity of the box. In other words, the volume or capacity tells you how much the box will hold. If length, width, and height are equal, the box is a **cube.**

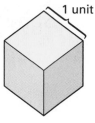

1 unit

The basic measurement for volume is a cube that measures 1 unit on an edge. This cube is 1 cubic unit. If the measure of the edge is in centimeters, the cube has a volume of 1 cubic centimeter (cm^3).

1 cubic unit

Volume is measured by the number of cubic units an object can hold. Here is how you find the number of cubic centimeters in a box.

Find the number of cubic centimeters in the base.	Multiply by the number of layers in the box.
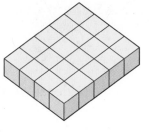	
$5 \cdot 4 = 20$	$5 \cdot 4 \cdot 3 = 60$
The base contains 20 cubes, each 1 cubic unit.	The volume is 60 cubic units.

The picture leads to a formula for finding the volume of a box.

Volume of a Rectangular Box
$$V = lwh$$

If the box is a cube ($l = w = h = e$),

$$V = e^3$$

The inside of a rectangular container is 50 inches in length and 20 inches in width. If you fill the container to a depth of 8 inches, what is the volume of the container?

Critical Thinking   The meter is considered a one-dimensional measurement. Why is the square meter considered a two-dimensional measurement? Why is the cubic meter considered a three-dimensional measurement?

Spheres

What do basketballs, globes, ball bearings, storage tanks, and weather balloons have in common? All of these objects have the familiar shape of a sphere. The volume (V) of a sphere depends on the length of its radius (r).

Volume of a Sphere

$$V = \frac{4}{3}\pi r^3$$

A technician for the National Weather Service is preparing to launch weather balloons. Each balloon is spherical in shape and has a radius of 3 feet. To calculate the number of cubic feet of helium in each inflated balloon, the technician uses the volume formula.

$$V = \frac{4}{3}\pi r^3$$
$$= \frac{4}{3}\pi (3)^3$$
$$= \frac{4}{3}\pi (27)$$
$$= 36\pi \approx 36(3.14) \approx 113.1$$

Each balloon uses about 113.1 cubic feet of helium.

Ongoing Assessment

Find the volume of a spherical tank with a radius of 9 meters.

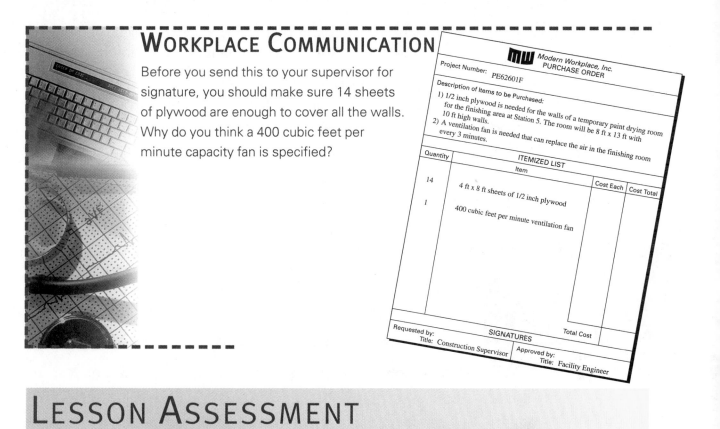

WORKPLACE COMMUNICATION

Before you send this to your supervisor for signature, you should make sure 14 sheets of plywood are enough to cover all the walls. Why do you think a 400 cubic feet per minute capacity fan is specified?

Modern Workplace, Inc.
PURCHASE ORDER

Project Number: PE62601F

Description of Items to be Purchased:
1) 1/2 inch plywood is needed for the walls of a temporary paint drying room for the finishing area at Station 5. The room will be 8 ft x 13 ft with 10 ft high walls.
2) A ventilation fan is needed that can replace the air in the finishing room every 3 minutes.

Quantity	ITEMIZED LIST Item	Cost Each	Cost Total
14	4 ft x 8 ft sheets of 1/2 inch plywood		
1	400 cubic feet per minute ventilation fan		
		Total Cost	

SIGNATURES
Requested by:
Title: Construction Supervisor
Approved by:
Title: Facility Engineer

LESSON ASSESSMENT

Think and Discuss

1 How are area and volume alike? How are they different?

2 How do you find the volume of a cube if you know the length of one edge?

3 How do you find the volume of a sphere if you know the diameter of the sphere?

Practice and Problem Solving

4. Find the volume of a box with length 3 meters, width 6.2 meters, and height 1 meter.

5. Find the volume of a cube with one edge 2.5 feet in length.

6. Find the volume of a sphere with a radius of 2.1 feet.

7. Find the volume of a sphere with a diameter of 6 meters.

8. Concrete costs $35 per cubic yard. If a concrete patio is to be 36 feet long, 12 feet wide, and 6 inches deep, how much will the concrete cost?

9. The diameter of the earth is about 7926 miles. Find the volume of the earth. Write your answer in scientific notation.

10. A steel ball has a diameter of 2 inches. It is placed inside a hollow sphere with a radius of 5 inches. How much water can be poured into the sphere?

11. A scoop of ice cream is placed on a cone. If the radius of the ice cream is 4 centimeters, what is the volume of ice cream?

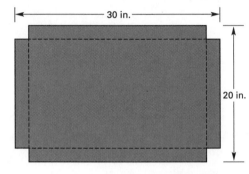

12. A rectangular piece of cardboard is 30 inches by 20 inches. A piece of cardboard 2 inches by 2 inches is cut from each corner. The resulting piece of cardboard is folded to make a rectangular container. Find the volume of the container.

13. How many cubic inches equal 2 cubic feet?

14. How many cubic centimeters equal one cubic meter?

Mixed Review

You are planning a vacation. You want to rent a car for four days and drive 250 miles. The On-Time Rental company charges a one-time cost of $45 plus 35 cents a mile. The Best Deal company charges a flat rate of $37 per day.

15. Which company should you use?

16. How much will you save by using this company?

A business shows a loss of $295 in January, a gain of $1655 in February, and a loss of $695 in March.

17. Write each amount as an integer.

18. In which month did the business have the greatest loss?

19. What is the total gain or loss for the three months?

20. The Mississippi River has a drainage area of about 1,150,000 square miles. Write this drainage area in scientific notation.

LESSON 3.5 INTEREST

During the next few years, you will either save money or borrow money. When you keep money in a savings account, the bank pays you *interest* for the use of the money. When you borrow money, you pay *interest* for using money. **Interest** is money paid for the use of money.

The money that is borrowed or saved is called the **principal.** The **rate of interest** is the percentage of the principal charged for the use of the money. The rate of interest is usually specified for one year. However, you can calculate interest for any period of time.

Simple Interest

Simple interest is only earned or charged on the principal. Simple interest (i) is found by multiplying the principal (p) by the annual rate of interest (r) and the period of time (t), in years, that the money is borrowed or saved.

Simple Interest

$$i = prt$$

Suppose you borrow $500 for one year at 12% simple interest. To find the interest charged, use the formula $i = prt$ and multiply 500 • 12% • 1.

Remember to change the percentage to a fraction or a decimal before you multiply. That is, 12% is $\frac{12}{100}$ or 0.12. Find the product on your calculator. The simple interest is $60.

When you borrow money, you must pay back the **total amount** (A) of the loan. The amount is the principal plus the interest. The amount of the payback is $500 + $60 or $560.

Critical Thinking   How is simple interest calculated if the money is borrowed for less than one year? More than one year?

Ongoing Assessment

If only one payment is made at the end of 3 years, what is the total payback on a loan of $750 borrowed at an annual rate of 9.5%?

Compound Interest

Suppose you save $500 for two years at 8% simple interest. After one year, you will have $540 in the bank. After the second year, you will get another $40 interest and have $580. If you save the money at a bank, however, you earn more than simple interest. **Compound interest** is interest paid on both the principal and the previously paid interest. Using compound interest (compounded yearly), the amount in the bank after one year is still $540. For the second year, however, the $540 becomes the principal. The interest for the second year is calculated using this new principal.

$$i = prt$$

$$540 \cdot 8\% \cdot 1 = 43.20$$

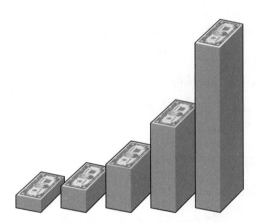

Instead of $580 after the second year, you now have $583.20. This is not a large increase over simple interest. However, over many years this amount will grow substantially. You can earn a great deal of money if you are paid compound interest instead of simple interest.

Total Amount with Compound Interest
$$A = p(1 + r)^n$$

A is the total amount and n represents the number of years.

Most banks compound interest on a daily rather than on a yearly basis. However, if interest is compounded even monthly, you receive much more money over a period of years than simple interest provides.

Using a Power Key

A bank is advertising a plan that encourages parents to invest money at 8% interest compounded yearly when their child enters first grade. If a parent invests $500 under this plan, how much money will the child have 12 years later?

The formula for compound interest is the following:

$$A = 500(1 + 0.08)^{12}$$

1 Use the power key on your calculator.

Enter 500   Press ⊠

Enter 1.08   Press y^x

Enter 12   Press ⊜

2 What is the result?

3 How much less is the same amount invested at simple interest?

LESSON ASSESSMENT

Think and Discuss

1 Compare the amounts you receive when you save $100 at 6% simple interest and $100 at 6% compounded annually for 10 years. Why is saving at compound interest important?

2 A credit card company charges 18% annual interest on any unpaid balance. Explain why this is not such a good way to borrow money.

Practice and Problem Solving

Find the difference for each amount of savings if simple interest or compound interest is used.

3. $1000 at 6% for 2 years.   **4.** $875 at 5% for 5 years.

5. $2,450 at 9% for 3.5 years.   **6.** $750 at 8.5% for 4 years.

7. Todd invested $750 at 6.5% compounded annually. How much did Todd have after 5 years.

8. Kesha borrowed $1800 at 5% simple interest. How much did Kesha owe after 6 months?

9. Angie put $2000 in a CD. The CD paid 6% compounded annually. How much is the CD worth after 3 years?

10. Suppose you invest $1000. How long will it take to double your money at 6% simple interest?

Mixed Review

A clothing company sells four lines of clothing. The chart shows the profit (P) and loss (L) in dollars for each line over one year.

~~Line A~~	~~Line B~~	~~Line C~~	~~Line D~~
220,000(P)	105,000(L)	92,500(L)	162,550(P)

11. Which line has the greatest loss?

12. Which line has the least profit?

13. Draw a number line and place each profit and loss in the correct position.

Write the product or quotient in scientific notation.
14. $(3.4 \times 10^3)(6.2 \times 10^{-5})$

15. $(1.25 \times 10^{-2}) \div (3 \times 10^{-3})$

16. What is the value of $m - (n - m)^2$, when m is 6 and n is 2?

17. A square picture is 3 feet on each side. It costs $15 to matte each square foot. How much does it cost to matte the picture?

18. A washer has an inside radius of 15 millimeters and an outside radius of 21 millimeters. What is the area of the washer?

19. A grain car is 30 feet long, 10 feet wide, and 8 feet high. One cubic foot of grain weighs 6 pounds. What is the weight of the grain needed to fill the car?

20. Find the volume of a sphere with a radius of 12 feet.

MATH LAB

Activity 1: Rate and Distance

Equipment   Calculator
Timer
Masking tape
Battery-powered cars
Tape measure

Problem Statement

You will measure the time required for battery-powered cars to travel a measured distance. You will then calculate the speed of the cars. Based on the comparison, you will predict the winner of a race. An actual race will test the validity of your prediction.

Procedure

a Place a 1-foot piece of tape on the floor as a starting line. Measure 15 feet and place another 1-foot piece of tape for a finish line.

b Form a team to work with each car. Each team should measure the time required for its car to go from the starting line to the finish line. Write this time on a sheet of data paper. Repeat and record the measurements for a total of five trial runs for each car.

c Use an appropriate formula to calculate the speed of the car for each trial run in feet per second (ft/sec). Round each answer to the nearest hundredth of a foot per second.

d Calculate the average speed of the car for the five trial runs to the nearest hundredth of a ft/sec.

e When all groups have finished recording results, compare the average speeds and predict which team's car will win a race.

f Measure a distance of 25 feet. Race the cars over the 25-foot distance. Record which cars take the first three places.

g How do the results of the race compare with the predictions made in Step **e**?

Activity 2: Radius and Volume of a Sphere

Equipment   Calculator
Micrometer caliper
100 BBs
Five $\frac{3}{8}$ inch ball bearings
10 ml graduated cylinder

Problem Statement

You will measure the diameter of two different spheres and calculate their volumes. Then, you will measure the volumes of a number of the spheres by the water-displacement method. From this data, you will calculate an average volume for each sphere and compare the average volume to the calculated volume. You will also calculate the radius of a sphere given its volume.

Procedure

a Use the micrometer to measure the diameter of a $\frac{3}{8}$ inch ball bearing. Convert this measurement from inches to centimeters. Record this measurement to the nearest 0.001 cm.

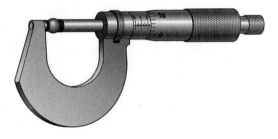

b Calculate the volume of the ball bearing using the formula for the volume of a sphere. Remember, the radius is one-half of the diameter.

c Fill the 10 ml graduated cylinder with water to about the 5 ml mark. Read the initial volume of the water. Record this initial volume.

d Place the ball bearings in the 10 ml graduated cylinder one at a time until all five are added. If any of the ball bearings are above the final water level, start over with more water. Read and record the final volume of the water.

e Subtract the initial volume from the final volume. Since a millimeter is the same as a cubic centimeter, write the volume change in both units. Divide this volume change by the number of ball bearings added. This is the average volume of one ball bearing.

f Compare the average volume measured in Step **e** to the calculated volume in Step **b.**

g Use the average volume calculated in Step **e** and the formula below to find the average radius of each ball bearing.

$$r = \sqrt[3]{\frac{3V}{4\pi}}$$

The symbol $\sqrt[3]{}$ means "cube root." For example, $\sqrt[3]{8}$ is the number that answers the question, "What are the three equal numbers whose product is 8?" In other words, $\sqrt[3]{8} = 2$ because $2 \times 2 \times 2 = 8$.

Most scientific or graphics calculators can be used to find a cube root.

See if either of these methods will find the cube root of 216.

1. If your calculator has a key, enter 216 and press .

2. If your calculator has a $\boxed{y^x}$ key, an $\boxed{INV}$ key, a $\boxed{\sqrt[x]{y}}$ or a key labeled $\boxed{y^{1/x}}$, enter 216 and press the necessary keys.

Try either method with 256. You should get 6.35 rounded to the nearest hundredth. Now continue with the Math Lab.

Compare this average radius to $\frac{1}{2}$ the diameter of the ball bearing measured with the micrometer calipers in Step **a.**

h Use the micrometer to measure the diameter of a BB. Convert this measurement from inches to centimeters. Record this measurement to the nearest 0.001 cm.

i Calculate the volume of the BB using the formula for the volume of a sphere.

j Fill the 10 ml graduated cylinder with water to about the 5 ml mark. Read and record the initial volume of the water.

k Add the BBs to the graduated cylinder, one at a time, until all 100 are added and all are below the final water level. Read and record the final volume of the water.

l Subtract the initial volume from the final volume. Divide this volume change by the number of BBs added. This is the average volume of a BB.

m Compare this average volume of a BB to the calculated volume from Step **h.**

n Use the average volume calculated in Step **l** and the formula for the radius of a sphere to calculate an average radius. Compare this average radius to $\frac{1}{2}$ the diameter of the BB you measured with the micrometer in Step **h.**

o How do the average measured volumes compare to the calculated volumes? How do the average radii compare to $\frac{1}{2}$ the measured diameters?

p Which method do you think is more accurate?

1. The one based on measurement with a micrometer.

2. The one based on water displacement.

Activity 3: Indirect Measurement of Height

Equipment Calculator
Protractor
PVC pipe-1 inch in diameter and 3 feet long
    with a swivel attached in the center.
Line level
Tape measure

Over the length of the pipe, draw a line through the swivel. Attach 20 feet of string to the swivel at point *A.*

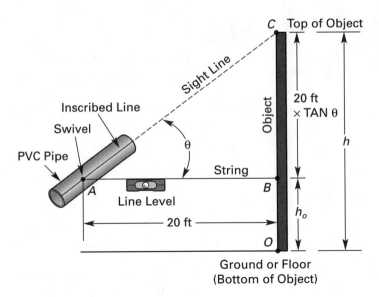

Problem Statement

You will use the device in the figure and a formula to determine the vertical height of an object perpendicular to the ground.

Procedure

a Form a five-member group and examine the figure.

- Member One should sight through the pipe along the line AC.

- Member Two should be ready to measure angle θ.

- Member Three needs to check the line level along AB to ensure the string AB is level.

- Member Four needs to hold the end of the string against the measured object at point B.

- Member Five needs to measure the height h_0 of the string AB above the base (point O) of the object.

b Study the drawing and examine the following formula:

$$h = 20(\tan \theta) + h_0$$

The expression $\tan \theta$ is read "tangent of theta." You will learn more about the tangent in a later chapter. For now, you just need to know that the tangent is the ratio of the side opposite angle θ to the side adjacent to angle θ in a right triangle. In the drawing, $\tan \theta$ is the ratio of the length of the segment $(h - h_0)$ to the length of the horizontal string.

Use a scientific calculator to find the value of h using the formula. Enter the measure of the angle θ in degrees and press the (TAN) key. The value for $\tan \theta$ will appear in the display.

Suppose θ is 45° and h_0 is 4 feet.

Enter 45° and press the (TAN) key. The display should read 1.0000000. If the display does not read 1.0000000, change your calculator to the degree mode. When you get a value of 1.0000000 for $\tan 45°$, multiply it by 20 and add 4 to this result. This gives a value of 24 feet for the height h.

c Select an object to measure. Some good choices outside your classroom are a flagpole, a wall of the school building, or a football goalpost. Inside the classroom, you can measure the height of a wall.

d Hold the free end of the 20-ft string against the selected object at point *B*. Stretch the string tight and attach the line level. Measure the length *AB* from the swivel to point *B* to verify that it is 20 ft. If *AB* is not 20 ft, use the true measurement *AB* instead of 20 in the formula for *h*.

e Hold the pipe at a comfortable height. Sight through the pipe to the top of the object.

f Adjust the end of the string at point *B* until the line level shows that the string is level. Measure and record the height *OB*. In the formula, this measure is called h_0.

g While sighting through the pipe to the top of the object, measure and record the angle of elevation, θ, between the string and the line inscribed on the pipe at the swivel.

h Use the appropriate formula to calculate the height of the object, *h*. Record this value.

i Describe other methods you could use to determine the height of the object. Which method do you think is most accurate?

j Describe how you would change the procedure to measure the width of a wide river or canyon.

MATH APPLICATIONS

The applications that follow are like the ones you will encounter in many workplaces. Use the mathematics you have learned in this chapter to solve the problems. Wherever possible, use your calculator to solve the problems that require numerical answers. A table of conversion factors can be found beginning on page A1.

An automotive technician needs to determine the resistance of an automobile starting motor that draws 90 amperes of current from a 12 volt battery.

1 Use the formula $R = \dfrac{V}{i}$ to calculate resistance, where R is the resistance (ohms), V is the voltage across the resistance (volts), and i is the current (amperes).

A cleaning solvent is purchased by the barrel. A useful formula for estimating the number of barrels in a cylindrical storage tank is

$$V = 0.14d^2 h$$

V = volume of solvent in barrels.
d = diameter of the storage tank in feet.
h = height of the storage tank in feet.

2 Approximately how many barrels of cleaning solvent can be stored in a 20-foot diameter tank that is 12 feet tall?

3 You are preparing a purchase order to refill the storage tank with solvent. The depth of the solvent remaining in the tank is 7 feet. How many barrels of solvent should you order?

Bricklayers can estimate the number of standard bricks needed to build a wall using the formula $N = 21LWH$.

N = number of bricks needed.
L = length of the wall in feet.
W = width of the wall in feet.
H = height of the wall in feet.

4 How many bricks are needed to construct a wall that is 150 feet long, 12 feet high, and 4 inches wide?

A construction team is pouring concrete for six cylindrical bridge supports. Each of the supports has a 2.4 foot radius and is placed into the ground to a depth of 12 feet and extends 15 feet above the ground.

5 What is the total length of each support in yards?

6 Use the formula

$$V = \frac{\pi d^2 L}{4}$$

where V is the volume of the cylinder, d is the diameter of the cylinder, and L is the length of the cylinder in yards to find the number of cubic yards of concrete needed to pour one of the supports.

7 A truck can deliver eight cubic yards of concrete with each load. How many trucks of concrete will the contractor need to pour the six supports?

Suppose that your time card shows that you worked seven hours each day, for five days. Your rate of pay is $7.45 per hour. You can determine your gross pay for a week by multiplying the total hours worked during the week by your hourly rate of pay.

8 Write a formula that determines weekly gross pay.

9 Use the formula to compute your gross pay.

10 If you work more than 40 hours during a week, you are paid 1.5 times the normal hourly pay rate for those hours worked in excess of 40 hours. Write a formula to determine your gross pay earnings for work in excess of 40 hours.

Your medical insurance policy requires you to pay the first $100 of your hospital expenses (this is known as a deductible). The insurance company will then pay 80% of the remaining expenses. The formula below expresses the amount that you must pay.

$$E = [(T - D) \times (1.00 - P)] + D$$

E is the expense to you (how much you must pay).
T is the total of the hospitalization bill.
D is the deductible you must pay first.
P is the decimal percentage that the insurance company pays after you meet the deductible.

11 Suppose you are expecting a short surgical stay in the hospital for which you estimate the total bill to be about $5000. Use the formula above to estimate the expense to you for the stay in the hospital.

AGRICULTURE & AGRIBUSINESS

Aerial photography is a useful tool in forestry surveys. A photograph can be used to estimate the diameter of a ponderosa pine. The diameter at chest height is found using the following formula:

$$D = 3.7600 + (1.3480 \times 10^{-2})\, HV - (2.4459 \times 10^{-6})\, HV^2 + (2.4382 \times 10^{-10})\, HV^3$$

D is the diameter at chest height in inches.
H is the height of the tree in feet.
V is the visible crown diameter from the photograph in feet.

12 An aerial photograph shows a visible crown diameter of 22 feet for 108-foot trees. Find the diameter of the trees.

When determining the usable log volume, deductions are made for slab, edgings, and saw kerf. The following formula is used to estimate the number of board feet that can be obtained from a log.

$$V = 0.0655\, L\, (1 - A)\, (D - S)^2$$

V is the usable volume of a log in board feet.
L is the length of the log in feet.
A is the decimal percent deduction for saw kerf.
D is the log diameter in inches.
S is the slab and edging deduction in inches.

13 Use the formula to estimate the usable volume of a 32-foot log having a diameter of 30 inches. Allow for 10% saw kerf and 3 inches of slab and edging deductions.

A commercial cherry grower estimates that when up to 30 trees are planted per acre, each tree can produce about 50 pounds of cherries per season. For each additional tree planted, the yield per tree is reduced by 1 pound.

14 Make a table of the yield per tree for several different planting densities: 30 per acre, 31 per acre, 32 per acre, and so on, up to 35 per acre.

15 Add a column to your table that shows how many total pounds of cherries would be produced per acre for each planting density.

16 Someone proposes that these data can be described by the formula

$$Y = 50 - (D - 30)$$

Y is the yield per tree in pounds.
D is the density of the planting in trees per acre.

Does this formula agree with the data in your table?

A floating electrical stock-tank heater uses 1500 watts of electrical power. The electric company charges you 12¢ per kilowatt-hour of usage. This means that if you use a one-kilowatt device for one hour, the electric company will charge you 12¢ for the electricity. If you use it for two hours, you will be charged 24¢, and so on.

17 During the winter, you use the stock-tank heater at night. A timer will control the heater, so that it operates for about 14 hours each night. Write an equation to compute the cost of electricity to run this heater for *d* nights.

18 Use your equation to find the electricity cost of using the heater for 90 nights.

BUSINESS & MARKETING

To consider an investment in certificates of deposit, you use the formula

$$A = P(1 + r)^n$$

Note: The units of time for *i*, *r*, and *n* must agree (for example, if you use a yearly percentage rate for *r*, you must use years for *n*).

19 A long-term certificate requires a deposit of $5000 for five years. Each year, the deposit earns 8.5% interest. Use the formula to determine the future value of the deposit at the end of the five-year term.

20 A short-term certificate requires a deposit of $1000 for six months. The deposit earns interest at an annual rate of 7%, or a monthly rate of $\frac{1}{12}$ of 7%. Use the formula to determine the total amount of the deposit at the end of the six-month term.

Series *EE* Savings Bonds can be purchased for one-half their face value. If they are held until maturity, they can then be redeemed for their face value, plus a bonus. A securities teller must explain to a customer how to compute the interest obtained on these bonds. The difference between the redemption value and the purchase price is the interest earned.

21 Write a formula that shows how to compute the purchase price of a Series *EE* Savings Bond, knowing only its face value.

22 Write a formula that shows how to compute the interest earned on a Series *EE* Savings Bond held until maturity, knowing only the purchase price and the redemption value.

A delivery service limits the packages it will handle to those weighing less than 70 pounds. In addition, the combined girth and length of the package must not exceed 108 inches. The length is the longest dimension of the package, and the girth is the total distance around the package at its widest part, not including the length.

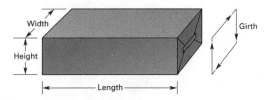

23 For a rectangular package with width *W*, length *L*, and height *H*, write a formula to compute the "combined girth and length."

24 You want to ship a small telescope in a box that weighs 54 pounds, and has a length of 54 inches, a width of 10 inches, and a height of 11 inches. Would this package pass the weight and size restrictions of the delivery service?

You can obtain a suggested retail price for used cars from a used-car guide. If the car has unusually low or unusually high mileage, the suggested price is adjusted accordingly. The following equation for cars driven between 20,000 and 70,000 miles seems to agree with your table of prices for a particular make and model of car.

$$P = -0.024\,M + 4850$$

P is the suggested retail price of the car.
M is the mileage shown on the car's odometer (in miles).

25 What is the suggested price for this type of car if the odometer shows 25,000 miles?

26 Does this equation suggest higher or lower prices for cars with unusually high mileage? What about very low mileage? Is this as you expect? What indicates this trend in the equation?

Some electric utilities offer an average billing plan to their customers. According to the plan, the amount a customer must pay on a given month's bill is determined as follows:

The current month's cost and the previous 11 months' costs are added, and this total is divided by 12. One customer's monthly utility costs are shown below.

Utility Costs for Customer #17662			
Month	Cost	Month	Cost
Current	$ 67.02	April	$76.15
September	92.66	March	70.28
August	126.90	February	55.23
July	127.43	January	58.18
June	112.76	December	49.72
May	89.19	November	59.14

27 Write a formula that allows you to compute this customer's bill using the average billing plan described above. The formula should require you to supply only the monthly costs to arrive at a bill for the current month. Use your formula to compute the bill for the current month.

You can make payments on an insurance policy either annually or monthly. If you choose to spread the premium payments over 12 months, the insurance company charges a handling fee of 11% of the annual premium. This handling fee is also spread over the 12 monthly payments.

28 Write a formula to determine the monthly payment for an annual premium plus the handling fee.

29 Use your formula to determine what the monthly payment would be if the annual premium is $288.

A client calls your investment firm to invest $20,000. She wants to earn at least $2400 interest per year by splitting the investment between two options. You can give her this choice: Option A pays 9% interest per year at relatively low risk, and Option B pays 17% per year at a higher risk. It is reasonable to keep the amount invested in the high-risk option as low as possible. The anticipated interest earned from each investment each year can be computed by multiplying the annual interest rate by the amount invested.

30 You advise your client to invest L dollars (L is less than $20,000) in the low-risk investment. Write an expression involving L that shows how much remains to invest in the high-risk investment.

31 Write a formula to compute how $2400 in interest can be earned from investing *L* dollars in the low-risk investment and the remaining amount in the high-risk investment.

32 How much of the $20,000 do you recommend the client deposit into each of the two investments?

You are performing laboratory tests on blood samples to find the amount of oxygen in the hemoglobin of the red blood cells. A blood test tells you that a 100 milliliter sample contains 12 grams of hemoglobin, and the hemoglobin is 91% oxygen. The following formula is used to detemine the amount of oxygen in the blood.

$$H = 1.36gs$$

H is the milliliters of oxygen in the hemoglobin from 100 ml of whole blood.
g is the number of grams of hemoglobin in 100 ml of whole blood.
s is the decimal value of oxygen per cent in the hemoglobin.

33 How much oxygen is in the hemoglobin of the sample (round your answer to 2 decimal places)?

As an exercise consultant, you have a formula that helps you estimate the percentage of body fat for men.

$$F = 0.49W + 0.45P - 6.36R + 8.71$$

F is the percent of body fat.
W is the waist circumference in centimeters.
P is the thickness of the skin fold above the pectoral muscle, in millimeters.
R is the wrist diameter in centimeters.

34 Compute the percent body fat for a male client who has a waist circumference of 87.3 cm, a skin fold thickness of 6.2 mm, and a wrist diameter of 6.5 cm (round your answer to one decimal place).

35 Suppose your brother's waist measures 34 inches, the skin fold above his pectoral muscle is about one-half inch, and his wrist diameter is two and one-half inches. Estimate your brother's percent body fat.

In a hospital, the doctor frequently orders a volume V (in cc) of IV (intravenous) fluid to be given over a specified period of hours, T.

The nurse must convert the rate from "cc per hour" to "drops per minute" using a conversion ratio of "15 drops per cc of fluid." A formula tells the nurse how many drops per minute to set the IV:

$$D = \frac{V}{4T}$$

D is the number of drops per minute.
V is the volume of fluid (in cc) ordered by the doctor.
T is the total time (in hours) which the fluid is to be given.

36 Suppose the doctor orders 1000 cc of D_5W to be given to a patient over eight hours. How many drops per minute should the IV supply to the patient?

Blood-sample analysis in the laboratory requires you to compute the mean corpuscular volume, or MCV.

$$MCV = \frac{H}{RBC} \times 10^7$$

MCV is the mean corpuscular volume in cubic centimeters.
H is the hematocrit value expressed in percent (for example, if the red blood amount is 52%, use 52—not 0.52—in the formula).
RBC is the red blood cell count in cells per cc.

37 Evaluate the equation when H is 45% and RBC is 5.3×10^6 cells per cc.

FAMILY AND
CONSUMER
SCIENCE

You have $8750 in fire and smoke damage to your house. Your insurance policy provides for 80% copayment. Your property is valued at $65,000, and the insurance policy covers $50,000.

38 Use the formula

$$P = \frac{i}{0.80V} \cdot L$$

P is the payment made by the insurance company.
i is the insurance coverage of the policy.
L is the loss due to fire.
V is the value of the property that is insured.

to determine the payment you will receive from the insurance company.

A furniture company rents a sofa and love seat combination for an initial fee of $79.80 and a monthly rental payment of $64.80. The company's "rent-to-own plan," lets you keep the furniture after the 24th payment.

39 Write a formula for the total cost to you of the furniture rental after any payment before the 24th payment.

40 Evaluate your equation for the 24th payment. How much will you have paid for the furniture when it is yours to keep?

INDUSTRIAL TECHNOLOGY

Greenshield's formula can be used to determine the amount of time a traffic light at an intersection should remain green: $G = 2.1n + 3.7$. G is the "green time" in seconds, n is the average number of vehicles traveling in each lane per light cycle.

41 Find the green time for a traffic signal on a street that averages 19 vehicles in each lane per cycle.

The *"WHILE"* formula is a simplified method to estimate the power of an air conditioning unit needed for a given room. The formula is

$$BTU = W \cdot H \cdot I \cdot L \cdot E \div 60$$

where *BTU* is the number of *Btu* (British thermal units) per hour of cooling power needed for the room.

> W is the width of the room in feet.
> H is the height of the room in feet.
> I is the insulation factor (10 for well insulated rooms and 18 for poorly insulated rooms).
> L is the length of the room in feet.
> E is the exposure factor.

You must determine the exposure factor from the orientation of the longest wall. If the wall faces north, $E = 16$; if east, $E = 17$; if south, $E = 18$; if west, $E = 20$. If two or more walls are equally the "longest," use the largest possible value for E.

42 A newly constructed room needs air conditioning. It is 18 feet wide, 24 feet long, and 9 feet high. The room is well insulated. One 24-foot wall faces south and one faces north. What size air conditioner does the formula predict is needed to cool this room (round your answer to the nearest 1000 *Btu* per hour)?

Chapter 3 Assessment

Skills

Evaluate each expression.

1. $9 + 16 \bullet 2$ **2.** $64 \div (16 \div 4)$ **3.** $8 - 3^2 + 9$

4. $5a - 7b$, when a is -3 and b is -4.

5. $m^3 + 6 - n^2$, when m is -2 and n is -3.

6. $pq^2 + r$, when p is 4, q is 3, and r is -10.

7. Give a numerical example of the distributive property.

Applications

8. The temperature on a bank thermometer shows 45°C. What is the temperature in degrees Fahrenheit?

9. A patient has a temperature of 100°F. What is the patient's temperature in degrees Celsius?

This table shows the number of plants growing per acre over five years.

Year	0	1	2	3	4	5
Plants	13.5	17.0	20.5	24.0	27.5	?

10. Write a formula to show the relationship between the number of plants growing per acre in a given year, and the number of plants growing per acre the following year.

11. Use the formula to complete the table for five years.

In Exercises 12–14, let A represent the area of a rectangle and P represent the perimeter. Let L represent the length and w represent the width of the rectangle.

12. Write the formula for finding the perimeter of the rectangle.

13. Write the formula for finding the area of the rectangle.

14. Write the formula for finding the length of the rectangle if you know its area and width.

15. What is the diameter of a circle that has a radius of 7.5 inches?

16. The path from the center of a circular garden to its edge is 8 feet long. What is the length of a path around the outer edge of the garden?

17. Suppose one quart of paint will cover 180 square feet. You need to paint a circular sign that has a diameter of 8 feet. How many quarts of paint must you purchase?

18. One cubic yard of concrete costs $48. A rectangular concrete pier has dimensions 4 feet by 2 feet. The pier is 18 feet high. How much will it cost to fill the pier with concrete?

19. The formula for the volume of a cylinder is $V = \pi r^2 h$, where r is the radius and h is the height. One cubic foot of water weighs about 62.4 pounds. What is the weight of the water in a cylindrical tank with a radius of 6 feet if the tank is filled to a depth of 12 feet?

20. What is the simple interest on a note for $2000 borrowed for 6 months at an annual rate of 8%?

Math Lab

21. Three battery-powered cars travel the following distances in the indicated times. Which car would win a three-way race?

Car	Distance (feet)	Time (seconds)
A	11	1.9
B	15	2.3
C	13	2.0

22. You measure the diameter of a sphere with a micrometer caliper. How would you use this measurement to find the volume of the sphere?

23. Find the height of a treetop if sighting through a pipe attached to a 20 foot string gives an angle (θ) of 60°. The string is 5 feet above the ground. [$h = 20(\tan \theta) + h_0$]?

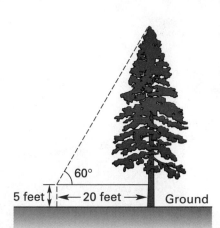

60°

5 feet ← 20 feet → Ground

CHAPTER 4

WHY SHOULD I LEARN THIS?

Equations are the backbone of today's technology. People use equations to control air pollution, design supersonic aircraft, and operate telecommunication systems. Learn how you will use equations in your future career.

SOLVING EQUATIONS

OBJECTIVES

1. Simplify and solve an equation.
2. Translate a problem into an equation.
3. Check the solution of an equation in terms of the problem.

In Chapter 3, you used equations in the form of formulas. In this chapter, you will solve formulas and equations for a specific variable. Who uses equations? Most people who work with money use equations—people in banks, government tax offices, real estate firms, credit unions, and payroll departments. People who work in factories, in hospitals, on farms, and around the home also use equations. Here are two ways people use equations on the job.

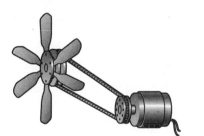

$B = 2L + 1.625\,(D + d)$
designing an exhaust fan

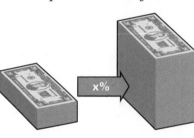

$A = P\,(1 + r)^n$
helping savings grow

How do equations help you solve problems? Often the problems you meet on the job are stated in words. If you can state a problem in the form of an equation, you can usually solve the equation and find an answer to the problem. Once you find an answer, it is important to check and analyze the solution in terms of the original situation or problem. This check will make sure your answer makes sense.

As you watch the video for this chapter, notice how people on the job use equations to solve problems.

LESSON 4.1 SOLVING MULTIPLICATION EQUATIONS

Understanding equations and how to rearrange them is an important skill in the workplace. Often a technician is asked to review technical reference books and find an equation for solving a particular problem.

Roland is an electronics technician. He knows the current and resistance specifications for a particular computer circuit but not the voltage. To find the voltage, he refers to a reference manual and finds the following equation:

$$V = I \bullet R$$

This equation is called Ohm's Law. In this equation, V represents the voltage, I represents the current, and R represents the electrical resistance. To find the voltage, Roland can multiply the measured values of the current by the resistance. The way this equation is arranged—with the V isolated on the left side—makes Roland's work very easy.

Suppose Roland is able to measure the voltage V and current I but needs to calculate the resistance R of the circuit. Can he still use Ohm's Law?

The answer is yes, but first Roland must rearrange the equation to "solve" for R.

Equations and Pan Balances

Remember, an equation states that two mathematical expressions are equal. For example,

$$V = IR$$

states that the expression V is equal to the expression IR.

One way to help you understand equations is to think about a pan balance. A pan balance is an instrument that measures the relative weights of objects.

Small weights are added to the right-hand pan until they balance the weight of the object in the left-hand pan. When "balanced," the weight of the object on the left equals the total weight on the right.

In the same way, an equation represents a balance between its two sides. The equal sign says the expression on the left side of the equation equals the expression on the right side. Consequently, even though one expression is represented by V and the other is represented by IR, they are equal, and both sides of the equation are balanced.

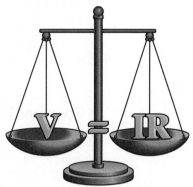

When solving equations, the most important thing to remember is that you must always keep the equation balanced. This balance is maintained by always performing the same mathematical operations on *both* sides of the equation.

> **Balancing Equations**
> Whatever mathematical operation is performed on one side of an equation must also be performed on the other side.

You can use this rule to solve multiplication equations.

Multiplication Equations

Solving Multiplication Equations

 1 Start with the equation $5 \cdot 3 = 15$.

2 Divide each side by 5.

3 What is the result?

Now try this:

4 Start with the equation $5x = 15$.

5 Divide each side by 5.

6 What is the result?

You have just solved the multiplication equation $5x = 15$ by dividing each side of the equation by 5. The resulting equation is $x = 3$. Thus, the solution is 3. To solve the equation, you used one of the basic properties of equality.

> ## Division Property of Equality
> If each side of an equation is divided by the same nonzero number, the results are equal; that is, the two sides of the equation stay equal or balanced.

The Division Property solves problems modeled by multiplication equations. When solving equations using division, it is easier to show the division in fraction form rather than with the division sign. Thus, $5x \div 5$ is written as $\frac{5x}{5}$.

When you solve an equation, give your reason for each step. This will help you solve equations as they become more complicated.

EXAMPLE 1 Solving a Multiplication Problem

A shipping clerk must ship several packages of chemicals to a laboratory. The container holds 18 pounds. If a packet weighs 3 pounds, how many chemical packets will each container hold?

SOLUTION

The equation $3x = 18$ models this problem. Here is how to solve $3x = 18$.

$3x = 18$	Given
$\dfrac{3x}{3} = \dfrac{18}{3}$	Division Property
$x = 6$	Simplify

The last step is an equation with *x isolated* on one side. The solution, 6, is on the other side of the equal sign. Thus, each container can hold 6 chemical packets.

Ongoing Assessment

a. Solve the equation $4x = -24$. Give the reason for each step.

b. Solve the equation $6x = 21$. Give the reason for each step.

Critical Thinking   Why do you choose the coefficient of the variable as the divisor when you are solving a multiplication equation?

Now return to the example of Ohm's Law.

Roland has measured the voltage and current in a computer circuit and needs to find the circuit resistance. He knows Ohm's Law relates these circuit measurements to each other through the following equation:

$$V = IR$$

Roland uses the following steps to solve this equation for *R*. Give the reason for each step.

$$V = IR \qquad\qquad ?$$
$$\frac{V}{I} = \frac{IR}{I} \qquad\qquad ?$$
$$\frac{V}{I} = R \qquad\qquad ?$$

You can write the isolated variable on either side of the equal sign. However, the final result is usually written with the isolated variable on the left side.

$$\frac{V}{I} = R \text{ means the same as } R = \frac{V}{I}$$

In solving $V = IR$ for R, I is divided by itself. When any nonzero number is divided by itself, the result is 1.

$$\frac{IR}{I} = \frac{I}{I} \cdot R$$

$$R = 1 \cdot R$$

When you multiply R by 1, the result is R. This is an important basic property.

Property of Multiplying by 1
For any number a,

$$a \cdot 1 = a$$

and

$$1 \cdot a = a$$

Ongoing Assessment

Solve the equation $C = \pi d$ for d. Show your steps. When did you use the Property of Multiplying by 1?

John is starting a new job. His employer has agreed to pay him $5.35 per hour. They also agree that John will work 15 hours per week after school and 8 hours on Saturdays.

To find his weekly pay, John uses the following equation:

weekly pay = $5.35/hr (15 hr + 8 hr)
         ↓   ↓
     hourly rate   total hours
                          worked

Let p represent John's weekly pay. You can rewrite the last equation as

$$p = 5.35 (15 + 8)$$

The solution is 123.05. This means that John will earn $123.05 each week.

Now suppose John wants to increase his weekly earnings to $150 per week so that he can afford a stereo. But he still wants to work the same number of hours. The only

way John can increase his earnings without working more hours is to earn a higher hourly rate. What new hourly rate does John need?

To answer this question, first modify the last equation by replacing p with \$150. Then let r represent the new hourly rate. The result is the following equation:

$$150 = r(15 + 8)$$

You can also write this equation in the following forms:

$$150 = r(23)$$

$$23r = 150$$

You can solve this equation by dividing both sides by 23. Why? The result is 6.52 rounded to the nearest hundredth. Thus, if John works 23 hours each week, he will have to earn \$6.52 per hour to make \$150 per week. Always check and analyze the solution to make sure it makes sense; that is, \$6.52 $\times$ 23 = \$149.96.

The Division Property of Equality is also useful in solving geometry problems.

EXAMPLE 2 Finding the Degrees in a Triangle

A sheet metal worker is making a warning sign in the shape of an equilateral triangle. How many degrees are in each angle of the triangle?

SOLUTION

It usually helps to draw a picture when solving geometry problems.

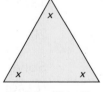
Equilateral Triangle

The sum of the angle measures of a triangle is 180°. Let x represent each angle measure. Since an equilateral triangle has three angles with equal measures, you can write the following equation:

$$3x = 180$$

If you divide each side by 3, the result is $x = 60$. Thus, each angle is 60°. Does this answer make sense?

LESSON ASSESSMENT

Think and Discuss

1 Explain how solving an equation is like using a pan balance.

2 How is the Division Property of Equality used to solve a multiplication equation?

3 Explain how to solve $ax = c$ for x.

4 How is the Property of Multiplying by 1 used in solving a multiplication equation?

5 Why is it necessary to check your answer to a problem after you solve the equation?

Practice and Problem Solving

Solve each equation for the given variable.

6. $8c = 64$ **7.** $8c = -64$

8. $-8c = 64$ **9.** $-8c = -64$

10. $5r = -27$ **11.** $12m = 80$

12. $15t = 140$ **13.** $-9p = 87$

14. $2.5y = 20$ **15.** $80 = -6z$

16. $-8 = -12b$ **17.** $7.5 = 2.5d$

18. $72 = -3.6d$ **19.** $-10q = -5$

20. $1.5e = 9$ **21.** $-0.3f = 9$

Solve each equation for the indicated variable. Give a reason for each step.

22. Solve for L in the equation $A = LW$.

23. Solve for i in the equation $V = ir$.

24. Solve for t in the equation $d = rt$.

25. Solve for p in the equation $i = prt$.

For each problem, write and solve an equation. Check your answer.

26. How many hours does it take a jet to fly 729 miles at a rate of 485 miles per hour?

27. A brick mason is building a patio in the shape of a parallelogram. The area of a parallelogram is the product of its base and height. Paul needs the patio to be 180 square feet. If the base of the patio is 12 feet, what is the height?

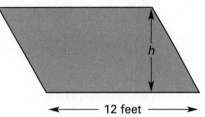

28. The sum of the angles in a square is 360°. Each of the angles in a square has the same measure. Show that each angle of a square measures 90°.

Mixed Review

29. The distance from Tom's apartment to the library is 2.4 kilometers. How far is this distance in meters?

30. A carton of butter weighs 2.25 kilograms. What is this weight in grams?

31. A container holds 520 milliliters of water. How many liters does the glass hold?

32. Paul weighs 90,250 grams. What is Paul's weight in kilograms?

LESSON 4.2 THE MULTIPLICATION PROPERTY OF EQUALITY

Knowing how to work with equations will help you solve everyday problems as well as problems on the job. Sometimes you will use multiplication to solve these problems.

Solving Division Equations

ACTIVITY | **The Multiplication Property of Equality**

1 Start with the equation $\frac{x}{5} = 3$.

2 Multiply each side of the equation by 5.

3 What is the result?

4 Use a pan balance to explain what happened.

This Activity leads to another basic property of equality.

> **Multiplication Property of Equality**
> If each side of an equation is multiplied by the same number, the two sides of the equation stay equal or balanced.

An auto assembly plant has just started a night shift. The plant operates with teams of 6 people each. How many people are needed to work the night shift if there are 7 teams? The equation that models this problem is $\frac{t}{7} = 6$, where t is the total number of people on the night shift. To solve this problem, use the Multiplication Property of Equality to isolate the variable.

$$\frac{t}{7} = 6 \qquad \text{Given}$$

$$7 \cdot \frac{t}{7} = 7 \cdot 6 \qquad \text{Multiplication Property}$$

$$t = 42 \qquad \text{Simplify}$$

The auto assembly plant will need 42 people to work the night shift.

Use the Multiplication Property of Equality to solve the equation

$$\frac{m}{-8} = -9$$

EXAMPLE 1 Finding the Mass of an Object

You can find the density (D) of an object by dividing its mass (M) by its volume (V). The density of water is 1 gram per cubic centimeter. What is the mass of 25 cubic centimeters of water?

SOLUTION

You are given that $D = \frac{M}{V}$. If you multiply each side of the equation by V, the result is $DV = M$ or $M = DV$. Thus, $M = 25$ grams.

Reciprocals

If the product of two numbers is 1, the numbers are called **reciprocals** or **multiplicative inverses**. Since $6 \cdot \frac{1}{6} = 1$, 6 and $\frac{1}{6}$ are reciprocals. What is the reciprocal of $\frac{3}{2}$?

Reciprocal Property
The product of any number and its reciprocal is 1.

Critical Thinking Which numbers are their own reciprocals? Explain why.

In Lesson 4.1, you solved $5x = 15$ by using the Division Property. Now, using reciprocals, you can solve the same equation with the Multiplication Property.

$$5x = 15 \qquad \text{Given}$$

$$\frac{1}{5} \cdot 5x = \frac{1}{5} \cdot 15 \qquad \text{Multiplication Property}$$

$$x = 3 \qquad \text{Reciprocal Property; Simplify}$$
$$\left(\frac{1}{5} \cdot 5 = 1 \text{ and } \frac{1}{5} \cdot 15 = 3 \right)$$

The solution is 3. Check to make sure the solution is correct.

$$5(3) = 15$$
$$15 = 15$$

The purchase price (P) of a series EE Savings Bond is found by the formula $\frac{1}{2}F = P$, where F is the face value of the bond. Use the formula to find the face value of a savings bond purchased for $500.

The Multiplication Property and the Reciprocal Property are often used together to solve a multiplication equation.

EXAMPLE 2 Finding the Speed of a Car

A technician at an automobile test track is calibrating a new speedometer. She drives the car 45 miles in 45 minutes ($\frac{3}{4}$ hours). What is the speed of the car?

SOLUTION

To find the distance (d), multiply the rate (r) and the time (t). You can express this relationship as the equation

$$d = rt \quad \text{or} \quad rt = d$$

Now substitute the values for d and t into the equation and solve.

$rt = d$	Distance Formula
$r\left(\dfrac{3}{4}\right) = 45$	Given
$\dfrac{4}{3} \cdot \dfrac{3}{4}\, r = \dfrac{4}{3} \cdot 45$	Multiplication Property
$r = 60$	Reciprocal Property; Simplify

The speed of the car is 60 miles per hour. Does this answer make sense?

Rearranging Equations

When you solve an equation, it is often necessary to rearrange the numbers and variables. Two basic number properties allow for this rearrangement.

One basic property allows you to change the *order* of the numbers or variables in an equation.

Commutative Property of Multiplication
For all numbers a and b,

$$ab = ba$$

The Commutative Property states that the order in which you multiply two numbers has no affect on the product.

For example, $3 \cdot 4 = 4 \cdot 3$ and $2(5 \cdot 8) = (5 \cdot 8)2$.

The other basic property allows you to change the *grouping* of the numbers or variables in an equation.

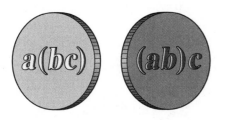

Associative Property of Multiplication
For all numbers a, b, and c,

$$a(bc) = (ab)c$$

The Associative Property states that the *grouping* of numbers you multiply has no affect on the product.

For example, $5(2 \cdot 8) = (5 \cdot 2)8$.

LESSON ASSESSMENT

Think and Discuss

1 How is the Multiplication Property used to solve a division equation? Give an example.

2 How do you find the reciprocal of a number?

3 How is the Multiplication Property used to solve a multiplication equation? Give an example.

4 How are the Commutative Property and the Associative Property used to rearrange the numbers and variables in an equation? Give an example of each.

Practice and Problem Solving

Find the reciprocal of each number.

5. 4   **6.** -6 **7.** $\dfrac{9}{10}$ **8.** $-2\dfrac{3}{4}$

Use the Multiplication Property to solve each equation. Check your answer.

9. $\dfrac{b}{12} = -7$ **10.** $8t = -128$

11. $125 = 5y$ **12.** $26 = \dfrac{m}{-2}$

13. $-9x = -66$ **14.** $-\dfrac{3}{5}a = 15$

15. $\dfrac{8}{5}k = -40$ **16.** $-36 = 1\dfrac{2}{3}d$

17. Solve the equation $A = LW$ for L. For W.

18. Write and solve an equation for this problem: Latisha has saved \$230 for a new compact stereo. So far, Latisha has $\dfrac{2}{3}$ of the amount she needs to buy the stereo. What is the cost of the stereo?

Mixed Review

Write each number in scientific notation.
19. 0.000056   **20.** 985,000   **21.** 143.95

From 3 AM until 8 AM, the temperature dropped two degrees an hour.

22. Write a multiplication expression to model the drop in temperature.

23. What is the change in temperature?

Twenty gallons of paint cost a painting contractor \$170.

24. Let a represent the cost of one gallon. Write a multiplication equation relating the cost of 20 gallons of paint to the total cost.

25. Use the equation you wrote to find the cost of one gallon of paint to the nearest cent.

LESSON 4.3 SOLVING PROPORTIONS AND PERCENT EQUATIONS

Ratio and Proportion

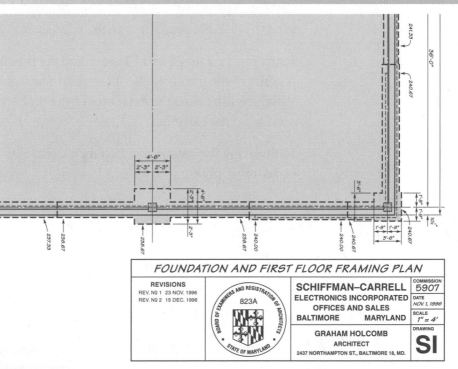

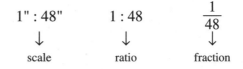

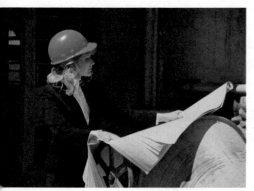

The scale drawing of an office building has a scale of 1 inch to 4 feet. This means that one inch on the drawing is the same as four feet in the actual office building. The scale is written as 1" : 4' or 1 : 4.

A **ratio** is a fraction that compares two quantities. Thus, you can also write the scale as $\frac{1}{4}$. However, it is better to keep the ratio in the same units. Since 4 feet equals 48 inches, this ratio can be written in one of three ways:

1" : 48"	1 : 48	$\frac{1}{48}$
↓	↓	↓
scale	ratio	fraction

The length of the office building on the drawing is 12 inches. How can you find the full-scale length? Let s represent the full-scale length of the laundry center. First, form the ratio $\frac{12}{s}$.

The numerator represents the scale drawing length → 12
The denominator represents the full-scale length ⟶ s

In this ratio, the units in the numerator and denominator are the same.

A **proportion** is an equation with a ratio on each side of the equal sign. Since the drawing is to scale, a proportion models the problem.

$$\frac{1}{48} = \frac{12}{s}$$

Critical Thinking   Show how the Multiplication Property can be used to solve the proportion in the scale drawing problem.

In the proportion, 48 and 12 are called the **means**. Thus, the product of the means is $48 \cdot 12$ or 576. The number 1 and the variable s are called the **extremes** of the proportion. What is the product of the extremes?

Multiplying the means and then the extremes is an example of **cross-multiplying**.

ACTIVITY   **Cross-Multiplication**

Find the product of the means. Then find the product of the extremes.

a. $\dfrac{3}{4} = \dfrac{6}{8}$ **b.** $\dfrac{4}{10} = \dfrac{6}{15}$ **c.** $\dfrac{-4}{8} = \dfrac{5}{-10}$ **d.** $\dfrac{10}{15} = \dfrac{4}{d}$

1   What do you notice about the results in **a, b,** and **c**?

2   Use your observation to solve the proportion in **d.**

Cross-multiplying is a method used to test whether two ratios are equal. For example, you can use cross-multiplication to show the two fractions, $\frac{3}{6}$ and $\frac{4}{8}$, are equal.

Find the product of the means: $3 \cdot 8 = 24$. Find the product of the extremes: $6 \cdot 4 = 24$. The cross-products are equal. Thus, the ratios are equal.

This leads to an important property for proportions.

Proportion Property
   In a true proportion, the cross-products are equal.

The Proportion Property is a tool for finding the actual length of the office building.

$$\frac{1}{48} = \frac{12}{s} \qquad \text{Given}$$

$$1 \cdot s = 12 \cdot 48 \qquad \text{Proportion Property}$$

$$s = 576 \qquad \text{Property of Multiplying by 1}$$

Thus, the actual length of the office building center is 576 inches or 48 feet.

Ongoing Assessment

Write and solve a proportion for this problem. A scale drawing for a rectangular tabletop measures 3 inches wide and 4 inches long. You want to make a table that is 5 feet wide. How long will it be? Show your work. Express your answer in feet and inches.

Percent

Percent means hundredths of a quantity. For example, 8% is just another way to write the ratio 8:100 or $\frac{8}{100}$. You can use proportions to change fractions or decimals to percents.

EXAMPLE 1 Changing Fractions to Percents

Change $\frac{5}{8}$ to a percent.

SOLUTION

Write a proportion
$$\frac{5}{8} = \frac{x}{100} \qquad \left(\frac{x}{100} \text{ represents a \%}\right)$$

and solve
$$8x = 500 \qquad \text{Proportion Property}$$

$$x = 62.5$$

Thus, $\frac{5}{8}$ is equal to 62.5%.

Although proportions solve percent problems, it is usually easier to use a *percent equation*. One way to write a percent equation is

$$a\% \text{ of } b = c$$

In this form, a is the percent, b is the base, and c is the percentage.

Critical Thinking Explain why the percent equation can also be written in the form:

$$\frac{a}{100} \cdot b = c$$

Tad earns a 6% commission on all of his sales. If Tad earns $150, what are his sales?

Use a percent equation. The percent is 6, and the percentage is 150. Let b represent the base (Tad's sales). Write a percent equation.

$$\frac{6}{100}b = 150$$

Multiply each side by $\frac{100}{6}$.

$$\frac{100}{6} \cdot \frac{6}{100}b = \frac{100}{6} \cdot 150$$

$$b = 2500$$

Tad has sales of $2500.

WORKPLACE COMMUNICATION

How many gallons of liquid nitrogen will Cryogenics Supply Corporation deliver?

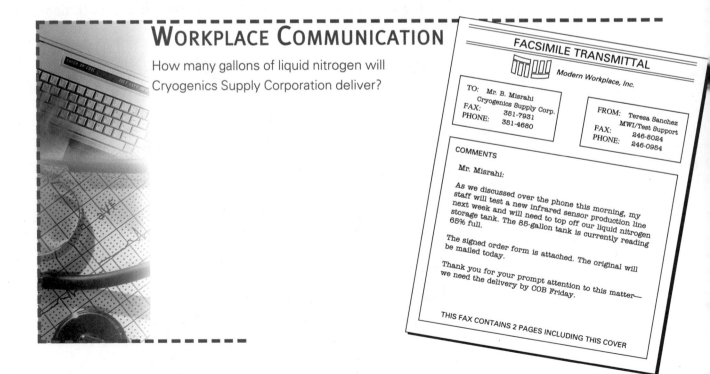

FACSIMILE TRANSMITTAL

Modern Workplace, Inc.

TO: Mr. B. Misrahi
Cryogenics Supply Corp.
FAX: 351-7931
PHONE: 351-4680

FROM: Teresa Sanchez
MWI/Test Support
FAX: 246-8024
PHONE: 246-0954

COMMENTS

Mr. Misrahi:

As we discussed over the phone this morning, my staff will test a new infrared sensor production line next week and will need to top off our liquid nitrogen storage tank. The 85-gallon tank is currently reading 65% full.

The signed order form is attached. The original will be mailed today.

Thank you for your prompt attention to this matter— we need the delivery by COB Friday.

THIS FAX CONTAINS 2 PAGES INCLUDING THIS COVER

LESSON ASSESSMENT

Think and Discuss

1 How are ratios and proportions related? Give an example.

2 Explain the meaning of the Proportion Property.

3 Explain how to use a proportion to change a fraction to a percent.

4 Explain how to use a percent equation to find what percent one number is of another.

Practice and Problem Solving

Solve each proportion.

5. $\dfrac{r}{5} = \dfrac{8}{10}$ **6.** $\dfrac{7}{3} = \dfrac{m}{9}$ **7.** $\dfrac{-6}{16} = \dfrac{15}{t}$ **8.** $\dfrac{-a}{12} = \dfrac{6}{9}$

Set up a proportion and solve. Round answers to the nearest tenth.

9. What is 25% of 120?   **10.** What percent of 10 is 6?

11. 64 is what percent of 256?   **12.** Find 12.5% of 72.

13. 6 is what percent of 9?   **14.** 9 is what percent of 6?

Set up a percent equation and solve. Round answers to the nearest tenth.

15. What percent of 300 is 13.5?   **16.** What is 300% of $20?

17. What is 0.5% of $120?   **18.** What is 15% of $225?

19. $2.50 is what percent of $10?

20. What percent of 12 is 2.4?

For each problem, set up a proportion or percent equation and solve.

21. Keesha bought a sweater for $50. If the sales tax rate is 8.5%, find the total amount Keesha paid the cashier.

22. Sami borrowed $1200 for one year. If Sami paid $90 simple interest, what was the interest rate?

23. Tom pays 6.2% of his wages for social security. If Tom paid $186 in social security taxes, what was his income?

24. Rhonda buys her plumbing supplies at wholesale. She buys pipe for $4/ft and sells it for $5/ft. What percent profit is she making?

Simplify.

25. $(4.2 \cdot 10^{-3})(5.9 \cdot 10^{-2})$

26. Water is stored in a rectangular container that is 8 inches by 12 inches by 4 inches. The water is transferred to a spherical container with a radius of 10 inches. Is the spherical container large enough to hold the water? By how much is it too small or too large?

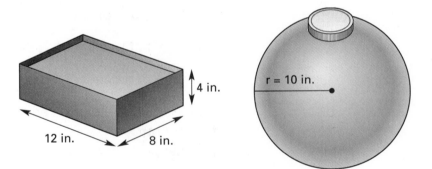

4 in.

12 in.   8 in.

r = 10 in.

27. Evaluate $-8c + 2d \div -4$ when c is 5 and d is -6.

28. Solve the equation $mx = b$ for x. Show your steps. What properties did you use?

LESSON 4.4 THE ADDITION PROPERTY OF EQUALITY

Subtraction Equations

The Blueberry Baking Company has announced a 48¢ reduction in price for its cookie products. Cookies that sold for $1.59 last week will now cost $1.59 − $0.48, or $1.11. Use your calculator to check whether this computation is correct.

You can model the drop in cookie price by an equation. Represent the original cookie price by p. Represent the new cookie price by n. The equation that models the relationship between the old and new cookie prices is the subtraction equation

$$n = p - 0.48$$

After the sale, the cookies return to their original price. If you add 0.48 to the right side of the equation, it will reverse the effect of subtracting 0.48. Thus, adding 0.48 will isolate p. However, to maintain the balance of the equation, you must also add 0.48 to the left side.

Addition Property of Equality
If the same number is added to each side of an equation, both sides of the equation stay equally balanced.

The Addition Property says that if you add 0.48 to the right side of the equation, you must also add 0.48 to the left side. Here are the steps that isolate p.

$n = p - 0.48$	Given
$n + 0.48 = p - 0.48 + 0.48$	The Addition Property
$n + 0.48 = p + 0$	Simplify
$n + 0.48 = p$	

Using the Addition Property, you have isolated p on one side of the equation while maintaining the "balance" of the equation.

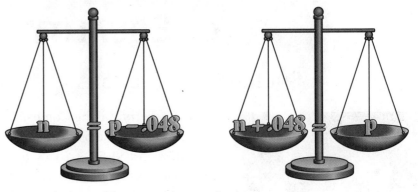

The pan balance illustrates the results.

Addition Equations

Now suppose the Blueberry Baking Company increases prices by $0.48 over the original price. The equation

$$n = p + 0.48$$

models this increase. In this equation, the variables p and n have the same meaning as before. Except this is now an addition equation. If you want to isolate p in this equation, you can subtract 0.48 from each side of the equation. When the same value is subtracted from each side of the equation, you are using the Subtraction Property of Equality.

Subtraction Property of Equality
If the same number is subtracted from each side of an equation, both sides of the equation stay equally balanced.

Before you use the Subtraction Property to solve the equation

$$n = p + 0.48$$

for p, another property is needed. Do you remember using this property earlier in this lesson?

Property of Adding Zero
For any number a,

$$a + 0 = a \text{ and } 0 + a = a$$

Now you can use the Subtraction Property to solve for p.

$n = p + 0.48$	Given
$n - 0.48 = p + 0.48 - 0.48$	Subtraction Property
$n - 0.48 = p$	Explain why

EXAMPLE 1 Using Sales Tax

The total price (p) of a car is equal to the sum of the sale price (s) and the sales tax (t). Write an addition equation and solve for the sale price.

SOLUTION

One way to write the addition equation is $s + t = p$. If you subtract t from each side of the equation, the result is

$$s = p - t$$

Ongoing Assessment

Solve the following equations for the indicated variable.

1. $n = p - 0.64$ for p

2. $y = x - 12$ for x

3. $k = m + 25$ for m

Sometimes you must rearrange the numbers and variables in an addition or subtraction equation. Remember the two properties you used in Lesson 4.2 to rearrange numbers and variables in a multiplication equation. The Commutative and Associative Properties are also true for addition.

Commutative Property of Addition
For all numbers a and b,
$$a + b = b + a$$

For example, $3 + 7 = 7 + 3$.

Associative Property of Addition
For all numbers a, b, and c,
$$a + (b + c) = (a + b) + c$$

For example, $6 + (4 + 8) = (6 + 4) + 8$.

Critical Thinking   Which basic property is illustrated?

a. $5(3 + 4) = (3 + 4)5$ b. $(1 + 2) + 3 = 3 + (1 + 2)$

EXAMPLE 2 Rearranging Equations

Solve the equation $15 + x = -32$.

SOLUTION

The Commutative Property states that you can add in any order. To rearrange the left side of the equation, replace $15 + x$ with $x + 15$.

$$x + 15 = -32$$

Now subtract 15 from each side,

$$x + 15 - 15 = -32 - 15$$
$$x = -47$$

The solution is -47.

The Associative Property states that you can regroup the numbers or variables in an equation. For example, to solve $(r + 5) + 10 = 25$, the Associative Property allows you to regroup to obtain $r + (5 + 10) = 25$. These properties are used together to rearrange the terms in an equation. This will simplify your work when solving problems in more complex situations.

EXAMPLE 3 Rearranging to Solve an Equation

Solve the equation $(-6 + n) + (3 - 2n) = 0$.

SOLUTION

You can use the definition of subtraction and at the same time rearrange the left side to rewrite the equation.

$$3 + (-6 - n) = 0 \qquad \text{Given}$$

$$-n + (3 - 6) = 0 \qquad \text{Rearrange the terms using the Commutative Property and the Associative Property}$$

$$-n - 3 = 0 \qquad \text{Simplify}$$

$$-n = 3 \qquad \text{Addition Property}$$

There are two ways to solve the equation $-n = 3$.

1. Think of the negative sign as an opposite sign. If the opposite of n is 3, n must be -3.

2. Think of $-n$ as a short way of writing $-1 \cdot n$. Multiply each side by -1. The result is $n = -3$.

By either method, the solution is -3. Check: $(-6 - 3) + (3 + 6) = 0$
$$-9 + 9 = 0$$

The solution to Example 3 leads to an important definition. You can find the opposite of a number by multiplying the number by -1.

Multiplying by -1
For any number a,

$$-a = -1a$$

EXAMPLE 4

Solve the equation $17 = 9 - n$.

SOLUTION

Subtract 9 from each side of the equation. The result is

$$-n = 8$$

The solution is -8 because $(-1)(-8) = 8$ and $-1(-n) = n$.

LESSON ASSESSMENT

Think and Discuss

1 Compare the Addition Property of Equality and the Subtraction Property of Equality. How are they alike and different?

2 How is the Property of Adding Zero used to solve an addition or subtraction equation?

3 Explain how to solve the equation $a + b = c$ for b.

Practice and Problem Solving

Solve each equation. Show each step.

4. $y + 5 = 9$ **5.** $22 = t + 18$ **6.** $q - 9 = -12$

7. $36 = c - 27$ **8.** $-14 = k + 36$ **9.** $18 + a = 17$

10. $-k = -8$ **11.** $18 - w = 20$ **12.** $36 = -15 - e$

Solve each equation. Check your answer.

13. $4.6 = z + 5.4$ **14.** $p - (-12) = 20$ **15.** $-1.9 = s - 5.1$

16. $f - 3.5 = -2.8$ **17.** $10 = 5.6 + n$ **18.** $-8.5 - d = -3$

19. $5.3 + x = -7.4$ **20.** $6 - x = -15.3$ **21.** $x - 2.78 = -5.9$

22. The formula for profit (p) is cost (c) minus overhead (o). Write an equation for profit and solve it for cost.

23. The equation $556 + n = 580$ represents the change in the number of computer workstations in a corporation over one year. Solve the equation for n to find the change in the number of workstations.

Ross wrote a check for $124 and received a notice from the bank stating that he was $51 overdrawn. His account now shows a negative balance.

24. Let *a* represent the original amount in Ross's account. Write an equation to model the bank's transaction.

25. Find the original amount in Ross's account.

Mixed Review

On three successive carries of the football, Todd lost 5 yards, gained 8 yards, and lost 3 yards. He started at the 25-yard line.

26. Use integers to show each carry.

27. On what yard line will the team start the next play?

On a vacation trip, Maria drove a total of 978 miles in 6 days.

28. Write an equation to model the number of miles Maria drove each day.

29. Solve the equation.

LESSON 4.5 SOLVING MULTISTEP EQUATIONS

Equations with One Variable

Many equations contain several terms. Often you must use more than one of the properties of equality to solve these equations. In the process of solving an equation with one variable, isolate the variable and simplify the constants on the other side of the equation.

EXAMPLE Solving a Linear Equation

Solve $3g - 7 = 11$.

SOLUTION

Start by identifying the variable. In this equation, it is the letter g. Then begin the process of isolating g.

Look at the equation $3g - 7 = 11$ and ask yourself these questions.

- How can I isolate the expression containing g from the other expressions on the same side of the equation?

- Once g and its coefficient are isolated, how can I isolate g on one side of the equation— that is, what is the next thing I must do to free g from $3g$?

As you carry out this process, always remember that whatever you do to one side of an equation—you must do the same to the other side. Remember to keep the equation balanced.

$$3g - 7 = 11 \qquad \text{Given}$$

$$3g - 7 + 7 = 11 + 7 \qquad \text{Addition Property}$$

$$3g = 18 \qquad \text{Simplify}$$

$$\frac{3g}{3} = \frac{18}{3} \qquad \text{Division Property}$$

$$g = 6 \qquad \text{Simplify}$$

Finding a solution is only half of the procedure when you solve an equation. You must also check to make sure the value you found is correct. To check your work, substitute your solution in the original equation. Since you are not sure that the two sides are really equal (remember you are checking your answer), write a question mark over the equal sign.

$$3g - 7 = 11 \qquad \text{Original equation}$$

$$3(6) - 7 \overset{?}{=} 11 \qquad \text{Substitute the value 6 for the variable } g$$

$$18 - 7 \overset{?}{=} 11 \qquad \text{Simplify by doing the arithmetic}$$

$$11 = 11 \qquad \text{Verify the equality}$$

Ongoing Assessment

Solve each equation. Show each step. Check your answers.

a. $4a + 5 = -15$ b. $8 - 3a = 9$ c. $\dfrac{2}{3}y - 1 = -13$

Using the Distributive Property

Recall how you used an equation to solve John's problem in Lesson 4.1. John had just started a job that paid an hourly rate of $5.35 per hour. His work schedule was 15 hours during the week and 8 hours on Saturday. This equation models John's weekly pay:

$$p = 5.35(15 + 8) = 123.05$$

John decides he needs to earn $150 per week while still working the same number of hours. Thus, he needs a new hourly rate r. He needs to replace p by 150 and 5.35 by r.

$$150 = r(15 + 8)$$

Solving this equation shows that John needs to make over $6 per hour. Since John probably will not make this rate in the near future, he must find a way to increase his weekday hours. To find the number of weekday hours, start with the original equation for weekly pay.

$$p = 5.35(15 + 8)$$

Again replace p with 150. Since 15 (the number of weekday hours) is going to change, replace 15 by the variable e. The 8 hours of weekend time remains the same. The new equation is

$$150 = 5.35(e + 8)$$

There are two ways to solve for e. One way is to solve for $(e + 8)$ by dividing both sides by 5.35 (Division Property of Equality). Then subtract 8 from both sides (Subtraction Property).

$150 = 5.35(e + 8)$	Given
$\dfrac{150}{5.35} = \dfrac{5.35(e + 8)}{5.35}$	Division Property
$28.04 = e + 8$	Simplify
$28.04 - 8 = e + 8 - 8$	Subtraction Property
$20.04 = e$	Simplify

Check to see if 20.04 is the correct solution. John must work just over 20 hours during the weekdays to earn at least $150 per week.

The other method for solving John's equation for e uses the Distributive Property. The following Activity will give you an opportunity to see how this property is used to isolate the variable e.

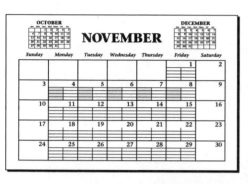

ACTIVITY 1   Using the Distributive Property First

Copy these steps on your paper. Give reasons for each step.

	Steps	Reasons
1	$150 = 5.35\,(e + 8)$	Given
2	$150 = 5.35e + 5.35 \cdot 8$	?
3	$150 = 5.35e + 42.8$	?
4	$150 - 42.8 = 5.35e + 42.8 - 42.8$	?
5	$107.2 = 5.35e$	?
6	$\dfrac{107.2}{5.35} = \dfrac{5.35e}{5.35}$	?
7	$20.04 = e$	?

Critical Thinking   How do the two methods for solving the equation $150 = 5.35\,(e + 8)$ compare? Which one do you prefer? Why?

Ongoing Assessment

Use two methods to solve the equation $5(t - a) = m$ for t. Show your steps and give a reason for each one.

You can write a general equation that describes John's salary situation. Let p be the weekly pay. Let r be the hourly rate. Let e be the hours worked weekday evenings. Let s be the hours worked on Saturdays. Then, the general equation is

$$p = r(e + s)$$

You can solve this equation for any of the variables. Complete the Activity to solve for r in terms of p, e, and s.

ACTIVITY 2   Using the Division Property First

Give reasons for each step.

	Steps	Reasons
1	$p = r(e + s)$	?
2	$\dfrac{p}{(e + s)} = \dfrac{r(e+s)}{(e+s)}$	?
3	$\dfrac{p}{(e + s)} = r$	?

The result of the activity is that r is now isolated on one side of the equation. Since you used only basic properties, the equation should be "balanced." To check your work, find numbers that make the original equation true. Then substitute these same numbers into the new equation and see if both sides are still equal.

WORKPLACE COMMUNICATION

Explain how you can use the ideas in Chapter 4 to reach this conclusion.

 File   Edit   View   Label   Special

mwi electronic mail

FROM: tsao@mwi.org
TO: anderson@mwi.org
          miller@mwi.org
SUBJECT: Power Restrictions for New Furnace

MESSAGE:

The new furnace at semiconductor production station 2 is scheduled for testing to begin next week. This furnace will require testing for the rest of the year; however, I'm concerned that it could cause us to exceed our budget for electricity for the year.

We have about $25,000 left in the budget to cover the electricity requirements for both furnaces—station 1 and station 2. The power company charges 3.5 cents per kW-hr of electrical energy used.

The furnace at station 1 must continue to operate at full capacity so that production is not affected. It will use about 4.76×10^5 kW-hr of electrical energy over the remainder of the year.

This means that for us to stay on budget station 2 can only operate at half power for the rest of the year (i.e., the new furnace can use about 2.58×10^5 kW-hr of electrical energy).

Please restrict the power of the new furnace to stay under this limit.

LESSON ASSESSMENT

Think and Discuss

1 Explain how to solve $2(a + b) = c$ for a in two different ways.

2 Why is it necessary to check the solution to an equation?

Practice and Problem Solving

Solve each equation. Show each step. Check your answer.

3. $2x + 15 = 29$ **4.** $16 = 4t - 20$ **5.** $7 + 3r = -11$

6. $-5x + 9 = 4$ **7.** $-6s - 12 = 24$ **8.** $3(x + 2) = -5$

Solve each equation. Check your answer.

9. $2.5(4 + b) = -20$ **10.** $3.7 = 1.8f - 3.9$

11. $-28 = 3.6(m - 6)$ **12.** $\frac{2}{3}c + 5 = 13$

13. $\frac{3}{4}(8 + 2y) = 3$ **14.** $\frac{5b + 9}{4} = -4$

15. Solve the equation $ax + b = c$ for x.

16. Solve the equation $a(x + b) = c$ for x.

The formula for the perimeter of a rectangle is $P = 2L + 2W$. P represents the perimeter, L the length, and W the width. Substitute the dimensions into the formula and solve the equation for the missing variable. (All measurements are in feet.)

17. length is 15; width is 12

18. perimeter is 60; length is 12

19. perimeter is 52; length is 15

20. perimeter is 36; width is 3

21. Solve the perimeter formula for L.

22. Rewrite the perimeter formula using the Distributive Property.

23. The sum of Lin's three test scores for the grading period is 264. There is one test remaining, and Lin would like to average 90 on his test scores. Lin uses the equation $90 = \frac{264 + s}{4}$ to find the fourth test score he must make. What is the test score?

24. A 60-liter solution is 30% acid. The formula $180 = 2(n + 60)$ can be used to find the number of liters of water (n) that must be added to make a 20% solution. How many liters must be added?

25. Natasha is running a "20% off" sale at her clothing shop. On one jacket, she is offering a savings of $25. What does Natasha usually charge for the jacket?

Mixed Review

Simplify
26. $-35 - (6 + 4 \bullet -5)$ **27.** $(2.6 \bullet 10^{-5})(5.12 \bullet 10^{7})$

The formula to find the volume (V) of a cylinder with radius (r) and height (h) is $V = \pi r^2 h$.

28. The Cup Company produces coffee mugs in the shape of cylinders. The mugs have a radius of 4 centimeters and a height of 9 centimeters. What volume of coffee will one of the mugs hold?

29. A restaurant wants to order mugs that hold 706.5 cubic centimeters of soup each. For stacking purposes, the cups must have a 10 centimeter diameter. What height should the cups be?

30. Solve the equation $\frac{c}{d} = \frac{m}{n}$ for d.

LESSON 4.6 EQUATIONS WITH VARIABLES ON BOTH SIDES

An electronics manufacturing plant produces telephones at the rate of 600 per day. A customer survey indicates that the demand for phones with built-in answering machines is twice as great as for those without them. If you are determining the production quota for a day, how many phones with answering machines would you schedule for production? Model the problem with an equation.

Let x represent the number of phones to be manufactured without the answering machines. Then $2x$ represents the number of phones with answering machines. The total number of phones must be 600.

$$\begin{matrix} \text{Phones without} \\ \text{machines} \end{matrix} + \begin{matrix} \text{phones with} \\ \text{machines} \end{matrix} = \begin{matrix} 600 \\ \text{phones} \end{matrix}$$

$$x \quad + \quad 2x \quad = \quad 600$$

To find the number of each type of phone, you need to solve the equation

$$x + 2x = 600$$

To see how to simplify the expression $x + 2x$, you can use a rectangular shaped block called an Algeblock.

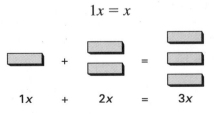

This Algeblock has a height of one unit and a length of x units. Remember that the coefficient of x is understood to be one.

That is,

$$1x = x$$

$$1x \quad + \quad 2x \quad = \quad 3x$$

Now use the Division Property to solve for x. If $3x = 600$, then $x = 200$ and $2x = 400$.

Thus, you will schedule the production of 200 phones without answering machines and 400 with answering machines. Does this answer make sense?

If all the terms in an expression have the same variable, the terms are called **like terms.** You solved the telephone problem by adding like terms using Algeblocks as a model. You can also add like terms by using the Distributive Property.

$$1x + 2x = (1 + 2)x$$

$$= 3x$$

EXAMPLE 1 Simplifying Expressions

Simplify the expression $3a + 2b + 4a + 4b$.

SOLUTION

$$3a + 2b + 4a + 5b = 3a + 4a + 2b + 4b \qquad \text{Rearrange terms}$$

$$= 7a + 6b \qquad\qquad\qquad \text{Add like terms}$$

You can also subtract like terms.

ACTIVITY   Subtracting Like Terms

1 Use a drawing of Algeblocks to simplify $5x - 3x$.

2 Use the definition of subtraction and the Distributive Property to simplify $5x - 3x$.

3 Explain why $6x - x$ is $5x$.

EXAMPLE 2 Equal Compensation

A salesperson in a stereo store is given a choice of two different compensation plans. One plan offers a weekly salary of $250 plus a commission of $25 for each stereo sold. The other plan offers no salary but pays $50 commission on each stereo sold. How many stereos must the salesperson sell to make the same amount of money under both plans?

Let x represent the number of stereos sold in one week.
Plan 1 offers a $25 commission ($25x$) plus $250 salary.
Plan 2 offers a $50 commission ($50x$).

Find when plan 1 ($25x + 250$) = plan 2 ($50x$).

$25x + 250 = 50x$	Given
$250 = 50x - 25x$	Subtraction Property
$250 = 25x$	Subtract like terms
$10 = x$	Division Property

Since $25(10) + 250 = 500$ and $50(10) = 500$, selling 10 stereos a week results in the same compensation.

Critical Thinking   Solve the equation $5 - 2x + 3 = 4x + 8 - 6x$. What does the solution tell you about the equation?

CULTURAL CONNECTION

In 1858, Henry Rhind bought an ancient Egyptian papyrus that was written by the Egyptian scribe Ahmes. The Ahmes, or Rhind Papyrus as it is known, contains 85 problems copied by Ahmes from earlier writings. One of the problems illustrates how the Egyptians used a method now called "guess and check" to solve an equation such as $5x + 3 = 28$. Here is how it works.

Guess a solution and substitute to see if you are correct. Try 10.

Since $5(10) + 3$ is 53, the guess is too high. Try 2.

$5(2) + 3$ is 13. This guess is too low. The next guess might be halfway between the high and the low. Try it and see if it works. Do you see how the Egyptians eventually found the right number? Use the guess and check method to solve these two equations.

1. $6b - 9 = 30$ **2.** $9w + 18 = 5w - 4$

3. What are the advantages and disadvantages of the guess and check method?

LESSON ASSESSMENT

Think and Discuss

 1 Explain how to use Algeblocks to add $7x + 3x$.

2 Explain how to use the Distributive Property to solve $7x + 3x = 20$.

3 Why can you add like terms?

4 Explain how to solve $3m - 2 = 5 - 4m$.

Practice and Problem Solving

Solve each equation. Show each step.

5. $3a + 9 + 4a = 15$ **6.** $4m - 8m - 10 = -24$

7. $5(n + 3) + 2n = -10$ **8.** $2(d - 3) - 5d = 18$

9. $w + 6 = w + 5$ **10.** $3r + 12 = 5r + 15$

Solve and check your solution for each equation.

11. $16 - p = 3p + 8 - 5p$ **12.** $5(y - 1) = -y$

13. $b - 8 = 4(2b + 3)$ **14.** $-4(3 + c) = 6(3 - c)$

15. $\frac{3}{4}(12x + 8) = 50x$ **16.** $\frac{t + 1}{9} = \frac{t}{-3}$

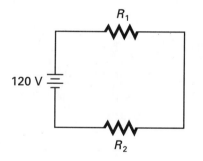

An electronics technician must connect two resistors in series with a 120-volt power supply. The total voltage across two resistors connected in series is the sum of the voltage across each resistor. The voltage across the second resistor is 60 volts greater than the voltage across the first resistor. Let x represent the voltage across the first resistor.

17. Write an expression for the voltage across the second resistor in terms of x.

18. Write an equation for the sum of the voltages across the resistors.

19. Find the voltage across each resistor.

A contractor is building a 4-foot circular sidewalk around a flower bed. The outside circumference of the sidewalk is 1.5 times greater than the circumference of the flower bed. Let x represent the radius of the flower bed.

20. Draw a picture to model the problem.

21. Write an equation that equates the outside circumference of the sidewalk to 1.5 times the circumference of the flower bed.

22. Solve the equation for the radius of the flower bed.

23. A manufacturer can produce 350 riding mowers per day. To meet the needs of distributors, the company must produce four times as many 38-inch mowers as 42-inch mowers. How many of each type of mower should be manufactured each day?

A person is using CAD software to draw rectangles. The software is programmed to produce rectangles as shown for each value of x.

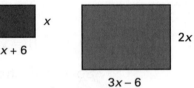

24. What value for x will produce rectangles with equal perimeters?

25. What are the dimensions of each of the rectangles in Exercise 24?

Mixed Review

The Marshall City Safety Department recorded the number of cars passing the intersection of Main and Jackson between 8 AM and 10 AM for five consecutive weekdays.

124   361   243   210   282

26. Find the average number of cars passing the intersection.

27. Use 250 as a benchmark. Subtract the benchmark from each recorded number. Write this deviation between the benchmark and each recorded number as an integer.

28. Find the average of the deviations. Add this number to the benchmark. Compare this sum to the average.

LESSON 4.7 EQUATIONS ON THE JOB

Solving an equation is a step-by-step process. However, finding a correct equation to model a problem is sometimes a difficult process. When you read a problem, try to understand the problem so that it makes sense to you. Follow these steps to help you understand the problem.

1. Look at all the words in the problem and make sure you understand what each word means. If you are not sure, ask someone or look up the word in a reference book, such as a dictionary. You must know what the words mean before you can begin to solve the problem.

2. After you understand the words, try to understand the situation. Picture in your mind the situation that the words describe. Draw a rough sketch of the situation.

3. Form a sentence (either in your mind or on your paper) that begins with what you are trying to find and continues with the verb *is*. For example, if you need to find the distance around a park, you might use this sentence: "The distance around the park *is* the sum of the lengths of all of the sides."

4. Rewrite your sentence in the form of a mathematical sentence or equation. Look for words in the problem that have specific mathematical meanings.

Explain how these steps are used in the following Activities.

ACTIVITY 1   A Soft Drink Problem

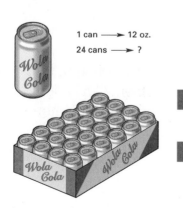

1 can ⟶ 12 oz.
24 cans ⟶ ?

A catering company uses a 24-can case of 12-ounce drinks to make ice cream sodas for a party. The caterer will adjust the other ingredients of the ice cream sodas based on the number of ounces of soft drink that are used. How many ounces are in a case of soft drinks?

1 Are all the words in the problem understandable? Does the drawing help you understand the problem better?

2 You know the weight in one can of soft drink. You want to find the weight in a case of 24 such cans. Complete this sentence.

"The weight of the soft drink in a case of 24 cans is . . . ."

3 Write an equation that models the problem.

ACTIVITY 2   Cutting a Metal Pipe

A plumber needs a 12.5 inch length of copper pipe. She has one piece that is 20.8 inches long. How much pipe must be cut off to leave a piece that is 12.5 inches long?

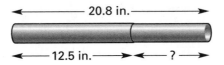

To "cut off" means to subtract. For this problem, use the sketch of the pipe to show how much to cut and how much to leave. You might begin your sentence in this way: "The piece cut off is . . . ." Finish the sentence and write an equation.

ACTIVITY 3   A Family Budget

The family budget has 10% of the take-home pay assigned to transportation. If the monthly take-home pay is $1200, how much is in the budget for transportation?

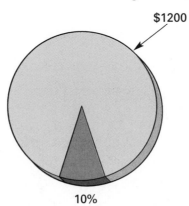

Do you know the meaning of the key words (budget, take-home pay, and transportation) in the problem? You might begin your sentence in this way: "The amount budgeted for transportation is . . . ." Finish the sentence and write an equation.

Compare your equations to the following ones. Often there is more than one correct way to write the equation for a problem. Each of the following equations shows just one of the possibilities.

1. $w = 24$ cans • 12 ounces per can, or $w = 24 \cdot 12 = 288$ oz

2. $l = 20.8$ inches $- 12.5$ inches, or $l = 20.8 - 12.5 = 8.3$ in.

3. $t = 10\%$ of $1200, or $t = 0.10 \cdot \$1200 = \$120$

When you use equations to solve problems on the job, the equations are not always written out for you. You may first have to restate the problem in equation form and then solve the equation.

Suppose the personnel manager of a large department store has just hired you. He explains that your weekly salary is $150. In addition, you will receive 5% of the total value of the sales you make during the week. How can you predict your gross pay for any week?

Remember that it is helpful to picture the problem as you begin to work toward a solution. In this case, you can picture your pay for a week as two stacks of money. The first stack is 150 dollar bills (or $150). The second stack represents 5% of your total sales of the week. The sum of the two stacks (in dollars) is your weekly pay.

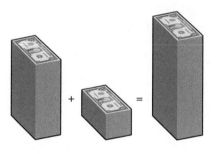

Now you can put your picture into words. Your weekly salary is $150 plus 5% of your total sales for the week. In equation form, this becomes

$$W = 150 + 0.05S$$

Here, W represents your weekly salary in dollars and S your total weekly sales in dollars. Notice that 5% is written as "0.05," the decimal form that your calculator understands. You can also rewrite your equation as

$$W = 0.05S + 150$$

5% = [0.05]

Suppose your total sales during the first week came to $1000. Substitute $1000 for S, and solve for W.

$$W = 0.05\,(1000) + 150$$

$$W = 50 + 150$$

$$W = 200$$

The solution is $200. First check to make sure the arithmetic is correct. Then analyze the solution in terms of the situation. In this problem, 5% means $5 for each $100 you sell. It makes sense that for $1000, you will get ten times as much, or $50. Thus, your total salary for this week is $50 + $150, or $200.

Use the equation $W = 0.05S + 150$ and complete the table.

W	S
200	1000
?	500
?	200
?	100

Write a sentence that describes the relationship between W and S.

Ongoing Assessment

What if your salary for the week is $500? What is the amount of sales you made for the week? The same formula models this situation.

$$W = 0.05S + 150$$

Substitute $500 for W and solve for S. Analyze your solution in terms of the situation.

Now you are ready to write and solve the equations needed in the Math Lab Activities and Math Applications.

MATH LAB

Activity 1: Balancing Equations

Equipment   Calculator
Soup can
1 pound coffee can
Wooden dowel about 1 inch in diameter
Vernier caliper
Cloth tape measure

Problem Statement

You will work with the formula for the circumference of a circle. Let C represent the circumference of a circle. Let d represent the diameter. Use 3.1416 as the approximate value of π.

$C = \pi d$ (this is Equation 1)

You will form new equations by solving the equation for d and for π. Then, you will check your work by using measured amounts to see if your equations balance.

Procedure

a   Measure the circumference of the coffee can, soup can, and wooden dowel. Record your results as the first three entries under the "Measured Left Side" column.

Equation	Left Side	Right Side	Object	Measured Left Side	Calculated Right Side
Equation 1	C	πd	Coffee can		
			Soup can		
			Dowel		
Equation 2	d		Coffee can		
			Soup can		
			Dowel		
Equation 3	π		Coffee can	3.1416	
			Soup can	3.1416	
			Dowel	3.1416	

   Measure the diameter of each object with the vernier calipers. Complete the "Measured Left Side" column by entering these measurements.

c Solve Equation 1 for d. Call this Equation 2. Record the right side of this equation in the "Right Side" column.

d Solve Equation 1 for π. Call this Equation 3. Record the right side of this equation in the "Right Side" column.

e Calculate and record the values for the right side of Equations 1, 2, and 3 using measured values for circumferences and diameters.

f Are both sides of all your equations equal? If the answer is yes, you have solved all the equations correctly. If not, check to be sure you correctly used the basic properties for solving equations correctly.

g Explain why the values might not be exactly the same even though you solved each equation correctly.

Activity 2: Creating Equations

Equipment	Ball of string (over 400 feet long)
	Masking tape
	Tape measure
	Calculator

Problem Statement

You will create an equation that determines the perimeter of an irregularly shaped area. You will then form several new equations by solving the perimeter equation for each of its variables. Finally, you will check the balance of these new equations.

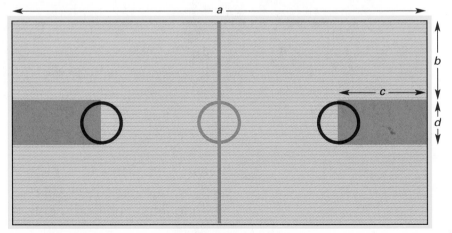

Procedure

 Measure the lengths of the irregular sections of the basketball court represented by *a, b, c,* and *d.* Record each value in Table 1.

Side	Length
a	
b	
c	
d	

Table 1. Perimeter

b Devise a way to use the string, tape, and tape measure to measure the highlighted perimeter in the illustration.

c Measure and record this perimeter in Table 1.

d Using the variables *a, b, c,* and *d,* create an equation that models the highlighted perimeter of the court. Record the right side of this equation in the first row of Table 2 under the "Right Side" column.

Left Side	Right Side	Measured Left Side	Calculated Right Side
p			
a			
b			
c			
d			

Table 2.

e Enter the measured value of the perimeter from Table 1 in the column labeled "Measured Left Side." Also calculate the value of the right side of the perimeter equation using the measures in Table 1. Record the result of your calculation in the first row of the "Calculated Right Side" column of Table 2.

f Solve the perimeter equation for the variable *a.* Record the right side of this new equation in the second row of the "Right Side" column.

g Complete the second row of Table 2 by using the measured values from Table 1.

h Complete Table 2 by repeating Steps **f** and **g** for the variables *b, c,* and *d.*

i What conclusions can you make about your calculations? What does the calculation tell you about the equations you created?

Activity 3: Solving Equations

Equipment   Algeblocks
               Algeblocks mat

If Algeblocks are not available, you can make paper or cardboard tiles in the dimensions shown here. Color half of each type of tile yellow and the other half green. Also make a mat like the one shown.

■ unit tile   ▭ *x*-tile

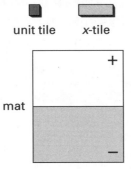

mat

Problem Statement

You will use Algeblocks or tiles to solve a simple equation.

Procedure

a Model the equation $3x - 2 = 7$ by placing the Algeblocks on your mat in this position:

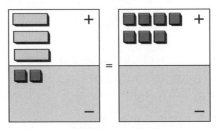

b Add two units to each side of the equal sign.

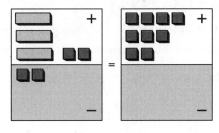

c Remove the two units on the positive section and the two units on the negative section of the left mat. What mathematical property is applied in this step?

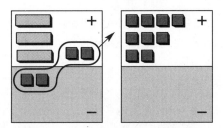

d Divide both sides into three groups. How many units are in each group?

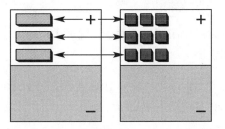

How does this show that 3 is the solution to $3x - 2 = 7$? Check your solution. Does it make the original equation true?

e Use Algeblocks to solve $5x + 3 = 2x - 3$. Record each step by drawing a picture. Write a summary of your findings.

The applications that follow are like the ones you will encounter in many workplaces. Use the mathematics you have learned in this chapter to solve the problems.

Wherever possible, use your calculator to solve the problems that require numerical answers.

Social security tax is deducted from your paycheck using the formula

$$T = P \times G$$

T is the tax amount to be deducted.
P is the decimal percentage rate for the tax.
G is your gross pay.

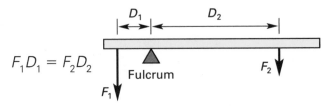

1 Write a sentence that describes how to compute the tax deducted from your paycheck.

2 Rewrite the formula isolating the variable *P*. In other words, write a formula for determining the value of *P*.

3 Suppose your paycheck stub shows gross pay of $301.60 and a tax deduction of $22.65. Use your formula to determine the percentage rate used for computing your deduction. Round the answer to two decimal places.

When a lever is used to lift or balance weights, as occurs in common patient scales seen in doctor offices, the sizes and positions of the weights are governed by the equation

$$F_1 D_1 = F_2 D_2$$

*F*₁ and *F*₂ are the weights (forces), and
*D*₁ and *D*₂ are the distances from the fulcrum.

4 You have a scale with a movable weight (F_2) that is 500 grams. The patient stands on a platform that is 0.2 cm (D_1) from the fulcrum. Write a formula you could use to compute the patient's weight (F_1) if the 500-gram weight is balanced at D_2 centimeters from the fulcrum.

5 What is the weight of the person on the scale if the scale balances when you place the 500-gram weight 18 centimeters from the fulcrum?

BUSINESS & MARKETING

It costs a recording company $9000 to prepare a CD for production. This represents a one-time fixed cost. Manufacturing, marketing, royalty payments, and so on are variable costs. Suppose the variable costs total $4.50 per CD. The CDs are sold for $6.75 each. How many CDs must be sold for the income from CD sales to equal the total costs.

6 Write an equation for the income (i), based on the number of CDs sold (N).

7 Write an equation for the total costs (C), based on the number of CDs produced (N).

8 Set the expression for i equal to the expression for C, and then isolate the variable N to find out what value of N is needed to make i equal to C.

9 You will receive a 5% commission on the sale of a condominium for a customer. The customer would like to obtain at least $60,000 from the sale. Use the formula below to determine the lowest selling price for the condominium that will allow the customer to receive the amount asked for. (Round your answer to the nearest $100.)

$$A = (100\% - c) \times s$$

A is the amount desired by the customer from the sale.
c is the commission that you will get from the sale.
s is the selling price.

The "Rule of 72" is a way to estimate the effect of compound interest on an investment. The rule states that you divide the annual percentage interest rate (expressed as a percent, not a decimal) into 72 to find the approximate number of years needed to double the value of your investment.

10 Write the "Rule of 72" as a formula. Be sure to identify the variables that you choose to use.

11 Use your formula to find out how many years it would take to double a $500 deposit in an account that paid an annual interest rate of 8.5%.

12 Use your formula to find out what interest rate would be needed to double an investment in five years.

You are told to "fax" (that is, electronically transmit a facsimile over the telephone lines) a document to your overseas branch office in Paris, France.

The phone company charges $1.94 for the first minute. After the first minute, the charge is $1.09 per minute plus a 3% tax on the total charges. On the other hand, a 2-day postal delivery can be made for $22.

13 Write a formula that can be used to compute the total cost of a phone call of *n* minutes, when *n* is greater than one. Be sure to define your variables. What does a 10-minute phone call cost?

14 Rewrite the formula, isolating the total length of the call, *n*.

15 Use this last formula to determine how long a phone call could be to match the cost of the 2-day delivery of the document. How do you interpret this value for *n?*

You need to estimate the copying cost for a job that requires a large volume of copying. After a phone call to the copy shop, you find that each copy on white paper costs $0.045. On colored paper, it costs $0.065. In addition, there is a $5 handling fee for jobs that exceed 500 copies. Each packet of materials that you will duplicate has 5 colored pages and 27 white pages.

16 Write a formula you can use to determine the total cost for producing *n* packets, when *n* is greater than 500.

17 Suppose your supervisor tells you that this project has only $2000 allotted for copying costs. Use your formula to determine how many packets you can produce with this allotted amount.

18 Is there enough money in the budget to produce 1000 packets?

19 You are told that the demand for a certain product can be described by the equation

$$D = 4000 - 44C \text{ (for values of } C \text{ between 10 and 90).}$$

D is the demand, in pounds, for the product.
C is the cost per pound in cents.

In addition, the supply of this product from the supplier is described by the equation

$$S = -1340 + 134C \text{ (for values of } C \text{ between 10 and 90).}$$

S is the supply, in pounds, of the product.
C is the cost per pound in cents.

When the supply exceeds the demand, the price will go down. On the other hand, when the demand exceeds the supply, the price will go up. At what cost per pound will the price stabilize and the supply equal the demand?

HEALTH OCCUPATIONS

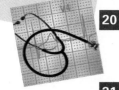

As children grow, dosages for medications gradually approach those for adults. Clark's rule is often used to determine the correct dosage for children.

$$\text{Child's dose} = \text{Adult dose} \times \frac{\text{Weight of child in pounds}}{150 \text{ pounds}}$$

20 What dosage should be administered to a child weighing 28 pounds if the adult dose of a certain drug is 80 milligrams? Round your answer to the nearest whole milligram.

21 According to Clark's rule, how much would a child have to weigh to receive the adult dosage of a medication?

A person's vital capacity is the amount of air that can be exhaled after one deep breath. An estimate of the vital capacity for women can be obtained by the formula

$$V = 0.041h - 0.018A - 2.69$$

V is the vital capacity in liters.
h is the woman's height in centimeters.
A is the woman's age in years.

22 Determine an estimated vital capacity for yourself or a friend using the formula.

23 Suppose that by using a spirometer, the lung capacity of a healthy woman is measured at 3.4 liters. The woman is 5 feet 6 inches tall. Isolate the variable A in the formula. Use the given data to find a value for A and complete the sentence that follows:

This woman exhibited the breathing capacity of a ___?___-year-old woman.

When a fair-skinned person comes into Marble's Tanning Salon, you normally advise him or her to start with 15 minutes' exposure, then gradually build up to a maximum of 23 minutes, increasing the exposure 1 minute every $2\frac{1}{2}$ days.

24 Which of the following equations would describe the exposure time, T (in minutes), for a fair-skinned client who has been tanning regularly for D days, prior to the point when he or she reaches the maximum time?

    a. $T = 23\,D - 15$
    b. $T = 2.5\,D^2 + 23$
    c. $T = 15\,D + 2.5$
    d. $T = 0.4\,D + 15$

25 How much exposure does the equation suggest you recommend on the 5th day of treatment? On the 10th day?

26 Solve the equation for D and use this new equation to determine how many days a client needs to reach the maximum 23 minutes of exposure.

The following table lists permissible sound exposures in the workplace established by the Occupational Safety and Health Administration (OSHA).

Hours of Exposure	Weighted per Day Sound Level (decibels)
8	90
6	92
4	95
3	97
2	100
1.5	102
1	105
0.5	110
0.25	115

A computer analysis of the data generates the equation

$$S = -2.817H + 108.9$$

S is the maximum permissible sound level in decibels.
H is the number of hours the person is in this environment.

27 Use the equation to compute the maximum possible sound levels for the hours of exposure listed in the table. Compare the computed sound levels to the values in the table.

28 Do you think the equation provides a good description of the table?

You are the manager of a drug testing laboratory. You want to compare the cost of breeding your own mice rather than purchasing them from a supply company. You can buy 6-week-old mice from the supply company for $4.90 each. Alternately, you estimate the cost of breeding the mice would be $830 each week in overhead and $0.60 per mouse each week for food. You need to find out how many mice would be needed each week to reach a point where the cost of purchasing equals the cost of breeding.

29 Write an expression for the cost of buying n mice from the supply company.

30 Write an expression for the cost of breeding n mice for one week.

31 You want to find how many mice it requires for the costs to be equal. You need to find the value of n that makes one expression equal the other. Set the two cost expressions equal to each other. Solve for the variable n. How many mice would the lab need per week for the decision to breed the mice to be cost-effective?

32 If more mice than this number are needed per week, would it be cheaper to purchase the mice or to breed them?

FAMILY AND CONSUMER SCIENCE

As an interior decorator, you frequently estimate costs for wallpapering a room. You always do the same set of calculations using different numbers for different rooms. For a given wall, you measure the height and the width. You divide the width of the wall by the width of the wallpaper to find how many strips will be needed. Then you multiply this number by the height of the wall to find the total length of paper needed.

33 Write a formula based on the above procedure that you could use to estimate the length of wallpaper needed to cover a wall. Be sure to define your variables.

34 If each roll contains 16 feet of wallpaper, modify your formula to provide an estimate of how many rolls of paper would be needed.

35 Try your formula for a wall that is 9 feet high and 14 feet wide, using wallpaper that is 27 inches wide. How many rolls would you estimate are needed to cover this wall?

36 What are some possible sources of error when using this formula?

Cosmetologists are frequently paid by the "salary and commission" method. The cosmetologist is paid a base salary plus a percentage of any money received that exceeds twice this base salary. The situation is modeled by the following equation:

$$G = S + C(i - 2S)$$

G is the gross pay (in dollars) for the period if i is greater than twice the salary amount.
S is the salary amount for the period.
C is the commission percentage rate (a decimal percent).
i is the gross income for services performed.

37 Suppose your weekly salary is $260, with a 15% commission on all sales over twice your salary. Use the formula to determine your gross pay if the gross income for your services totaled $685 during this week.

38 How much income for services is needed to have a gross pay of $300.

39 You examine the cost of bus travel for your vacation. After analyzing the fare schedule, you make a table of costs and mileages.

Mileage	One-way Fare
35	$ 5
100	10
145	17
180	23

Use a proportion to estimate the cost of a one-way fare for travel to a city that is 50 miles away.

Suppose you consider two job offers. One has a starting salary of $12,900 with raises of $1000 per year. The second offer is for a starting salary of $15,000 with raises of $650 per year.

40 Will these two salaries ever be the same? If so, when?

41 You hope to qualify for a promotion and a significant increase in salary after 5 years. Which of the two starting jobs would give you the higher salary at the end of five years?

INDUSTRIAL TECHNOLOGY

You are designing an exhaust fan for a small garage by using an electric motor connected to a large fan blade by a fan belt and two pulleys.

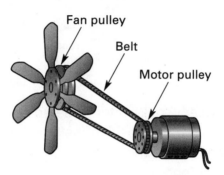

Fan pulley

Belt

Motor pulley

The approximate length of belt needed to connect the two pulleys is given by the formula

$$B = 2L + \frac{\pi}{2}(D + d)$$

B is the length of the belt.
L is the distance between the two pulley centers.
D is the diameter of the larger pulley.
d is the diameter of the smaller pulley.

42 The formula with definitions for the variables does not specify any units. You should assume the equation is true as long as the units from the variables are the same. If *L, D,* and *d* are measured in meters, what is the unit for *B*?

43 Compute the approximate length of belt needed to connect the two pulleys if the two pulley centers are 3 feet apart. The larger pulley is 8 inches in diameter, and the smaller pulley is 3.5 inches in diameter.

In close-up photography, the distance of an object from the lens determines how far the lens must be from the film. This requires special lenses and focusing mechanisms. These distances (all measured in the same units) are related by the formula

$$\frac{1}{f} = \frac{1}{p} + \frac{1}{q}$$

f is the focal length of the lens.
p is the distance of the object being viewed from the center of the lens.
q is the distance of the image formed on the film from the center of the lens.

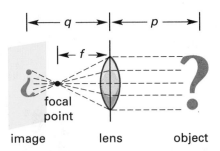

image   lens   object

You have a 50-millimeter close-up lens. With this lens, the maximum distance you can position the lens from the film plate is 6.2 centimeters. You need to find the closest you can position an object to the lens to get a sharp photograph.

44 Rewrite the formula, isolating the variable representing the object distance.

45 Determine the object distance predicted by the equation for a lens with a focal length of 50 millimeters and a maximum image distance of 6.2 centimeters.

You need to make a wire-wound resistor with a resistance of 15 ohms using 22-gauge copper wire. From a table, you find that 22-gauge copper wire has a resistance of 0.0135 ohm per centimeter length. Thus, you can determine the resistance of a wire by multiplying the length of the wire by the number of ohms per centimeter.

46 Write a formula to determine the resistance of any length of 22-gauge copper wire. Be sure to define your variables and their units.

47 Use your formula to determine the resistance of 75 feet of 22-gauge copper wire.

48 Rewrite your formula, isolating the variable for length. Use this new formula to determine what length of 22-gauge copper wire you need to create the 15-ohm wire-wound resistor.

49 A particular electronics circuit lists several resistances R_2, R_3, and R_4 that are dependent on R_1. You cannot measure R_1 but you *can* measure R_2—it measures 110 ohms. Use this measurement to determine what the resistance for R_1 must be. Also determine what R_3 and R_4 must be.

Resistor	Resistance (ohms)
R_2	$1.5\,R_1$
R_3	$10\,R_1$
R_4	$0.25\,R_1$

Ohm's Law states that the voltage (in volts) is equal to the current (in amperes) times the resistance (in ohms).

50 Write Ohm's Law as an equation. Use V for the voltage, i for the current, and r for the resistance.

51 Suppose you observe various currents through a 3300-ohm resistor. Rewrite your equation for this condition.

52 Use the equation from Exercise 51 to determine what voltage you should measure if the current in the 3300-ohm resistor is 0.0175 ampere.

The National Fire Codes have guidelines for storing flammable liquids. An amount of 2 gallons per square foot is permitted in a storeroom with no fire protection (sprinkler or other type) and a rated fire resistance of 1 hour. The storeroom cannot be larger than 150 square feet.

53 Write an equation to express the relationship between the gallons of flammable liquid that can be stored and the area of the storeroom.

54 Identify the variables and the constants in your equation.

55 Use the equation to calculate the floor area required to store a maximum of 165 gallons.

56 How many gallons can you store in the largest storeroom?

You need to calibrate a variable, wire-wound resistor. You find that when the contact is at the minimum position, your resistor has a resistance of 4 ohms. Then, for every centimeter of travel away from this minimum position, the resistance increases by 2.5 ohms. This relationship continues up to a maximum of 29 ohms. At this resistance, the contact is at the farthest position—10 centimeters from the minimum.

57 Write a formula to compute the resistance of the resistor based on the distance from the minimum position.

58 How can you be sure of the validity of your formula?

59 Isolate the variable for the distance of the contact. Where should you place the contact to obtain a resistance of 22 ohms?

You are writing a contract for commercial sandblasting before repainting a building. To estimate the cost of the sandblasting, you determine the sand cost and the labor cost (to operate the sandblaster). You find that the sand costs $4 per bag and that it takes about 10 minutes to load and use each bag of sand. The labor cost for the operator is $40 per hour plus a one-time $110 equipment fee. Find an equation to estimate the total cost of the sandblasting, depending on the time.

60 Write an equation for the cost of the equipment used by an operator who works T hours.

61 Determine how much sand would be used in an hour (60 minutes) and hence how much it would cost per hour. Write an equation for the cost of sand when sandblasting for T hours.

62 Add the cost of the labor and equipment to the cost of the sand to find the total cost of sandblasting for *T* hours.

63 Use your equation to estimate the total cost if the job takes 2 hours. What is the cost for a 4-hour job and for a 6-hour job?

You work in a metal shop and must often drill equally spaced holes along a length of stock. The holes on the ends of the stock must be centered exactly 1 inch from the ends. The remaining holes will be equally spaced. A few examples are shown below.

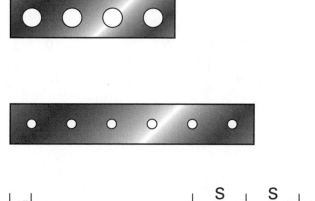

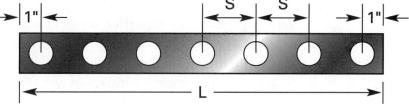

64 Write a formula for the spacings of the holes on a piece of stock that is *L* inches long and that must have a total of *N* equally spaced holes.

65 What spacing between holes (their centers) would be needed for 9 holes on a bar that is 10 feet long?

66 Suppose you use a $\frac{3}{4}$-inch drill bit (i.e., which has a $\frac{3}{4}$ inch diameter). Describe the layout of the holes on the bar if your equation yielded a spacing value of 0.75 inches. For nine holes in the bar, what length of stock would yield this value for the spacing?

Skills

Solve each equation. Check your answer.

1. $10a = -100$

2. $c + 25 = 16$

3. $3q + 5 = 19$

4. $-0.5m - 7 = 12$

5. $\dfrac{-20}{n} = \dfrac{12}{15}$

6. 15% of $x = 45$

7. 35% of $800 = y$

8. $\dfrac{1}{2} - r = -\dfrac{3}{4}$

9. $8.5 = 2.6 - 2d$

10. $5(3 + x) = 20$

11. $-9 = \dfrac{1}{2}(4 - y)$

12. $\dfrac{16 - x}{4} = -20$

13. $8y = 10 - 6y$

14. $-3(d + 6) = 2d$

15. $4(3 - d) = -(d + 5)$

Applications

Write and solve an equation for each situation.

16. A company pays one-half of the monthly premium on life insurance policies for its employees. If the company pays $140 a month for each employee, what is the total premium?

17. A TV and VCR together sell for $700. The price of the TV is $2\frac{1}{2}$ times the price of the VCR. What is the price of the VCR?

18. A maintenance department employs six workers. Each employee earns the same hourly rate. Last week each employee worked a total of 40 hours. The total payroll for the month was $2,640. What is the hourly pay for each worker?

19. One Saturday, Rhonda earned $10 more than Anne. If Rhonda and Anne earned a total of $124, how much did Anne earn?

20. One plumber charges $90 for a call and then $35 an hour. Another plumber charges $40 for a call and then $40 an hour. How many hours will the plumbers work if they make the same amount of money in the same amount of time?

Math Lab

21. Use the equation $C = \pi d$ and the measured values for C and d in the table to calculate an average value for π.

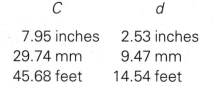

C	d
7.95 inches	2.53 inches
29.74 mm	9.47 mm
45.68 feet	14.54 feet

22. The perimeter of the irregular figure is 250 feet. What is the length of side x?

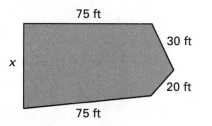

75 ft

30 ft

x

20 ft

75 ft

23. Use Algeblocks to solve the equation $4x - 2 = 10$

CHAPTER 5

WHY SHOULD I LEARN THIS?

The right solution can mean the difference between success and failure in a business. Successful problem solvers often use linear equations. Find out how you can use linear equations to increase your problem solving skills and make you more valuable to your employer.

GRAPHING DATA

OBJECTIVES

1. **Graph data as points on a coordinate system.**
2. **Graph an equation.**
3. **Find the slope of a graphed line.**
4. **Find the intercepts of a graphed line.**

In this chapter, you will draw the graphs of equations that represent straight lines. These equations are called **linear equations**.

A linear equation with two variables tells how those variables relate to each other. For example, the linear equation $d = 2r$ tells you that the diameter for any circle is twice the radius of that same circle.

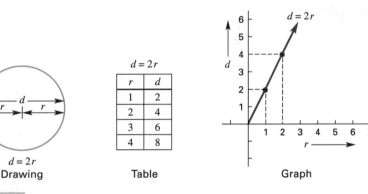

r	d
1	2
2	4
3	6
4	8

$d = 2r$

Drawing   Table   Graph

You can see this relationship if you *make a table* of values that satisfy the equation. Each pair of values in the table makes the equation true.

You can also see this if you *draw a graph.* The graph of a linear equation tells you how the two variables relate to each other, but in picture form. This graph is a line that shows how the radius and diameter are related.

As you watch the video for this chapter, notice how people on the job use equations to solve problems.

LESSON 5.1 COORDINATES AND GRAPHS

Coordinates on a Line

In Chapter 4, you used whole numbers, integers, and rational numbers to solve equations. A **rational number** is any number that can be written in the form $\frac{a}{b}$ where a and b are integers and b does not equal zero. *Terminating* decimals such as 0.5 are rational numbers because you can write them as fractions.

$$0.5 = \frac{5}{10} = \frac{1}{2}$$

Repeating decimals such as 0.333 . . . are also rational numbers. For example,

$$0.333 \ldots = \frac{1}{3}$$

There are also decimals that do not terminate or repeat. Such decimals are called **irrational numbers**. The number π is an example of an irrational number.

$$\pi = 3.14159264 \ldots$$
$$\text{never terminates or repeats}$$

Together, the rational and irrational numbers make up the set of **real numbers**. You use a number line to model the set of real numbers. Every point on the line is represented by exactly one real number, and every real number is represented by exactly one point.

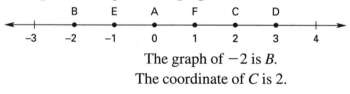

The numbers on the number line are referred to as the **coordinates** of the points. The points are **graphs** of the coordinates.

The graph of -2 is B.
The coordinate of C is 2.

Ongoing Assessment

What is the graph of 3? What is the coordinate of E?

For the rest of your study in Algebra this year, you will use the real number system. So, from now on, when the word *number* is used, it will mean *real number.*

Critical Thinking   One of the basic properties of the real number system is called **density**. That is, between any two real numbers there is always another real number. How can you show that the set of integers is not dense? How can you show that the set of rational numbers is dense?

Coordinates on a Plane

A plane is a flat surface that extends in all directions. The floor in your classroom is a model of a plane. You can also think of a piece of paper as a model for a plane.

From early civilizations, we know that people pictured pairs of numbers as points graphed on a plane. About 350 years ago, a French soldier, René Descartes, developed a system that used horizontal and vertical number lines to picture points and lines on a plane. His ideas worked so well that today we call this system the **Cartesian coordinate system** in his honor.

To draw a Cartesian coordinate system, begin by drawing a number line horizontally in the middle of your paper (if you want to use the whole sheet for the graph). This is the *x*-**axis**. Place the zero point about halfway across the page. This point is the **origin**. Label this point with the letter *O*.

Now draw another number line perpendicular (making a right angle) to the first, crossing the first line at the origin. This is the *y*-**axis**. Notice that the positive numbers are to the right of the origin for the horizontal line and up from the origin for the vertical line.

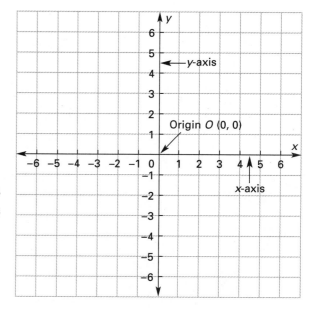

The size of the unit (the distance from zero to one) does not have to be the same on both number lines. But it is generally a good idea to draw the units the same unless you have a special reason for making them different. For example, you might need to have many small units on one line and only a few large units on the other.

The size of the unit is always the same on the positive and negative sides of a number line.

The horizontal number line and the vertical number line are both called **axes**. The horizontal axis is perpendicular to the vertical axis, and they cross at the origin, or **zero point**.

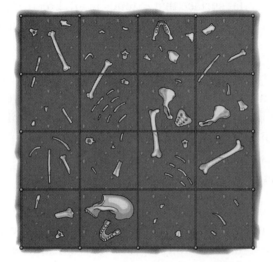

Pairs of numbers called **coordinates** represent every point on the Cartesian coordinate system. The first coordinate always tells how far to the right or left of the origin the point is located. The second coordinate in the pair always tells how far up or down from the origin the point is located. The number pair that names the origin or zero point is (0,0). Archaeologists use Cartesian coordinate systems to identify the location of artifacts.

EXAMPLE

Write the number pair that represents point A.

SOLUTION

To find the number pair that identifies point A, begin at the origin where the axes cross. Move to the right along the x-axis until you are under point A. As you move, count the number of units from the origin. Write down this number. Then move up along a line parallel to the y-axis until you reach point A. As you

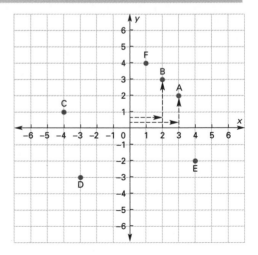

move, count the number of units from the *x*-axis. Write this number to the right of the first number. Place a comma between the pair of numbers and enclose them in parentheses. The pair (3,2) identifies point *A*.

Write the number pair that represents point *B*.

Did you write (2,3)? Notice that it *does* make a difference which number you write first. Point *A* is (3,2) and point *B* is (2,3). They are *not* the same point.

The pair of numbers that identifies a point in the Cartesian coordinate system is called an **ordered pair**. For example, the pair (3,2) that identifies point *A* is the ordered pair (3,2).

Ordered Pairs
The ordered pair of numbers that identifies any point is (*x,y*). The *x*-value of a point is the **x-coordinate** and is always written first. The *y*-value is the **y-coordinate** and is always written second.

ACTIVITY 1   Naming Points with Ordered Pairs

Write an ordered pair for each of the points *C, D, E,* and *F.* Remember, if the point lies to the left of the *y*-axis, its *x*-value is negative. If it lies below the *x*-axis, its *y*-value is negative.

Quadrants

The Cartesian coordinate system divides a plane into four quadrants.

Notice that the quadrants are numbered in a counter-clockwise direction.

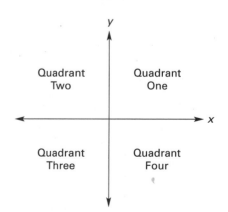

Use the ordered pairs you wrote for Activity 1 to help you choose the correct word (positive or negative) to complete these statements.

1 All the points in the first quadrant have x-values that are ___?___ and y-values that are ___?___.

2 All the points in the second quadrant have x-values that are ___?___ and y-values that are ___?___.

3 All the points in the third quadrant have x-values that are ___?___ and y-values that are ___?___.

4 All the points in the fourth quadrant have x-values that are ___?___ and y-values that are ___?___.

LESSON ASSESSMENT

Think and Discuss

1 Explain how the rational and irrational numbers are alike .

2 Explain the difference between a coordinate on a line and a coordinate on a plane.

3 How is a point on the Cartesian coordinate system named?

4 Why are the coordinates of a point written as an ordered pair?

Practice and Problem Solving

Write the ordered pair that represents each point.

5.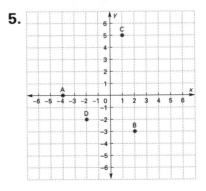

6.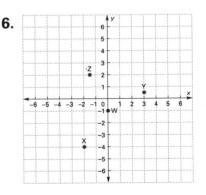

7. Identify the coordinates of each city.

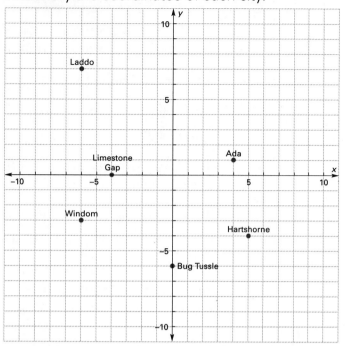

8. The data from a manufacturing process are shown on a graph. Identify the coordinates relating time deviation (*x*) to the length deviation (*y*).

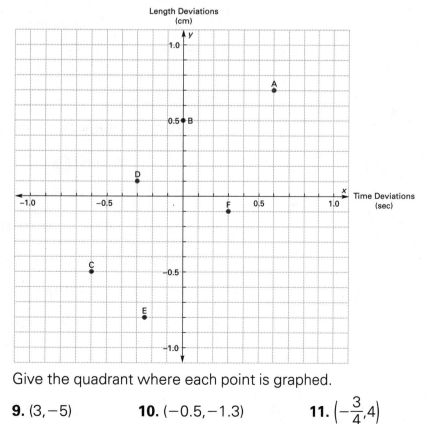

Give the quadrant where each point is graphed.

9. $(3, -5)$ **10.** $(-0.5, -1.3)$ **11.** $\left(-\dfrac{3}{4}, 4\right)$

Tell if the sentence is true or false. If it is false, give a counter-example. Let *a* and *b* represent integers.

12. If $a < b$, then $-a > -b$ **13.** $a + b = a - (-b)$

There are 10^2 centimeters in a meter. There are 10^3 meters in a kilometer.

14. How many centimeters are in a kilometer? Write your answer as a power of ten and in decimal notation.

15. How many kilometers are in a centimeter? Write your answer as a power of ten and in decimal notation.

The radius of a front wheel of a car is 42.3 centimeters. The radius of a rear wheel is 47.9 centimeters.

16. How far does the car travel for one turn of the rear wheel?

17. How far has the rear wheel traveled when the front wheel makes 500 turns?

The freshman class voted on the school color. The final vote was as follows:

Red 76   Blue 45   Green 90   Black 89

Write an equation and solve.

18. What percent voted for red?

19. How many votes would make 40% of the total vote?

20. What percent voted for black and blue combined?

21. What percent of the green vote is the blue vote?

Solve each equation for the given variable.
22. $6x - 33 = 15$ **23.** $4(y - 3) = -8$

LESSON 5.2 GRAPHING POINTS AND LINES

Ordered pairs of numbers identify all points located on the Cartesian coordinate plane. If you turn this idea around, you can say that every point on the Cartesian coordinate plane identifies an ordered pair of numbers.

The Cartesian Coordinate System
Every point on the Cartesian coordinate plane has a matching ordered pair of numbers. Conversely, every ordered pair of numbers has a matching point on the Cartesian coordinate plane.

There are many different coordinate systems. Any reference to a "coordinate system" in this book refers to the Cartesian coordinate system.

Graphing a Point

Draw a Cartesian coordinate system. Mark six equal units in all four directions. To locate the point $(-3, -3)$, begin at the origin. The x-value (-3) tells you to move three units to the left. The y-value (-3) tells you to then move three units down. Mark this as point G. You have located the point $G(-3, -3)$ as shown. To

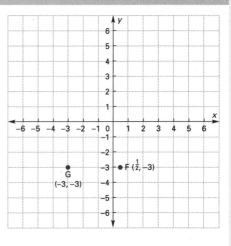

locate point F, with coordinates $(\frac{1}{2}, -3)$, move $\frac{1}{2}$ unit to the right along the x-axis. Now move 3 units down. The correct location of point $F(\frac{1}{2}, -3)$ is shown.

ACTIVITY 1   Locating Points

1 Draw a Cartesian coordinate system.

2 Label the horizontal axis the x-axis. Label the vertical axis the y-axis.

3 Locate and label the the ordered pairs: $A(1,5)$, $B(-1,5)$, $C(-1,-5)$, and $D(1,-5)$.

4 Draw the following line segments: AB, BC, CD, and DA.

5 Name the figure you have drawn.

Range and Scale

Choosing the range and scale units for the axes is an important step in making a graph. You can find the range needed for an axis by simply finding the difference between the smallest and largest values to be graphed along that axis. Once you know the range (and the size of the graph paper),

you then choose a scale unit so that the correct number of intervals fits along the axis.

Suppose you want to make a graph showing the distance traveled versus time for a race car moving at a constant speed of 200 miles per hour. You have the following data to graph:

Time	Distance
0.0 hr	0 mi
0.5 hr	100 mi
1.0 hr	200 mi
1.5 hr	300 mi
2.0 hr	400 mi
3.0 hr	600 mi
4.0 hr	800 mi

Also suppose you choose to plot time along the x-axis and distance along the y-axis. To establish the range for each axis, do the following:

- For *time along the x-axis,* find the difference between the largest and smallest values of the data for time. The range is 4 hours.

- For *distance along the* y-*axis,* find the difference between the largest and smallest values of the data for distance. The range is 800 miles.

To establish the scale unit, do the following:

- For the *scale unit along the* x-*axis* (the time axis), divide the range (4 hours) by the length of the x-axis. You should always allow margins of about 1.5 inches for labeling. For a standard page that is 8.5 inches wide, subtract about 3 inches for margins. This leaves about 5 inches of page space. Next, divide 4 hours by the 5 inches of page space. This gives a scale unit of 0.8 hour per inch, or about 1 hour per inch. Choose the scale unit of "one inch equals one hour." If the graph paper is marked off in small squares, you may want to count the nearest number of squares in an inch. Let the scale be, for example, "5 squares equal one hour."

- For the *scale unit along the* y-*axis* (the distance axis), divide the range (800 miles) by the y-axis length. For a standard page length of about 11 inches, you might choose 8 inches (leaving 3 inches for top and bottom margins) and get a scale of 100 miles per inch. Again, if you prefer to work with squares and there are about 2 squares to the inch, choose your scale unit as "1 square equals 50 miles."

With the range and scale chosen as outlined, the axes of your graph and your plotted points will look like those shown here. This graph is reduced in size to save space.

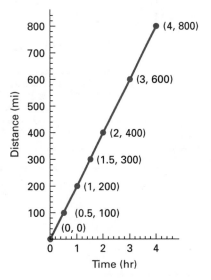

For the graph on the previous page, note the following:

- The scale units are large enough to give a good visual display of the graphed data.

- The scale units are large enough to let you determine distances traveled at the half-hour points.

The overall graph fits nicely on the 8.5-by-11-inch page. The graph provides ample room for the title and for the labels and scale units along each axis. Every graph should have a title, and scale units should always be labeled. That way, anyone who examines the graph will know what it represents.

ACTIVITY 2   **Changing Scale**

Use the same data given in the distance-versus-time table. Leave ample margin space. Use the given scale units to draw a graph on 8.5-by-11-inch paper.

1 Case 1. x-axis: 1 inch = 4 hours
                 y-axis: 1 inch = 400 miles

Can this be done? Why or why not?

2 Case 2. x-axis: 1 inch = $\frac{1}{2}$ hour
                 y-axis: 1 inch = 50 miles

Can this be done? Why or why not?

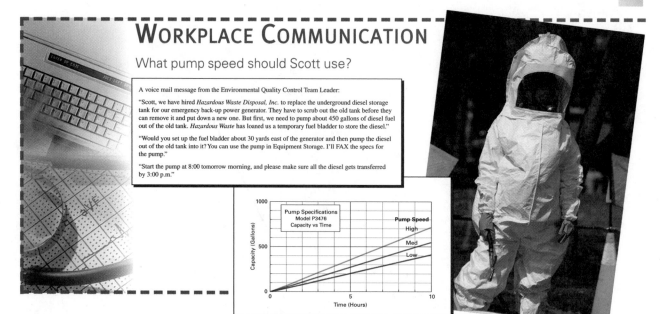

WORKPLACE COMMUNICATION

What pump speed should Scott use?

A voice mail message from the Environmental Quality Control Team Leader:

"Scott, we have hired *Hazardous Waste Disposal, Inc.* to replace the underground diesel storage tank for our emergency back-up power generator. They have to scrub out the old tank before they can remove it and put down a new one. But first, we need to pump about 450 gallons of diesel fuel out of the old tank. *Hazardous Waste* has loaned us a temporary fuel bladder to store the diesel."

"Would you set up the fuel bladder about 30 yards east of the generator and then pump the diesel out of the old tank into it? You can use the pump in Equipment Storage. I'll FAX the specs for the pump."

"Start the pump at 8:00 tomorrow morning, and please make sure all the diesel gets transferred by 3:00 p.m."

LESSON ASSESSMENT

1 Explain how to graph a point on the Cartesian coordinate plane.

2 Why is it important to understand how to determine the range and scale for the axes of a graph?

Practice and Problem Solving

Draw a Cartesian coordinate system. Locate and label each point.

3. $G(-3,4)$ **4.** $H(3,-4)$ **5.** $J(-3,-4)$ **6.** $K(3,4)$

7. Connect the points in order starting from quadrant one and ending in quadrant four. Describe the figure drawn.

Draw a Cartesian coordinate system. Locate and label each point.

8. $G\left(0,-3\frac{1}{2}\right)$ **9.** $H(-2,-2)$ **10.** $K(4,-2)$ **11.** $L\left(6,-3\frac{1}{2}\right)$

12. Connect the points in order starting from quadrant one and ending in quadrant four. Describe the figure drawn.

Draw a Cartesian coordinate system. Locate and label each point.

13. $S(0,0)$ **14.** $T(1,2)$ **15.** $U(3,6)$ **16.** $V(-2,-4)$

17. Connect the points. Describe the figure drawn.

18. Robots in a manufacturing plant move parts along an assembly line. Their movements are described in a coordinate plane. The range of Robot A is from $(7,8)$ to $(12,8)$. The range of robot B is from $(5,3)$ to $(5,-2)$. The range of Robot C is from $(7,-7)$ to $(12,7)$. Draw a Cartesian coordinate system. Plot the points for the range of each robot.

19. A jig-boring operation uses rectangular coordinates to establish each hole location. Draw a Cartesian coordinate system. Plot the centers of the holes specified by the ordered pairs
$A(6,-2)$, $B(6,2)$, $C(10,1)$, and $D(10,-2)$.

Write each number in scientific notation.
20. 0.000385

21. $(1.89 \cdot 10^2)(7.4 \cdot 10^4)$

22. What is the value of $3m^3 - 2n$ when m is -3 and n is 12?

23. What is the value of $-2(r - t) + 5(t - r)$ when r is 2 and t is 4?

24. A carpenter is cutting circular blocks from plywood. If the diameter of the circle is 2.1 meters, what is the area of the plywood section being cut out?

25. If you borrow $1800 for 3 years at 6.5% simple interest, how much will you owe in interest?

Solve each equation for c.
26. $7c + 2 = -37$

27. $-2c - 8 = 12$

28. $\frac{2}{3}c + 11 = 5$

29. $\frac{c}{-4} = \frac{9}{3}$

30. 30% of $c = 120$

31. c% of $90 = 150$

LESSON 5.3 THE SLOPE OF A LINE

ACTIVITY 1   The *y*-axis

1 Graph these three points on the same Cartesian coordinate system.

$(0,-2)$, $(0,0)$, $(0,3)$

2 Describe the location of the three points.

All three points lie on the *y*-axis (the vertical axis). The *x*-value of each point is zero. However, the *y*-values are different. Any ordered pair that has an *x*-value equal to zero *must* identify a point somewhere along the *y*-axis. Thus, the equation $x = 0$ describes the *y*-axis.

ACTIVITY 2   The *x*-axis

1 Graph these three points on the same Cartesian coordinate system.

$(-2,0)$, $(0,0)$, $(3,0)$

2 Describe the location of the three points.

3 What is true about the *y*-value for each of these points? What is the equation that describes the *x*-axis?

ACTIVITY 3   Equal Coordinates

1 Graph these three points on the same Cartesian coordinate system.

$(-2,-2)$, $(0,0)$, $(3,3)$

2 Describe the location of the three points. Do all of these points lie on one straight line? What is true of the *x*-values and *y*-values of all these points?

3 Write an equation that describes this line.

You have used equations to describe three lines. You can also graph the lines on the same coordinate system.

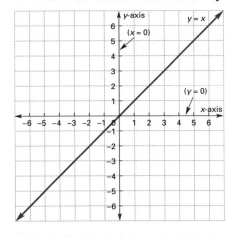

Slope

Imagine that each of the three lines ($x = 0$, $y = 0$, and $y = x$) is a hill. You have to climb each line. Which line is easiest to "walk up"? The "$y = 0$" line (or x-axis) is the easiest—it has no rise at all. You can say that the line $y = 0$ has a steepness of zero. This steepness is called the **slope** of the line. Thus, the line $y = 0$ has a slope of zero. Which line represents a hill so steep that it is impossible to climb? You cannot "walk up" the "$x = 0$" line (or y-axis) at all. The slope of this hill is so great that you cannot assign a number to it. The slope of the line $x = 0$ is undefined.

Which line represents a hill that is fairly steep, but one you could still climb? The "$y = x$" line has a slope somewhere between the x-axis (with a slope of zero) and the y-axis (with a slope that is undefined). How can you find the slope of the $y = x$ line?

The slope of a line is a measure of its steepness or "tilt." The steepness of a line (or a hill) is found by comparing its vertical rise to its horizontal run. A very steep road has a large amount of vertical rise for a given amount of horizontal run.

A road with a gentle slope has a small amount of vertical rise for the same amount of horizontal run.

Slope
The slope of a line is the ratio of the rise to the run.

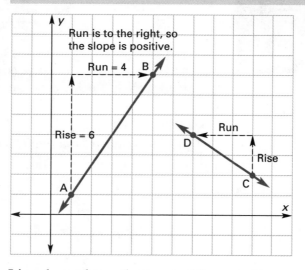

Lines have slopes that are positive, negative, or zero. To find the slope of a line, you need two points on the line. Imagine yourself walking from point *A* to point *B*. However, in this imaginary walk, you must first move up, and then go right or left; you cannot go diagonally.

Count the steps (or units) going up to find the rise. Then count the steps (or units) either left or right (the run) to reach *B*. If you move to the right, the number for the run is positive; if you move to the left, the number for the run is negative.

Once you know the value of the rise and run, write a fraction with the rise as the numerator and the run as the denominator. This fraction representing the ratio $\frac{rise}{run}$ is the slope of the line. The slope of line *AB* is

$$\frac{6}{4} \text{ or } \frac{3}{2}$$

Critical Thinking Why is the slope of the line passing through *C* and *D* negative?

EXAMPLE 1 Finding Slope

A surveyor places two stakes, *A* and *B,* on the side of a hill. Stake *A* is 20 feet lower than Stake *B*. If the horizontal distance between the stakes is 200 feet, what is the slope of the hill?

SOLUTION

Draw a diagram. The rise is 20 feet. The run is 200 feet. The slope is $\frac{20}{200}$ or $\frac{1}{10}$ or 0.1.

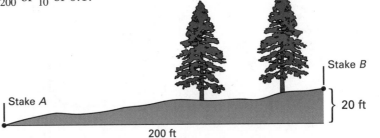

Stake *B*

Stake *A*

20 ft

200 ft

Look below at the graph of *y* = *x*. Pick two points, such as *P*(1,1) and *Q*(3,3). To move from *P* to *Q,* you can move up two units and then to the right two units. The slope is the ratio of the rise to the run.

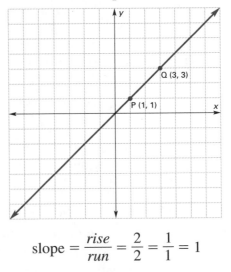

$$\text{slope} = \frac{rise}{run} = \frac{2}{2} = \frac{1}{1} = 1$$

Thus, the slope of the line *y* = *x* is 1.

The Slope Formula

The coordinates of any two points on a line determine its slope. The difference between the *y*-coordinates is the rise. The difference between the *x*-coordinates is the run. This gives a formula for finding slope.

Since the slope is the ratio $\frac{rise}{run}$, the slope can also be written in the following way:

$$\text{slope} = \frac{\text{difference of } y\text{-coordinates}}{\text{difference of } x\text{-coordinates}}$$

To find the slope of a line between two points, you can use the slope formula.

Slope Formula

If $A(x_1, y_1)$ and $B(x_2, y_2)$ are two points on line AB, then the slope of $AB = \frac{y_2 - y_1}{x_2 - x_1}$

When you use the slope formula to find the slope of a line between two points, be sure to subtract the coordinates in the same order.

EXAMPLE 2 The Slope Between Two Points

Find the slope of the line that joins $A(-2, -3)$ and $B(1, 4)$.

SOLUTION

Method 1 Make a sketch.

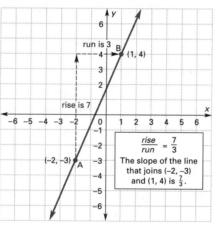

Start at the lower point, A. Move up to find a rise of 7 units. To reach point B, move to the right 3 units. This is a run of 3. Thus, the slope is $\frac{7}{3}$.

Method 2 Use the slope formula.

The difference in the y-coordinates is $4 - (-3) = 7$.

The difference in the x-coordinates is $1 - (-2) = 3$.

The slope is the ratio $\frac{7}{3}$.

LESSON ASSESSMENT

1 Describe the points that are located on the *y*-axis and on the *x*-axis.

2 Explain what the slope of a line means.

3 Explain how to find the slope of a line if you know its rise and its run.

4 Explain how to use the slope formula to find the slope of a line.

5 Explain why a slope can be positive or negative, and describe the line that models each slope.

Practice and Problem Solving

Find the slope of a line for the given rise and run.

6. rise 3, run −5   **7.** rise −2, run 4 **8.** rise 8, run 0

9. rise −10, run −2   **10.** rise 0, run 3   **11.** rise −2, run −10

A fencing contractor uses a scale drawing on a coordinate plane to calculate a bid on a job. Part of this calculation includes finding the slopes of the lines in a drawing. Find the slopes of the line segments with the given endpoints below.

12. $A(-3,4)$, $B(2,-1)$ **13.** $M(2,-5)$, $N(-1,-1)$

14. $C(2,3)$, $D(4,3)$ **15.** $X(0,0)$, $Y(2,4)$

16. $S(-6,2)$, $T(-6,5)$ **17.** $R(-8,-1)$, $S(3,-2)$

18. $L(-4,2)$, $M(3,-2)$ **19.** $P(7,10)$, $Q(-7,10)$

20. $G(-5,2)$, $H(5,-2)$

21. It is 220 miles from Johnson City to Putnam. The elevation of Johnson City is 3000 feet. The elevation of Putnam is 3500 feet. What is the average rate of increase in elevation per mile from Johnson City to Putnam?

22. What is the slope of the roof below?

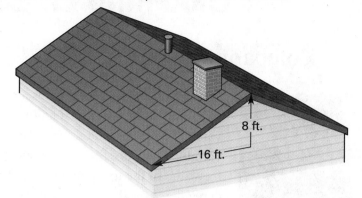

8 ft.

16 ft.

23. In a landing approach, an airplane maintains a constant rate of descent of 50 feet for every 500 feet traveled horizontally. What is the slope of the line that represents the landing approach of the plane?

Mixed Review

24. One light year is the distance light travels in one year. Light travels about 5.88×10^{12} miles in one year. Our galaxy is approximately 100,000 light years in diameter. What is the diameter of our galaxy? Write your answer in scientific notation.

For each situation, write and solve an equation.

25. The amount of water flowing over a dam at noon is 3.5 gallons per hour more than its rate at mid-morning. When the water flow was tested at noon, it had reached 48 gallons per hour. What was the rate of the water flow at mid-morning?

26. Keshia sells her inventory for twice what she pays. After expenses of $240 are deducted, Keshia finds she has $680 left. What did Keshia pay for her initial inventory?

27. Ramon is carpeting a rectangular room with a perimeter of 110 feet. One side of the room is 5 feet longer than the other. Find the length of the longer side.

Solve each equation. Check your answer.

28. $3d + (-4) = -1$ **29.** $x\%$ of $50 = 12$ **30.** $\dfrac{-5}{6} = \dfrac{10}{r}$

LESSON 5.4 GRAPHING LINEAR EQUATIONS

Comparing Slopes

The equation $y = 2x$ expresses a relationship between two variables x and y. In this relationship, the value of y *depends* on the value that is substituted for x. Thus, x is called the **independent** variable and y is called the **dependent** variable. If you select 1 as the value of the independent variable x, then 2 is the value of the dependent variable y. The equation $y = 2x$ is solved when you find the ordered pairs that make the equation a true statement. One solution of the equation is the ordered pair (1,2). Since the solution of $y = 2x$ is a set of ordered pairs, a table of values or a graph can represent $y = 2x$.

In a table, the first column is usually labeled as the independent variable. The second column is labeled as the dependent variable. Since you cannot list all the ordered pairs that make the equation a true statement, three to five pairs are usually enough.

x	y
1	2
0	0
-1	-2
-2	-4

After you complete the table, graph each pair of (x,y) values as a point. Draw the line that passes through all the points. Write the original equation next to or along the line.

Examine the figure that shows the graphs of $y = 2x$ and $y = x$ on the same pair of axes. Because the graphs are straight lines, $y = 2x$ and $y = x$ are called **linear equations**. Since the graph of the equation is determined by the table, you can find the slope of $y = 2x$ from the table.

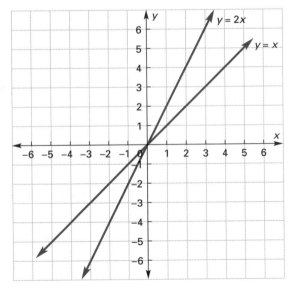

Critical Thinking Explain how to use the table to find the slope of the linear equation $y = 2x$.

ACTIVITY 1   The Slope of a Line

1 Use a table to graph each of the three equations on the same coordinate axes. Use four ordered pairs for each table.

a. $y = x$ **b.** $y = 3x$ **c.** $y = 5x$

2 Use the slope formula to determine the slope of each line. Compare the slope of each line to the coefficient of the independent variable. What do you notice?

3 Guess the slope of $y = 6x$. Use a table to check your guess.

Notice that the graph of each equation is a straight line that passes through the point (0,0).

Linear Equation

The equation $y = mx$ is a linear equation. The graph of $y = mx$ is a line with slope m. The line passes through point (0,0).

Critical Thinking   What happens to the slope of the line $y = mx$ as m increases?

EXAMPLE 1 Negative Slope

Compare the slope of $y = -2x$ with the slope of $y = 2x$.

SOLUTION

Make a table of values and graph $y = -2x$. First, choose four values for x. Then, use the equation to find the values for y.

x	y
-1	2
0	0
1	-2
2	-4

After you complete the table, graph each pair of (x,y) values as a point. Draw the line that passes through the points. Write the equation along the line.

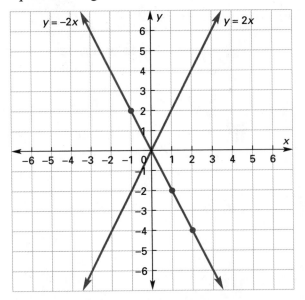

The steepness of these two lines is the same. However, they rise in opposite directions. What is the slope of each line? Notice that the slopes are opposite but equal values.

Positive and Negative Slope
 A line that rises to the right has a positive slope. A line that rises to the left has a negative slope.

Look at your table of values for $y = -2x$. Notice that as x increases in value, y decreases. In this case, the graph of the equation has a negative slope.

Slope-intercept Form of an Equation

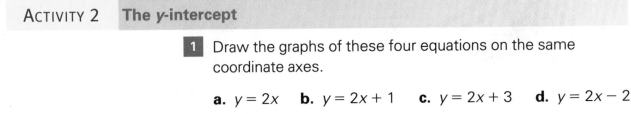

ACTIVITY 2   **The y-intercept**

1 Draw the graphs of these four equations on the same coordinate axes.

   a. $y = 2x$ **b.** $y = 2x + 1$ **c.** $y = 2x + 3$ **d.** $y = 2x - 2$

2 What is the slope of each line?

3 Where does each line cross the y-axis?

4 Where do you think the graph of $y = 2x + 5$ crosses the y-axis?

The y-value of the point where the line crosses the y-axis is called the **y-intercept**. The equation $y = 2x + 5$ is written in **slope-intercept form.** The slope is 2, and the y-intercept is 5.

> ## Slope-intercept Form of a Linear Equation
> $y = mx + b$ is a linear equation. The slope is m and the y-intercept is b. The line crosses the y-axis at the point $(0, b)$.

EXAMPLE 2 Graphing a Linear Equation

Graph the equation $2y = -3x + 6$ using slope-intercept form.

SOLUTION

First, write the equation in slope-intercept form. That is, solve the equation for y.

$$2y = -3x + 6 \qquad \text{Given}$$

$$\left(\frac{1}{2}\right)2y = \frac{1}{2}(-3x + 6) \qquad \text{Multiplication Property}$$

$$y = -\frac{3}{2}x + 3 \qquad \text{Distributive Property and Simplify}$$

The slope is $\dfrac{-3}{2}$, and the line crosses the y-axis at $(0,3)$.

Now draw the coordinate axes. Locate the point where the line will cross the y-axis. Since the slope is $\frac{-3}{2}$, you can find another point by moving down 3 and to the right 2. Locate that point. Connect the two points with a line.

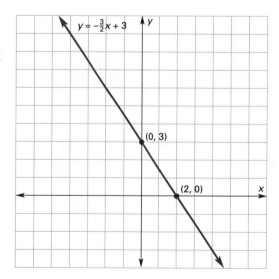

Write the equation $3x - y = -2$ in slope-intercept form. Graph the equation.

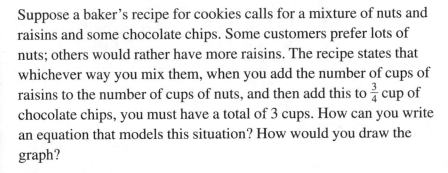

Linear equations model many problems. Once an equation is determined, you can use the graph to interpret the solutions.

Suppose a baker's recipe for cookies calls for a mixture of nuts and raisins and some chocolate chips. Some customers prefer lots of nuts; others would rather have more raisins. The recipe states that whichever way you mix them, when you add the number of cups of raisins to the number of cups of nuts, and then add this to $\frac{3}{4}$ cup of chocolate chips, you must have a total of 3 cups. How can you write an equation that models this situation? How would you draw the graph?

- First, translate the problem into sentence form:

  The number of cups of raisins plus the number of cups of nuts plus $\frac{3}{4}$ cup of chocolate chips must equal 3 cups.

- Second, let the letter r represent the number of cups of raisins and n represent the number of cups of nuts. Now you can change the sentence to the equation: $r + n + \frac{3}{4} = 3$.

To graph the equation, choose one of the variables as the independent variable. For example, let n be the independent variable. Then r takes the position of y, and n takes the position of x. To write the equation in slope-intercept form, isolate r on the left side of the equation.

$$r + n - n + \frac{3}{4} = 3 - n \qquad \text{Subtraction Property}$$
$$r + \frac{3}{4} - \frac{3}{4} = 3 - n - \frac{3}{4} \qquad \text{Subtraction Property}$$
$$r = -n + 2\frac{1}{4} \qquad \text{Simplify}$$

By comparing the equation $r = -n + 2\frac{1}{4}$ with the slope-intercept form $y = mx + b$, you can see that the slope (m) is -1. So you would expect this line to be higher on the left. What numbers do you want to use on the r and n axes? Since the values for both will be small positive numbers, including fractions, you might choose a fairly large (1 inch or so) distance for your unit. Draw the coordinate system on your paper, mark the units, and label the axes.

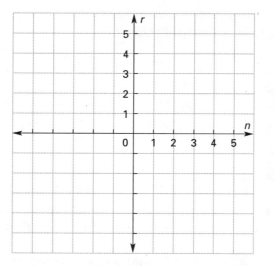

You can use the *y*-intercept to locate one point and use the slope to find another. Or you can make a table of several values. With only two ordered pairs, you can draw a straight line, but with several additional values, you can check your work. If all points are not on the same straight line, you have an error. If this happens, go back and check your arithmetic—and the way you made the graph. Graph your values and draw the line.

The graph of the equation $r = -n + 2\frac{1}{4}$ is a line that lies in quadrants one, two, and four. But the situation in the problem makes sense only in the first quadrant, where both r and n are positive numbers. As a result, the line you graph to fit the problem is limited to the first quadrant. Use a solid line in the first quadrant to emphasize where the data in the problem apply.

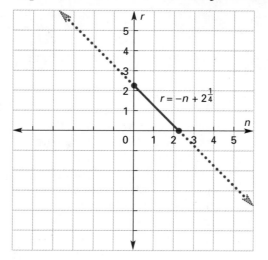

You can tell roughly what the temperature (T) is on a summer evening if you hear a cricket chirping and count how many times (N) it chirps in one minute. The formula to find the temperature in degrees Fahrenheit is the following:

$$T = \frac{1}{4}N + 40$$

1 Draw a graph of the formula. Use your graph to find the temperature if you count 100 chirps per minute. Find how many chirps you might hear if the temperature is 95 degrees.

2 What are reasonable values for T and N in this problem? Compare your graph to the one here.

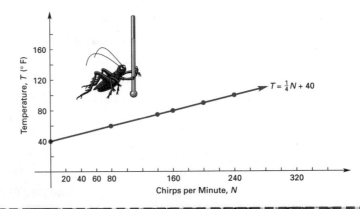

CULTURAL CONNECTION

Hypatia, the first woman mentioned in the history of mathematics, wrote about the work of a mathematician known as Diophantus of Alexandria. Sometime in the third century BCE, Diophantus wrote a text about arithmetic. In his text, Diophantus worked with equations that have more than one whole number solution. The equations are called Diophantine equations. Consider this problem.

In a pet shop there are several kittens and birds. The shopkeeper counted exactly 20 legs in the shop. How many kittens and birds does the shopkeeper have?

Let x represent the number of birds and y the number of kittens. The linear equation $2x + 4y = 20$ models this situation.

There are an infinite number of solutions to the equation. But the solution must make sense. How many kittens and birds can the shopkeeper have?

LESSON ASSESSMENT

Think and Discuss

1 How can you use an equation to make a table?

2 How can you use a table to graph an equation?

3 Explain why *m* is the slope of the equation $y = mx$.

4 Explain why the graph of the linear equation $y = mx + b$ crosses the *y*-axis at $(0, b)$

5 How can you graph the equation $3y + x = 9$ using slope-intercept form?

Practice and Problem Solving

Use a table to graph each equation. Give the slope and *y*-intercept of each graph.

6. $y = x + 2$ **7.** $y = -3x + 5$ **8.** $y = 2x - 4$

9. $x + y = 8$ **10.** $2x - y = 6$ **11.** $5x = 2y - 3$

12. $y = -5x - 3$ **13.** $y = \dfrac{1}{5}x - \dfrac{2}{3}$ **14.** $y - x = 4$

An asphalt company has found the equation $G = 10 + 20L$ approximates the number of trucks of gravel required for surfacing three-lane city streets. Let *G* represent the number of trucks of gravel required and *L* represent the length of the street in kilometers.

15. What are the slope and *y*-intercept of the equation?

16. Complete a table for the street lengths and the number of trucks of gravel. Use street lengths between 1 and 10 kilometers.

17. During one week, the asphalt company used 130 truckloads of gravel. How many kilometers of street did the company resurface?

In an experiment on plant growth, a certain species of plant is found to grow 0.05 centimeters per day. The plant measured two centimeters when the experiment started. Let H represent the ending measurement of the plant. Let d represent the number of days during which the experiment takes place.

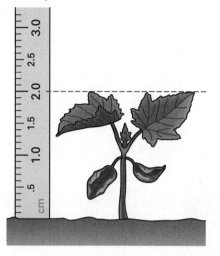

18. Write an equation that models the growth of the plant.

19. What are the slope and *y*-intercept for the equation?

20. At the end of the experiment, the plant was 3.3 centimeters tall. How many days did the experiment last?

Mixed Review

Write and solve an equation for each situation.

21. Paul is going to use 30% of his savings to make a down payment on a car. He can make a down payment of $2500. How much does Paul have in his savings account?

22. Tamara charges $6 per hour to clean windows. She also receives $3 for transportation to the job. One Saturday, Tamara earned $39. How many hours did Tamara work?

23. Nat has $50 and is saving money at $15 per week. Marti has $15 and is saving money at $20 per week. In how many weeks will they have the same amount of money?

24. At 7 AM, Jared notices that the temperature is 15°C. The weather forecaster just reported that from 5 AM to 7 AM, the temperature rose 5 degrees. What was the temperature at 5 AM?

LESSON 5.5 THE INTERCEPTS OF A LINE

Finding the Intercepts

The equation $y = mx + b$ is written in slope-intercept form. You know that the line representing this equation has a slope of m and crosses the y-axis at $(0,b)$. That is, the y-intercept is b. What is the y-intercept of $2x + y - 6 = 0$? To find the y-intercept, rewrite the equation in slope-intercept form.

$2x + y - 6 = 0$ Given

$\qquad y = -2x + 6$ Why?

The graph of $y = -2x + 6$ crosses the y-axis at $(0,6)$. Thus, the y-intercept is 6. What is the x-intercept of the graph?

The x-intercept is the point where the graph crosses the x-axis. All along the x-axis, the y-values are zero. Substitute zero for y and solve for x to find the x-intercept.

$\qquad y = -2x + 6$ Given

$\qquad 0 = -2x + 6$ Substitute zero for y

$\qquad 2x = 6$ Addition Property

$\qquad x = 3$ Division Property

The line crosses the x-axis at the point $(3,0)$. The x-intercept is 3.

Since the x- and y-intercepts represent two points on the line, you have another way to graph a linear equation. Graph the intercepts and connect them with a line.

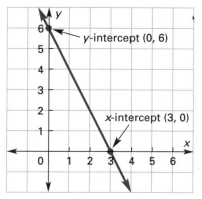

To check your graph, calculate one more point and graph it to be sure it is on the same line. For example, choose 1 for x. Then substitute in the equation $y = -2(1) + 6$. The result is the point (1,4). Does this point lie on the same line determined by the intercepts?

It is easy to find the intercepts for the graph of a linear equation. To find the y-intercept, replace x by zero and solve for y. To find the x-intercept, replace y by zero and solve for x.

The equation $y = -2x + 6$ models this workplace situation. A tank is filled with 6 gallons of water. A pump begins to remove the water at a rate of 2 gallons per minute. How long will it take the pump to empty the tank? Here, the slope represents the pumping rate. What does the y-intercept represent? What does the x-intercept represent?

Ongoing Assessment

What are the intercepts for the equation $3x - 5y = 15$?

Change of Scale

Sometimes the slope and the intercepts of an equation make it necessary to change the scale of the graph.

EXAMPLE Changing Scale

Use the intercepts to graph $y = 20x + 25$.

SOLUTION

To find the y-intercept, replace x by zero. The result is a y-intercept of 25.

To find the x-intercept, replace y by zero. The result is the x-intercept $\frac{-25}{20}$ or $\frac{-5}{4}$.

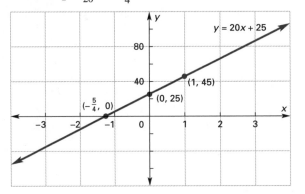

To check the graph, find one more point. If x is replaced by 1, y is equal to 45. Make sure the line passes through (1,45).

Since the slope of the line (20) is very steep, each axis has a different scale. This scale keeps the graph for $y = 20x + 25$ from being too close to the y-axis. By compressing the numbers along the y-axis (so that 40 units cover the same distance as one unit along the x-axis), the graphed line appears to have a very gentle rise.

Critical Thinking   Explain how changing the scale can make it hard to interpret the graphed data.

ACTIVITY   Rate of Change

The A-1 Cable Company installs underground cable. They charge $650 for the first 50 feet and $6 per foot for each additional foot. Let L represent the length of the cable installed (after the first 50 feet). Let d represent the number of dollars in the total cost.

1 Write an equation to describe the cost of installing cable.

2 Which variable is the

   a. independent variable?   **b.** dependent variable?

   Explain your answer.

3 Write the cost equation so that it is in slope-intercept form with the dependent variable (d) alone on one side of the equal sign.

4 Graph the equation.

5 What is the slope of the line? What is the y-intercept? What is the meaning of the slope and y-intercept in this situation?

6 Which part of your graph makes sense for the workplace situation?

7 Use your graph to read the approximate cost of installing a total of 175 feet of underground cable.

8 Check your answer. Show your check. Does your answer make sense?

Lesson Assessment

Think and Discuss

1 Explain how to find the *y*-intercept of a line.

2 Explain how to find the *x*-intercept of a line.

3 Give an example of a situation where the slope or intercepts would result in a change of scale of the graph.

Practice and Problem Solving

Find the *x*- and *y*-intercept for each line.

4. $y = 2x + 4$

5. $y = -3x + 5$

6. $y = 5x - 3$

7. $y = \frac{1}{2}x - 4$

8. $y = \frac{3}{4}x + \frac{1}{2}$

9. $y = 1.5x - 2.5$

10. $3x + y = -12$

11. $4x - y = -10$

12. $y - x = 5$

13. $2x - 3y = 12$

14. $5x + 3y = 15$

15. $x - 3y = -8$

Choose the independent and dependent variables. Then, write and graph an equation for each situation.

16. To convert Celsius to Fahrenheit degrees, multiply the Celsius temperature by 1.8 and add 32.

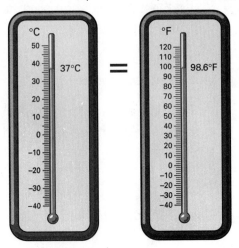

17. The distance from the start line after a certain amount of time if a runner is given a 2-mile head start and runs 4 miles per hour.

18. Lengths and widths of rectangular lawns with perimeters of 100 meters.

19. Amount earned if Anna mows lawns for $8 per hour and charges $6 for cleaning each yard.

20. Total cost if a seed company charges a fixed rate of $2.50 per pound plus a one-time $0.50 packaging charge.

Mixed Review

Write the answer as a power of ten.
21. $10^{-3} \cdot 10^{-2}$ **22.** $10^5 \cdot 10^{-5}$ **23.** $10^6 \div 10^{-2}$

24. Tim invested $5250 at 5.5% simple interest for 18 months. How much interest did Tim receive?

25. Draw a circle so that it fits into a square exactly. If the length of one side of the square is 625 centimeters, what is the area of the circle?

Evaluate each expression when x is -4 and y is -2.
26. $5x - 3y$ **27.** $x(y - 5) + 3$ **28.** $x^2 - y^3$

Catenya keeps a record of her earnings from the utility company. There are several ways Catenya can predict how much she will earn for any number of hours that she works. For example, she can write an equation or draw a graph.

Hours	Earnings
7	$52.50
4	$30.00

To draw the graph, plot the two points and draw a line through them. What is the scale? Why is the graph drawn only in the first quadrant?

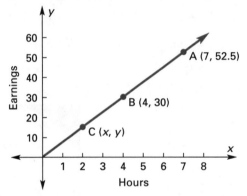

Using the Slope Formula

To write an equation, you will need to use the slope formula. The points $A(7, 52.5)$ and $B(4, 30)$ are on the line. Thus, the slope (m) of the line is

$$m = \frac{y_2 - y_1}{x_2 - x_1} = \frac{30 - 52.5}{4 - 7}$$

$$= \frac{-22.5}{-3}$$

$$= 7.5$$

You can represent every point on the line by an ordered pair (x, y). The slope of the line between $C(x, y)$ and $B(4, 30)$ is also 7.5 or $\frac{7.5}{1}$. Use the slope-intercept formula again. This time, use points C and B and solve for y.

$$\frac{7.5}{1} = \frac{y - 30}{x - 4} \qquad \text{Slope Formula}$$

$$7.5(x - 4) = y - 30 \qquad \text{Proportion Property}$$

$$7.5x - 30 = y - 30 \qquad \text{Distributive Property}$$

$$y = 7.5x \qquad \text{Addition Property}$$

The equation $y = 7.5x$ models Catenya's earnings.

In this equation, y is the dependent variable, and x is the independent variable. The equation shows that Catenya receives $7.50 for each hour she works. The amount of earnings (y) is a function of the time Catenya works (x).

The linear equation $y = 7.5x$ is a **linear function**. In this situation, the slope of the line represents a **rate of change** of $7.50 in the linear function. When x increases by 1, y increases by 7.5. This type of a rate of change is called **direct variation**.

EXAMPLE 1 Finding the Equation of a Line

Find the equation of the line passing through $M(3,1)$ and $N(1,-3)$. Describe the graph connecting the points.

SOLUTION

Find the slope.
$$m = \frac{y_2 - y_1}{x_2 - x_1}$$

$$= \frac{-3 - 1}{1 - 3}$$

$$= 2$$

Let $P(x,y)$ be any point on the line. Use $M(3,1)$ as the second point.

$$2 = \frac{y - 1}{x - 3} \qquad \text{Given}$$

$$2(x - 3) = y - 1 \qquad \text{Proportion Property}$$

$$2x - 6 = y - 1 \qquad \text{Distributive Property}$$

$$y = 2x - 5 \qquad \text{Simplify}$$

The graph of the linear function $y = 2x - 5$ is a line with slope 2 and y-intercept -5.

> **Linear Function**
> An equation written in the form
>
> $$y = mx + b$$
>
> is a linear function. The rate of change of the function is m.

Find the equation of the line passing through $D(4,3)$ and $E(1,2)$ and describe its graph.

Horizontal and Vertical Lines

ACTIVITY 1 Horizontal Lines

1 Find the equation of the line passing through $(1,4)$ and $(2,4)$.

2 Find the equation of the line passing through $(-2,-3)$ and $(1,-3)$.

3 What do the slopes of the lines have in common?

4 Graph each equation on the same coordinate system.

5 What do the graphs have in common?

You can write each equation in Activity 1 in the form $y = b$. The equation can also be written as $y = mx + b$, where m is 0 and b is a constant. Thus, $y = b$ is a constant function. Its graph is a horizontal line, parallel to the x-axis. The value of b tells you how far above or below the origin the graph is drawn.

ACTIVITY 2 Vertical Lines

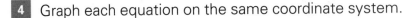

1 Find the equation of the line passing through $(4,1)$ and $(4,2)$.

2 Find the equation of the line passing through $(-2,-2)$ and $(-2,1)$.

3 What do the slopes of the lines have in common?

4 Graph both equations on the same coordinate system.

5 What do the graphs have in common?

You can write each equation in Activity 2 in the form $x = c$. However, the equation $x = c$ is not a function. In the next chapter you will find out why. The graph of $x = c$ is a vertical line, parallel to the y-axis. The value of c tells you how far to the right or left of the origin the graph is drawn.

Critical Thinking   Compare these graphs to the equations in the Activities. How are they alike or different?

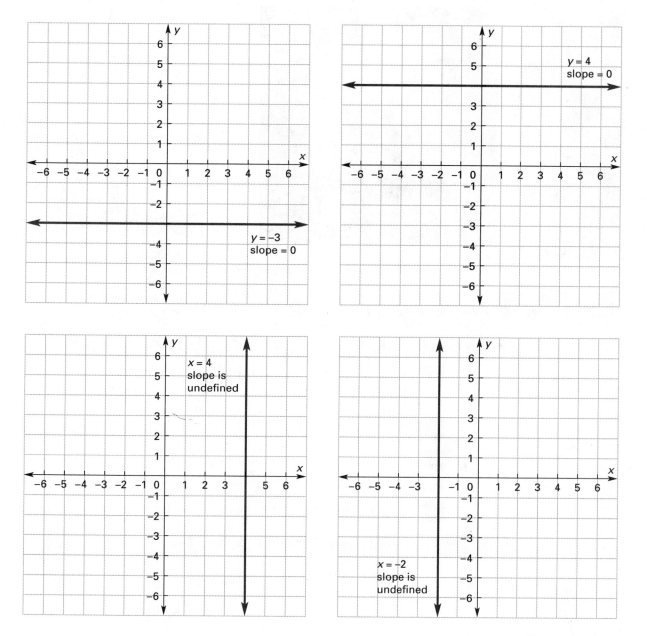

$y = -3$
slope = 0

$y = 4$
slope = 0

$x = 4$
slope is undefined

$x = -2$
slope is undefined

WORKPLACE COMMUNICATION

How did Sara use linear functions to estimate the amount of optical fiber needed for the next job? What slope did she use? What rate of change does this slope represent?

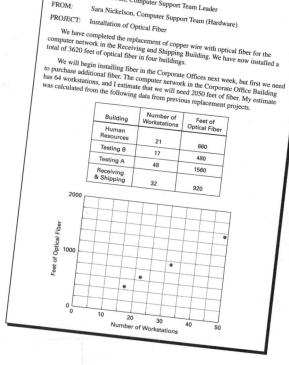

PROGRESS REPORT

TO: Tim Howell, Computer Support Team Leader
FROM: Sara Nickelson, Computer Support Team (Hardware)
PROJECT: Installation of Optical Fiber

We have completed the replacement of copper wire with optical fiber for the computer network in the Receiving and Shipping Building. We have now installed a total of 3620 feet of optical fiber in four buildings.

We will begin installing fiber in the Corporate Offices next week, but first we need to purchase additional fiber. The computer network in the Corporate Office Building has 64 workstations, and I estimate that we will need 2050 feet of fiber. My estimate was calculated from the following data from previous replacement projects.

Building	Number of Workstations	Feet of Optical Fiber
Human Resources	21	
Testing B	17	660
Testing A	48	480
Receiving & Shipping	32	1560
		920

LESSON ASSESSMENT

Think and Discuss

1 Explain how to find the equation of a line passing through two points if you know the coordinates of each point.

2 What is a linear function?

3 How are the rate of change of a function and the slope of a line related?

4 Explain why a horizontal line represents a linear function.

Practice and Problem Solving

Find the equation of the line passing through the given points.

5. (4,5), (2,3)

6. (3,−2), (6,0)

7. (−1,8), (−3,5)

8. (−2,4), (4,4)

9. (−7,−2), (−6,−1)

10. (−9,6), (−9,−6)

A moving van travels 325 miles in 5 hours. The next day, the same van travels 520 miles in 8 hours. Let x represent the time and y represent the distance traveled.

11. Write the two ordered pairs from the times and distances given. Find the rate of change in the distance the car travels.

12. Write a linear function that models the situation.

Todd is selling tickets for the school play. It costs $5 for an adult ticket and $3 for a child's ticket. Todd needs to sell a total of $65 in tickets. Let x represent the number of adult tickets sold. Let y represent the number of child's tickets sold.

13. Write a linear function that models the number of adult and child's tickets Todd must sell to reach his goal.

14. Graph the linear function.

15. Suppose (10,5) is a point on your graph. What does the ordered pair represent in the problem?

Hook's Law states that the distance (d) a spring is stretched is directly proportional to the force (F) applied. The formula is often written in the form $F = kd$, where k is called the spring constant. Suppose a force of 500 pounds stretches a truck's suspension spring one inch.

16. What is the spring constant of the truck spring?

17. Write a linear function that models the spring displacement as a function of applied force.

18. How far will the spring stretch under a force of 700 pounds?

19. How much force is needed to stretch the spring 3 inches?

Represent each situation with a positive or negative number.

20. A profit of $1250

21. A weight loss of 12 pounds

22. A salary increase of $2/hr

23. A temperature fall of 9 degrees

Solve each equation. Check your answer.

24. $5t - 12 = -t$ **25.** 45% of $m = 90$ **26.** $\dfrac{-20}{d} = \dfrac{7.5}{15}$

Write and solve an equation for each situation.

27. Winston finished 20% of his job assignment the first day. In that time, he was able to paint 24 outside window casings. How many window casings will Winston paint at the end of his assignment?

28. A draftsman is completing a design that uses isosceles triangles. In one of the triangles, the measure of the vertex angle is represented by a. The measure of each base angle is represented by the expression $2a + 20$. Find the size of each angle in the triangle.

LESSON 5.7 USING A GRAPHICS CALCULATOR

Most graphics calculators graph linear functions written in the form $y = mx + b$. Check out the graphics calculator you will be using. Although the instructions given in this text might differ somewhat from how your calculator works, the ideas should be similar. Here is one way to graph $y = 3x + 1$.

Press the Y= key. Then enter $3x + 1$. Now press the GRAPH key. You should see the graph of $y = 3x + 1$ appear on the screen.

The graphics calculator window is very small. Thus it is important that a "friendly window" (one with a readable scale) is selected before you graph a function. The WINDOW or RANGE key is used to set the maximum (max) and minimum (min) range and the scale (scl) division. The standard setting for this text will be

$X\text{min} = -10$ $X\text{max} = 10$ $X\text{scl} = 1$

$Y\text{min} = -10$ $Y\text{max} = 10$ $Y\text{scl} = 1$

Ongoing Assessment

Use your graphics calculator to graph $y - 2x = -3$.

Critical Thinking   Explain how to use your graphics calculator to graph $y = 2x + 2$ and $y = -5x + 3$ on the same coordinate plane.

The TRACE key is used to find the coordinates of any point on the graph of the linear function.

ACTIVITY 1   Graphing a Linear Function

1 Graph the function $y = 3x + 1$ on your graphics calculator.

2 Press the TRACE key.

3 What is the result?

4 Press the left and right arrow keys.

5 What is the result?

There is another important key on your graphics calculator.

1 Press the (ZOOM) key.

2 What is the result?

3 Explain how the (ZOOM) key results in a more accurate reading of the coordinates of a point.

LESSON ASSESSMENT

Think and Discuss

1 What are the advantages of using a graphics calculator to graph a linear function? What are the disadvantages?

Explain how to graph each equation on your graphics calculator.

2 $y = 3x - 1$

3 $2x + y = 6$

4 $y = \frac{1}{2}x + 5$

5 Reread the situation presented at the beginning of Lesson 5.6. Explain how to change the range or scale settings to get a friendly window for this situation.

Practice and Problem Solving

Use your graphics calculator.
Graph both equations on the same coordinate axis.

    a. $y_1 = 3x + 2$ **b.** $y_2 = 3x - 1$

6. What is the relationship between the lines?

7. What is the relationship between the slopes?

8. Write another equation with the same relationship. Graph the equations on the same coordinate axis to check your result.

Graph both equations on the same coordinate axis.

    a. $y_1 = -2x + 2$ **b.** $y_2 = \frac{1}{2}x - 1$

9. What is the relationship of the lines?

10. What is the relationship of the slopes?

11. Write the equation of a line perpendicular to the graph of $y = 4x - 3$. Graph the equation to check your result.

Mixed Review

Solve each equation.

12. $5a - 12 = -27$

13. $\dfrac{-6}{9} = \dfrac{r}{-12}$

14. 25% of $d = 80$

15. $\dfrac{2}{3}m + 7 = -15$

16. $-24 = 18 - q$

17. $5s + 4 = 13 - 4s$

Write each equation in slope-intercept form.

18. $5x + 2y = 20$ **19.** $3x - y = 2$ **20.** $2x = 9 + 3y$

Write and solve an equation for each situation.

21. Latisha invests $2500 at 6.5% simple interest. How much does she have at the end of 18 months?

22. Shareef sells model cars for $8. At one craft show, Shareef earned $1032. How many model cars did he sell?

23. The combined total company sales for this year and last year are $3,500,000. This year, sales are 1.5 times last year's sales. What are the sales for this year?

24. Mr. Williams is putting a fence around a rectangular garden. The perimeter of the garden is 300 feet. The length of the garden is twice the width. How long is the garden?

MATH LAB

Equipment   Calculator
1 lb coffee can
Graduated cylinder, 500 ml capacity
Vernier caliper
Ruler

Problem Statement

You will examine the relationship between the volume (V) and height (h) of a cylindrical can and plot data of volume vs. height.

The formula for the volume of a cylinder is $V = \pi r^2 h$.

Procedure

a Measure the inside diameter of the 1 pound coffee can. Calculate and record the radius in centimeters. To convert from inches to centimeters, use 2.54 centimeters equal to 1 inch. Calculate and record the area of the base of the coffee can.

b Measure 100 ml of water in the 500 ml graduated cylinder and pour it into the coffee can. Be sure the can is level. To check if it is level, measure the height of the water in the can at three different positions around the inside edge of the can. If all three measurements are equal, the can is level. Measure and record the height in centimeters of the water in the can. Record the volume of the added water.

c Repeat Step **b** by adding 100 ml of water four more times for a total of 500 ml of water. Be sure to record the height of the water and the total volume of water in the can after each addition of water. When you are finished, you should have five pairs of values for V and h in your data table.

d Graph your data. Graph the volume on the vertical axis and the height on the horizontal axis. Be prepared to defend your answers to these questions.

1. Is the relationship between volume and height linear?

2. Is the equation $V = \pi r^2 h$ linear if r is constant?

3. If the equation is linear, what is the slope of the line represented by the equation?

4. How does this slope compare to the calculated value from Step **a**?

5. Does the graphed line pass through the origin if you extend it? Should it?

6. What does the point on the line at the origin mean?

7. How would the slope change if you used a larger diameter coffee can?

Activity 2: Measuring in Inches and Centimeters

Equipment   Calculator
Tape measure marked in inches and centimeters

Problem Statement

The relationship that converts measurements in inches (i) to centimeters (C) is

$$C = 2.54i$$

You will measure several lengths in inches and centimeters and plot corresponding pairs of measurements on a graph. You will interpret the graph to verify the value of the coefficient 2.54.

Procedure

a Measure and record the width of a sheet of paper in inches and in centimeters.

b Measure and record the length of a sheet of paper in inches and in centimeters.

c Measure and record the width of the classroom door in inches and in centimeters.

d Measure and record the height of the classroom door in inches and in centimeters.

e Measure and record the width of the teacher's desk in inches and in centimeters.

f Measure and record the length of the teacher's desk in inches and in centimeters.

g Graph your data. Graph the measurement in centimeters on the vertical axis. Graph the measurement in inches on the horizontal axis. *Caution:* Study the range of the data. Then choose the scales for the *x* and *y* axes. Make certain all the data will fit on the graph.

h Draw an unbroken line that best connects the six points on your graph. Is the graph a straight line? If a point is not on the straight line, double-check your measurements and graph for that point.

i Choose any two points on the graphed line—such as *A* and *B* in the drawing shown here. These points need not include the points you plotted to draw the graph. Based on the values of these points, subtract the smaller centimeter value from the larger centimeter value. The result is the difference in centimeter values for the two points. Label this on the graph as Δcm. [Note: Δ is the Greek letter delta. It is often used to indicate a *difference* in values.]

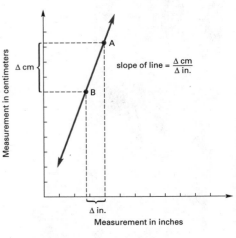

j For the same two points, and in the same order, find the difference in inch values. Label this on the graph as Δin.

k Divide the Δcm value by the Δin. value. This is the slope of the graphed line and is the value of *m* in the slope-intercept form of a linear equation $y = mx + b$. Compare your calculated slope to the value 2.54 in the equation 1 (cm) = 2.54 · 1 (in.).

l For your graphed line, what is the *y*-intercept? Does this value make sense?

Equipment   Calculator
1000 ml beaker
Graduated cylinder, 500 ml capacity
Spring scale with 500 gm capacity
String

Problem Statement

If the temperature is constant, the relationship between the weight and volume of liquid is a linear function. You will study this relationship by weighing different volumes of water and then calculating the slope and y-intercept.

Procedure

a Tie a string around the top of the graduated cylinder. Weigh the graduated cylinder using the spring scale. Record this weight.

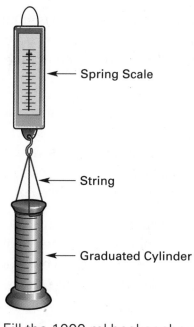

Spring Scale

String

Graduated Cylinder

b Fill the 1000 ml beaker about half full of water.

c Pour about 100 ml of water from the beaker into the graduated cylinder and weigh the water and graduated cylinder with the spring scale. Record the volume of water and the weight.

d Repeat Step **c** four more times until about 500 ml (but no more) is added to the graduated cylinder.

e Graph your data. Graph the weight on the vertical axis and the volume of water in the graduated cylinder on the horizontal axis.

f Draw an unbroken line connecting the points on your graph. Is the graph a straight line? If any point is not on the straight line, check your data and graph to be sure the point is correctly plotted.

g Choose two points next to each other. Subtract the smaller weight from the larger weight. This is the difference in weights. Label this difference as Δw. (Refer to Activity 2 for a drawing that shows a similar graph with similar measurements.)

h For the same two points, subtract the smaller volume from the larger volume. This is the difference in volumes. Label this difference as ΔV.

i Divide the difference in weight, Δw, by the difference in volume, ΔV. This is the slope of the graphed line and is the value of m in the slope-intercept form of a linear equation $y = mx + b$.

j What is the weight on the graph when the volume of water is zero? Compare this value to the weight of the empty graduated cylinder. This value is the y-intercept of the graph and corresponds to the b value in the slope-intercept form of a linear equation $y = mx + b$.

k Write a linear equation in slope-intercept form that shows the relationship between the weight and volume of given amounts of water as established by your data. Your equation should include the value for the constant m and the constant b.

MATH APPLICATIONS

The applications that follow are like the ones you will encounter in many workplaces.
Use the mathematics you have learned in this chapter to solve the problems.
Wherever possible, use your calculator to solve the problems that require numerical answers.

At a constant temperature, the weight of water increases as the volume increases. The graph below shows the relationship between water's weight and volume at 60°F.

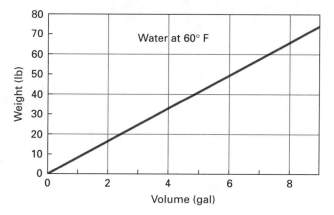

1 What is the weight of 2 gallons of water at 60°F? What is the weight of 6 gallons of water at 60°F?

2 What is the slope of the graphed line? What is the *y*-intercept of the graphed line?

3 Write the equation of the graphed line in slope-intercept form.

For each range of data below, use the maximum length of the axis given for a graph to determine a scale unit for that axis. For example, for a data range of 0 to 4 hours to fit an axis 4 inches in length, you can use a scale unit of 1 hour per inch.

Data Range	Maximum Length of Axis	Scale Unit
4. 0 to 800 miles	8 inches	
5. 0 to 500,000 persons	5 inches	
6. 0 to 60 seconds	12 centimeters	
7. 25 to 30 minutes	10 centimeters	
8. 80°C to 120°C	10 centimeters	

Pressure changes with depth in water and with altitude in air. The following tables list values of pressures at different depths below and altitudes above sea level.

Water Pressure vs Depths		Air Pressure vs Altitude	
Depth (ft)	Pressure (Atm)	Altitude (1000 ft)	Pressure (Atm)
0	1	0	1.00
33	2	10	0.69
66	3	20	0.46
100	4	30	0.30
133	5	40	0.19
166	6	50	0.11
200	7	60	0.07
300	10		
400	13		
500	16		

9 Plot graphs of the two sets of data.

10 Does either graph show a linear relationship? If so, which graph(s)?

11 If any of the graphs are linear, identify the slope of the graph and the *y*-intercept of the graph.

12 Use your values of the slope and the intercept to write an equation for the linear relationship(s).

Steers weighing 800 to 900 pounds consume an average of 22.3 pounds of feed per day. These steers should gain weight at an average rate of 2.7 pounds per day.

13 Write an equation in slope-intercept form showing the number of days for a steer in this weight range to gain W pounds of body weight.

14 Write another equation in slope-intercept form to express the number of days needed for an average steer in this weight range to consume F pounds of feed.

15 Both equations from Exercises 13 and 14 define the number of days. Thus, you can set them equal to each other to obtain a new equation. Do this and obtain an equation that relates the pounds of feed consumed (F) to the pounds of weight gained (W). Isolate the variable for the pounds of feed consumed.

16 Use your equation from Exercise 15 to calculate the feed consumption needed for a 50-pound weight gain.

A plant's water requirement is defined as the pounds of water that must pass from the roots and out of the leaves to produce one pound of dry plant matter. The water requirement for corn is 368 pounds of water per pound of dry plant matter produced.

17 Write an equation in slope-intercept form that expresses the relationship between the yield of plant matter and the water requirement for corn.

18 Use your equation to calculate the water required to produce 8,000 pounds of dry corn plant matter.

19 An acre-inch of water is equal to 226,512 pounds of water. Use your equation to calculate the yield of dry corn plant matter that consumes one acre-inch of water.

A seed corn company buys foundation seed from a seed stock company for $4.00 per 1,000 viable kernels. The planting rate for seed corn is 16,000 kernels per acre.

20 Write an equation in slope-intercept form for the relationship between the acres planted and the number of kernels of seed corn.

21 Write an equation in slope-intercept form that shows the relationship between the acres planted and the cost of the seed corn.

22 Use your equation from Exercise 21 to calculate the cost of the seed corn used to plant 120 acres.

A certain farming area has found that the cost per acre of loading hay to be closely related to the total labor hours required to complete the job. The following graph shows a plot of cost per acre versus labor hours per acre.

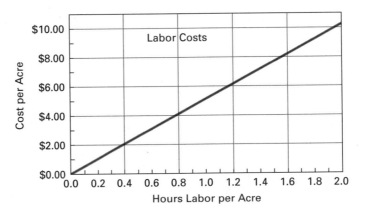

23 What is the slope of the graph? What rate of change does this slope represent?

Farmers can use various methods to load hay, each requiring different labor hours per acre. The labor required for three methods is as follows:

Trailed wagon and chute	1.54 hours per acre
Bale thrower	1.16 hours per acre
From ground by hand	1.97 hours per acre

24 What is the cost for loading 100 acres of hay from the ground by hand?

25 What are the costs for loading hay with a bale thrower and with a wagon and chute for 100 acres?

The various food supplies in a zoo are used at different rates. Suppose the current quantity and usage for three different types of foods are as listed below.

Food	Current Quantity (in pounds)	Usage per Day (in pounds)
Type A	700	40
Type B	640	60
Type C	480	30

26 Let x represent the number of days. Let y represent the number of pounds of food. Let the rate *usage per day* be the slope and *current quantity* be the y-intercept. Write linear equations for the food types that represent the number of pounds of food remaining after x days. (Put the equations in slope-intercept form.)

27 Choose a scale and draw a graph of the three lines. Label each line according to the food type. The slopes are negative. Why?

28 Use your graph to determine which of the three types of food will be the first to be consumed. How many days in the future will this occur? (Hint: This is the x-intercept for one of your graphed lines.)

A farmer can estimate the cost of transporting crops to market as a sum of the cost of fuel and the cost of the vehicle, as they relate to the distance to market. Fuel costs for the vehicle are about $0.095 per mile. The cost of the vehicle is a fixed value of $2500 per year.

29 Write an equation for the annual transportation cost as a function of total annual miles traveled.

30 What is the slope of your equation? What does this rate represent?

31 What is the meaning of the y-intercept of the equation?

32 Draw a graph of the equation.

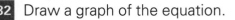

A researcher is analyzing past trends in consumer spending. The graph shown here displays some changes in consumer spending that took place over a period of ten years.

33 The line representing spending for durables and the line for nondurables appear similar. What feature of these two lines is nearly the same?

34 The two lines for services and nondurables are similar but have one feature that distinguishes them. What feature of these two lines is notably different?

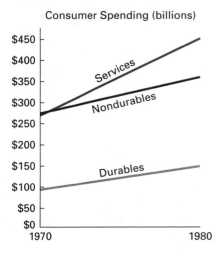

A department store pays each of its senior sales clerks a weekly salary of $150 plus a 3% commission on the clerk's gross sales. This graph shows the pay scale.

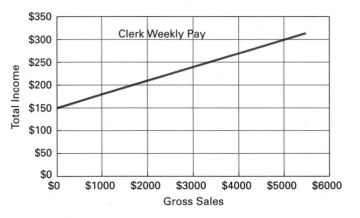

35 What is the y-intercept of the graph? What quantity does this intercept represent?

36 What is the slope of the graph? What rate does the slope represent?

37 Write the equation for the graphed line in slope-intercept form.

38 Suppose the store increased the commission to 5% but decreased the weekly salary to $100. How would this change affect the graph? Write the equation for these new conditions.

The financial department of a production firm presented the graph shown below of the production costs for a certain item. The graph shows the fixed costs—the costs for tooling and overhead that are required to produce even one item—and the variable costs—the costs due to materials, labor, marketing, transportation, and so on.

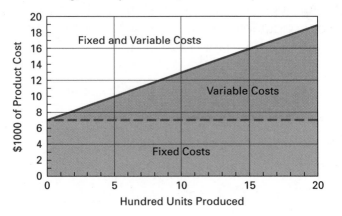

39 Identify the slope and *y*-intercept of the graph for total production cost.

40 How are these (slope and *y*-intercept) related to the fixed costs and the variable costs?

You must have some brochures printed, so you make a telephone call to a local print shop. You are told that there is no setup charge. The cost for printing various quantities is given to you in a table.

Quantity	Cost
50	$ 7.68
100	8.91
150	10.14
200	11.36
250	12.59
500	18.73
750	24.86
1000	31.00

41 Choose a scale and graph the cost data. Draw a line that seems to fit the data.

42 Determine the slope and *y*-intercept of your line.

43 What is the meaning of this nonzero *y*-intercept?

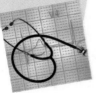

HEALTH OCCUPATIONS

Research indicates that blood flow through the kidneys decreases with age, as shown in this simplified graph of experimental findings.

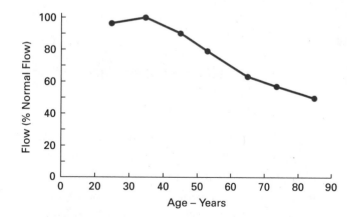

44 Would you say that the changing blood flow shows a linear relationship, or close to it? If not, could you say that it was linear for some ages?

45 A computer analysis of these data reported the following equation for relative blood flow.

$$F = -1.18A + 141$$

A is the age in years, from 35 to 85 years.

Evaluate the equation for several ages to see if it agrees with the graphed data. (Round your answers to the nearest whole number before comparing to the graph.)

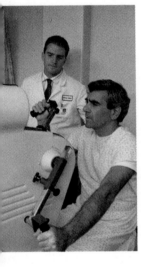

A "stress test" is a method of evaluating the health of a patient's cardiovascular system. In this test, a technician monitors the patient's pulse rate during an exercise session on a treadmill or stationary bicycle. A patient's pulse rate is not allowed to exceed a maximum rate. The maximum rate is based on the patient's age, from the following graph.

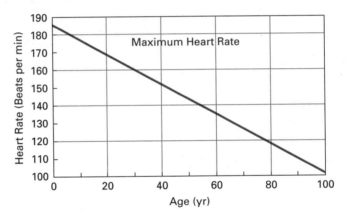

46 What is the *y*-intercept of the graph?

47 What is the slope of the graph?

48 Write an equation for the graphed line in the slope-intercept form.

49 Use your equation to find the maximum heart rate in a stress test for a 44-year-old patient.

Weight loss occurs when more calories are expended in exercise than are consumed in the daily diet. When 3500 calories more are used in exercise than are consumed in the diet, about 1 pound of fat is lost. The following graph shows the weight loss versus calories expended in exercise that can be expected by a 5 foot 6 inch, 132-pound woman who consumes a 2400-calorie daily diet. (Her weight remains stable with the 2400-calorie daily diet and normal exercise.)

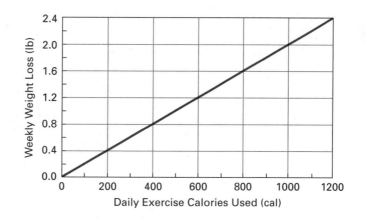

50 What is the slope of the line?

51 What is the equation of the line?

52 How much weight will the woman lose each week if her exercise uses 350 calories per day?

53 If the woman decreases her dietary calories from 2400 calories to 2150 calories per day, how will this affect the graphed line? What is the equation for the new line?

A pickle recipe requires $\frac{3}{4}$ ounce of salt per pound of cucumbers.

54 Write an equation in slope-intercept form that shows the relationship between the pounds of cucumbers and the ounces of salt.

55 How much salt do you need to pickle 1 ton of cucumbers? Remember that 1 ton = 2000 pounds.

56 How many pounds of cucumbers can you pickle with 100 pounds of salt? Remember that 1 pound = 16 ounces.

INDUSTRIAL TECHNOLOGY

When you use a thermocouple to measure the temperature at a location, you must convert a voltage reading from the thermocouple into a temperature. You recorded the following voltages from a voltmeter and the corresponding temperatures from a thermometer.

Temperature (°C)	Voltage (mV)
0	0.00
100	6.32
200	13.42
300	21.03
400	28.94
500	37.00
600	45.09

57 Graph the data. Based on your graph, is the relationship between temperature and thermocouple voltage a linear one? Explain your answer.

58 Write a formula for a line, in slope-intercept form, that will very closely describe the relationship between the two variables T (temperature in °C) and V (voltage in millivolts). This is the calibration equation for the thermocouple.

59 How can you use this equation for temperature measurements?

60 What is the meaning of the y-intercept in the calibration equation? What rate does the slope represent?

You have a standard assembly used to construct large exhaust fans for industrial warehouse facilities. The assembly is designed so that the motor pulley and the fan pulley are 28 inches apart. A small adjustment is available for belt tension. The motor pulley has a diameter of 3 inches. Based on the required fan speed and fan size, you can vary the size of the fan pulley. To help you determine the belt sizes, you have the formula

$$B = 2L + 1.625(D + d)$$

B is the length of the belt needed.
L is the distance between the two pulley centers.
D is the diameter of one pulley.
d is the diameter of the other pulley.

Fan Pulley
Belt
Motor Pulley

61 Substitute the known values given above into the formula and rewrite the formula as a linear equation in slope-intercept form. Identify the variables, the coefficient, and the constant. In linear equations, what are other names for the coefficient and the constant?

62 Try your equation for fan pulley diameters of 6 inches and 10 inches. Check your results with the original formula.

You can use a flow test to determine if a radiator is clogged. In a flow test, you measure the volume of water that flows through a radiator by gravity in a measured time period. Automotive manufacturers establish the standards for these tests. If less water than the standard flows through a radiator, it is clogged. The standard established by Ford Motor Company is 42.5 gallons per minute.

63 Write an equation in slope-intercept form that shows the relationship between the time in seconds and the volume in gallons measured by the flow test.

64 Use the equation to calculate the water that should flow through a Ford radiator that is not clogged in 20 seconds.

65 Use the equation to calculate the time needed for a Ford radiator that is not clogged to fill a 5 gallon bucket.

66 Suppose you test a Ford radiator and measure a flow of 25 gallons of water in 40 seconds. Should you consider the radiator clogged?

The following table gives selected values of wire gauge numbers and the corresponding wire diameters.

Wire Gauge	Diameter (in inches)
0	0.3249
5	0.1819
10	0.1019
15	0.05707
20	0.03196
25	0.01790
30	0.01003
35	0.005615
40	0.003145

67 Draw a graph of the data.

68 Is the relationship between wire gauge number and diameter a linear relationship?

CHAPTER 5 ASSESSMENT

Skills

Write the number pair represented by each point on the graph below.

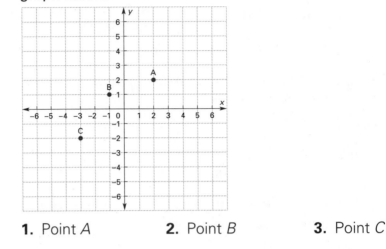

1. Point *A*　　　　**2.** Point *B*　　　　**3.** Point *C*

Name the quadrant where each point is graphed.
4. $(-4,-3)$　　　**5.** $(2,-6)$　　　**6.** $(-1,7)$

Find the slope of the line through each pair of points.
7. $(-2,3)$ and $(5,2)$　**8.** $(2,4)$ and $(7,-1)$　**9.** $(6,0)$ and $(0,6)$

Name the slope and y-intercept of the graph for each equation.
10. $y = 3x + 7$　　**11.** $y = -2x + 5$　　**12.** $y = \frac{2}{3}x - 4$

Write each equation in slope-intercept form.
13. $x + 2y = 6$　　**14.** $2x - y + 5 = 0$　　**15.** $3y - 5x = 6$

Applications

16. If $(2,3)$, $(-2,3)$, and $(2,-3)$ are vertices of a rectangle, what are the coordinates of the fourth vertex? Explain your answer.

17. A department store sells suit jackets and skirts. The jackets are always twice as much as the matching skirt. Write an equation to represent the relationship of jacket cost to skirt cost.

18. Mr. Garcia drives at a constant speed for 5 hours. At the end of two hours, he had driven 90 miles. After four hours, he had driven 180 miles. How many miles had he driven after five hours? Write an equation and draw a graph to determine your answer. At what rate is Mr. Garcia driving?

19. A cable TV company charges $30 for a basic installation plus $5 for each additional television it hooks up. If the total cost is a linear function, write an equation to represent this function and find the cost of having a cable installation including four additional televisions.

20. A company has developed a new product. Market research has determined that the demand for the product is a linear function of the price. The company can sell 70 items per month if it charges $150 each, but only 20 items if it charges $200 each. How many items of the new product will the company sell if it charges $180 each?

Math Lab

21. The volume of a cylinder is $V = \pi r^2 h$. If you graph the volume of water in a can as a function of the height of the water, you see a straight line. What is the slope of the line?

22. If one inch is equal to $2\frac{54}{100}$ centimeters, write a linear equation to convert centimeter measures to inches. What is the y-intercept of your equation?

23. You use a graduated cylinder to measure the volume and weight of various amounts of water. When you plot the weight on the horizontal axis and the volume on the vertical axis, you see a straight line. But the line does not pass through the origin—there is a y-intercept. What does the y-intercept represent?

CHAPTER 6

WHY SHOULD I LEARN THIS?

The next wave of technological breakthroughs will involve nonlinear systems. Industry needs people who can use nonlinear mathematical models to create new products that will improve our lives. This chapter shows how nonlinear systems are applied to the workplace and are important to your future.

NONLINEAR FUNCTIONS

OBJECTIVES

1. Represent relations and functions as tables of data, ordered pairs, graphs, and equations.
2. Identify the domain and range of a function.
3. Identify and graph nonlinear functions involving absolute value, squares, square roots, exponents, and reciprocals.
4. Solve problems involving nonlinear equations either by graphing or by using algebraic methods.

In chapter 5 you studied the patterns of linear functions. You represented these patterns with tables of ordered pairs, with graphs, and with equations of the form $y = mx + b$.

In this chapter you will see how patterns of nonlinear functions can also be represented by tables, graphs, and equations.

Automotive engineers and technicians use nonlinear functions to design and test electric cars. Structural engineers model the motion on bridges with nonlinear functions.

As you watch the video, look for graphs that are not straight lines. These graphs will represent nonlinear functions.

LESSON 6.1 RELATIONS AND FUNCTIONS

You use relationships between two variables every day. One example of a *relation* is the phrase "is a friend of."

In the sentence,

x is a friend of *y*

any two friends can replace the variables.

Jamie is a friend of Roberto

is an example of the relationship.

Sets of ordered pairs, tables, and graphs are all useful tools used to represent relationships between two variables. For example, you can estimate the distance of an approaching thunderstorm by counting the seconds between a flash of lightning and the resulting sound of thunder. This works because sound travels about $\frac{1}{3}$ of a kilometer every second. Consequently, if 3 seconds go by between the flash and the sound, the storm is 1 kilometer away, and so on.

A set of ordered pairs is called a **relation**. You can use ordered pairs to represent the data for an approaching thunderstorm.

$$R = \{(3, 1), (6, 2), (9, 3), (12, 4), \ldots\}$$

The braces { } mean *the set of.* You can list a relation in a table or draw its graph on a coordinate plane.

Time in Seconds	Distance in Kilometers
3	1
6	2
9	3
12	4
⋮	⋮

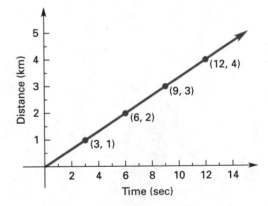

There is a pattern in the ordered pairs describing the thunderstorm data. Represent time with the independent variable *t* and the distance with the dependent variable *d*. The following equation represents the relationship:

$$d = \frac{1}{3}t$$

This equation is a shortcut way of saying that *the distance sound travels depends on time.*

Sequences

Examine the following sequence of numbers:

$$88, 96, 104, 112, 120, \ldots$$

The numbers in this sequence record the speed of an electric car each time it passes the timing line. The first speed measured is 88 feet per second and the car is steadily increasing its speed. Do you see a pattern? The difference between any two numbers is 8. You can order the numbers in the sequence with the set of whole numbers and list them in a table.

Lap Number	Speed	Difference in Speed
0	88	
1	96	$96 - 88 = 8$
2	104	$104 - 96 = 8$
3	112	$112 - 104 = 8$
4	120	$120 - 112 = 8$

If you graph the ordered pairs, you can see that the graph is a line with slope 8 and *y*-intercept 88. The equation representing the relationship between the ordered pairs is

$$y = 8x + 88$$

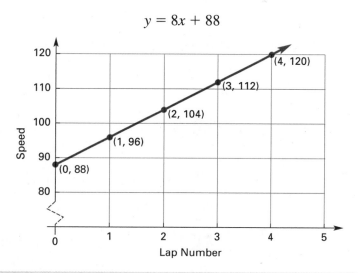

Ongoing Assessment

What is the speed of the car after 5 laps? By how much does the car increase its speed each lap?

EXAMPLE 1 Charges for Service Calls

The XYZ Plumbing Company has a schedule of charges for making service calls. The company charges a flat rate of $40 for each call plus $20 per hour on the job. Show the schedule of charges as a table, a graph, and an equation.

SOLUTION

Time in Hours	Charge in Dollars
0	40
1	60
2	80
3	100
4	120

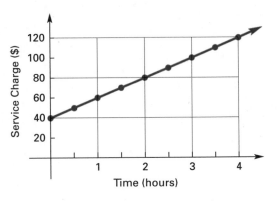

The equation $c = 20t + 40$ represents the relationship between the time worked (t) and the charge for the work (c).

ACTIVITY Sequences and Equations

Use the sequence 3, 7, 11, 15, 19, . . . to complete the following:

1 Use whole numbers to write the sequence as a set of ordered pairs beginning with (1, 3) and ending with (5, 19).

2 Use the ordered pairs to make a table.

3 Graph the set of ordered pairs.

4 Find the slope and y-intercept of the graph.

5 Find an equation that expresses the relationship between the numbers in the sequence.

6 Extend the sequence to three more numbers.

Function Notation

A **function** is a relation with one additional condition.

> **Function**
>
> A function is a set of ordered pairs (x, y) such that for any value of x, there is exactly one value of y.

A function machine can explain how a function operates. If you put in a certain sequence of numbers, you will get out a definite sequence of numbers. Sometimes you can determine the rule by examining the input and output numbers.

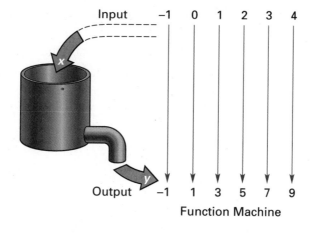

Function Machine

The equation $y = 2x + 1$ describes the rule used in the function machine. In this case, x is the independent variable and y is the dependent variable. Since a *unique* value for y is determined for each value of x, y is a function of x. A relation is a function if for each value of the independent variable x, there exists exactly one value of the dependent variable y. In function notation, you write

$$f(x) = 2x + 1$$

The symbol $f(x)$ is read "f of x" or "f is a function of x." In function notation, an input x determines the output $f(x)$ according to the rule $2x + 1$. Usually the letters f, g, and h are used to represent functions. The set of input values is the **domain** of the function. The set of output values is the **range**.

EXAMPLE 2 Evaluating a Function

Let $f(x) = 3x - 2$. What is the value of $f(x)$ when $x = 5$?

SOLUTION

Substitute 5 for x in the rule and simplify.

$$f(5) = 3(5) - 2$$
$$= 15 - 2$$
$$= 13$$

Thus, $f(5) = 13$.

Ongoing Assessment

a. If $h(x) = -5x - 6$, what is $h(-4)$?

b. If $g(x) = x^2 - 5x + 6$, find $g(2)$ and $g(3)$.

When you use function notation where $y = f(x)$, graph x on the horizontal axis and $f(x)$ on the vertical axis. The graph of the function in Example 2 looks like this.

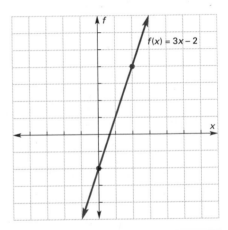

$f(x) = 3x - 2$

LESSON ASSESSMENT

Think and Discuss

1 Give examples of mathematical relations that are not functions. Explain why they are relations but not functions.

2 How many ways can you describe a function? Give examples.

3 Explain why a sequence of numbers might be considered a function.

4 Explain what the notation $g(x) = 3x + 5$ means.

5 Explain how to find the value of a function.

For each sequence, write a rule that gives the relationship between the numbers and find the next three numbers in the sequence.

6. 3, 4, 5, 6   **7.** 0, 2, 4, 6   **8.** −1, 2, 5, 8

9. 6, 11, 16, 21   **10.** −1, 1, 3, 5   **11.** 1, $\frac{3}{2}$, 2, $\frac{5}{2}$

If $h(x) = 2x - 7$, find
12. $h(4)$ **13.** $h(0)$ **14.** $h(-3)$

If $f(x) = -5x + 3$, find
15. $f(0.2)$ **16.** $f(-1)$ **17.** $f\left(\frac{3}{5}\right)$

For each situation, make a table, find and graph the equation, and give the slope and *y*-intercept.

18. Roberto started walking from a point 3 miles from his house and walked away from his house at a constant speed of 4 miles per hour. What is Roberto's distance (*d*) from home as a function of time (*t*)?

19. In Carlette's appliance repair business, she charges $15 for making a house call. She also charges $8 per hour. What are Carlette's total charges (*c*) as a function of time (*t*)?

20. Videos rent for $3 per day. To join the video club, you pay a one-time fee of $12. What is the cost (*y*) of renting videos for *x* days?

21. Write 0.000085 in scientific notation.

22. Write 100,000,000 as a power of ten.

Solve each equation.
23. $5y - 4 = 25$ **24.** $2x + 9 = -7$ **25.** $3(r - 1) = -9$

26. 20% of $g = 20$ **27.** $95 = 15x + 20$ **28.** $\frac{m}{-5} = \frac{16}{-10}$

29. A store pays $650 for a stereo system. It charges its customers $1050 for the system. What is the profit from the sale of a stereo system written as a percent of the store's cost?

LESSON 6.2 THE ABSOLUTE VALUE FUNCTION

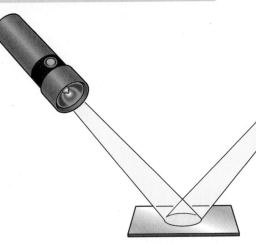

In the "function machine" illustrated in Lesson 6.1, each input number resulted in exactly one output number. In some functions, however, two or more input numbers might result in the same output number. The absolute value function is one such function.

Recall that the absolute value of a number is its distance from zero on the number line. For example, $|-2| = 2$ and $|2| = 2$. Both -2 and 2 are two units from zero on the number line. This leads to a definition of absolute value expressed with variables.

Absolute Value
If $x \geq 0$, then $|x| = x$
If $x < 0$, then $|x| = -x$

The absolute value function is written

$$y = |x| \qquad \text{or} \qquad f(x) = |x|$$

The domain of this function is the set of all numbers, but the range is restricted to the set of positive numbers and zero.

Critical Thinking Why are functions that involve the measurement of time or distance usually restricted? Describe a function that has a restricted domain and range. In which quadrants does its graph belong?

EXAMPLE 1 Graphing the Absolute Value Function

Graph $y = |x|$.

SOLUTION

1. Make a table.

x	$\lvert x \rvert$
-2	2
-1	1
0	0
1	1
2	2

2. Draw a graph.

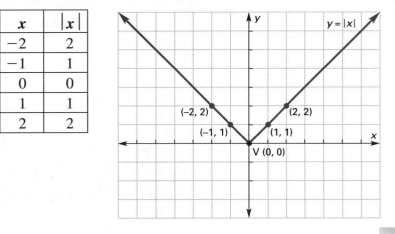

The graph of the absolute value function looks like a V. The point V(0,0) is called the **vertex** of the graph.

Critical Thinking   Compare the graph of the absolute value function $y = \lvert x \rvert$ with the graphs of $y = x$ and $y = -x$. How are the graphs alike and different?

A graphics calculator will make the following activities easier to do. Check your calculator to see how to graph the absolute value function. If you do not have access to a graphics calculator, make a table for each function and draw its graph.

ACTIVITY 1   Vertical Slides

1 Graph $y = \lvert x \rvert$.

2 Graph each function on the same coordinate axes.

a. $y = \lvert x \rvert + 3$ **b.** $y = \lvert x \rvert + 1$ **c.** $y = \lvert x \rvert - 1$

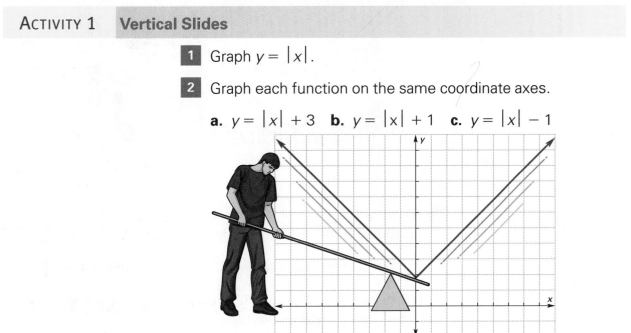

3 The function $y = |x|$ is called the **parent function** for absolute value. Each of the other absolute value graphs results from a movement of the parent graph. Explain how adding or subtracting a constant value can affect the graph of the parent function.

When the graph of a parent function slides up or down, the slide is called a **vertical translation**.

> **Vertical Translation**
> If $y = |x| + c$,
>
> then the vertex of $y = |x|$ is translated up or down by c units.

ACTIVITY 2   Horizontal Slides

1 Graph $y = |x|$.

2 Graph each function on the same coordinate axes.

      a. $y = |x - 3|$ **b.** $y = |x + 1|$ **c.** $y = |x - 1|$

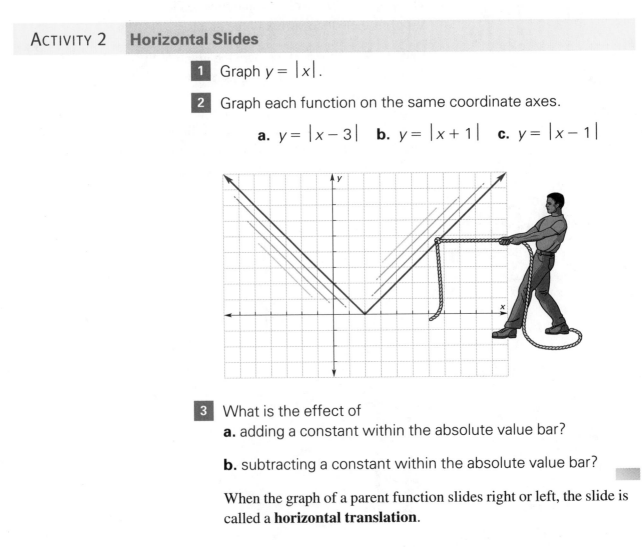

3 What is the effect of
a. adding a constant within the absolute value bar?

b. subtracting a constant within the absolute value bar?

When the graph of a parent function slides right or left, the slide is called a **horizontal translation**.

Horizontal Translation

If $y = |x + c|$,

then the vertex of $y = |x|$ is translated right or left by c units.

Ongoing Assessment

Describe the graph of the function $h(x) = |x + 2| - 5$.

ACTIVITY 3   Stretching and Shrinking

1 Graph each function on the same coordinate axes.

    a. $y = 2|x|$ **b.** $y = 3|x|$ **c.** $y = 5|x|$

2 Compare each function to $y = |x|$.

3 Graph each function on the same coordinate axis.

    a. $y = \frac{1}{2}|x|$ **b.** $y = \frac{1}{3}|x|$ **c.** $y = \frac{1}{5}|x|$

4 Compare each function to $y = |x|$.

5 What effect does the constant a in $y = a|x|$ have on the parent function when a is less than 1 and greater than zero?

When the parent function $y = |x|$ is multiplied by a constant, the constant is called the **scale factor** of the function.

What happens when each value of $y = |x|$ is multiplied by the scale factor 3? Each point of each arm of the V is moved up by a factor of 3. This has the effect of pulling the arms of the V together. In other words, the angle between the arms has shrunk. On the other hand, if the scale factor is $\frac{1}{2}$, each point on each arm of the V is moved down by a factor of $\frac{1}{2}$.

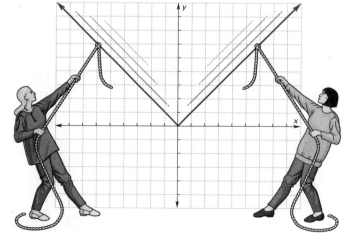

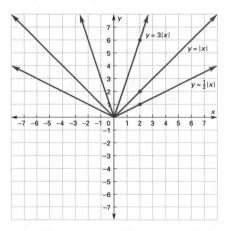

Scale Factor

If $y = a|x|$, then a is the scale factor of the function.

If a is greater than 1, the angle between the lines of the graph becomes smaller or shrinks.

If a is greater than zero, but less than 1, the angle becomes greater or stretches.

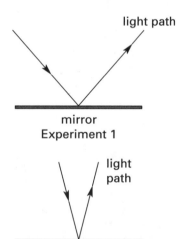

light path

mirror
Experiment 1

light path

mirror
Experiment 2

A technician is performing an experiment with a laser. He is aiming the beam at a mirror and checking its reflection.

Can the technician use an absolute value function to describe the path of the light?

Suppose the laser is moved to another position to perform another experiment. How can the absolute value function described in the first experiment represent the light path in the second experiment?

EXAMPLE 2 Combining Moves

Describe the graph of $f(x) = 4|x| - 1$.

SOLUTION

The scale factor 4 shrinks the graph of the parent function by 4 units. Since the constant 1 is subtracted, the graph is moved down 1 unit.

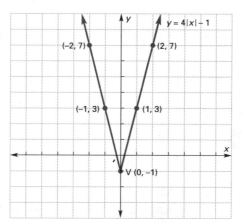

$y = 4|x| - 1$

$(-2, 7)$ $(2, 7)$

$(-1, 3)$ $(1, 3)$

V $(0, -1)$

ACTIVITY 4   A Reflection

1 Repeat Activity 3 with each scale factor changed to a negative number.

2 Explain what happens to the parent graph under these circumstances.

3 Create a definition for reflecting the absolute value function over the *x*-axis.

4 Check with your classmates. How do your definitions compare?

WORKPLACE COMMUNICATION

Use the absolute value function to model the struts at the current position and at the suggested new position. What scale factor can be used for the absolute value function for the new position of the struts? If the suggestion is implemented, what length of material is needed to fabricate the new struts?

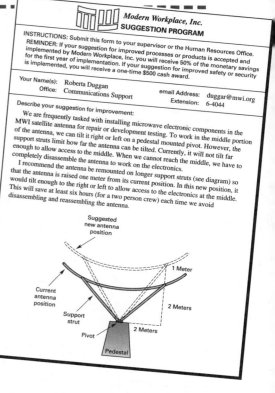

Modern Workplace, Inc.
SUGGESTION PROGRAM

INSTRUCTIONS: Submit this form to your supervisor or the Human Resources Office.
REMINDER: If your suggestion for improved processes or products is accepted and implemented by Modern Workplace, Inc. you will receive 50% of the monetary savings for the first year of implementation. If your suggestion for improved safety or security is implemented, you will receive a one-time $500 cash award.

Your Name(s): Roberta Duggan
Office: Communications Support

email Address: duggar@mwi.org
Extension: 6-4044

Describe your suggestion for improvement:

We are frequently tasked with installing microwave electronic components in the MWI satellite antenna for repair or development testing. To work in the middle portion of the antenna, we can tilt it right or left on a pedestal mounted pivot. However, the support struts limit how far the antenna can be tilted. Currently, it will not tilt far enough to allow access to the middle. When we cannot reach the middle, we have to completely disassemble the antenna to work on the electronics.

I recommend the antenna be remounted on longer support struts (see diagram) so that the antenna is raised one meter from its current position. In this new position, it would tilt enough to the right or left to allow access to the electronics at the middle. This will save at least six hours (for a two person crew) each time we avoid disassembling and reassembling the antenna.

LESSON ASSESSMENT

1 Explain why $y = |x|$ is a function.

2 What happens to the vertex of $y = |x|$ when 3 is added to or subtracted from the function? Give an example.

3 What happens to the shape of $y = |x|$ when the function is multiplied by 10? Give an example.

4 What happens to the shape of $y = |x|$ when it is multiplied by $\frac{1}{4}$? Give an example.

5 Explain what happens to the graph of $y = |x|$ when it is multiplied by a negative scale factor.

Practice and Problem Solving

Graph each function. Describe the graph in relation to the parent function $y = |x|$.

6. $y = |x| + 7$ **7.** $h(b) = 2|b|$

8. $r = \frac{3}{4}|t|$ **9.** $g(s) = -3|s| + 1$

10. $y = \frac{1}{2}|x| - 3$ **11.** $a = 4|b| + 6$

12. $f(x) = \frac{2}{3}|x + 4|$ **13.** $b = 3|c| + 2\frac{1}{2}$

14. $y = -\frac{5}{2}|x| - 2$

Mixed Review

15. The science club treasury had a balance of $350.45. Over the next four weeks, the following withdrawals (W) and deposits (D) are made. Find the final balance.

$114.36(W); $39.57 (W); $205.15(D); $185.59 (W)

Write each expression as a power of ten.
16. 10,000,000   **17.** 0.00001

Solve each equation.
18. $5a - 9 = 26$ **19.** 30% of $150 = c$ **20.** $\frac{-6}{9} = \frac{m}{15}$

21. 6% of $q = 96$ **22.** $\frac{3}{4}x - 7 = -19$ **23.** $7 = 0.5x - 3$

Write and solve an equation for each situation.

24. The yearly simple interest rate paid on an investment is 7.5%. If $15,000 is invested, what is the interest paid for two years?

25. One angle of a triangular support is three times as large as the smallest angle. The largest angle is 80° greater than the smallest angle. Find the measurement of each angle.

Find the slope and y-intercept.

26. $3y - 9 = 5x$ **27.** $3x + 2y = 12$

LESSON 6.3 THE QUADRATIC FUNCTION

Area of a Square

Is the equation $y = x^2$ a function? To find out, first make a table. Then draw a smooth curve connecting the points.

x	x^2
-2	4
-1	1
0	0
1	1
2	4

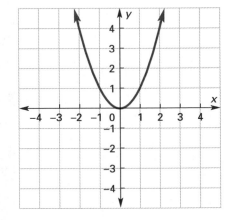

Note that for each input value in the domain, there is exactly one output value in the range. The equation $y = x^2$ is a function. This function models the area of a square. Replace the dependent variable (y) with the symbol A to represent area. Replace the independent variable (x) with the symbol s to represent the length of a side. The equation for the area of a square is

$$A = s^2$$

Critical Thinking What, if any, are the restrictions on the domain and range of the function $A = s^2$?

The Parabola

An equation written in the form $y = x^2$ or $f(x) = x^2$ is called a **quadratic equation** or **quadratic function**. A quadratic equation or function has exactly one term in the expression that is raised to the second power. The area formula is an example of a quadratic function. The graph of a quadratic function is a curve called a **parabola**. The point V(0,0) is the **vertex** of the parabola. Many real world objects are described by parabolas.

Quadratic functions can model the relationship between the distance (d) and time (t) for a free-falling object.

EXAMPLE 1 Quadratic Functions

How far will a bungee jumper free-fall in 5 seconds? What are the domain and range for the function?

SOLUTION

The function is $d = 16t^2$, where d is in feet and t is in seconds. Evaluate the function when t is 5 seconds. Thus, the distance is $d = 16(5^2)$ or 400 feet. The domain is measured in time and the range in distance. Thus, the domain and the range are both restricted to numbers greater than or equal to zero.

Ongoing Assessment

Graph the function $y = x^2$. Describe the graph.

Critical Thinking Compare the graph of $y = x^2$ with $y = |x|$. How are they alike and different?

ACTIVITY 1 Vertical Translations

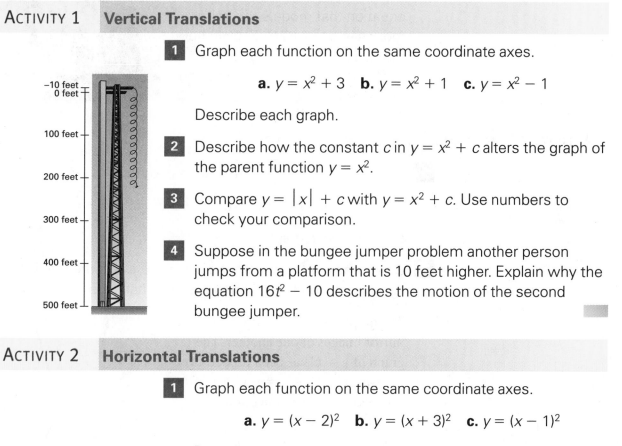

1 Graph each function on the same coordinate axes.

a. $y = x^2 + 3$ **b.** $y = x^2 + 1$ **c.** $y = x^2 - 1$

Describe each graph.

2 Describe how the constant c in $y = x^2 + c$ alters the graph of the parent function $y = x^2$.

3 Compare $y = |x| + c$ with $y = x^2 + c$. Use numbers to check your comparison.

4 Suppose in the bungee jumper problem another person jumps from a platform that is 10 feet higher. Explain why the equation $16t^2 - 10$ describes the motion of the second bungee jumper.

ACTIVITY 2 Horizontal Translations

1 Graph each function on the same coordinate axes.

a. $y = (x - 2)^2$ **b.** $y = (x + 3)^2$ **c.** $y = (x - 1)^2$

Describe each graph.

2 How does the constant c in $y = (x + c)^2$ alter the graph of the parent function $y = x^2$?

3 How are $y = |x + c|$ and $y = (x + c)^2$ alike?

ACTIVITY 3   Stretching and Shrinking a Parabola

1 Graph each function on the same coordinate axes.

   a. $y = 3x^2$ **b.** $y = 2x^2$ **c.** $y = \frac{1}{2}x^2$

Describe each graph.

2 Describe how the constant a in $y = ax^2$ alters the graph of the parent function $y = x^2$ when a is greater than 1. What happens when a is greater than zero, but less that 1?

3 The moon's gravitational force is only about $\frac{1}{6}$ that of the Earth. If a ball is dropped on the moon, the equation that models the ball's motion is $d = 2.67t^2$. Explain how the constants in the equations $16t^2$ and $2.67t^2$ affect the motion of the ball.

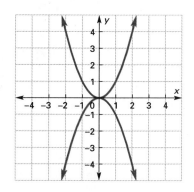

ACTIVITY 4   Reflecting a Parabola

1 Graph each function on the same coordinate axes.

   a. $y = -3x^2$ **b.** $y = -2x^2$ **c.** $y = -\frac{1}{2}x^2$

Describe each graph.

2 Describe how the constant a in $y = ax^2$ alters the graph of the parent function $y = x^2$ when a is less than zero.

The graphs of $y = x^2$ and $y = -x^2$ are mirror images of one another. The graph of $y = x^2$ has been flipped over the x-axis to graph $y = -x^2$. This flip is called a **vertical reflection**.

Vertical Reflection

The graph of $y = -x^2$ is called the vertical reflection of $y = x^2$ and appears as the mirror image of $y = x^2$ across the x-axis.

EXAMPLE 2 Moving a Parabola

Graph the function $y = -2x^2 + 1$.

SOLUTION A Without a Graphics Calculator

Graph $y = x^2$.

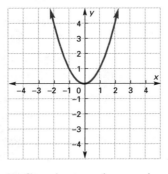

Vertical stretch by the scale factor 2.

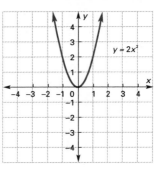

Reflect it over the x-axis.

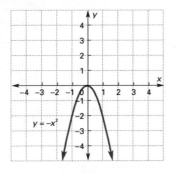

Move the vertex up 1 unit.

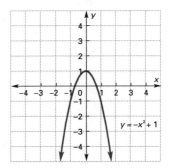

SOLUTION B With a Graphics Calculator

Set an appropriate Range.

Input the function.

Press the (GRAPH) key.

Critical Thinking   Explain how to move the graph of the parent function $y = |x|$ to graph $y = -\frac{1}{2}|x| - 3$.

EXAMPLE 3 Braking Distance

A driver education course teaches that the higher the speed of a car, the greater the distance needed to stop. You might think that if you go twice as fast, you need twice the distance to stop. Is this true? Let b represent the braking distance in feet. Let v represent the speed of the car in miles per hour when you start to brake. The function

$$b = 0.037\, v^2$$

is the estimated braking distance of a car on dry concrete. Make a table of car speed and braking distance. Examine the pattern in the table. If a car moves twice as fast, does it need twice the distance to stop?

SOLUTION

The function is a quadratic. The domain and range are restricted to numbers greater than zero. The scale factor (0.037) will horizontally stretch the parent function. Thus, the faster you drive, the more distance you need to brake. You can illustrate this by graphing the ordered pairs formed by speeds of 20, 30, 40, 50, and 60 miles per hour.

Speed (mph)	Breaking Distance (ft)
20	15
30	33
40	59
50	93
60	133

The braking distances are rounded to the nearest whole number to simplify the graph. You can see that braking distance does not double as you double your speed.

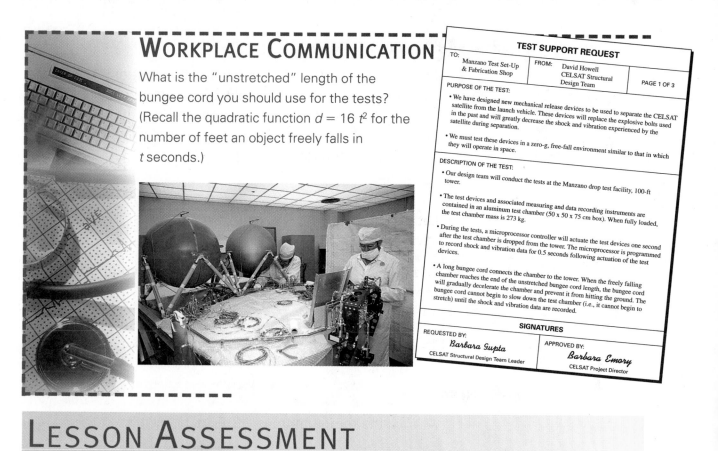

WORKPLACE COMMUNICATION

What is the "unstretched" length of the bungee cord you should use for the tests? (Recall the quadratic function $d = 16\,t^2$ for the number of feet an object freely falls in t seconds.)

TEST SUPPORT REQUEST

| TO: Manzano Test Set-Up & Fabrication Shop | FROM: David Howell CELSAT Structural Design Team | PAGE 1 OF 3 |

PURPOSE OF THE TEST:

• We have designed new mechanical release devices to be used to separate the CELSAT satellite from the launch vehicle. These devices will replace the explosive bolts used in the past and will greatly decrease the shock and vibration experienced by the satellite during separation.

• We must test these devices in a zero-g, free-fall environment similar to that in which they will operate in space.

DESCRIPTION OF THE TEST:

• Our design team will conduct the tests at the Manzano drop test facility, 100-ft tower.

• The test devices and associated measuring and data recording instruments are contained in an aluminum test chamber (50 x 50 x 75 cm box). When fully loaded, the test chamber mass is 273 kg.

• During the tests, a microprocessor controller will actuate the test devices one second after the test chamber is dropped from the tower. The microprocessor is programmed to record shock and vibration data for 0.5 seconds following actuation of the test devices.

• A long bungee cord connects the chamber to the tower. When the freely falling chamber reaches the end of the unstretched bungee cord length, the bungee cord will gradually decelerate the chamber and prevent it from hitting the ground. The bungee cord cannot begin to slow down the test chamber (i.e., it cannot begin to stretch) until the shock and vibration data are recorded.

SIGNATURES

REQUESTED BY:
Barbara Gupta
CELSAT Structural Design Team Leader

APPROVED BY:
Barbara Emory
CELSAT Project Director

LESSON ASSESSMENT

Think and Discuss

1 Explain why $y = x^2$ is a function.

2 What happens to the vertex of the parent quadratic function when you add or subtract a positive constant?

3 What happens to the vertex of the parent quadratic function

a. when you add a positive constant (c) to the independent variable?

b. when you subtract a positive constant (c) from the independent variable?

4 What happens when you multiply the parent quadratic by a constant greater than 1?

5 What happens when you multiply the parent quadratic by a constant greater than zero but less than 1?

6 Explain what happens when you multiply the parent by a number less than zero.

The power, in watts, of a microwave transmitter is given by the function $P = I^2 R$. I is the electrical current passing through the transmitter in amperes, and R is the resistance of the transmitter in ohms. Find the power of a 4-ohm transmitter when the current is

7. 5 amps   **8.** 10 amps   **9.** 20 amps   **10.** 30 amps

Let $g(x) = -3(x + 2) - 4$. Find $g(x)$ when x is
11. -2 **12.** 4 **13.** 0 **14.** $-\dfrac{1}{3}$

Graph each function. Describe the graph in comparison to the parent graph $y = x^2$.
15. $y = x^2 - 1$ **16.** $g(a) = -a^2 + 3$ **17.** $d = 3(t + 2)^2 + 1$

18. $h(n) = -2n^2 - 3$ **19.** $A = \pi r^2$ **20.** $f(x) = \dfrac{1}{2}x^2 - 5$

At 4 AM the temperature is $-8°$F. At 11 AM it is $-12°$F.

21. At what time is the temperature greater?

22. What is the difference in temperature between the two readings?

23. During the next three hours, the temperature increases 5 degrees. What is the new temperature?

24. Convert the new temperature to Celsius.

Write an equation and solve.
25. What percent of 80 is 15?   **26.** 35 is what percent of 10?

Solve each equation.
27. $4a - 6 = 9$ **28.** $3(2x - 3) = -5$

Graph each equation.
29. $2y + x = 4$ **30.** $3y - 5x = 8$

You have already used the formula $A = s^2$ to find the area (A) of a square when the length of a side (s) is known. If you know the area, the following formula provides a means for finding the length of one side:

$$s = \sqrt{A}$$

This formula is read "s is equal to the square root of A." The equation $s = \sqrt{A}$ is an example of the general equation $y = \sqrt{x}$.

What is the meaning of $y = \sqrt{x}$? Since $2^2 = 4$, 2 is a square root of 4. Thus, $\sqrt{4} = 2$. However, $(-2)^2 = 4$. Thus, $\sqrt{4} = -2$. Every positive number except zero has two square roots. Since $\sqrt{4} = 2$ or -2 (written as $\sqrt{4} = \pm 2$), one input number (4) has two output numbers (2 or -2). Thus, the equation $y = \sqrt{x}$ does *not* represent a function. To find out more about this equation, you can make a table and draw a graph.

x	$\sqrt{x}$
0	0
1	± 1
4	± 2
9	± 3

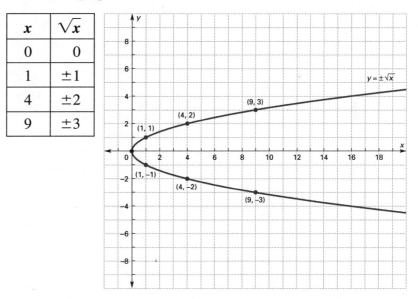

You can tell by the table that "finding the square root of a number" is not a function. However, if you look at the graph, you can see that the top curve represents positive square roots and the bottom curve represents negative square roots. Furthermore, the curves are vertical reflections of each other across the x-axis. If you restrict the range to positive values and zero, $y = \sqrt{x}$ becomes a function. This function is the principal square root function. From now on *the square root sign will mean positive square root.* If the negative square root is needed, you write $-\sqrt{}$.

Square Root
If $x \geq 0$, $\sqrt{x^2} = x$.

Solving Equations

We have now defined $y = \sqrt{x}$ as a function.

Critical Thinking   Explain how to find the domain and range of the function $f(x) = \sqrt{x}$.

The equation $y = \sqrt{4}$ has exactly one solution. The solution is 2. However, the equation $x^2 = 4$ has two solutions, 2 and -2.

Solving $x^2 = c$
If $x^2 = c$ and $c \geq 0$,

1. $x = \pm\sqrt{c}$.

2. The solutions are $\sqrt{c}$ and $-\sqrt{c}$.

EXAMPLE 1 Falling Objects

A ball falls from the top of a 250-foot building. The equation that models the distance (d) it falls in time (t) is $d = 16t^2$. How long does it take the ball to hit the ground?

SOLUTION

Let d equal 250 and solve the equation for t.

$16t^2 = 250$ Given

$t^2 \approx 15.6$ Division Property (calculator)

$t \approx \pm\sqrt{15.6}$ Solving $x^2 = c$

Since t represents time, only the positive square root makes sense. Using your calculator, you will find that it takes about 3.9 seconds for the object to reach the ground.

The Allied Storage Company has a new storeroom that provides 6000 square feet of floor space. If the storeroom has a square shape, find the approximate length of each side.

The Pythagorean Theorem

A **right triangle** contains exactly one right angle. Many early civilizations found an interesting relationship between the three sides of a right triangle.

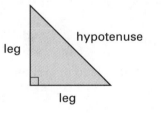

If you place squares on each of the sides of a right triangle and measure the area of each square, you will find that

| the area of the square on the hypotenuse | = | the sum of the areas of the squares on the other two sides. |

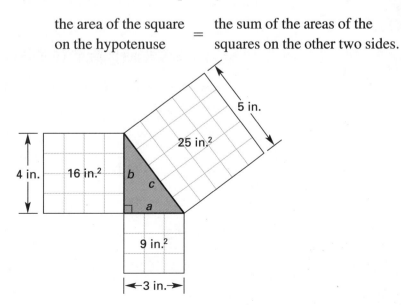

This relationship is called the **Pythagorean Theorem** or Pythagorean Formula. (To find out why, read the Cultural Connection at the end of this Lesson.) The Pythagorean Theorem is usually written in the form

$$c^2 = a^2 + b^2$$

In this form, c is the length of the hypotenuse, and a and b are the lengths of the legs. Applying the Pythagorean Theorem to the right triangle in the illustration gives

$$c^2 = a^2 + b^2$$

$$5^2 = 3^2 + 2^2$$

$$25 = 9 + 16$$

EXAMPLE 2 Finding Distance

An escalator takes you from the second floor to the third floor of a building. During the ride, the escalator rises a distance of 30 feet and moves horizontally a distance of 27 feet. What is the length of the escalator?

SOLUTION

Use a right triangle to model the situation. Let c represent the length of the escalator.

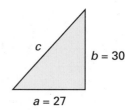

Use the Pythagorean Theorem. Substitute 27 for a, 30 for b, and solve for c.

$c^2 = a^2 + b^2$	Given
$c^2 = (27)^2 + (30)^2$	Substitution
$c^2 = 729 + 900$	Simplify
$c^2 = 1629$	Simplify
$c = \pm \sqrt{1629} \approx \pm 40.36$	Solve for c

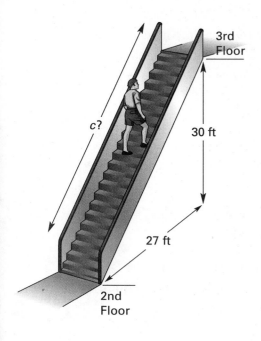

3rd Floor

c?

30 ft

27 ft

2nd Floor

Since only the positive square root makes sense, the escalator is about 40.36 feet in length.

You can write the Pythagorean Theorem in terms of a^2 and b^2.

Thus,

$$a^2 = c^2 - b^2 \qquad \text{and} \qquad b^2 = c^2 - a^2$$

EXAMPLE 3 Finding a Leg of a Right Triangle

Suppose the hypotenuse of a right triangle is 20 feet long. If one of the legs is 16 feet long, what is the length of the other leg?

SOLUTION

Use the formula $a^2 = c^2 - b^2$. Replace c with 20 and b with 16.

$$a^2 = (20)^2 - (16)^2$$

Use your calculator to find the length is 12 feet. Check your answer to be sure.

CULTURAL CONNECTION

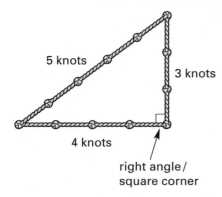

5 knots

3 knots

4 knots

right angle/
square corner

12 knots

The Pythagorean Theorem, in one form or another, was known by the ancient Chinese, Babylonians, and Egyptians. The Babylonians and Egyptians probably used the rule to survey lands after the flooding of their rivers. People also used the rule along with a knotted rope to mark off a square corner.

The earliest known informal proof of the rule is a diagram found in a Chinese manuscript called the *Chiu Chang.* However, a formal proof was not developed until many years later. During the sixth century BCE, a secret society called the Pythagoreans gathered to study philosophy, music, and mathematics. It is not known if the members of the society actually devised a formal proof, but the formula is named in honor of Pythagoras, the founder of this secret group.

LESSON ASSESSMENT

1 Explain why every positive number can have two square roots.

2 What is meant by the principal square root?

3 Why is $f(x) = \sqrt{x}$ considered a function?

4 Explain how to find the solution to $3x^2 = 48$.

5 Explain how to use the Pythagorean Theorem to find the length of the hypotenuse of a right triangle.

Practice and Problem Solving

Use your calculator to find each square root. If necessary, round to the nearest hundredth.

6. $\sqrt{225}$ **7.** $-\sqrt{2500}$ **8.** $-\sqrt{196}$ **9.** $\sqrt{615}$

Solve each equation. If the answer is not an integer, round it to the nearest hundredth.

10. $5a^2 = 500$ **11.** $2d^2 = 48$ **12.** $3x^2 - 8 = 100$

Use the Pythagorean Theorem to find the missing dimension of the triangle. If the answer is not an integer, round it to the nearest hundredth.

	side A	side B	hypotenuse
13.	3 meters	4 meters	?
14.	5 inches	7 inches	?
15.	?	12 feet	13 feet
16.	11 feet	35 feet	?
17.	21 inches	?	230 inches
18.	2 kilometers	5 kilometers	?

19. You ride your bike north 8 miles. Then you turn west and ride 7 miles. To ride directly home, how far do you have to travel?

20. A ladder is leaning against a wall at a point 30 feet above the ground. The bottom of the ladder is 40 feet from the wall. What is the length of the ladder?

21. You need to bury a water pipe in your backyard. Use the dimensions in the drawing to find what length pipe to buy.

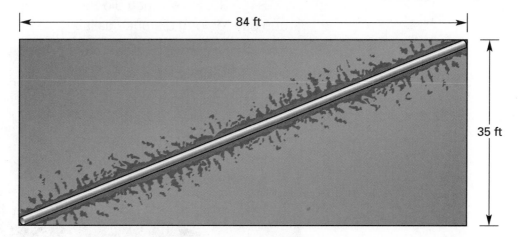

84 ft

35 ft

22. A baseball infield is shaped like a square. It is 90 feet between bases. How far is a throw from third base to first base?

Mixed Review

23. What is the value of $x(y + z)$ when x is -3, y is -5, and z is 8?

24. What is $f(-3)$ when $f(t)$ is $5t^2 - 3$?

25. What are the slope and y-intercept of $4x + y = -9$?

26. A manufacturer of lawn watering systems produces a sprinkler head that covers a circular area 12 feet in diameter. How many square feet of lawn should the manufacturer advertise the sprinkler head will cover?

27. A rectangular tank is 6 feet long, 8 feet wide, and 5 feet high. The tank is completely full of water. If one cubic foot is equivalent to about 7 gallons of water, about how many gallons of water will the tank hold?

Write and solve an equation for each situation.

28. Sharon and Jack are washing the windows in a large apartment building. Sharon has washed 36 more windows than Jack. If 148 windows have been washed so far, how many windows has Sharon washed?

29. The Top Notch Marketing Group spent three times as much on long distance calls as on local calls. The total phone bill is $415.36. How much did the Top Notch Marketing Group pay for long distance calls?

LESSON 6.5 EXPONENTIAL FUNCTIONS

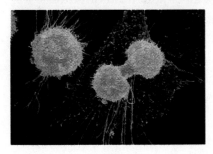

Plants and animals grow through a process called *cell division*. Life begins with one cell and progressively doubles so that 1 cell becomes 2 cells, 2 cells become 4 cells, 4 cells become 8 cells, and so on. This process gives rise to the following sequence:

$$1, 2, 4, 8, \ldots$$

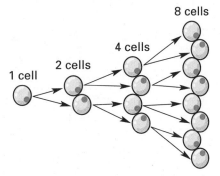

If you let y represent the number of cells present and x represent the number of cell divisions that have occurred, the sequence can be described by

$$y = 2^x$$

Does this equation describe a function?

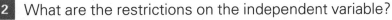

ACTIVITY 1   Cell Division

Use the equation $y = 2^x$ to complete this activity.

1 Let the independent variable be x. Let the dependent variable be y. Make a table with values of x from zero through 4. Remember that $2^0 = 1$.

2 What are the restrictions on the independent variable?

3 What are the restrictions on the dependent variable?

4 Draw coordinate axes. Plot the points from your table. Connect the points with a smooth curve.

5 Explain why the equation $y = 2^x$ describes a function.

Exponential Growth

The equation $y = 2^x$ is an example of an **exponential function**. You can recognize an exponential function by observing that the independent variable is an exponent. Any function in the form $y = a^x$ where a is greater than 1 models **exponential growth**. Compare the graph of $y = 2^x$ with the graph of $y = 4^x$. As x gets larger, the graph of $y = 4^x$ rises faster.

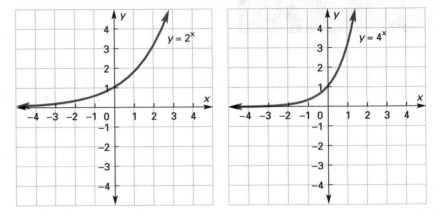

EXAMPLE 1 Increasing Bacteria

At the beginning of an experiment, a laboratory culture dish contains 500 bacteria. The number of bacteria increases by 50% each hour. After h hours, the number of bacteria (B) present is given by the formula:

$$B = 500(1.5)^h$$

1. What are the restrictions on the domain and range of the function?

2. Graph the function.

3. Estimate the number of bacteria in the culture dish after three-and-a-half hours.

SOLUTION

1. The domain includes zero and all the positive numbers. The range includes zero and all the positive numbers.

2. In this problem, the independent variable is h, and the dependent variable is B. Make a table using whole numbers as replacements for h. (Use the [Yˣ] key on your calculator.)

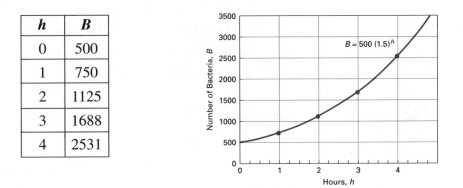

h	B
0	500
1	750
2	1125
3	1688
4	2531

The numbers representing B are rounded to the nearest whole number. There are no fractional bacteria.

3. To find the number of bacteria after 3.5 hours, substitute 3.5 for h, and solve for B.

$$B = 500(1.5)^{3.5}$$

The exponent 3.5 is between 3 and 4. The answer should lie between 1688 and 2531. If you use the exponent key on your calculator, the result is 2066.75697. This is rounded down to about 2066 bacteria after three-and-a-half hours. Use the graph to check this approximation. Does the answer make sense?

Ongoing Assessment

Use your calculator to estimate the number of bacteria in the culture after three-and-three-quarters hours. Use the graph to check your estimate.

Exponential Decay

Some carbon atoms are radioactive. The radioactive carbon atom, carbon-14, has a half-life of about 5700 years. The half-life is the time it takes for one-half the radioactive atoms to decay.

Suppose that 22,800 years ago, a tree died. At the time of its death the tree contained 50 grams of carbon-14. After one half-life of time (5700 years) passes, the remains of the tree contain half the original amount of carbon-14 (25 grams). After two half-lives (11,400 years), the tree's remains contain one fourth the original amount of carbon-14 (12.5 grams). After three half-lives, one-eighth remains, and so on.

The following sequence describes this carbon-14 decay process:

$$1, \frac{1}{2}, \frac{1}{4}, \frac{1}{8}, \dots$$

The equation $y = \left(\frac{1}{2}\right)^x$ models this sequence. The independent variable, x, represents the number of half-lives. Does this equation describe a function?

ACTIVITY 2   Radioactive Decay

Use the equation $y = \left(\frac{1}{2}\right)^x$ to complete this Activity.

1 Let the independent variable be x. Let the dependent variable be y. Make a table for the first four values of x. Remember that $\left(\frac{1}{2}\right)^0 = 1$.

2 What are the restrictions on the independent variable?

3 What are the restrictions on the dependent variable?

4 Plot the points in your table.

5 Connect the points with a smooth curve.

6 Explain why the equation $y = \left(\frac{1}{2}\right)^x$ describes a function.

Any function in the form $y = a^x$ where a is between zero and 1, is called an **exponential decay function**. Describe the effect on the graph $y = a^x$ as x increases in value.

Critical Thinking Suppose the decay function is multiplied by a constant. How is the graph of the parent function affected?

EXAMPLE 2   Population Change

The population of a city is 100,000. The population is declining by 5% each year. What is the population after 3 years?

SOLUTION

Each year the population is 95% of the year before. The decay function that models this situation is

$$y = 100{,}000(0.95)^x$$

Substitute 3 for x and solve for y. If you use your calculator, you will find that the population is 85,738 after three years.

Critical Thinking Explain how to use negative exponents to make the exponential decay function look like the exponential growth function.

LESSON ASSESSMENT

Think and Discuss

1 Explain how to tell if an exponential function is a growth function.

2 Explain how to tell if an exponential function is a decay function.

3 Explain why exponential growth and decay functions have restrictions on their domain and range.

4 Explain why increasing a number by a% is the same as taking $(100 + a)$% of the original number.

5 Explain why decreasing a number by b% is the same as taking $(100 - b)$% of the original number.

Practice and Problem Solving

Sketch the graph of each function. Describe the graph.

6. $y = 3^x$ **7.** $y = 5(2)^x$ **8.** $y = 8\left(\frac{1}{2}\right)^x$

9. $y = 2^x + 1$ **10.** $y = 2^x - 1$ **11.** $y = \left(\frac{1}{2}\right)^x + 1$

12. What effect does adding or subtracting a constant to the parent exponential function have on its graph? Give an example.

13. What can you do to the parent exponential function to reflect its graph over the x-axis? Give an example.

14. Describe the equation that models this sequence:

$$1, 3, 9, 27, \ldots$$

Write the function that models each situation. Solve each problem.

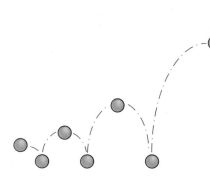

15. Suppose a ball bounces to one-half of its previous height on each bounce. Make a table of values for the first four bounces. Draw the graph.

16. The population of a city is 50,000. It is increasing at an annual rate of 3%. What is the population at the end of 4 years?

17. An investment compounds annually at 6% per year is kept for 5 years. Write the exponential function that models the amount in the investment after each year. If the initial amount invested was $1000, how much is in the account after 5 years?

18. Suppose you start with a penny and someone doubles the amount you get every day. How much will you receive on the tenth day?

19. The value of a car depreciates by 8% each year. If a new car is purchased for $20,000, what is the car worth after 4 years?

Mixed Review

20. A light-year (5.88×10^{12} miles) is the distance light travels in one year. The star Vega is 23 light-years from Earth. Write the distance (in miles) from Vega to Earth in scientific notation.

Write an equation and solve.

21. You make 6% commission on all of your sales. If your commission is $250, how much are your sales?

22. A spherical tank has a radius of 4 meters. Find the volume of the tank.

23. If $f(x) = 5x + 9$, what is $f(-3)$?

24. If $g(r) = r^2 - r$, what is $g(-5)$?

25. Find the equation of the line passing through the points $A(1,3)$ and $B(-2,5)$.

LESSON 6.6 THE RECIPROCAL FUNCTION

In Lesson 4.2 you used the reciprocal of a number to solve equations where the coefficient of the variable is a rational number.

For example,

$$\frac{3}{4}x = 12 \qquad \text{Given}$$

$$\frac{4}{3} \cdot \frac{3}{4}x = \frac{4}{3} \cdot 12 \qquad \text{Multiplication Property of Equality}$$

$$x = 16 \qquad \frac{4}{3} \cdot \frac{3}{4} = 1 \text{ and } \frac{4}{3} \cdot 12 = 16$$

The Multiplication Property of Equality is effective in solving this equation because $\frac{4}{3}$ is the reciprocal of $\frac{3}{4}$. Remember, the product of a number and its reciprocal is always one. Multiplying by the reciprocal of the coefficient of x isolates x on one side of the equation.

You can extend the idea of a reciprocal to functions. The equation that describes the **reciprocal function** is

$$y = \frac{1}{x} \ (x \text{ cannot equal zero})$$

Notice that in a reciprocal function the independent variable is in the denominator of the function.

ACTIVITY 1 | **Graphing Part of the Function**

1 Complete the table for positive values of x.

x	$\dfrac{1}{x}$
4	?
2	?
1	?
$\dfrac{1}{2}$	?
$\dfrac{1}{4}$	?

2 Plot the points and connect them with a smooth curve.

3 What are the restrictions on the domain and range?

4 What happens to y when x is very large?

5 What happens to *y* when *x* is very small?

6 What happens if you try to substitute zero for *x*?

ACTIVITY 2 **The Rest of the Story**

1 Repeat the same steps used in Activity 1. But this time, use the opposite of each value in the domain.

2 Put the two graphs together. Describe what you see. Why is this a representation of a function?

The graph of the function $y = \dfrac{1}{x}$ is called a *hyperbola*. The graph is made up of two branches, one in the first quadrant and the other in the third quadrant.

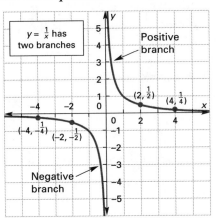

Critical Thinking Explain why the branches of the graph of a hyperbola do not touch each other. Explain why the branches of the graph get closer and closer to the axes but never touch.

Inverse Variation

The reciprocal function is an example of a relation called inverse variation. In **inverse variation**, as one variable gets larger, the other variable gets smaller.

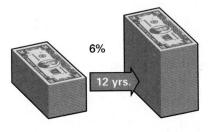

A formula called the Rule of 72 is an inverse variation used to approximate how fast money will double when it is invested at a given compound interest rate. The number of years (*y*) to double an investment is equal to 72 divided by the annual interest rate (*r*) expressed as a percent.

$$y = \frac{72}{r}$$

If you invest at a rate of interest of 6%, it will take about $\frac{72}{6}$, or 12 years, for your money to double.

How long will it take to double your money if the interest rate is 7%?

Critical Thinking   If $y = \dfrac{1}{x}$ is the parent reciprocal function, what effect does multiplying the function by a positive constant have on its graph?

For example, light intensity follows an *inverse square law*. The intensity (I) of light that falls on an object at a given distance (d) from a light source of power (P) is given by this formula:

$$I = \frac{P}{d^2}$$

If the distance is measured in meters and the power is measured in lumens, then the intensity is measured in lumens per square meter.

EXAMPLE Brightness of Light

A photographer takes pictures for a store catalog. The item being photographed is 3 meters from a 72 lumen light source. Where should the object be placed to receive twice as much light?

SOLUTION

To find the intensity when the object is 3 meters from the light source, substitute 72 for P, 3 for d, and solve for I. Since 3^2 is 9, and 72 divided by 9 is 8, the intensity is 8 lumens per square meter.

But you need to find the distance at which the luminosity is twice as much, or 16 lumens per square meter. Replace I by 16, and solve for d.

$16 = \dfrac{72}{d^2}$ Given

$16d^2 = 72$ Multiplication Property

$d^2 = 4.5$ Division Property

$d = \pm\sqrt{4.5}$ Solving $x^2 = c$

The distance is about 2.1 meters. Why is the negative solution not used?

The Pencil or Vertical-Line Test

You can use a technique called the "pencil test" or "vertical-line test" to distinguish between relations that are functions and those that are not. With this test, move a pencil (oriented as a vertical line) from left to right across the graphed curve of the relation. As the pencil moves, check to see how many points of the curve it intersects at any one time. If the pencil crosses the graph of the relation one point at a time, the relation is a function.

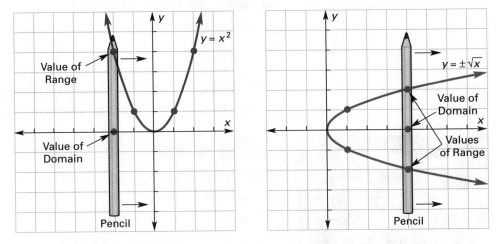

Critical Thinking   Explain how the vertical-line test shows that $y = x^2$ is a function but $y = \pm\sqrt{x}$ is not a function.

LESSON ASSESSMENT

Think and Discuss

1 How can you find the reciprocal of a number?

2 Why is there no reciprocal for zero?

3 How are hyperbolas and parabolas alike? How are they different? Draw a picture to illustrate your explanation.

4 Explain the meaning of inverse variation. How is inverse variation different from direct variation?

5 Explain how the vertical-line test can be used to test for a function. Give an example.

Graph each function. Use a graphics calculator if possible.

6. $y = 2\left(\dfrac{1}{x}\right)$ **7.** $y = 5\left(\dfrac{1}{x}\right)$ **8.** $y = 10\left(\dfrac{1}{x}\right)$

9. What effect does multiplying the parent reciprocal function by a number greater than 1 have on its graph?

Graph each function. Use a graphics calculator if possible.

10. $y = -2\left(\dfrac{1}{x}\right)$ **11.** $y = -5\left(\dfrac{1}{x}\right)$ **12.** $y = -10\left(\dfrac{1}{x}\right)$

13. What effect does multiplying the parent reciprocal function by a number less than -1 have on its graph?

Graph each function. Use a graphics calculator if possible.

14. $y = \dfrac{1}{2x}$ **15.** $y = \dfrac{1}{5x}$ **16.** $y = \dfrac{1}{10x}$

17. What effect does multiplying the parent reciprocal function by a number greater than zero but less than 1 have on its graph?

In computer programming, a convenient function for rounding is the *greatest integer function*. This function, written $y = \text{INT}(x)$, is defined as follows:

A number (x) written in decimal form is rounded down to the nearest integer (y). For example,

$$\text{INT}(3.1459) = 3 \qquad\qquad \text{INT}(-3.1459) = -4$$

$$\text{INT}\left(\dfrac{3}{4}\right) = 0 \qquad\qquad \text{INT}(-1) = -1$$

18. What is the domain of the INT function?

19. What is the range of the INT function?

20. Graph the INT function for the numbers in the domain that are between -5 and 5.

21. Describe the graph of the INT function.

22. How can you tell by looking at the graph that the INT function is a function?

23. How many 32-cent stamps can you buy for $1? Explain why this situation can be modeled by an INT function.

24. Explain how to evaluate the greatest integer function using your calculator.

Mixed Practice

When taking a 1-minute timed test, you stop at 55 seconds.

25. Are you over or under the allotted time?

26. Use an integer to show how much you are over or under.

27. What is the absolute value of your miss?

Suppose you are going to tile the floor shown here.

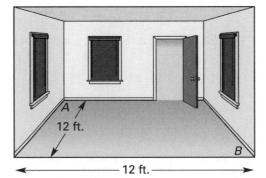

28. If you want to use tiles that are 9 inches by 9 inches, how many tiles will you need?

29. What is the distance from corner *A* to corner *B*?

Thomas needs to cut a 72-inch board into three pieces with the following specifications. One piece is five inches longer than the shortest piece and five inches shorter than the longest piece.

30. Write an equation to model the situation.

31. Find the length of the shortest piece.

32. Find the length of the longest piece.

MATH LAB

Equipment   Calculator
Pizza price list showing prices of at least four different pizzas

Problem Statement

A price list for pizza usually shows the prices for pizzas of different diameters. The price increases as the diameter increases. You will examine a price list for several different size pizzas with the same ingredients to determine the relationship of the price to the diameter.

Procedure

a Make a table with columns for diameter, area, and price.

b Record the diameter and price for each pizza in the table.

c Calculate the area of each size of pizza. Record the area in the table.

d Use your data to make a graph. Use the x-axis for the diameter. Use the y-axis for the price.

e Now make a second graph using the x-axis for the diameter and the y-axis for the area. Compare the two graphs you have drawn.

f Now make a third graph using the x-axis for area and the y-axis for price. What do you notice?

g Is the graph of pizza area versus price *linear?* If so, what is the slope of the graphed line? What are the units of this slope? What is the significance of this slope?

h Explain why pizza shops would lose money if their prices were proportional to the diameter of their pizza instead of the area.

Activity 2: Exponential Growth

Equipment   Calculator
                Checkerboard
                1 pound of kitty litter

Problem Statement

You will examine how organisms such as bacteria increase in number if they double at regular intervals. You will do this by recording how quantities of objects increase as they are doubled when you move from one checkerboard square to the next. The equation that describes this doubling process is

$$N = 2^{(x-1)}$$

N is the number of objects present, and x is the number of doubling intervals. You will use grains of cat litter as objects, and each square of a checkerboard will represent a doubling interval.

1	2	3	4	5	6	7	8
9	10	11	12	13	14	15	16
17	18	19	20	21	22	23	24
25	26	27	28	29	30	31	32
33	34	35	36	37	38	39	40
41	42	43	44	45	46	47	48
49	50	51	52	53	54	55	56
57	58	59	60	61	62	63	64

Procedure

a Place one grain of cat litter on square 1, two grains on square 2, four grains on square 4, and so on until you reach square 8.

b For squares 1 through 8, make a table showing the number of each square and the corresponding number of grains of cat litter on that square.

c Use your table to draw a graph. Use the x-axis for the number of the square and the y-axis for the number of grains. Is the graph linear? Why or why not?

d Use the formula $N = 2^{(x-1)}$ and your calculator to determine the number of grains on each of the next 8 squares. Continue your table.

e Make a graph of the new data. Is this graph similar to the one you have already drawn?

f Find the sum of the grains on the first 7 squares. How does the sum compare to the number of grains on square 8?

g Find the sum of the grains on the first 15 squares. How does the sum compare to the number of grains on square 16?

h What pattern do you see in the last two steps? Use this pattern to determine how many grains will be placed on square 64.

i Based on the pattern you have discovered, decide which is the better deal:

1. You can have ten million dollars today; or

2. You can start with $1 and double the amount every day for a month.

Activity 3: Height and Time of Bouncing Ball

Equipment   Calculator
Tape measure
Timer
Rubber ball

Problem Statement

When a ball is dropped to the floor from a given height, the time it takes to stop bouncing depends on the height from which it is dropped.

Procedure

a Locate an area where there is a hard, smooth floor. Drop the ball onto the floor from a height of 6 feet. Measure the length of time that the ball bounces. Record the drop height and time.

b Repeat the drop three more times from the same height. Average the four trials and record the average in a table.

c Repeat steps **a** and **b** using heights of 5 feet, 4 feet, 3 feet, 2 feet, and 1 foot.

d Graph your data. Use the x-axis for the average drop height and the y-axis for the average time the ball bounces.

e Is the relationship between drop height and time linear or nonlinear?

You have a 12-foot extension ladder. This ladder, when fully extended, is about 12 feet long. When retracted, it is about 6 feet long. Suppose you place the base of the ladder 2.5 feet away from a wall.

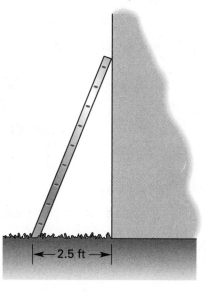

2.5 ft

 1 Use the Pythagorean formula to obtain an equation for the reaching height of the ladder for extensions from 6 feet to 12 feet.

2 Is this equation linear or nonlinear?

 3 Make a table of reaching heights for several values of the ladder extension between 6 feet and 12 feet. Draw a graph of the reaching heights versus ladder extensions. Does your graph confirm your answer to Exercise 2? Explain why or why not.

You can predict the balance of your savings account each month. This is easy with a scientific calculator using the formula

$$B = P(1 + i)^n$$

B is the balance after n months.

P is the starting balance (assuming you make no deposits or withdrawals).

i is the percent interest paid each month, as a decimal value.

n is the number of months.

4 Is the formula a linear or nonlinear equation? Explain your answer.

5 For an annual interest rate of 6%, the monthly interest rate would be $\frac{1}{12}$ of 6%, or 0.5%. For a starting balance of $1000, make a table showing the growth of this deposit. Select several different numbers of months between zero and 120 months (that is, a span of 10 years). (Hint: Be sure to use the decimal value for the interest rate.)

6 Add another column to your table, computing the balances for the same values of *n* for an annual interest rate of 7%.

7 Draw a graph of the balances you calculated. Connect the sets of points from each percentage rate with a smooth curve. Does the appearance of the graph agree with your answer to Exercise 4?

When you travel at a constant speed of 0.75 mile per minute (that is, 45 miles per hour), the total mileage *D* you travel during *t* minutes can be computed by the formula

$$D = 0.75t$$

On the other hand, if you accelerate and change your speed by 0.0167 mile per minute (that is, one mile per hour) each minute, the distance *D* you can travel in *t* minutes can be computed by the formula

$$D = \frac{1}{2}(0.0167)t^2$$

8 Make a table of values for *D* for each case above, (**a**) with a constant speed of 0.75 mile per minute, and (**b**) accelerating 0.0167 mile per minute each minute. Use values of *t* from 0 to 120 minutes.

9 Draw a graph of the values in your table and connect each set of points with a smooth line or curve. Label each line (or curve) as either "constant speed" or "constant acceleration."

10 Suppose you travel for 60 minutes with each of the methods described above. With which method would you be farther from your starting position? If you continued on for another 120 minutes, with which method would you be farther along?

11 Describe a method to find at what time *t* both methods would have you at the same distance from the starting point.

12 On a certain golf course, the traditional play is to take two strokes to go around a small lake. Each stroke requires a drive of about 250 yards. Determine how long a drive is needed to cross the lake in one stroke.

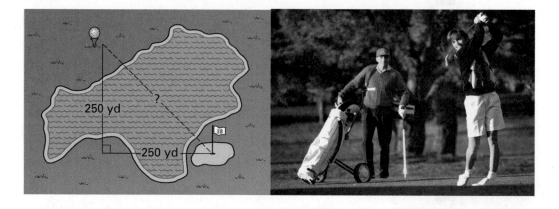

The water content of soil affects its weight. The results of certain soil tests produce the results shown below as a graph (*called a compaction curve*).

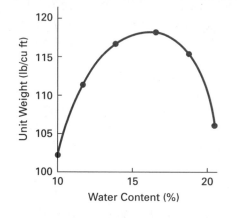

13 Examine the curve that is drawn through the test data. Is the trend in the data linear or nonlinear?

14 The most important feature of the curve is the maximum (that is, the high point of the peak). The maximum weight shown on the graph is called the *maximum dry density*, or the M.D.D. The percent water content at which this maximum occurs is called the *optimum moisture*, or O.M., of the soil. Approximately what are the M.D.D and the O.M. of this soil sample?

One of the basic tests performed on soils is the sieve test. The soil sample is allowed to settle through a stack of sieves that are of progressively finer and finer mesh. The amount of soil that remains in each sieve is measured. The percent of the soil sample that passes through each sieve is computed. The results from a typical soil sample are tabulated below.

Sieve Size	Soil Particle Size (mm)	% Finer
3 in.	76.2	92%
2 in.	50.8	90%
1 in.	25.4	89%
1/2 in.	12.7	82%
No. 4	4.76	73%
No. 10	2.00	66%
No. 20	0.84	56%
No. 40	0.42	41%
No. 60	0.25	25%
No. 100	0.15	12%
No. 200	0.07	4%

15 Draw a graph of the percentage finer versus the soil particle size for percentages less than 75%. Sketch in a curve that seems to fit the points on your graph.

16 Does the graph seem to be linear or nonlinear?

17 Using your sketched curve, estimate the particle sizes that correspond to percentages of 10% and 60%. (The particle size corresponding to 10% is called the *effective size*.)

18 Compute the ratio of the two particle sizes determined above: the size for 60% to the size for 10%. (This ratio is called the *uniformity coefficient*.)

Lumber mill operators estimate usable log volume from the formula below, which allows for saw kerf, slab, and edging deductions:

$$V = 0.0655L (1 - A) (D - S)^2$$

V is the usable volume of a log in board feet.

L is the length of the log in feet.

A is the decimal value of the percent deduction for saw kerf.

D is the log diameter in inches.

S is the slab and edging deductions in inches.

19 Rewrite the equation above for logs that are 32 feet long, cut with a 10% saw kerf, and have 3 inches of slab and edging deductions.

20 Make a table of the usable board feet of lumber that can be obtained under these conditions for logs with diameters of 10 in., 15 in., 20 in., 30 in., and 40 in. (Round your answers to the nearest 10 board feet.)

21 Draw a graph of this data.

22 What is the meaning of the graph as it approaches the x-axis? At what diameter does the usable number of board feet equal zero, based on this equation?

BUSINESS & MARKETING

The graph below shows the future value of a savings account earning 6% annual interest when $100 monthly deposits are made.

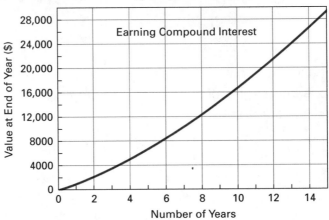

23 Does the graph show a linear relationship between the number of years and the savings account balance?

24 Compute the slope of the line that is formed by using the origin and the balance at 10 years as the endpoints.

25 Compare this slope with the slope of the line representing $100 per month deposits. Explain this comparison. Use the graph to estimate the amount of interest accumulated in the account after 10 years.

26 Suppose you determined the slope of the line formed by using the origin and the balance at 15 years. Would this slope be greater or less than the slope you calculated in Exercise 24? Explain your answer.

The graph below illustrates the changing labor costs in the American automotive industry for the years from 1960 through 1984.

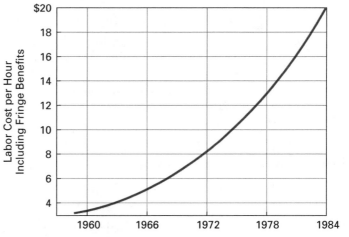

Labor Cost per Hour Including Fringe Benefits

27 Is this a graph of a linear trend or a nonlinear trend?

28 The shape of the curve most resembles which relationship?

   a. $y = x^2$ **b.** $y = \sqrt{x}$

   c. $y = \dfrac{1}{x}$ **d.** $y = |x|$

29 Compute the slope of the small section of the graph for the period of the 1960s, and then for the 1970s. Round your answer to the nearest $0.01 per hour. Did you get about the same values? Explain.

The "Rule of 72" is a simple way to observe the effect of compound interest. With this rule you can estimate the number of years needed to double an investment:

$$Y = \frac{72}{i}$$

Y is the years needed to double an investment.

i is the annual percentage at which the account earns interest, such as 8.5 for 8.5%.

30 The equation above most closely resembles which of the following nonlinear forms:

   a. $y = x^2$ **b.** $y = \sqrt{x}$

   c. $y = \dfrac{1}{x}$ **d.** $y = |x|$

31 Make a table of values for *Y* for interest rates of 3%, 4%, 5%, 6%, 7%, and 8%.

32 Draw a graph of the data in your table and connect the points with a smooth curve.

33 Is the slope of your curve steeper for smaller interest rates or for larger interest rates? Explain what this means.

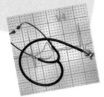

HEALTH OCCUPATIONS

When the count of red blood cells increases, there is more friction between the layers of cells. This friction determines the thickness, or viscosity, of the blood. A common measure of the blood—cell count—is the *hematocrit*, the percentage of the blood by volume that is cells. A graph of the effect of various hematocrit values is shown below. (Note that the viscosity of water is shown as equal to 1 on the graph, for comparison with whole blood and with plasma.)

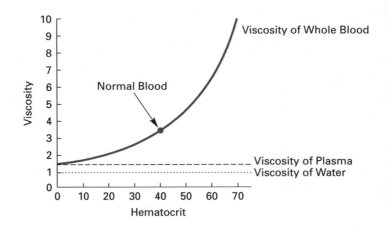

34 Which of the curves in the graph appears to be nonlinear?

35 Which of the following relationships does the nonlinear graph appear to resemble:

a. $y = x^2$ **b.** $y = \sqrt{x}$

c. $y = \dfrac{1}{x}$ **d.** $y = |x|$

36 Investigate the nonlinear curve for low values of hematocrit as well as for higher values. In which case does a small change in the hematocrit have more effect on the viscosity?

The vapor pressure of water is an important factor in inhalation therapy. A table of the vapor pressures of water (in units of millimeters of mercury) for various temperatures is shown below.

Temperature (°C)	Pressure (mm Hg)
0	4.6
10	9.1
20	17.4
30	31.5
40	54.9
50	92.0
60	148.9
70	233.3
80	354.9
90	525.5
100	760.0

37 Draw a graph of the vapor pressure of water for the temperatures listed in the table above.

38 Which of the following relationships does the nonlinear graph resemble:

a. $y = x^2$ **b.** $y = \sqrt{x}$

c. $y = \dfrac{1}{x}$ **d.** $y = |x|$

Doctors will frequently order IV (intravenous) fluids for patients needing additional body fluids. The hospital staff adjusts the IV drip rate so that the patient receives the prescribed amount of fluid in the prescribed amount of time. The drip rate can be determined using the formula

$$D = \frac{V}{4T}$$

D is the drip rate, in drops per minute.

V is the prescribed volume of fluid, in cubic centimeters (cc).

T is the prescribed amount of time for the fluids to be given to the patient, in hours.

39 A common volume for IV fluids is 1000 cc. Rewrite the equation that would be used to determine the drip rate for this volume, for various times T.

40 A burn victim may require such a volume to be given rapidly—perhaps in 3 to 4 hours. For a more stable patient, you may only wish to keep the vein open by a very slow flow—a total time span as long as 24 hours. Make a table of the drip rates indicated by your equation for times ranging from 3 hours to 24 hours.

41 Draw a graph of this equation using your table of data from above.

INDUSTRIAL TECHNOLOGY

Police officers frequently investigate the scenes of automobile accidents. By measuring the length of the skid marks, they can make a reasonable estimate of a car's speed before the skid began. Below is a table of average skid distances for an automobile with good tires on dry pavement.

Speed (mph)	Skid Distance (ft)
20	19
30	38
40	70
50	110
60	156
70	215
80	276

42 A graph of these data might make it easier to translate skid marks to estimated speeds. Choose a scale for your axes and make a graph of the data.

43 Draw a line or a curved line that seems to match the data. Should your line (or curved line) pass through the origin? Explain. Which of the following does the graph resemble?

a. $y = x^2$ **b.** $y = \sqrt{x}$ **c.** $y = \dfrac{1}{x}$

d. $y = |x|$ **e.** $y = x$

44 Suppose you measured a skid mark of 90 feet. Use your graph to estimate the speed the car was traveling at the time the skid began.

While adjusting the spark advance on a high-performance engine, you monitor the exhaust valve temperature. Below is a table of the data you collect.

Spark Advance (degrees)	Exhaust Valve Temperature (°F)
10	1350
20	1280
30	1255
40	1255
50	1300

45 Choose an appropriate scale and graph the data.

46 Is this graph linear or nonlinear? Draw a straight or curved line that seems to match the data.

47 If you want to set the spark advance to obtain the lowest temperature, what setting would you choose? Show this on your graph.

48 A surveyor is establishing lot lines for a housing development. To find the distance across a lake between two points A and B, she measures the two perpendicular distances shown. Determine the distance across the lake to the nearest foot.

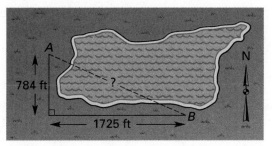

Below are two graphs that show a relative comparison of the speed versus torque (or turning power) for a direct-current (dc) motor and an alternating-current (ac) motor.

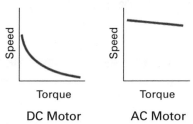

DC Motor   AC Motor

49 Identify whether each graph appears linear or nonlinear.

50 For the dc motor, describe what happens to the motor speed when the torque on the motor is very small.

51 How does this trend compare to the ac motor?

52 Select one of the graphs above that is nonlinear. Which of the following relationships does the graph appear to resemble?

a. $y = x^2$ **b.** $y = \sqrt{x}$

c. $y = \dfrac{1}{x}$ **d.** $y = |x|$

The thermistor, the RTD, and the thermocouple are three common electronic measuring devices. A graph of the measurement sensitivity (either voltage or resistance) is shown for each of these types of devices.

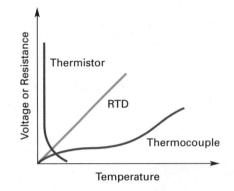

53 Identify which of the three curves are nonlinear and which are linear, or very nearly linear.

54 The thermistor curve is very different from the RTD and thermocouple curves. This curve most closely resembles which of these relationships:

a. $y = x^2$ **b.** $y = \sqrt{x}$

c. $y = \dfrac{1}{x}$ **d.** $y = |x|$

55 What advantage would there be in using a thermistor for temperature measurements?

A technical handbook for temperature measuring equipment shows the following graph of response times for various types of thermocouple constructions.

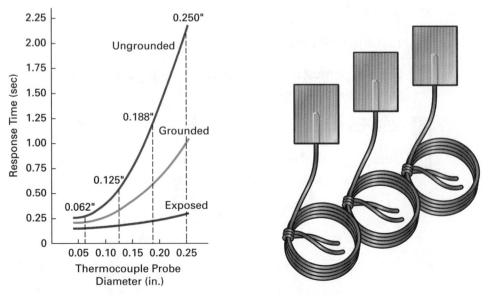

56 Identify the three types of thermocouple probes shown by the three curves in the graph.

57 Based on this graph, which type of thermocouple has a response time that is most sensitive to the diameter of the probe? Explain your answer.

58 If you needed a probe with a short response time (less than half a second) and a diameter of about 0.05 inch, would one type of probe perform significantly better than another? What if you were using probes with diameters of 0.25 inch?

59 Could you use a linear equation to approximate any of the regions of any of these curves? If so, describe those regions.

With the rising utility of superconductors, measurement of very low temperatures has become critical. The temperature scale of Fahrenheit is no longer convenient, and so the Kelvin temperature scale is used. In the Kelvin scale, 0°K is called *absolute zero* (the lowest possible temperature), and the freezing point of water is at 273°K. Some superconducting devices operate very close to 0°K. To measure these low temperatures requires special equipment. Below is a voltage response curve for one type of sensor.

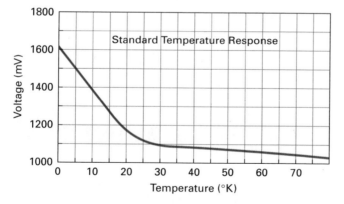

60 Identify two ranges of temperatures in the graph for which the response curve is very nearly linear.

61 For each of the ranges above, determine the approximate slope of the graph in the region. (Round to the nearest 1 mV per °K.)

62 What advantage is it for the slope to be steeper in the region close to 0°K?

The force of drag on a car is an important design consideration. By measuring how much the speed drops during a 10-second interval of coasting, an estimate of the drag force on the car can be found. The force of drag is related to the change in speed divided by the amount of time. The table lists data from such a test.

Speed(km/hr)	Drag Force (kg force)
20	29
30	30
40	34
50	38
60	43
70	50
80	59
90	68
100	75
110	86

63 Draw a graph of the data listed above and draw a smooth curve through the plotted points.

64 Does the relationship appear to be linear or nonlinear?

65 Based on your graph, can you predict what the drag force would be at a speed of zero? Explain.

The reference manuals for many automobiles specify how hard you should tighten bolts in various places on the car. This is done by specifying the "torque" to apply to the bolt. The torque is equal to the force multiplied by the length of the lever arm.

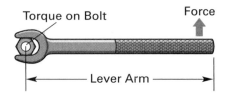

Torque on Bolt   Force

Lever Arm

You can compute the force needed when using wrenches of different lengths by rearranging the relationship to obtain the expression below.

Torque = Force • Lever Arm

Torque is measured in foot-pounds.

Lever Arm is measured in feet.

Force is measured in pounds.

66 Rewrite the equation to show *Force* as a function of *Lever Arm*. For a torque of 45 foot-pounds, make a table and graph of the dependent variable when the independent variable ranges in value between 0.5 and 3.

67 Is this relationship linear or nonlinear?

68 What is the difference in required force if you change from a small wrench that is 6 inches long to one that is 12 inches long? What difference is there when going from a wrench 2 feet long to a wrench 3 feet long?

Ethylene glycol is commonly used as an antifreeze. When added to the water in a car radiator it lowers the freezing point. The table below shows the freezing point for various concentrations of this antifreeze.

Concentration (%)	Freezing Point (°C)
0	0
9.2	−3.6
18.3	−7.9
28	−14
37.8	−22.3
47.8	−33.8
58.1	−49.3

69 Draw a graph of the freezing points for the various concentrations listed above. Connect the points with a smooth curve or line. Does your curve suggest a linear relationship or a nonlinear one?

70 Use your graph to estimate what concentration would provide protection down to − 40°C.

71 Someone suggests simply using pure antifreeze, with a concentration of 100%, reasoning that this would provide the lowest temperature protection. Can you use your graph to estimate what freezing point protection this would give you? Explain your answer. (Hint: The freezing point of 100% ethylene glycol is actually about −17°C.)

A pile driver is a large mass used as a hammer to drive pilings into soft earth to support a building. The hammer is lifted to a certain height and allowed to drop freely, striking the piling and driving it into the ground. The velocity with which the pile driver strikes the piling is related to the height from which it is dropped by the equation

$$v^2 = 64.4h$$

v is the final velocity of the hammer, in feet per second.

h is the distance that the hammer falls, in feet.

72 Solve the equation above for the variable *v*.

73 Draw a graph of your equation for various values of *h* from zero to 25 feet.

74 This is clearly a nonlinear equation because of the radical. Can any portion of your graph be approximated by a linear relationship? If so, what portion?

You are designing an electric circuit that might have many voltage cells (batteries). Each cell has its own resistance of 5 ohms in addition to the 16-ohm resistance of the rest of the circuit. Each cell has a voltage of 1.4 volts. Below is an equation that predicts the current for such a circuit, depending on the number of cells used.

$$i = \frac{nV}{R + nr}$$

i is the current in the circuit, in amperes.

n is the number of cells used.

V is the voltage supplied by each cell, in volts.

r is the internal resistance of each cell, in ohms.

R is the resistance of the rest of the circuit, in ohms.

75 Substitute the known values into the equation above to obtain an equation for *i* dependent on the number of cells used in the circuit.

76 Make a table of the current predicted by the equation for values of *n* from zero to 10 cells.

77 Draw a graph of the values in your table. Explain why it would not make sense to connect the points in this graph.

Rooftops of houses are often built in the shape of a pyramid with a square base. The *lateral area* of a pyramid is the total surface area of the pyramid minus the area of the base. The lateral area of this pyramid can be expressed by the formula

$$L = 2s \sqrt{0.25s^2 + h^2}$$

L is the lateral area of the pyramid-shaped surface.

s is the width of the base.

h is the height of the peak.

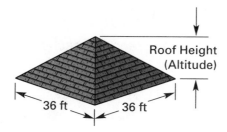

Roof Height
(Altitude)

36 ft   36 ft

78 For a house that has each base width equal to 36 feet, as shown in the figure, rewrite the formula to obtain an equation for *L* in terms of the height of the roof peak, *h*.

79 Suppose each bundle of shingles covers 25 square feet of roof. Rewrite the equation to allow you to compute the number of bundles, *n*, needed to cover any pyramid-shaped room with a square base.

80 Make a table of the number of bundles of shingles needed for each height.

81 Draw a graph of your data. Sketch a smooth curve through your points. Compare your graph to the graph of $y = \sqrt{x}$. How are they alike and how are they different?

Skills

If $f(x) = 3x - 5$, find

1. $f(2)$ **2.** $f(-4)$ **3.** $f\left(\dfrac{5}{3}\right)$

If $f(x) = -3x^2 + 2$, find

4. $f(3)$ **5.** $f(-1)$ **6.** $f\left(\dfrac{1}{2}\right)$

Graph on a coordinate system:

7. $y = 4|x|$ **8.** $y = |x| - 4$ **9.** $y = \dfrac{1}{4}|x|$

10. $y = x^2 + 4$ **11.** $y = -2x^2 - 1$ **12.** $y = \dfrac{1}{2}x^2 + 3$

13. $y = 2^x + 3$ **14.** $y = \left(\dfrac{1}{3}\right)^x$ **15.** $y = |x + 3|$

16. Give the domain and range of each function.

     a. $y = x^2$ **b.** $y = \sqrt{x}$

Applications

17. The length of a rectangular room is 12 feet, and the width is 16 feet. What is the length of the diagonal of the room?

18. A Navy plane is 2000 feet above and at a 5,000 foot horizontal distance from the end of an aircraft carrier deck. How far from the end of the deck is the plane?

19. A rectangular carton is 45 inches wide and 60 inches long. Can a skateboard that is 70 inches long fit in the box? Why or why not?

20. What is the width of a rectangle that has an area of 93 square feet and a length of 10 feet?

Math Lab

21. Which pizza is the better bargain?

   a. a10 inch pizza for $5.00
   b. a 12 inch pizza for $7.00

22. Which gives the larger ending balance?

   a. an initial amount of $100,000 that compounds at an annual interest rate of 10% for 30 years
   b. an initial amount of $0.01 that doubles every year for 30 years.

23. You drop a ball from a height of one foot and observe that it takes 11 seconds for the ball to stop bouncing. You drop the same ball from 2 feet and observe the ball bounces for 25 seconds. The next drop will be from 3 feet. How long will the ball bounce?

   a. 39 seconds
   b. less than 39 seconds
   c. more than 39 seconds

SKILLS REVIEW

SKILLS REVIEW

Congratulations. You have finished the first half of *CORD Algebra 1*. Before you begin Part B, stop and review some of the important skills presented in Part A. First, read the Skill Review; then, complete the exercises. This will give you a good foundation for a successful journey through the second half of *CORD Algebra 1*.

Skill 1 Adding and Subtracting Integers

When two integers have the same sign, add their absolute values. Use the common sign.

$$5 + 8 = 13 \qquad\qquad (-6) + (-7) = -13$$

When two integers have opposite signs, find the difference of their absolute values. Use the sign of the integer with the greater absolute value.

$$9 + (-4) = 5 \qquad\qquad -8 + 3 = -5$$

To subtract an integer, add its opposite.

$$12 - (-4) = 12 + 4 = 16 \qquad\qquad -7 - (-6) = -7 + 6 = -1$$
$$15 - 3 = 15 + (-3) = 12 \qquad\qquad -9 - 6 = -9 + (-6) = -15$$

Perform the indicated operations from left to right.

1. $-12 - 9$ **2.** $15 - 22$ **3.** $-14 + (-17)$

4. $3 + 8 - 6$ **5.** $-6 + 4 - (-3)$ **6.** $-5 - 7 - (-6)$

7. $23 - 48 + 35$ **8.** $6.1 - 4.8 - 4.2$ **9.** $-\$5.60 + \$3.33 - \$8.65$

Skill 2 Multiplying and Dividing Integers

When two integers have the same sign, their product or quotient is positive.

$$8 \times 9 = 72 \qquad\qquad -6 \times -7 = 42$$
$$36 \div 12 = 3 \qquad\qquad -48 \div -6 = 8$$

When two integers have opposite signs, their product or quotient is negative.

$$7 \times (-3) = -21 \qquad\qquad -15 \times 4 = -60$$
$$35 \div -5 = -7 \qquad\qquad -50 \div 2 = -25$$

Perform the indicated operations from left to right.

1. 17×-4 **2.** $-108 \div -6$ **3.** $-6.2 \div -3.1$

4. -2.5×-1.2 **5.** $39 \div -1.3$ **6.** -0.5×-3.24

7. $-8 \times -3 \times -2$ **8.** $-45 \div 9 \times -6$ **9.** $-100 \div -5 \div -5$

SKILLS REVIEW

Skill 3 Writing Large Numbers in Scientific Notation

To convert a large number from decimal notation to scientific notation:

1. Start from the left. Place a pointer to the right of the first nonzero digit.

16,500,000.
↑

2. Count the spaces as you move from the counter to the decimal point.

16,500,000.
↑ 7 spaces

3. Move the decimal point to the pointer and multiply the number by the power of ten equal to the spaces you moved.

1.65×10^7

Convert each amount to scientific notation.

1. 6,200,000 **2.** 425,000 **3.** 360,000,000

4. 19,256,000 **5.** 158,600,000 **6.** 8,000,000,000

7. The average distance from the Earth to the sun is thirty-three million miles.

8. The national debt is over four trillion dollars.

Skill 4 Writing Small Numbers in Scientific Notation

To convert a small number from decimal notation to scientific notation:

1. Start from the left. Place a pointer to the right of the first nonzero digit.

0.00035
↑

2. Count the spaces as you move from the decimal point to the pointer.

0.00035
↑ 4 spaces

3. Move the decimal point to the pointer and multiply by the power of $\frac{1}{10}$ equal to the spaces you moved.

3.5×10^{-4}

Convert each amount to scientific notation.

1. 0.016 **2.** 0.0004 **3.** 0.000007

4. 0.000315 **5.** 0.0000009 **6.** 0.000024

CORD Algebra

SKILLS REVIEW

Skill 5 Evaluating Expressions

To evaluate an expression, substitute the given value for the variable and perform the operations according to the order of operations rule.

- First, perform all operations within parentheses.
- Second, perform all operations involving exponents.
- Third, multiply and divide in order from left to right.
- Finally, add and subtract in order from left to right.

$$-3 \times (4 \div 2)^2 + 1 = -3 \times 2^2 + 1 = -3 \times 4 + 1 = -12 + 1 = -11$$

Evaluate each expression.

1. $10 \div (y + 1)$ when y is -6

2. $-5(b \div 3)$ when b is -6

3. $4 + 3(m - 1)$ when m is -2

4. $2 + (t + 4)^2 \div 5$ when t is 1

5. $5(3a + 2)^2 + 9$ when a is -3

6. $8 + -4(c - 5)^3 \div 2$ when c is 3

Skill 6 Using Formulas

A formula is a general rule expressed with letters written in the form of an equation.

The formula for the circumference of a circle:	$C = 2\pi r$
The formula for the area of a circle:	$A = \pi r^2$
The formula for the volume of a rectangular box:	$V = LWH$
The formula for the volume of a sphere:	$V = \frac{4}{3}\pi r^3$
The formula for simple interest:	$I = PRT$

Find

1. The circumference of a circle if the radius is 2.4 meters.

2. The area of a circle if the radius is 3 inches.

3. The volume of a rectangular box if the length is 2 feet, the width is 3 feet, and the height is 5 feet.

4. The volume of a sphere if the radius is 5 centimeters.

5. The simple interest if the principal is $500, the rate is 9.5%, and the time is 3 years.

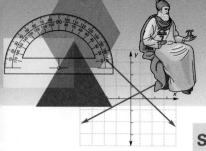

Skill 7 Solving Multiplication and Division Equations

If each side of an equation is multiplied by the same number or divided by the same nonzero number, the results are equal; that is, the equation is still balanced.

$$\text{Solve } 3b = 9$$
$$\frac{3b}{3} = \frac{9}{3}$$
$$b = 3$$

$$\text{Solve } \frac{c}{-8} = 2$$
$$-8 \cdot \frac{c}{-8} = -8 \cdot 2$$
$$c = -16$$

Solve each equation.

1. $5q = -30$

2. $\dfrac{k}{4} = -3$

3. $\dfrac{2}{3}m = 12$

4. $\dfrac{d}{7} = 1$

5. $1\dfrac{1}{2}r = -6$

6. $-8p = -40$

7. Solve $I = PRT$ for T.

8. Solve $\dfrac{d}{r} = t$ for d.

Skill 8 Solving Addition and Subtraction Equations

If the same number is added to or subtracted from each side of an equation, the results are equal; that is, the equation is still balanced.

$$\text{Solve } a - 6 = 9$$
$$a - 6 + 6 = 9 + 6$$
$$a = 15$$

$$\text{Solve } b + 4 = -6$$
$$b + 4 - 4 = -6 - 4$$
$$b = -10$$

Solve each equation.

1. $24 + q = 35$

2. $-15 + v = -25$

3. $9 = s + 15$

4. $p - 4.4 = 8.3$

5. $t - 8.1 = -5.3$

6. $-p + 3.4 = 1.7$

7. Solve $a + b = c$ for a.

8. Solve $a - b = c$ for a.

Skill 9 Solving Multi-Step Equations

Many equations require more than one operation to solve. Isolate the variable and simplify the constants on the other side of the equal side. Use the order of operations.

Solve $5n + 4 = 9$

$$5n + 4 - 4 = 9 - 4$$
$$5n = 5$$
$$\frac{5n}{5} = \frac{5}{5}$$
$$n = 1$$

Solve $-3(n - 4) = -6$

$$-3n + 12 = -6$$
$$-3n + 12 - 12 = -6 - 12$$
$$-3n = -18$$
$$n = 6$$

Solve each equation.

1. $3m - 4 = 11$

2. $6(b + 2) = 12$

3. $1\frac{1}{2}x + 5 = 179$

4. $-24 = -3(2m + 1)$

5. $\frac{3}{4}(x + 2) = -6$

6. $\frac{8c + 2}{2} = -15$

Skill 10 Solving Equations with Variables on Both Sides

The Distributive Property can be used to add and subtract like terms.

$$5a + 6a = (5 + 6)a = 11a \qquad 8b - 3b = (8 - 3)b = 5b$$

To solve an equation with variables on both sides, isolate the variables on the same side. Combine like terms and solve.

Solve $6m = 2m + 4$

$$6m - 2m = 2m - 2m + 4$$
$$4m = 4$$
$$m = 1$$

Solve each equation.

1. $5t + 2t = -21$

2. $-6r = 3r + 18$

3. $12 - 2y = 4y$

4. $3(n + 2) = 2n$

5. $b - 4 = 3(b + 2)$

6. $4(b - 5) = 2(b + 3)$

SKILLS REVIEW

Skill 11 Locating Points on a Coordinate System

An ordered pair of numbers (x,y) called coordinates are used to locate points on a coordinate system. The x-value gives the distance to the left or right of the origin. The y-value gives the distance above or below the origin. The point A (2,3) is 2 units to the right of the origin and 3 units above the origin.

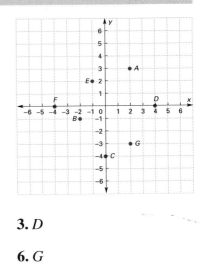

Name the coordinates of each point.

1. B **2.** C **3.** D

4. E **5.** F **6.** G

7. Explain how to locate S $(-2,4)$.   **8.** Explain how to locate T $(-5,-5)$.

Skill 12 Finding the Slope of a Line

The slope of a line is the ratio of its rise to its run. If A (a,b) and B (c,d) are two points on a line, the slope of the line is

$$\frac{b - d}{a - c} \quad \text{or} \quad \frac{d - b}{c - a}$$

The slope of line AB is

$$\frac{3 - 1}{4 - 2} = \frac{2}{2} = 1$$

Find the slope given the rise and the run.

1. rise 4; run -2 **2.** rise -1; run -5 **3.** rise 3, run -12

Find the slope of the line that passes through each pair of points.

4. A $(5,3)$ and B $(2,6)$ **5.** C $(-4,1)$ and D $(2,-6)$ **6.** E $(-3,-5)$ and F $(0,-3)$

7. Explain how to find the slope of the line passing through G $(-s,t)$ and H $(u,-v)$.

8. The slope of a line is $\frac{2}{3}$. One point on the line is X $(3,5)$. List two more points on the line.

SKILLS REVIEW

Skill 13 Finding the Intercepts of a Line

The graph of an equation written in slope-intercept form $y = mx + b$ is a line with slope m that intersects the y-axis at the point $(0,b)$. The value of b is called the y-intercept of the line. To find the slope and y-intercept for $4x - 5y = -20$, write the equation in slope intercept form

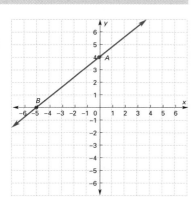

$$4x - 5y = -20$$

and solve for y:

$$y = \frac{4}{5}x + 4$$

The slope is $\frac{4}{5}$, and the y-intercept is 4. To find the x-intercept, substitute 0 for y and solve for x. If $4x - 5(0) = -20$, $x = -5$. The x-intercept is -5. On the graph, A is $(0,4)$ and B is $(-5,0)$.

Find the slope and y-intercept of each line.

1. $y = 5x + 8$ **2.** $2x + y = 6$ **3.** $3x = 5y - 10$

Find the x- and y-intercept of each line.

4. $3x - 4y = 12$ **5.** $2y + 3x = 10$ **6.** $x - 8y = -4$

Skill 14 Finding the Equation of a Line

The equation of a line can be determined if you know any two points on the line. First, find the slope of the line through A and B.

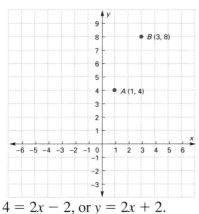

$$m = \frac{8 - 4}{3 - 1} = \frac{4}{2} = 2$$

Let $P(x,y)$ be another point on the line passing through A and B. Find the slope determined by P and either of the other two points.

$$m = \frac{y - 4}{x - 1}$$

Thus, $\frac{y - 4}{x - 1} = 2$, and the equation of the line is $y - 4 = 2x - 2$, or $y = 2x + 2$.

Find the equation of the line through the given points.

1. $(1,2)$ and $(2,4)$ **2.** $(0,4)$ and $(2,6)$ **3.** $(1,4)$ and $(6,12)$

4. $(-2,0)$ and $(3,-4)$ **5.** $(-1,3)$ and $(2,2)$ **6.** $(-3,-4)$ and $(3,4)$

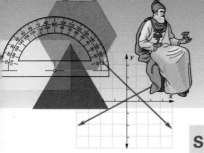

SKILLS REVIEW

Skill 15 Graphing the Absolute Value Function

The absolute value of a number is its distance from zero on a number line. Thus,

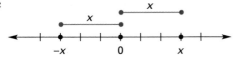

$$\text{If } x \geq 0, \ |x| = x$$

$$\text{If } x < 0, \ |x| = -x$$

The absolute value function is written $y = |x|$ or $f(x) = |x|$. The graph of the parent absolute value function looks like a V with its vertex at V(0,0).

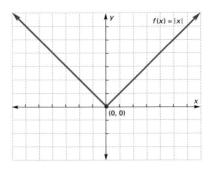

Describe the changes in the graph of the parent function for each equation.

1. $y = |x| + 1$ **2.** $y = |x + 1|$ **3.** $y = |x| - 3$

4. $y = 2|x|$ **5.** $y = -3|x|$ **6.** $y = |x - 1| + 2$

Skill 16 Graphing Quadratic Functions

The parent quadratic function is written

$$y = x^2 \text{ or } f(x) = x^2$$

The graph of $y = x^2$ is a parabola with its vertex at V(0,0).

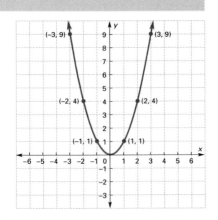

Describe the changes in the graph of the parent function for each equation.

1. $y = x^2 + 1$ **2.** $y = (x + 1)^2$ **3.** $y = x^2 - 3$

4. $y = 2x^2$ **5.** $y = -3x^2$ **6.** $y = (x - 1)^2 + 2$

CORD ALGEBRA 1

PART B 1

Mathematics in Context

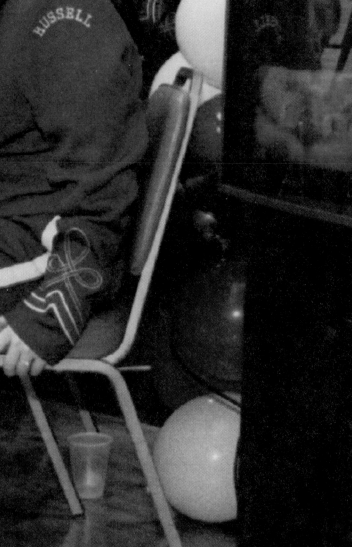

CHAPTER 7

WHY SHOULD I LEARN THIS?

Information will continue to grow as technology expands. In this chapter, you will explore how the principles of statistics and probability are used to work with large amounts of data. Learn how statistics and probability can help you manage information and expand your career options.

STATISTICS AND PROBABILITY

OBJECTIVES

1. Calculate the mean, median, and mode for a set of data.
2. Draw a histogram from a frequency distribution table.
3. Find the probability of a single event.
4. Find the probability of mutually exclusive events.
5. Find the probability of dependent and independent events.

Computer networking and the Internet have saturated the world with numerical data. People must have quick ways to analyze that data and extract the information it contains. **Statistics** is the branch of mathematics that helps organize and find meaning in numerical data.

For example, teachers collect a large amount of numerical data each time a test is given. They use statistics to analyze this data and provide insights into how individual students are performing with respect to the rest of the class.

Probability is useful in predicting outcomes of a future event. For example, on election night, news reporters often make predictions

on which of the candidates will win an office. To make these predictions, telephone surveys are conducted to collect data on how small, specially selected groups of people voted. These voter responses are analyzed and compared with past voting patterns to determine the probability of a certain candidate winning. From these probabilities, political analysts forecast winners in elections.

As you watch the video for this chapter, notice how people on the job use statistics and probability to solve problems.

LESSON 7.1 MEASURES OF CENTRAL TENDENCY

You are the production supervisor in a small manufacturing plant that produces handmade skateboards. To measure the productivity of three workers, you make a table to record how many skateboards each worker completes in 10 successive work days.

Production Data

					Day of the Week					
Worker	M	T	W	Th	F	M	T	W	Th	F
A	1	2	2	3	3	4	5	4	5	5
B	6	1	2	5	3	2	3	2	7	1
C	7	6	5	4	2	3	2	3	2	2

Now that you have organized the data in a table, what are some of the typical values you can find that give meaning to the data? For example, for each worker,

a. Which number occurs most often?

b. If a worker's production numbers are arranged in order, what is the middle number?

c. What is the average number of skateboards produced?

The answer to each of these questions is a statistic called a **measure of central tendency**.

> ### Measures of Central Tendency
> A measure of central tendency is a number used to represent all the numbers in a data set.

The mode, the median and the mean are the most commonly used measures of central tendency. Use the questions **a**, **b**, and **c** above to write a definition of mode, mean, and median. Give some examples where you would use each measure during a typical day.

The Mode

Look at the data in the table for Worker A. Which number occurs most often? You can see from the data set that 5 occurs three times. The other numbers occur two times or less.

The number that occurs most frequently in a set of numbers is called the **mode** of the set. The mode is 5 skateboards per day for Worker A.

Mode
The mode is the number that occurs most often in a set of numbers.

Ongoing Assessment

Use the data table to find the mode for Workers B and C.

EXAMPLE 1 Zero Mode or Two Modes

Find the mode for data set **a** and **b**.

a. 5   6.2   7   8   9.4

b. 4   4   5   6   6
        6   7   8   8   8
        9   10

SOLUTION

Data set **a** does not have a mode. No one number in this set occurs more frequently than any of the other numbers.

Data set **b** has two modes, 6 and 8. Both 6 and 8 occur most frequently (three times).

A set of numbers may have no mode, one mode, or more than one mode. You can usually find the mode by quickly looking at the data.

The mode is meaningful when there is a large number of values in the data set. The mode represents the *most typical* value in a large set of data.

Suppose you are responsible for inventory control at an office supply store. You may be very interested in what kind of paper people are using the most; that is, you might want to know the modes for paper weights and sizes.

The Median

Another measure of central tendency is the **median**.

To find the median of the data for Worker A, first list the data in order from greatest to least.

5   5   5   4   4   3   3   2   2   1

The median divides the data into two equal groups. In other words, there are as many numbers above the median point as there are below it. Since there are ten numbers in all (for ten days), look for the point that has five numbers above it and five numbers below it. Draw a line to show where this happens.

5   5   5   4   4 | 3   3   2   2   1

Where does the line fall? It falls halfway between 3 and 4. Therefore, the median production for Worker A is 3.5 skateboards.

Notice that the median represents the midpoint on a scale of measurement. The median does not necessarily equal an actual data value. Thus, the median for Worker A's daily production is 3.5 skateboards. Yet, Worker A never actually produced 3.5 skateboards on any of the ten days.

Median
The median is the middle number of a data set, when the data are arranged in numerical order. If there is no middle number, the median is the average of the two middle numbers.

Ongoing Assessment

Use the data table to find the median for Workers B and C.

The median is a useful way to show the central tendency for data that may have some extreme values. These values are not really typical of the rest of the data. These extreme values are called **outliers**. An outlier can result from unusual circumstances during the collection of data. For example, you are collecting data on the heights of students in your class, and one of your class members is the center of the basketball team. This height may be an outlier in your data set.

A person serving tables in a restaurant recorded these tips on 7 different days.

Day 1	Day 2	Day 3	Day 4	Day 5	Day 6	Day 7
$24	$15	$22	$80	$16	$21	$19

1 What is the mode for the data?

2 What is the median for the data?

Critical Thinking   What are outliers in the data set in Activity 1. What effect does an outlier such as $80 have on using the median to report the daily tips received?

The Mean

A third measure of central tendency (the one used most often) is called the **mean**, or average.

> **Mean**
>   The mean or average of a set of data is the sum of all the values of the data divided by the number of elements in the data set.

EXAMPLE 2 Finding a Mean

Find the mean of the tips from Activity 1.

Day 1	Day 2	Day 3	Day 4	Day 5	Day 6	Day 7
$24	$15	$22	$80	$16	$21	$19

SOLUTION

First, add all the tips. Use your calculator to help. The total is $197.

Second, divide the total ($197) by the number of days (7)

The mean value for the data is about $28.14 per day.

You can now find three measures of central tendency for a set of data. Each measure gives useful information about the data.

Critical Thinking   A friend is thinking about taking a job at the same restaurant and asks, "How much in tips do you usually get for a day?" For the data set given in Example 2, which measure of central tendency would you use to answer this question? Explain your answer.

ACTIVITY 2   Comparing Measures of Central Tendency

1 Complete the table using the numbers from the production data table.

Three Measures of Central Tendency			
	Mode	**Median**	**Mean**
Worker A	5	3.5	?
Worker B	?	?	?
Worker C	?	?	?

2 Circle the largest number in each column.

3 Which worker(s) has the highest mode?

4 Which worker(s) has the highest median?

5 Which worker(s) has the highest mean?

6 Suppose you must move two workers to another job. Which worker would you not move? Explain your answer.

Critical Thinking   Here are three examples where the mode, median or mean can be used to describe a situation. Which measure of central tendency would you use for each one, and why?

1. A supermarket manager wants to know how many boxes of cereal customers purchase most often.

2. A real estate agent wants to set the selling price of a three-bedroom house so that there are as many comparable houses above as below the price.

3. Students want to know the central tendency that tells the average of all their test scores in a certain grading period.

LESSON ASSESSMENT

Think and Discuss

1 Explain what is meant by "measures of central tendency."

2 Describe a situation in which the mode is the best measure of central tendency.

3 Describe a situation in which the median is the best measure of central tendency.

4 Describe a situation in which the mean is the best measure of central tendency.

5 Create a set of data using five numbers in which the mean, the median, and the mode are the same number.

Practice and Problem Solving

Mary's first semester pop quiz scores are:

5, 7, 7, 9, 10, 10, 10, 12, 14

Find the

6. Mode.   **7.** Median.   **8.** Mean.

Mary's test scores for the semester are:

85, 75, 70, 95, 55, 60, 100, 70, 85, 70, 65, 70, 85

Find the

9. Mode.   **10.** Median.   **11.** Mean.

The semester grade point averages for Mary and her friends are:

2.9, 2.3, 2.5, 2.4, 2.9, 2.7, 2.4, 2.4, 2.3, 2.4

Find the

12. Mode.   **13.** Median.   **14.** Mean.

Shelly Lynn records the number of cellular phone subscriptions she sells in different cities:

355, 262, 285, 237, 262, 258, 275

Find the

15. Mode.   **16.** Median.   **17.** Mean.

At the beginning of an aerobics class, the instructor recorded the weights of the participants.

Tom 58 kg, Joe 60 kg, Ann 60 kg, Jaren 49 kg, Nancy 63 kg, Kim 60 kg, Chris 67 kg, Megan 58 kg

18. Record these names and weights in a table from the least weight to the greatest weight.

19. Find the mode for this data.

20. Find the median for this data.

21. Find the mean or average weight for the class.

22. Suppose John, who weighs 65 kg, joins the class. Without computing, estimate how John's weight will affect the mean you found in Exercise 21.

23. Will John's weight affect the mode or median? Explain why or why not.

Mixed Review

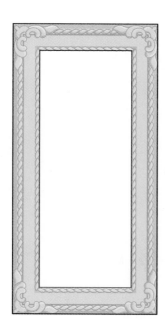

24. The speed of light is approximately 299,000,000 meters per second. Write this number in scientific notation.

25. Laura is going to loan her brother $5000 to use as a down payment on a new car. If her brother borrows the money at 8% simple interest, how much money will he owe Laura at the end of 6 months?

26. The formula $C = \dfrac{A}{46}$ gives the number of circuits required to wire a room of area A. Solve the formula for A.

Write and solve an equation for each situation.
27. After Thomas wrote a check for $249, he had $878.45 in his checking account. How much did Thomas have in his account before he wrote the check?

28. The length of a rectangular picture frame is twice its width. The perimeter of the frame is 108 inches. Find the length and width of the picture frame.

LESSON 7.2 FREQUENCY DISTRIBUTIONS

In Lesson 7.1, you used a data table to organize production data for three workers.

Skateboard Production Data										
	Day of the Week									
Worker	M	T	W	Th	F	M	T	W	Th	F
A	1	2	2	3	3	4	5	4	5	5
B	6	1	2	5	3	2	3	2	7	1
C	7	6	5	4	2	3	2	3	2	2

Organizing Data

A **frequency distribution** is another useful tool for organizing information. To make a frequency distribution table for Worker A, start by separating the data into five classes. Each class represents a different production level. For example, the class labeled 5 means a worker produces 5 skateboards in a day.

Next, tally (count) the number of times Worker A meets the production levels specified by the classes. For example, on the first Monday Worker A produced 1 skateboard. Therefore, place a tally mark in the class labeled 1. Continue through the rest of the day's list, placing tally marks in the appropriate classes. Finally, determine the class frequency by counting the number of tallies in each class. These steps will produce the following frequency distribution table.

Production Data for Worker A		
Classes	Tally	Frequency
5	///	3
4	//	2
3	//	2
2	//	2
1	/	1

To find the mean from a frequency distribution table, first multiply the class number by the frequency. If you let c represent the class and f represent the frequency, this product is

$$c \cdot f$$

Next, add each of the products. The Greek symbol Σ (sigma) means to add a set of numbers. Thus, the second step is to find

$$\Sigma(c \cdot f)$$

Finally, divide the sum by the total number n of tallies. The result is the mean and is usually denoted by $\overline{x}$. The formula for finding the mean of a frequency distribution is written as follows.

Formula for the Mean

Let c represent the class, f the frequency, and n the number of tallies. To calculate the mean $\overline{x}$ from a frequency distribution table, use this formula:

$$\overline{x} = \frac{\Sigma(c \cdot f)}{n}$$

EXAMPLE 1 Finding the Mean

Use the formula to find the mean of Worker A's production.

SOLUTION

Complete the table.

Classes	Tally	Frequency	$(c \cdot f)$
5	///	3	15
4	//	2	8
3	//	2	6
2	//	2	4
1	/	1	1
		10	34

$$n = 10 \qquad \Sigma(c \cdot f) = 34$$

Thus, $\overline{x} = \dfrac{34}{10} = 3.4$.

Critical Thinking How can you find the mode and the median by using the tally marks?

Make a frequency distribution table for Worker B and Worker C. Use the formula to find the mean for each worker. Use the tally marks to find the mode and median for each worker.

EXAMPLE 2 Using Data from a Frequency Table

Use the frequency distribution table in Example 1.

a. On how many days did Worker A make fewer than three skateboards?

b. On what percent of the days did Worker A make more than the mean of the distribution?

SOLUTION

Examine the frequency distribution table in the columns labeled "classes" and "frequency." The number in the frequency column is the number of days the worker equaled the production level in the classes column.

a. Worker A made 1 skateboard on one day and 2 skateboards on two days. Thus, Worker A made fewer than 3 skateboards three days.

b. The mean is 3.4. Worker A made 4 skateboards on two days and 5 skateboards on three days. Thus, Worker A made more than the mean on five out of ten days, or 50% of the days.

Histograms

A graph of the frequency distribution is a very useful means of displaying large sets of data. Consider the weights (in pounds) of thirty football players.

286, 234, 211, 268, 227,
273, 276, 250, 184, 230,
228, 202, 260, 205, 193,
250, 248, 234, 234, 197,
246, 224, 218, 235, 235,
253, 219, 241, 226, 246

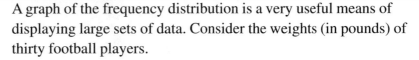

1 Identify the largest and smallest numbers in the group, 286 and 184. The difference between these numbers is the **range** of the data. What is the range for this data?

2 Group the data into weight intervals that each contain ten pounds. This is how the table begins:

Weight	Tally	Frequency
280–289	/	1
270–279	//	2

3 Complete the table on your paper.

4 Draw a positive *x*-axis and a positive *y*-axis.

5 Label the *x*-axis as "Weights of Players". The values on the *x*-axis begin with the weight interval 180–189 and end with 280–289.

6 Label the *y*-axis as "Frequencies." The frequencies begin with zero and go up to 6.

7 For each weight interval, draw a rectangle to show the frequency. Since each weight interval has equal value, the widths of the rectangles are the same.

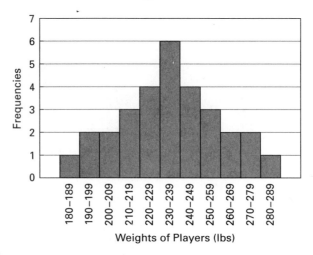

8 The height of each rectangle represents the frequency. The tallest rectangle represents the weight interval that occurs most often. Which weight interval is the mode for the weights of the football players?

Scientific or graphics calculators can evaluate many of the formulas used in statistics. Use your calculator to find the mean in the Activity.

The graphics calculator uses a menu key with the abbreviation $\boxed{\text{VARS}}$. You can use the variable statistics key to find the mean ($\bar{x}$), the sum (Σ), and the number of data values (n).

WORKPLACE COMMUNICATION

For the July 15 data, which time interval represents the mode for bandwidth usage? If the rate of increase in the use of the Internet continues, how long do you think it will be until MWI exceeds the capacity (100% of available bandwidth) of its T-1 connection? Why is there a "dip" in the middle of the histograms?

ALERT   ALERT   ALERT   ALERT   ALERT   ALERT

THIS IS AN AUTOMATIC ALERT MESSAGE FROM
ServPro
YOUR INTERNET SERVICE PROVIDER

Modern Workplace, Inc.
Attention: Computer Network System Administrator

Dear System Administrator:

This automatic alert message has been generated because your system has exceeded 70% of the capacity of your T-1 connection to *ServPro* and the Internet.

We detected your system above 70% of available bandwidth on July 15, during the hours of operation from 7AM to 7PM. Data for July 15, as well as six months prior, are shown below.

The number of Internet users on the MWI computer network has increased significantly over the past six months. If the increase continues, we highly recommend you install another T-1 connection to *ServPro*. We are standing by to assist you in this matter. Please call your *ServPro* technical representitive, Ms. Janet Porter, at 1-800-*SERVPRO* for further advice and assistance.

MWI T-1 Bandwidth Usage for January 15
(Percent of T-1 Bandwidth Being Used vs. Time of Day: 7am 9 11 1pm 3 5 7pm)

T-1 Bandwidth Usage for July 15
(Percent of T-1 Bandwidth Being Used vs. Time of Day: 7am 9 11 1pm 3 5 7pm)

LESSON ASSESSMENT

Think and Discuss

1 What is a frequency distribution?

2 How is the expression $\Sigma \, (c \cdot f)$ used?

3 Explain how to find the mean of a frequency distribution.

4 Explain how to find the mode of a frequency distribution from a histogram.

Practice and Problem Solving

In one month a landscape company trimmed trees on 20 different days. The supervisor recorded the number of trees trimmed each day.

	M	Tu	W	Th	F
Week 1	10	30	7	5	25
Week 2	24	28	11	17	16
Week 3	19	28	21	26	35
Week 4	3	4	21	13	9

5. Use the data to make a frequency distribution table.

6. Make a histogram of the data.

For the data, find the

7. Range.

8. Mode.

9. Median.

10. Use the formula to find the mean of the data.

The Johnson Company manufactures linen goods. In one week, the company packed and shipped the following number of linen packages to 25 retail customers.

80, 95, 50, 85, 100, 50, 75, 75, 80, 85, 70, 90, 95, 80, 85, 75, 80, 85, 70, 80, 100, 85, 60, 80, 95

11. Set up a frequency table and tally the number of packages.

12. Total the frequencies.

13. Make a histogram.

14. Find the mode, median, and mean number of packages sent.

Use the heights in inches of the students in your class to make a table.

15. Make a frequency distribution for the data.

16. Use the table to make a histogram.

17. What is the range of the data?

18. What is the mode of the data?

19. What is the median of the data?

20. What is the mean of the data?

21. Which measure of central tendency best describes the height of the students in your class?

22. Are there any outliers in your data? If so, what effect does the outlier have on the mean of your data?

Mixed Review

Write and solve an equation for each problem.

23. Todd receives a commission of 3% on all sales over $5000. What is Todd's commission on sales of $8500?

24. A local survey shows that 7 out of 12 people wear passenger seat belts. How many people out of 216 passengers would you expect to wear seat belts?

The equation $y = 40x + 100$ models the cost in dollars (y) for renting a backhoe for a number of hours, x.

25. What is the slope of the equation?

26. What is the y-intercept?

27. Graph the equation.

28. What is the value of y when x is 2.5?

LESSON 7.3 SCATTER PLOTS AND CORRELATION

The ABC Video Game Company has recently noticed an increase in the number of defective video games produced. The production manager has asked a quality control technician the reason for this increase. The technician has a hunch that the increase is related to the absentee rate of the workers. She gathers the following data.

Defective Video Games Test 1										
	M	T	W	Th	F	M	T	W	Th	F
Absentee Workers	9	11	5	4	2	7	7	11	10	5
Defective Video Games	9	10	11	6	3	6	8	9	7	4

To check the technician's hunch, plot the data as individual points on a coordinate graph. This graph is called a **scatter plot**.

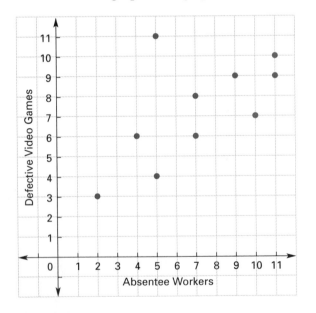

ACTIVITY   **Information from a Scatter Plot**

1 Look at the days that have a low number of absentee workers. Do these days also have a low rate of defective video games?

2 Describe how the points are scattered. Do they appear to cluster around a straight line?

3 Does the line rise or fall as you move from left to right?

4 How does the line represent the relationship between the number of absentee workers and defective video games?

5 Complete Steps 1–4 with the following:

Defective Video Games Test 2										
	M	T	W	Th	F	M	T	W	Th	F
Absentee Workers	7	5	4	5	7	3	7	9	6	2
Defective Video Games	6	4	6	3	4	10	2	3	7	9

6 Complete Steps 1–4 with the following:

Defective Video Games Test 3										
	M	T	W	Th	F	M	T	W	Th	F
Absentee Workers	8	6	8	10	4	3	10	6	5	9
Defective Video Games	6	3	9	8	4	10	4	5	7	5

Correlation

In Test 1 the data show a pattern that resembles a line rising from the left to the right. Since the slope of this line is positive, there is a **positive correlation** between the two sets of data. A positive correlation indicates that both variables in a test are increasing. As the number of absentee workers increases, so does the number of defective video games.

If the slope is negative, a **negative correlation** exists; that is, as one variable increases, the other decreases.

Sometimes there is a positive or negative correlation between data sets. Then you can predict the behavior or trend of one variable if you know the behavior of the other.

If the plot is scattered in such a way that it does not approximate a line, there is no correlation between the sets of data.

What would you conclude about the correlation between the number of defective video games and absentee workers over the six-week period? Are there any other factors than can explain the increase? Just because a positive correlation exists between the two variables does not mean one of the variables *causes* the other to increase.

Ongoing Assessment

Describe the correlation in Test 2 and Test 3. Explain what the correlation means in terms of an increasing or decreasing trend in the data.

Line of Best Fit

A **line of best fit** is a straight line that best represents the data on the scatter plot. This line may pass through some of the points, none of the points, or all of the points.

You can use a measure called the **correlation coefficient**, represented by the variable r, to describe how close the points cluster around the line of best fit. If all the points are on a line and the line has a positive slope, the correlation coefficient r is 1. If all the points are on a line and the line has a negative slope, r is -1. If the data is scattered at random, r is close to zero.

If r is close to 1 or -1, the trend of the data is well represented by a line, and there is a strong relationship between the data and the line. If r is close to zero, there is a weak relationship between the data and the line. In the real world, few relationships show a perfect correlation of 1 or -1.

The scatter plot on the next page shows the relationship of weight (x-axis) to height (y-axis) of the first 12 students to try out for the varsity basketball team. You can use a graphics calculator to plot the data points. In the statistics menu of the graphics calculator, the line of best fit is called the **regression line**. The correlation coefficient from the calculator is about 0.75.

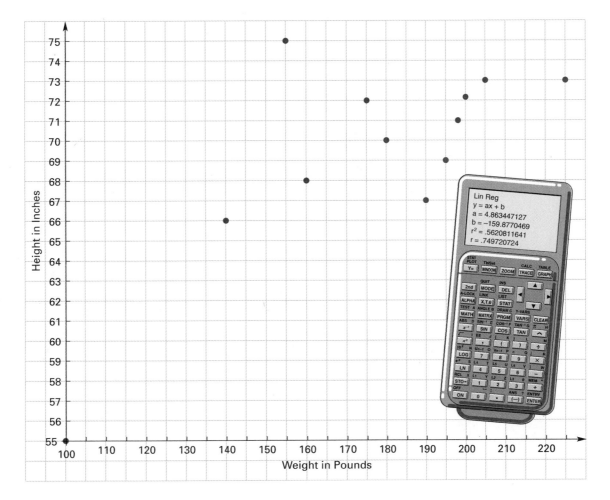

The scatter plot shows data points with axes labeled "Height in Inches" (vertical, ranging 55–75) and "Weight in Pounds" (horizontal, ranging 100–220).

Calculator display reads:

Lin Reg
y = ax + b
a = 4.863447127
b = −159.8770469
r² = .5620811641
r = .749720724

Critical Thinking Explain what the correlation coefficient of 0.75 means with respect to the basketball players.

LESSON ASSESSMENT

Think and Discuss

1 How is data displayed using a scatter plot?

2 Compare the meaning of a positive correlation and a negative correlation between two sets of data.

3 Explain how a "line of best fit" is used with a scatter plot.

Practice and Problem Solving

Jenny works in a state health agency. She is measuring the relationship between the amount of physical exercise per week and age. She records the data as ordered pairs. The first number in each ordered pair is a person's age and the second number is the number of hours of physical activity per week for that person.

(20, 15), (22, 11), (30, 6), (30, 7), (34, 6.1), (26, 13), (26, 8.5), (18, 16), (36, 3), (36, 5.8), (28, 11), (30, 9), (40, 3)

4. Plot the points on a scatter plot.

5. Does the data show any correlation between age and the number of hours of weekly exercise? If so, what kind of correlation?

6. Draw a line of best fit.

7. Use the slope and a point on the line to write an equation to approximate the line of best fit.

8. Enter the data in a graphics calculator and find the regression line.

9. Compare your line and the graphics calculator line and explain any differences.

The relationship between the number of hours studied and the score on an exam is represented by the ordered pairs listed below. The first number of the pair represents the number of hours studied and the second number represents the test score.

(10, 100), (0, 50), (1, 60), (3, 70), (5, 75), (4, 80), (7, 95), (9, 80), (9, 90), (7, 70), (9, 80), (5, 85), (10, 85), (4, 70), (6, 85), (6, 90), (9, 80), (10, 95), (3, 75)

10. Draw a scatter plot for the data.

11. Find the mode of the test scores.

12. What was the test score of the student who studied zero hours?

13. How many students got a 90% on the test? How many hours did they study?

14. Does the scatter plot indicate a negative or positive correlation?

15. Draw a line of best fit.

16. Use the slope and a point on the line to write an equation to approximate the line of best fit.

17. Enter the data on a graphics calculator and find the regression line.

18. Compare your line with the graphics calculator line and explain any differences.

Mixed Review

The fax bill for the Captain Tire Company averages $285 each month. The comptroller asked one of the assistants in the accounting office to compare the fax bills for the last five months. The monthly bills are as follows:

Jan.	Feb.	Mar.	Apr.	May
$240	$210	$325	$290	$305

19. Use integers to show the deviation between the average fax bill and the bill for each month.

20. Draw a number line with $205 at the center point. Use the results of Exercise 19 to place each amount on the number line in the correct order.

21. Boa is pouring a concrete walkway around a circular swimming pool. The radius of the pool is 10 feet. Boa wants the walkway to be 3 feet wide at each point. What is the area of the walkway Boa is planning?

Monique starts walking at a location 2 miles from her office. She walks at a constant speed of 3.5 miles per hour in a direction away from her office.

22. Find a function that models the distance Monique is from her office. Write the function in $f(x)$ notation, where $f(x)$ is distance and x is time.

23. Find $f(x)$ when x is 5.

24. Make a table of ordered pairs for the walk and graph the ordered pairs.

LESSON 7.4 PROBABILITY

Every evening after work, Carlos puts all of his change in a jar. One evening Carlos counts his coins and finds he has the following:

10 quarters   15 dimes   25 nickels

ACTIVITY 1   Part of a Whole

1 How many coins does Carlos have in the jar?

2 What fraction of the coins are

    a. quarters?   **b.** dimes?   **c.** nickels?

3 What percent of the coins are

    a. quarters?   **b.** dimes?   **c.** nickels?

4 Suppose Carlos reaches into the jar and, without looking, removes a coin. What are his chances that the coin will be a quarter?

Removing a coin from a jar is an example of an **event**. In Activity 1, there are 50 ways for an event (removal of a coin) to occur. Each occurrence is referred to as an **outcome**. For example, removing any one of the fifty coins is an outcome.

In Activity 1, Step 4, the outcome of removing a quarter is called the **favorable outcome**. Since there are 10 quarters in the jar of 50 coins, the chance of the favorable outcome (removing a quarter) occurring is 10 out of 50, or 20%.

When Carlos reaches into the jar and removes a coin without looking, he is *randomly* selecting a coin. A random selection means that all coins are *equally likely* to be selected. Whenever all the outcomes are equally likely, the probability is called **theoretical probability**.

Theoretical Probability
The theoretical probability $P(E)$ that an event E will happen is:

$$P(E) = \frac{\text{Number of favorable outcomes}}{\text{Total number of outcomes possible}}$$

Thus, the probability of removing one quarter, $P(\text{Quarter})$ from Carlos' jar of 50 coins is found in the following way:

$$P(\text{Quarter}) = \frac{10}{50} = \frac{1}{5} = 20\%$$

ACTIVITY 2   All or None

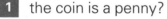

Carlos keeps his pennies in a separate jar. So far he has saved 120 pennies. He removes one coin from this jar. What is the probability

1   the coin is a penny?

2   the coin is a dime?

The questions in Activity 2 lead to the following definitions:

Probability of 1 and 0
If an event is sure to happen, the probability is 1.

If an event is impossible, the probability is zero.

Critical Thinking   Complete the following sentence. The probability of any event is greater than or equal to ___?___ but less than or equal to ___?___. Explain why.

A bowl contains 5 red marbles, 3 blue marbles, and 12 green marbles. You remove one of the marbles. What is the probability the marble will be

a. red?   **b.** blue?   **c.** green?   **d.** orange?

e. What is the sum of these four probabilities?

Sondra is a botanist. She is developing new plants for a local nursery. She is crossing plants that have red flowers with plants that have white flowers. A red-flowering plant has two red genes (RR). A white-flowering plant has two white genes (WW). Each parent contributes one gene to the seedlings.

Use a chart to model the possible color combinations for the seedlings. The top row shows the genes for the red-flowering plant. The left column contains the genes for the white-flowering plant. Each seedling has one gene from each parent and produces a pink flower (RW).

		Red-flowering Plant	
		R	R
White-flowering Plant	W	RW	RW
	W	RW	RW

RW = Pink

Sondra now crosses two of the pink-flowering plants. To see what the results might be, Sondra makes a new chart for the second generation.

		Pink Parent Plant	
		R	W
Pink Parent Plant	R	RR	RW
	W	RW	WW

RR = Red
RW = Pink
WW = White

EXAMPLE 1 Probability of an Event

What is the probability that a plant in the second generation is

a. pink?   **b.** red?

SOLUTION

Crossing two pink-flowering plants has the following possible outcomes:

$$(RR) \quad (RW) \quad (RW) \quad (WW)$$

a. There are two ways to get a pink-flowering plant: (RW), (RW). Thus,

$$P(RW) = \frac{2}{4} = \frac{1}{2} = 50\%$$

b. There is one way to get a red-flowering plant: (RR). Thus,

$$P(RR) = \frac{1}{4} = 25\%$$

Since there is one way to get the favorable outcome of a red flower and three ways to fail, the **odds** of getting a red flower are 1 to 3, or $\frac{1}{3}$.

Odds of an Event

$$\text{Odds of an event} = \frac{\text{Number of favorable outcomes}}{\text{Number of unfavorable outcomes}}$$

Critical Thinking   How do you find the probability that an event will not happen? What are the odds that an event will not happen?

EXAMPLE 2 Tossing Coins

Suppose you toss two quarters into the air. What are the odds that neither quarter will show heads?

SOLUTION

Heads/Heads   Heads/Tails

Tails/Heads   Tails/Tails

The coins can show (HH), (HT), (TH), (TT). For this example (TT) is the favorable outcome. There is one way to get tails on both quarters and three ways to fail. The odds are 1 to 3 neither quarter will come up heads.

LESSON ASSESSMENT

1 Describe how you can find the probability of buying the winning raffle ticket out of the total number sold.

2 Discuss the circumstances under which the probability of an event is zero or 1.

3 What is meant by the odds of an event occurring?

4 Discuss the difference between probability and odds.

Practice and Problem Solving

A tennis bag contains four white, five orange, and eleven green balls. If the tennis coach takes one ball from the bag, find the probability that it is

5. Orange.　　**6.** White.　　**7.** Green.

8. The faces of a number cube are numbered 1, 2, 3, 4, 5, and 6. If the cube is tossed, what are the odds that a 5 will turn up?

9. Past records show that a machine's probability of producing a defective bearing is 12%. Find the probability that a bearing from that machine will not be defective.

10. A store running a sales promotion has a box containing 60 envelopes. Twenty of the envelopes hold $5 bills. The other envelopes are empty. If one envelope is drawn, what is the probability that it will contain $5?

The chart gives a current inventory of T-shirts. If you choose one shirt from the inventory, what is the probability that it will be

	Large	Small
Blue	11	6
Red	14	9
Green	5	10

11. a red T-shirt?

12. a small T-shirt ?

13. a large, green T-shirt?

14. Write a problem in which the probability of an event happening is $\frac{3}{4}$.

15. Write a problem in which the odds of an event occurring are $\frac{2}{3}$.

The personnel office randomly schedules two employees, Leo and Vince, to work during two shifts, A and B.

16. Write all the possible outcomes for shift assignments as ordered pairs. For example, let (LA, VB) represent Leo on the first shift and Vince on the second shift.

17. What is the probability that Leo and Vince will be assigned to the same shift?

18. What is the probability Vince will be assigned to shift A?

Suppose you toss a penny, a nickel, and a dime. A head on the penny, a tail on the nickel, and a head on the dime can be recorded as the ordered triple (h, t, h). There are eight possible ways to record the heads and tails of a toss.

19. Use ordered triples to indicate all the possible outcomes of tossing the three coins.

20. What is the probability that all the coins will land heads up?

21. What are the odds that only one of the coins will land heads up?

22. A rectangular garden is 12 feet by 16 feet. A path is built diagonally from one corner to the opposite corner. How long is the path?

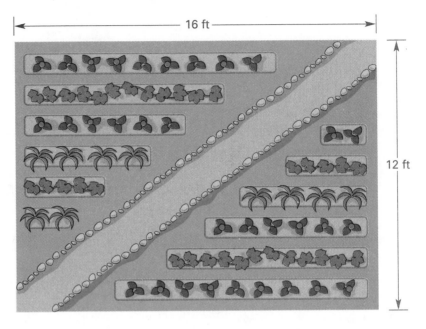

Write and solve an equation for each situation.

23. Each base angle of an isosceles triangle is 30 degrees less than the vertex angle. Find the measure of the vertex angle.

24. A board 600 centimeters long is cut into two pieces. One piece is three times the length of the other. How long is the shorter section?

25. The sum of three consecutive integers is −57. What are the integers?

The function $f(x) = x^2$ is an example of a nonlinear function.

26. Graph f for the values of x such that $0 \le x \le 3$.

27. If the graph of f is translated two units to the right, what are the new coordinates of the vertex?

28. How is the graph of $g(x) = (x + 1)^2$ related to the graph of f?

LESSON 7.5 EXPERIMENTAL PROBABILITY

In Lesson 7.4 you used theoretical probability to predict the outcome of events where all the outcomes were equally likely to occur. For example, when you toss a coin there is a theoretical probability of 50% that it will come up heads.

However, not all events are random. Assume there are 10,000 cars in your city and last night an accident occurred. You would not expect that the probability of being in the accident to be $\frac{1}{10,000}$ for each car in the city. In this situation, the probability depends on many factors, such as the weather and how other people drive.

When events are not random, you use **experimental probability**. In experimental probability, sample data or observations are used to estimate the probability of a specific event occurring. For example, statisticians for auto insurance companies collect data to determine the probability that people of various ages will have an accident. These probabilities are the key factors in setting insurance rates.

You can calculate experimental probability for both random and nonrandom events.

ACTIVITY 1   An Experiment with a Penny

1 Make a table like this for 20 trials.

Trial	Heads	Tails
1	?	?
2	?	?
3	?	?

2 Toss a penny 20 times. Record how the coin comes up each time. For each trial, record *H* under the Heads column and *T* under the Tails column.

3 How many times did the coin come up heads? What percent of the tosses were heads?

4 If you toss a penny 10 more times, how many times do you think it will come up heads?

Heads   Tails

5 Try 10 more tosses and record the result. Is the result the same as your guess?

6 Based on your results of this experiment of randomly tossing a penny, what is the probability that you will get 5 heads if you toss a penny 10 times?

Activity 1 shows how data is used to find experimental probability. The experimental probability of an event is found by collecting data and evaluating the data to predict the outcome of a specific event.

Usually, you observe a large number of trials and count how many times the event occurs. Once the count is completed, you can use the following formula to calculate the experimental probability.

Experimental Probability
Based on sample data or observations, the experimental probability $P(E)$ that an event E will happen is

$$P(E) = \frac{\text{Number of times the event happens}}{\text{Total number of tries}}$$

Ongoing Assessment

Set up an experiment with a number cube that has one number (1 through 6) on each face. What is the experimental probability of rolling a 4 in one try? Compare your results with the theoretical probability that the same event will occur.

As you found in Activity 1, you can use experimental probability to predict the outcome of a random event. However, experimental probability is most useful when the possible outcomes are *not* equally likely.

EXAMPLE 1 Make a Prediction

A consumer group surveyed its members and found that many of them had problems with a certain brand of tire. Out of a total of 14,537 tires purchased, 422 developed problems within the first 1000 miles. Unfortunately, you have just had one of these tires installed on your car. What is the probability that your tire will have a problem in the first 1000 miles of use?

SOLUTION

The only data you have is experimental. Thus, use the experimental probability formula.

$$P(\text{tire failure}) = \frac{422 \text{ tire failures}}{14{,}537 \text{ tires purchased}} \approx 2.9\%$$

Using experimental probability, you can predict that there is about a 2.9% chance (or about 1 chance in 34) that you will have a tire problem in the first 1000 miles.

Random Numbers

In **inferential statistics**, you use random sampling of a group to predict trends or behaviors of the group. Random sampling assumes that every possible sample of the group has the same probability of being selected.

Under these circumstances, you can use **random numbers** to choose sample selections. Calculators or spreadsheet programs can generate random numbers. Many statistics books contain lists of randomly generated numbers.

EXAMPLE 2 A Random Number Problem

What is the probability that a pair of random numbers ranging in value from 1 to 9 contains a 3?

SOLUTION

Generate 20 random numbers from 1 through 9. Use the numbers to form ten pairs. Count the number of pairs that contain a 3.

The table lists ten pairs of random numbers using the numbers from 1 through 9. In three of the pairs, one of the numbers is a 3. Thus, for this sample the experimental probability is $\frac{3}{10}$, or 30%.

1,6	2,2
2,1	4,**3**
3,6	7,1
9,7	9,9
8,5	2,**3**

Critical Thinking   In Example 2 find the experimental probability that at least one pair will add up to 10 or less.

Simulations

When it is impossible to set up an experiment to test an entire event you can sometimes use a **simulation** instead of the real experiment.

Suppose a potential wholesale supplier of tomato seeds for your garden supply store claims that half its seeds germinate within ten days of being planted. To verify the claim, you plant 20 seeds according to instructions. At the end of ten days, you dig up the seeds and test them for germination. The first four seeds you test have not germinated. If the supplier's claim is correct, what is the probability of having dug up four nongerminated seeds in a row?

Activity 2 presents a way of using random numbers to answer this question. The coin flipping simulates the germination rate of tomato plants.

ACTIVITY 2 A Coin Toss Simulation

1 Look at your record in Activity 1. Write your data so that the results of the 20 coin tosses are in the order they happened. For Example, H, H, T, H, ...

2 Examine the sequence to see if there are any places where four heads in a row occur.

3 How many times does the sequence H, H, H, H occur? Write a ratio using this number as the numerator and 20 as the denominator. What experimental probability did you get?

4 Combine your data with the rest of your class. Calculate the experimental probability from the larger data set. How does the probability from the larger set of data compare with that of your experiment with twenty data points?

There are many ways to generate random numbers to simulate experiments. You can use coins, number cubes, spinners, calculators, computers, or tables. The important thing to remember is that the more trials you take, the closer the simulation approximates the results of random events in the real world.

CULTURAL CONNECTION

Every culture plays games where winning is decided by the luck of the toss or the luck of the draw. Early American Indian children played a game that consisted of tossing five flat sticks with pictures on one side. After the toss, points were awarded based on the number of sticks showing the pictures.

In the seventeenth century, two famous French mathematicians, Pierre Fermat and Blaise Pascal, were given a problem similar to the one that follows.

The Situation:

Two players with equal skills are involved in a game of chance. The game is stopped before the game can be completed. Player A needs two points to win and Player B needs three points to win.

The Problem:

How should the amount of money already wagered be split between Player A and Player B?

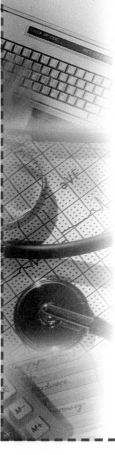

WORKPLACE COMMUNICATION

You need to do some preplanning before you set up the equipment for the $P(BE)$ verification test: Assume the CELSAT data communications system achieves the specified $P(BE)$. For this probability of error, how many bits (tries) will be required to record the required 10^4 errors (events) in the verification test? At the data rate of the commercial modem, how many hours do you need to schedule to complete the test?

			TECHNICAL REQUIREMENTS DOCUMENT CELSAT Cellular Communications Satellite	Page 9 of 13
Requirement Identification Number	Name	Requirement Specification	Description	Test Verification Requirement
36	Modem Data Rate	28.8 kbps (kilobits per second)	Data is transmitted as a long, coded set of bits (zeros and ones). The satellite and earth terminals shall utilize commercial, off-the-shelf modems that transmit and receive data bits at a rate of 28.8 kilobits per second. (28.8×10^3 bits per second)	Operation of all modems to be used in space shall be verified by testing in the MWI space radiation simulator.
37	Probability of Bit Error $P(BE)$	10^{-5}	Bit errors can be created when data bits are reversed (zeros changed to ones and ones changed to zeros) at random, due to electronic and atmospheric noise interference with the radio signal. $P(BE)$ is the probability that any given bit has been reversed and is in error. $P(BE)$ shall be no greater than 10^{-5}.	$P(BE)$ for the data communications system shall be tested in the MWI anechoic chamber by transmitting and receiving a known data stream and recording errors. The duration of the test shall be such that at least 10^4 errors are recorded.

LESSON ASSESSMENT

Think and Discuss

1 Compare experimental probability with theoretical probability.

2 Explain how random numbers can be used to simulate a probability problem.

Practice and Problem Solving

Toss a coin 100 times. Record each toss as heads (H) or tails (T).

3. What is the difference between the number of heads and the number of tails?

4. Record the total number of heads and tails tossed by your class.

5. Using theoretical probability, how many heads and how many tails would you expect for the data from the entire class?

6. Compare your class's experimental results and the theoretical probability.

Toss a nickel and a penny 100 times and record the heads (H) and tails (T) as ordered pairs. Record the nickel as the first number in the pair.

7. What does (T, H) represent?

8. What is the experimental probability that both coins will come up heads on a single toss?

9. List the four possible outcomes as ordered pairs.

10. What is the theoretical probability that both coins come up heads?

11. Compare the theoretical probability and the experimental probability.

12. Based on your experimental results, how many times would you expect two heads to come up if you toss the two coins 100 more times?

Roll a number cube 100 times. Record each number as it comes up.

13. Use your data to estimate the probability of rolling a 6 on a single roll.

14. Based on your results, how many 4s would you expect to get in another 100 tosses?

15. Combine the class results from tossing a number cube. Use this data to estimate the probability of rolling a 6 on a single roll.

16. What is the theoretical probability that a 6 will come up on one roll of the number cube? Compare the theoretical probability to the experimental probability.

Describe a simulation using random numbers to estimate the answer to each of the following situations.

17. Music practice is held once a week, Monday through Friday. What is the probability it will be held on Wednesday?

18. You take a true-false test with 5 questions. What is the probability of guessing 4 questions correctly?

19. You have a 50-50 chance to make a free throw. What is the probability you will make 2 in a row if you take 5 shots?

Mixed Review

20. A storage tank holds 9000 cubic meters of water. Use scientific notation to write the volume of the tank in cubic centimeters.

Write and solve an equation for each situation.

21. Paul is responsible for mowing 40 miles of highway medians for the state. If he has completed $\frac{2}{5}$ of his mowing, how many miles does he have left to mow?

22. In a random sample of 400 voters, 90 people said they planned to vote in the next city election. If there are 10,000 registered voters in one district, how many of those people can be expected to vote?

23. Becky is a quality control supervisor. On Monday she found several defective parts. On Tuesday she found 25 more defective parts than she did on Monday. If Becky found 67 defective parts altogether, how many did she find on Monday?

24. The distance around a rectangular room is 72 feet. If the length of the room is 16 feet, what is the width?

25. An airplane is flying at an altitude of 25,000 feet. To land, the plane must descend at a rate of 750 feet per second. Write an equation for the altitude (in feet) as a function of time (in seconds) to model the descent of the airplane from its current altitude.

LESSON 7.6 THE ADDITION PRINCIPLE OF COUNTING

Suppose you roll a red number cube and a green number cube. What is the probability that a red 3 and a green 4 will come up?

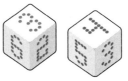

Probability of A and B

Since any of the six faces of a number is equally likely to come up, use theoretical probability. The probability that you toss a red 3 and a green 4 is the ratio of the number of favorable outcomes to the number of the possible outcomes.

Use a set of ordered pairs to represent all the possible outcomes of rolling different colored number cubes. You can graph these outcomes on a coordinate system. Use the *x*-axis to represent the possible outcomes for the red cube. Use the *y*-axis to represent the possible outcomes for the green cube.

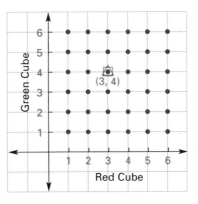

Each ordered pair on the graph represents one possible outcome of the event. The entire set of possible outcomes is called a **sample space** for the event.

The sample space contains 36 possibilities for tossing a red cube and a green cube. But there is only one way to toss a red 3 and a green 4. Thus,

$$P(\text{R3 and G4}) = \frac{1}{36}$$

Probability of A or B

What is the probability that a red 3 *or* a green 4 will occur in one toss of the cubes?

In probability, the words *and* and *or* have special meanings. The word *and* means both events occur. The word *or* means one of the events or both of the events occur.

The ordered pairs representing a red 3 are graphed using a square. The ordered pairs representing a green 4 are graphed using a triangle. There are 11 ordered pairs representing a toss that contains a red 3 or a green 4.

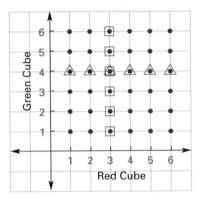

Thus, the probability of tossing a red 3 or a green 4 is

$$P(\text{R3 or G4}) = \frac{11}{36}$$

or about 31%.

Be careful when you count the events on a graph. The combination of a red 3 and a green 4 is a single event and should only be counted once.

EXAMPLE 1 $P(\text{Sum 7})$

If you roll the two number cubes once, what is the probability that the sum of the numbers will be 7?

SOLUTION

Make a sample space for the cube toss. Circle all the points that represent a sum of 7. Since there are 6 such points.

$$P(\text{sum 7}) = \frac{6}{36} = \frac{1}{6}$$

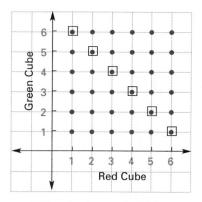

Critical Thinking   What is the probability of tossing a sum of 5 or less on one toss of the two number cubes.

Counting Choices

EXAMPLE 2 Art and Music

The chart shows enrollment for art and music classes at a local high school.

	Music	Art	Total
Boys	10	18	28
Girls	15	12	27
Total	25	30	55

a. How many students are boys?

b. How many students are in music?

c. How many students are boys *and* are in music?

d. How many students are boys *or* are in music?

SOLUTION

a. The first row shows 28 students are boys.

b. The first column shows 25 students are in music.

c. The intersection of the first row and first column shows 10 boys are in music.

d. 28 of the students are boys. There are 25 students in music. If you add 28 and 25, then you must subtract 10, because 10 of the boys have been added twice. Thus, 43 students are boys or are in music.

Use the chart from Example 2. Find the number of students who are girls or are in art.

The examples lead to an important principle of counting.

> ### Addition Principle of Counting
> Suppose there are *m* ways to make a first choice, *n* ways to make a second choice, and *s* ways to count the choices twice. Then the number of ways to make the first choice *or* the second choice is
>
> $$m + n - s$$

The Addition Principle of Counting applies to the probability problems in this lesson. For example, there are six combinations in which the red cube can come up 3. There are six combinations in which the green cube can come up 4. But as stated before, the one combination of a red 3 *and* green 4 is counted twice. Thus, there are $6 + 6 - 1$ (or 11) favorable outcomes for tossing a red 3 or a green 4 out of 36 possible choices.

$$P(\text{R3 or G4}) = \frac{11}{36}$$

Mutually Exclusive Events

If two events cannot occur at the same time, the events are **mutually exclusive**. If you remove a coin from a jar of coins, you cannot pick one coin that is *both* a quarter and a dime. You can use the fact that the results are mutually exclusive to determine the probability that either one event or the other event will occur.

Recall how the coins in Carlos' jar are distributed.

10 quarters　　　　15 dimes　　　　25 nickels

If Carlos removes one coin from the jar, what is the probability that the coin is either a quarter or a dime? There are too many events to build a sample space and graph the favorable outcomes. Instead, look at the probability that each event can happen and find the sum of those probabilities.

The probability of removing a quarter is $\frac{10}{50}$.

The probability of removing a dime is $\frac{15}{50}$.

Thus, the probability of removing a quarter or a dime is

$$\frac{10}{50} + \frac{15}{50} = \frac{25}{50}, \text{ or } 50\%$$

The probability of removing a quarter or a dime is 50%. If Carlos returns the coin to the jar and repeats the experiment 100 times, he will get either a quarter or a dime on about 50 of the tries.

> **Probability of Mutually Exclusive Events**
> If A and B are mutually exclusive events,
> $$P(A \text{ or } B) = P(A) + P(B)$$

LESSON ASSESSMENT

Think and Discuss

1 What is a sample space?

2 Explain the difference between the use of *and* and *or* in finding probabilities.

3 When is the Addition Principle of Counting used in finding probability?

4 What are mutually exclusive events?

Practice and Problem Solving

Refer to the sample space for rolling two number cubes. Find the probability of each event.

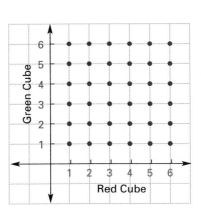

5. A red 4 and a green 6.   **6.** A red 4 or a green 6.

7. A red prime number and a green even number.

8. A red composite number or a green multiple of 2.

9. Either cube shows a 1.   **10.** Neither cube shows a 1.

11. Both cubes show a 1.   **12.** The sum is 5.

13. The sum is less than 7.   **14.** The sum is greater than 12.

15. There are six entrances to a building. Find the probability that two firefighters, entering at random and at different times, will choose the same door.

Four tickets in a box are numbered from 1 to 4. If two tickets are drawn, find the probability that:

16. the sum of their numbers is odd.

17. the product of their numbers is even.

The chart gives the current inventory of T-shirts. If you choose two shirts from the inventory, what is the probability that the T-shirts will be

Color	Large	Small
Blue	11	6
Red	14	9
Green	5	10

18. red and small?   **19.** blue or large?   **20.** large and not red?

21. The probability that you will buy an American brand of car is $\frac{3}{5}$. The probability that you will buy a German brand of car is $\frac{1}{4}$. What is the probability that you will buy an American or German brand of car?

22. The probability that it will rain is 0.6. The probability that it will be partly cloudy is 0.15. What is the probability that it will rain or be partly cloudy?

23. Solve the formula $i = prt$ for r.

24. A wire 24 inches long is bent to form a rectangle. If $x + 2$ represents the length of the rectangle, write an expression for its width in terms of x.

25. Mario is the supervisor for production in a cellular phone plant. In May Mario's shift made 200 more phones than in April. In June the number of phones made was double that of May. If Mario's shift production totaled 3800 items, what were each months production figures?

26. A garden is in the shape of a triangle with a perimeter of 37 feet. The longest side of the triangle is 5 feet greater than the shortest side. The remaining side is 2 feet greater than the shortest side. Find the dimensions of the garden.

LESSON 7.7 THE FUNDAMENTAL COUNTING PRINCIPLE

Suppose you have the following clothes in your closet.

Shirts: plaid, red, blue, tan
Pants: brown, black
Shoes: plastic sandals, canvas shoes, leather shoes

In how many different ways can you choose the three articles of clothing?

You can answer a question by drawing a **tree diagram**.

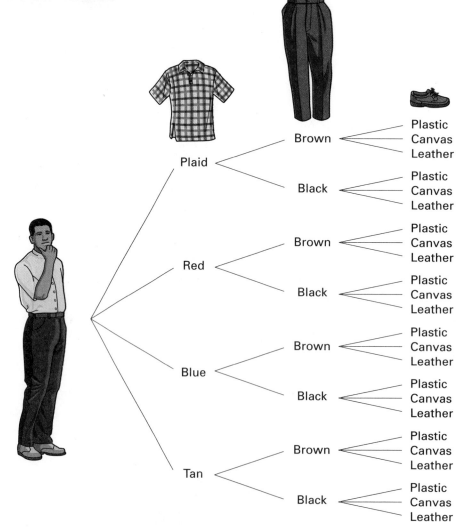

Start at the left side of the tree and follow one path all the way to the right. For example, start by choosing the plaid shirt. You can choose either black or brown pants. If you choose the brown pants, you can choose either plastic, canvas, or leather shoes. If you choose the black pants, you can choose either the plastic, canvas, or leather shoes. Thus, there are 6 ways to choose a combination of clothing if you start with a plaid shirt.

Ongoing Assessment

Begin with a different shirt and follow the paths to the right. How many different paths can you take?

The easiest way to count all possible paths is to count the final branches at the right side of the tree diagram. The final count for all the possible clothing combinations is 24.

ACTIVITY 1   Making Choices

To order a radio, you can make one choice from each of these three categories:

Colors: green, pink, yellow
Type: AM, FM, AM/FM
Accessories: battery, tape
  player

1 Create a tree diagram to show the possible combinations of color, type and accessory.

2 How many combinations are there?

Creating a tree diagram is a good way to picture choices. However, it takes time, and if there are many choices, there may not be enough room on your paper. Another way to organize and count the possible combinations in the clothing problem is found in Activity 2.

ACTIVITY 2   Multiplying to Find Choices

1 Draw three boxes, one for each kind of clothing.

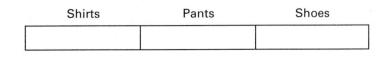

Shirts	Pants	Shoes

2 In the first box, write the number of choices you can make in selecting a shirt. There are 4 (plaid, red, blue, and tan). Complete the boxes for the other possible choices.

3 Find the product of the numbers in the three boxes. How does this product compare to the number of combinations you found using the tree diagram?

Ongoing Assessment

Use the method in Activity 2 to find the number of choices in Activity 1. Compare the results of the two methods.

Activity 2 is an application of the **Fundamental Counting Principle**.

Fundamental Counting Principle
To find the number of ways to make successive choices in a series of different categories, multiply together the number of choices in each category.

For every choice of shirt in the clothing problem, there were two choices of pants. For every choice of pants, there were three choices of shoes. A choice in one category of clothing did not affect the number of choices in any other category.

Similarly, in the radio problem, a choice of color, type, or accessory did not affect the number of choices in either of the other two categories. However, in some problems, a choice in one category eliminates choices from other categories.

EXAMPLE 1 Filling Officer Positions in a Club

The computer club elects three officers—president, secretary, and treasurer. Five members of the club (Akers, Chen, Miller, Vincent, and Wohl) are eligible for election as officers. How many different ways can these 5 people fill the 3 positions?

SOLUTION

For president (first position) there are 5 choices (Akers, Chen, Miller, Vincent, and Wohl). The first position can be filled in 5 different ways.

After one person is chosen to fill the first position, there are only 4 choices left to fill the second position of secretary. The second position can be filled in 4 different ways.

For treasurer (third position) there are only 3 choices left. The third position can be filled in 3 different ways. Use the Fundamental Counting Principle to find the number of ways to choose.

President		Secretary		Treasurer	
5	•	4	•	3	= 60

There are 60 different ways to fill the 3 positions given 5 choices.

EXAMPLE 2 Sitting at a Table

If 5 people can sit at the head table for a banquet, and there are 10 people to choose from, how many different seating arrangements can there be?

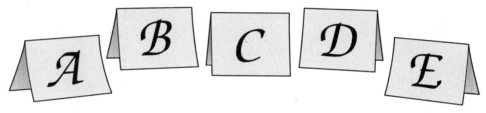

SOLUTION

Draw a picture to show the 5 positions. There are 10 ways to fill position A. After A is filled, there are 9 ways to fill position B. Use the Fundamental Counting Principle.

$$10 \cdot 9 \cdot 8 \cdot 7 \cdot 6 = 30{,}240$$

Thus, there are 30,240 possible ways for 5 of the 10 people to sit at the head table.

LESSON ASSESSMENT

Think and Discuss

1 Explain how to make a tree diagram to find the total number of ways to make two choices if each choice has three different possibilities.

2 When is it easier to use the Fundamental Counting Principle rather than to create a tree diagram? Give an example.

Practice and Problem Solving

In each of the following situations, use the Fundamental Counting Principle. Then, check your work by drawing a tree diagram.

3. If Nancy has 3 skirts, 2 blouses, and 4 scarves, how many different outfits can she wear?

4. A penny, a nickel, and a dime are tossed. How many different outcomes are possible?

5. You are placing decals on model race cars. Your choices consist of blue, white, green, or black decals in sizes small, medium, and large. How many choices do you have?

6. How many ways can you choose a team captain and a co-captain from a group of 5 players?

7. The day care center at a semiconductor plant serves breakfast and lunch Monday through Friday. How many meals are served each week?

8. Brian plays clarinet and saxophone, Hector plays the oboe, and Gina plays piano, flute, and trumpet. If Brian, Hector, and Gina form a trio, how many combinations of instruments can they play?

Use the Fundamental Counting Principle to answer the following questions.

9. Tickets for 10 performances of a play are sold for 6 different prices. How many combinations of performances and prices must be printed?

10. While Jeff is buying new tires for his car, he finds that they come in 16 different sizes, 4 different grades, 3 different types of construction, and either white sidewall or black sidewall. How many tire choices does Jeff have?

11. The Down Home Cafe serves 3 kinds of soup, 3 kinds of salad, and 5 choices of salad dressing. How many different orders of soup and salad (with dressing) are possible?

12. How many outfits are possible with 4 shirts, 2 pairs of slacks, 4 sweaters, and 3 ties?

13. There are 6 members of a club. How many ways can they elect a president and a secretary?

14. There are 5 runners in the 100-meter dash. Find the number of different ways the runners can finish the race in first through fifth place.

15. Seven students are in the finals of a skateboard competition. Only first, second, and third places are awarded. How many ways can the students finish in the competition?

16. Each school lunch package contains a ham, a chicken, or a turkey sandwich and a lemonade or orange drink. If the packages are randomly distributed, what is the probability that you will get a chicken and lemonade combination?

Mixed Review

Write and solve an equation for each situation.

17. Amber has saved $\frac{3}{5}$ of the cost for her first year at community college. If Amber has saved $690, what is the first year cost of college?

18. Ann and Kerry worked a total of 42 hours at their part-time jobs. Ann noticed that she worked 6 more hours than Kerry. How many hours did Kerry work?

19. Gilberto works part-time and Kyle works full-time. Kyle earns four times as much as Gilberto. If the difference in their pay is $150, how much does Gilberto earn?

20. The length of one side of a triangle is $\frac{1}{2}$ that of the second side. The length of the third side is $\frac{4}{5}$ the length of the second side. If the perimeter of the triangle is 138 inches, what is the length of each side?

Graph each function.

21. $f(x) = |x|$ **22.** $h(x) = |x| + 1$ **23.** $g(x) = |x| - 1$

24. Give the coordinates of the vertex for each absolute value function.

LESSON 7.8 INDEPENDENT AND DEPENDENT EVENTS

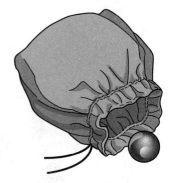

Margo has a bag that contains two red marbles, a blue marble, and a green marble. She draws one marble at random from the bag. She replaces the marble and draws another marble from the bag.

The probability that Margo removes a red marble on her first draw is $\frac{1}{2}$. Since she replaced the marble, the probability of removing a red marble on the second draw is still $\frac{1}{2}$. This is an example of events that are **independent**. Two events are independent if the outcome of the first event does not affect the second event.

ACTIVITY   Multiplying Probabilities

1 If a quarter is tossed once, what is the probability it will come up heads?

2 If a dime is tossed once, what is the probability it will come up heads?

3 Make a sample space for tossing a quarter and a dime.

4 Use the sample space to find the probability both tosses are heads.

5 What is the product of the probabilities from Steps 1 and 2? How does this result compare with your result from Step 4?

Tossing a quarter and tossing a dime are independent events. In the Activity, you showed that the probability of two independent events can be found by multiplying the probabilities of each event.

> **Probability of Independent Events**
> If A and B are independent events, then
>
> $$P(A \text{ and } B) = P(A) \bullet P(B)$$

For Margo, this means that the probability of removing a red marble on the first draw and removing a red marble on the second draw is
$$\frac{1}{2} \bullet \frac{1}{2} = \frac{1}{4}$$

EXAMPLE 1 Number Cubes

A red number cube and a green number cube are tossed. What is the probability that the number on the red cube is even and the number on the green cube is less than 5?

SOLUTION

Tossing an even number on the red cube and a number less than 5 on the green cube are independent events.

The probability that the red cube comes up with an even number is 3 out of 6, or $\frac{1}{2}$. The probability that the green cube comes up with a number that is less than 5 is $\frac{4}{6}$, or $\frac{2}{3}$.

Since the tosses of the red cube and the green cube are independent events,

$$P(\text{even red and green} <5) = \frac{1}{2} \cdot \frac{2}{3} = \frac{1}{3}$$

Check this answer by using a sample space of ordered pairs.

Ongoing Assessment

Eight cards are numbered 1, 2, . . ., 8. The cards are placed in a bag. One card is drawn and replaced. Then another card is drawn. What is the probability that the first card is greater than 5 and the second card is divisible by 2?

Suppose Margo does not replace the first marble before she draws a second marble. If Margo removed a red marble on the first draw there are 3 marbles left and only one of them is red. On the second draw, the probability that Margo will draw a red marble is $\frac{1}{3}$. This is an example of events that are **dependent**. Two events A and B are dependent when the first event A affects the probability of the second event B. If B follows A, then the probability of B occuring is written $P(B|A)$ and is read, "the probability of B *given that* A has occurred."

Probability of Dependent Events
If A and B are dependent events, and B follows A, then

$$P(A \text{ and } B) = P(A) \cdot P(B|A)$$

Thus, for Margo the probability of removing a red marble on the first draw and a red marble on the second draw is

$$\frac{1}{2} \cdot \frac{1}{3} = \frac{1}{6}$$

EXAMPLE 2 Dependent Events

You have a bag containing 3 red marbles, 2 blue marbles, and 5 green marbles. You draw one marble *but do not replace it* . Then you draw another marble. What is the probability of drawing a red marble on the first draw and a green marble on the second draw?

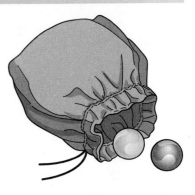

SOLUTION

For the first draw, there are 10 marbles in the bag. Each of the 3 red marbles is equally likely to be drawn. Thus,

$$P(\text{red on first draw}) = P(A) = \frac{3}{10}$$

After the first marble is drawn, there are 9 marbles left. If a red marble was taken on the first draw, there are 2 red marbles, 2 blue marbles, and 5 green marbles left in the bag.

$$P(\text{green on second draw}) = P(B|A) = \frac{5}{9}$$

Since removing a red marble and then removing a green marble without replacing the first red marble are dependent events,

$$P(A \text{ and } B) = P(A) \cdot P(B|A)$$

$$P(\text{red on first draw and green on second draw}) = \frac{3}{10} \cdot \frac{5}{9} = \frac{1}{6}$$

Ongoing Assessment

Recall the coins in Carlos' jar: 10 quarters, 15 dimes and 25 nickels. Suppose Carlos removes 3 coins from his jar without replacing any of them. What is the probability that he will remove a quarter on the first draw, a dime on the second draw, and a nickel on the final draw?

LESSON ASSESSMENT

 1 What is the difference between a dependent event and an independent event? Give an example.

2 How can you find the probability of tossing a two on a red number cube and a different number on a green cube

    a. using a sample space?

    b. using the independent events formula?

Practice and Problem Solving

Use these two spinners. What is the probability you will spin

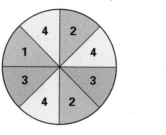
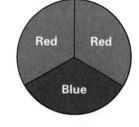

Spinner 1   Spinner 2

  3. a 4 and a blue?

  4. a 2 and a red?

  5. a 1 or a red?

  6. a prime number and a blue?

  7. a number greater than 3 or a red?

Three coins are tossed. What is the probability that

  8. 3 heads come up?

  9. exactly two heads come up?

  10. exactly 1 head comes up?

  11. no heads come up?

  12. Why must the answers to Exercises 8 and 11 be the same?

A quality control schedule for bicycle gears requires an accuracy check of a sample consisting of three large gears, four intermediate gears, and five small gears. In the first check, one gear is pulled out of the sample and checked. Then a second gear is pulled out and checked. What is the probability that

13. both gears are small?

14. the first gear is small and the second is large?

In the second check, one gear is pulled and then replaced. Then the second gear is pulled. What is the probability that

15. both gears are small?

16. the first gear is small and the second is large?

Circle B is drawn inside circle A. The area of circle B is $\frac{1}{4}$ the area of circle A. Suppose someone has randomly placed two dots within circle A. What is the probability that

17. both dots are in circle B?

18. one dot is inside circle B and one dot is outside circle B?

19. both dots are outside circle B?

20. both dots are inside circle A?

21. both dots are outside circle A?

The letters $M A T H E M A T I C A L$ are written on notecards and placed in a hat. One card is drawn and not replaced. Then a second card is drawn. What is the probability that

22. the first card is a T and the second is an M?

23. the first card is an A and the second is an A?

24. the first card is a vowel and the second is a consonant?

25. the cards can be used to spell AT? (Hint: This includes an *or* situation.)

26. A three-person production team is chosen randomly from a group of three men and three women. What is the probability that the team will consist of three women and one man?

27. The top four sales representatives, Mark, Samantha, Rosa, and Ron, are in a random drawing. Only first and second prize will be awarded. What is the probability that Mark will win first prize and Rosa will win second prize?

Mixed Review

Alicia sells lamps at craft fairs. She uses the function $f(x) = 25x - 250$ to project her profit for selling x lamps.

28. Graph the function and identify the slope and y-intercept.

29. At what point does Alicia begin to make a profit?

30. If Alicia makes a profit of $125, how many lamps did she sell?

Write and solve an equation for each situation.

31. A certain new car depreciates at the rate of $2400 per year. A new car that cost $26,000 is now worth $16,400. How old is the car?

32. A plumber needs to buy three sections of pipe for a new home construction project. The length of the longest pipe is twice the length of the shortest pipe. The other pipe is one-and-three-fourths the length of the shortest pipe. In all, the plumber needs 57 feet of pipe. How long is each section of pipe?

MATH LAB

Activity 1: Comparing Reaction Time

Equipment   Calculator
Meter Stick

Problem Statement

A reaction time is how long it takes to do something (such as step on the brakes) after you receive a certain signal (such as seeing a child start across the street in front of you). Reaction times vary from one person to the next. Reaction times, from one measurement to the next, also vary for the same person.

You will measure the reaction time for stopping a meter stick that falls between your thumb and forefinger. Based on the data for your group and your class, you will calculate measures of central tendency, make a frequency distribution table, and draw a histogram.

Procedure

a One person should hold the meter stick at the 100 cm end. Another member of your group (whose reaction time is being measured) should place his or her hand at the bottom of the stick with thumb and forefinger each two centimeters to either side of the 0 cm mark. When the person holding the meter stick drops it, the other person, looking only at the bottom of the stick, catches the stick as quickly as possible by pressing thumb and forefinger together.

b The length of the meter stick from the 0 cm end to the "catch point" is related to the elapsed time from the moment of "drop" to the moment of "catch." Thus, you can use this length as a measure of the reaction time. For each drop, read the millimeter mark that is just visible above the thumb. Use this reading as an indication of the reaction time. A lower reading (75 mm) represents a faster reaction time than a higher reading (92 mm).

c Have each member of the group catch three drops. Record the meter stick readings for each drop.

d When all groups have completed their measurements of reaction times, write the meter stick readings for the entire class on the chalkboard. These readings become the data set for the class.

e Divide the reaction times for the class data into five or six convenient intervals (for example, 50–59 mm, 60–69 mm, 70–79 mm, and so on). Make a frequency distribution table.

f Draw a histogram of the class data.

g Determine the mode and the median of the reaction times for the entire class.

h Find the mean of the class data.

i Compare your group data to the class histogram.

Activity 2: Games That Use Dice

Equipment   Calculator
                 Two dice

Problem Statement

You will examine the possible outcomes that occur from throwing a single die and a pair of dice. Many board games, such as Monopoly™, use dice to advance players around the playing board. On any throw, each face on a die is equally likely to come up. Since only one face can come up at a time, the results are mutually exclusive.

When you throw two dice, the sum of the dots on the faces can be anything from 2 through 12. These sums are not equally likely. Even though there is only one way to throw a 2 (1+1), there are six ways to throw a 7 (1+6, 2+5, 3+4, 4+3, 5+2, and 6+1). Thus, the events throwing a 2 and throwing a 7 are not equally likely.

Procedure

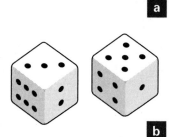

a Have each member of the group throw a single die 12 times and a pair of dice 12 times. Record the results. For a single-die throw, record the dots on the face that come up. For a two-die throw, record the sum of the dots on the two faces that come up. For example, if you throw a 5 and a 3, record an 8.

b Make a table similar to that shown below for the data you collect. The single-die throw should list the "face showing" and the "corresponding frequency" for the 12 throws. The two-dice throw should list the "sum of the two faces" and the "corresponding frequency," also for 12 throws.

Single-die Throw		Two-dice Throw	
Face Showing	Corresponding Frequency	Sum of Faces	Corresponding Frequency
1	2	2	0
2	3	3	0
3	1	4	3
4	2	5	2
5	2	6	1
6	2	7	2
		8	1
		9	1
		10	1
		11	1
		12	0

c Draw histograms for both the single-die and two-dice data.

d Make a table similar to the one above for all your group's data. Draw histograms of the single-die data for the group and the two-dice data for the group.

e Make a data table for the entire class. Draw separate histograms of the single-die data and the two-dice data for the class.

f Are there similarities in the three histograms for the single-die data? Are there similarities in the three histograms for the two-dice data? How are the histograms of single-die data different from the histograms of the two-dice data?

Activity 3: Predicting Physical Traits

Equipment   Calculator

Problem Statement

You will examine the frequency distributions for three physical traits. These traits are hair color, eye color, and dominant hand (left or right). Based on these experimental probabilities, you will predict the frequency of these traits in another class. You will also make a tree diagram for dominant hand, eye color, and hair color that shows the number of different combinations possible.

Procedure

a Record the frequencies of right-handed people and left-handed people for your group. Record these frequencies in a table similar to the one shown below for an assumed group of five members.

> Right-handed   4
> Left-handed   1

b Record the frequencies of different eye colors for your group. Use the four choices of brown, blue, green, and other. Record these frequencies in a table similar to the one shown below.

> Brown eyes   3
> Blue eyes   1
> Green eyes   0
> Other   1

c Record the frequencies of different hair colors for your group. Use the four choices of brown, blonde, red, and black. Record these frequencies in a table similar to the one below.

> Brown hair   3
> Blonde hair   1
> Black hair   1
> Red hair   0

d Write each of the three frequency tables for your group on the chalkboard. Combine all your groups' data to make a frequency table for the entire class. The table should show dominant hand, eye color, and hair color frequencies.

e Make a tree diagram showing the different branches for dominant hand, eye color, and hair color. How many branches are in the far-right column? What does this number represent?

f Use the Fundamental Principle of Counting to calculate the number of different combinations of dominant hand, eye color, and hair color. Compare your answer to Step **e**.

g From the frequency table for the entire class, calculate the experimental probability for each trait.

h Based on the experimental probabilities you have calculated, predict how many students in your high school should:

- be left-handed
- have red hair
- have blue eyes

(Your teacher can get the enrollment figures for you.) How can you improve the reliability of your prediction?

i Based on the experimental probabilities you have calculated, determine the probability that a student in your school will be left-handed, *and* have blonde hair, *and* have blue eyes. (You can assume each of these traits is independent for this calculation.)

j How does your experimental probability of being left-handed compare with the commonly accepted probability of 13% for the overall population?

MATH APPLICATIONS

The applications that follow are like the ones you will encounter in many workplaces. Use the mathematics you have learned in this chapter to solve the problems.

Wherever possible, use your calculator to solve the problems that require numerical answers.

Most teachers issue term grades based on the average grade for the period. Some teachers suggest that using the median of a student's grades for the period is a better method. Suppose these are your test grades.

Test Grades:
D	F	D	C
C	B	B	C
B	B	A	B
B			

1 Use the numerical scale of 4 for an A, 3 for a B, 2 for a C, 1 for a D, and 0 for an F. Convert your letter grades to number grades.

2 Determine your mean grade for the period.

3 Determine your median grade for the period.

4 Based on your findings, would you rather have a *mean* teacher or a *median* teacher?

The players on a professional basketball team have the following free-throw percentages at the middle of the season:

Player #	Free-throw %	Player #	Free-throw %
1	73.7	2	86.6
3	85.2	4	86.9
5	71.2	6	75.0
7	77.2	8	85.2
9	86.7	10	73.1
11	99.5	12	80.2

5 Determine the median free-throw percentage for this team.

6 Determine the mean free-throw percentage for this team.

7 A competing team has a higher median free-throw percentage of 85.0% but a lower mean of 77.0%. How can this be possible?

8 Suppose your team trades Player #5 for a player with a free-throw percentage of 79.0%. How will this affect the team's median free-throw percentage? How will it affect its mean free-throw percentage?

You are shopping for an inexpensive room thermometer in a discount store. There are 7 thermometers of the particular style you need on the shelf. However, you notice that they do not seem to agree on the present temperature in the store. Below are the temperatures you read from each of the thermometers.

Temperature (°F)

72 73 74 76
74 73 74

9 Determine the mode of the data.

10 Determine the mean of the data.

11 What is your estimate of the true temperature in the store?

12 Which thermometer would you choose to buy?

A security company has developed a lie detector that analyzes the stress in a person's voice to determine whether the person is telling the truth. The company claims the device to be accurate 94.7% of the time.

13 Suppose that this device is applied to a panel of candidates running for political office. If the claim is true, what is the probability that the device is correct in analyzing the truth of one candidate's statements?

14 Suppose 4 candidates are on the panel. What is the probability that the device can correctly analyze the integrity of all the candidates' statements?

15 What is the probability that the device is *not* able to correctly judge all four candidates' statements?

United States telephone numbers consist of a 3-digit area code, a 3-digit exchange, and a 4-digit station number.

16 In how many ways can you arrange the 3-digit exchange and 4-digit station number to form a telephone number?

17 Suppose that someone randomly dials the area code in a long distance call (in the United States). The exchange and station numbers match yours. What is the probability that the number dialed is your phone number? (Note: Even though some area codes are not valid, such as, 000, it is still possible to dial these numbers.)

During times of national crisis, the U.S. government can draft young men into the armed services. This has been done in the past by randomly assigning numbers from 1 to 366 to the calendar days. Traditionally, two drums of capsules are used. One drum contains number capsules identified 1 to 366. The second drum contains birth-date capsules, one for each of the 366 days of the year (assuming a leap year). One capsule is randomly drawn from each drum so that a number capsule is matched with a date capsule. For example, a number capsule of 128 and a date capsule of April 4 means that those 18-year-olds born on April 4th would be the 128th group to be drafted.

18 What is the theoretical probability that your birth date will be the first date capsule to be drawn?

19 What is the theoretical probability that the number 1 will be the first date capsule drawn from the other drum?

20 What is the probability that your birth date will be drawn first *and* assigned the draft number 1? Express this number as a percent.

21 Consider all the possible draft numbers that can be assigned to your date of birth. What is the probability that your date of birth is assigned the number 1?

22 Why is the probability for Exercise 21 higher than the probability for Exercise 20?

Your company orders a large shipment of scientific calculators from an office supply store. Since the supply house does not have enough of any one brand, it substitutes other brands with similar functions. The packing list reports this distribution of brands.

Brand	Quantity Shipped
A	18
B	30
C	12
D	10

23 If all the calculators are randomly distributed to the company staff, what is the probability that you will receive a Brand A calculator?

24 Still assuming a random distribution of the calculators to the company staff, identify the possible outcomes of such a handout of calculators. Are the possible outcomes resulting from this random distribution of calculators mutually exclusive?

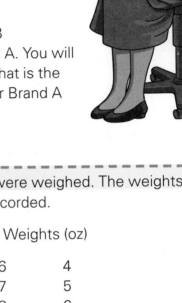

25 Suppose you decide that the Brand B calculator is of equal quality as Brand A. You will be satisfied if you received either. What is the probability that you will receive either Brand A or Brand B?

Samples of apples from an orchard were weighed. The weights were rounded to the nearest ounce and recorded.

AGRICULTURE & AGRIBUSINESS

Sample Apple Weights (oz)

6	6	4
6	7	5
5	8	6
5	5	6
4	5	7
6	5	7
5	5	

26 Construct a tally of the frequencies using one through ten ounces as weight classes.

27 Determine the percent frequency for each class.

28 Construct a percent frequency histogram of the data.

29 What is the mode, median, and mean of the apple weights sampled?

The weights of grain loaded in the bins of a river barge were recorded as follows:

Load	Weight per Bin (ton)
Barley	52.8
Corn	57.0
Flaxseed	49.8
Oats	43.4
Rice	49.1
Rye	52.6
Sorghum	56.6
Soybeans	54.8
Wheat	57.8

30 Compute the mean weight in a bin.

31 Compute the median weight of a bin.

32 Compute the range of weights for these bins.

The table below lists the average wheat yields for several countries during a certain time span.

Average Wheat Yields (pounds per acre)

Brazil	678	Canada	1401
France	2766	India	723
Japan	1999	Mexico	2150
Spain	1008	USA	1552
Russia	946	United Kingdom	3613

33 What is the mean wheat yield?

34 What is the median wheat yield?

35 What is the range of the wheat yields?

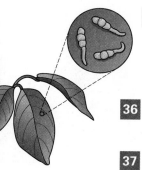

You inspect the trees in your orchard for a certain parasite that has appeared in some of your trees. You collect samples from 30 randomly selected trees. Your analysis shows that 12 of the samples are infested and the rest of the samples are clean.

36 What are the two possible events that can occur in your orchard trees with respect to the parasitic infestation?

37 Are these events mutually exclusive? Are they equally likely?

38 What is the experimental probability that a randomly selected tree from your orchard is infested with this parasite?

You read in the newspaper that a man purchased 12 eggs and found every one to be double-yolked. One source stated that the random chance of getting an egg that is double-yolked is 1 in 531.

39 What are the odds of getting a double-yolked egg?

40 If the number of yolks in each of the twelve eggs is independent of one another, what is the probability of getting 12 double-yolked eggs?

41 Suppose your family purchases 12 eggs every week. Approximately how many dozen eggs would your family have to purchase at this rate to expect to find one double-yolked egg inside a carton?

42 What other factors might affect the probability of getting even one double-yolked egg?

BUSINESS & MARKETING

You have asked twenty of your customers to fill out a questionnaire about the quality of your product. One question asks how satisfied they are with their purchase, on a scale of 1 (dissatisfied) to 5 (very satisfied).

Customer Ratings

3	4
1	3
4	4
2	5
4	5
5	4
3	1
4	4
2	3
3	4

43 Make a frequency distribution table for this data.

44 Identify the mode and the median.

45 Compute the mean rating of these customers. Would the mean be a good way to summarize such ratings? Explain your answer.

46 Which measure of central tendency helps you determine if at least half of your customers are reasonably satisfied with their purchase (that is, they give a rating of 4 or 5)?

Your retail toy business frequently experiences short-lived fads. A recent fad involved a particular type of stuffed animal. Your records reflect the rapid rise and decline of sales of this item.

Week #	Items Sold
1	14
2	32
3	68
4	122
5	77
6	39
7	17
8	5
9	2

47 Pick a convenient interval size and draw a histogram to reflect the changing sales for each week.

48 Find the mode for the frequency distribution.

49 Why would the mode of a distribution like this be of interest to a store owner?

50 Suppose the mode occurred later for the next fad. What might the store owner assume about the sales for the weeks that will follow? Is this a valid assumption?

You have recorded the end-of-the-month inventories for your clothing boutique this past year.

End-of-the-month Inventory

18	22	14
20	18	21
28	23	22
17	22	19

51 Determine the mode, median, and mean of this data.

52 Your goal for the end-of-the-month inventory is at least 19 items. How is the median the most useful measure of central tendency in evaluating this goal?

Your company is going to have an open house. A three-member committee is needed to plan the activities. Since all of the workers in your office want to be on the committee, a drawing will be held to elect members. Three names will be selected randomly from a hat containing the names of the seven workers in your office. The first name drawn will be the chairperson of the committee, the second will act as secretary, and the last will be the treasurer.

53 How many different ways are there to construct this committee?

54 Is each of these arrangements of three workers mutually exclusive? Is each equally likely?

55 What is the probability that you will be selected to be chairperson of the committee?

Mortality tables are used by life insurance companies to set insurance premiums. After age 30, mortality rates increase. Thus, insurance companies charge increasingly higher premiums for age groups after 30. The sample mortality table shows the number of persons out of a sample group of 100,000 who are still alive after the indicated age.

Age	Number Living	Age	Number Living
0	100,000	50	089,700
10	098,200	60	082,100
20	097,100	70	065,200
30	096,300	80	039,200
40	094,400		

56 What is the probability that one of these people randomly chosen at birth will live to be 40 years old?

57 What is the probability that one of these people randomly chosen at birth will live to be 80 years old?

58 If one of these people lives to be 60 years old, what is the probability that this person will live another 10 years?

59 If one of these people lives to be 70 years old, what is the probability that this person will live to be 80 years old? Compare this to the answer for Exercise 57.

60 If you were an insurance agent, how would numbers like these influence your decision to accept or reject clients for life insurance, and the premiums that you would ask them to pay?

The drip rate of a patient's IV changes depending on the position of the patient's arm. Over the course of a couple of hours, you count and record the number of drops per minute several times.

Number of Drops per Minute

10	12	10	8
11	11	10	9
11	11		

61 Draw a frequency distribution of the drip rates.

62 What are the values of the mean, mode, and median of this data?

63 What is the range of these data?

A consumer magazine performed a study on the leading brands of black tea, analyzing the caffeine content of a cup of each brand. The table lists the findings, in milligrams of caffeine per cup of brewed tea.

Brand	Caffeine (mg/cup)	Brand	Caffeine (mg/cup)
1	43	14	77
2	39	15	66
3	75	16	51
4	69	17	52
5	10	18	41
6	63	19	44
7	62	20	52
8	61	21	65
9	58	22	74
10	42	23	74
11	69	24	77
12	74	25	44
13	80		

64 Group the data by caffeine content into classes of 10 to 19, 20 to 29, and so on, and draw a histogram of the data.

65 Identify the mode and median of this data.

66 Determine the mean of the data and compare it to the mode and median you determined above.

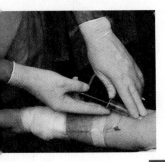

A person's blood is one of four possible types, A, B, AB, or O, depending on the person's gene pair. The gene pair is the result of the combination of the gene pairs from the person's parents. Type A, for example, results from a combination yielding gene pairs of AA. Type B results from a combination yielding gene pairs of BB. And type AB results from a combination yielding the mixed gene pair of AB. You can chart the combinations of gene pairs for blood type, as you did for flower genes earlier in this chapter.

67 Suppose one parent has a gene pair of AA and the other parent has AB. Make a chart of the possible outcomes of gene pairs for their offspring.

68 List the probabilities for each of the possible blood types.

69 Out of a population of 100 children whose parents have the gene pairs above, approximately how many would you expect to have blood type A? How many would you expect to have blood type AB? How many would have blood type B?

A simple test to determine left-handedness involves spreading apart the ring and middle fingers of each hand to form a "V." A right-handed person can stretch the fingers of his or her left hand farther than the right hand. A left-handed person will have a larger spread on his or her right hand. If there is no difference in the spread, the person tends to be right-handed and has good coordination in both hands. A quick test of this theory found that it was correct in 22 out of 25 cases.

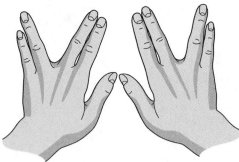

70 A test outcome is correct or incorrect. How many outcomes are possible with the 25 test cases? Are they all equally likely? Does the test of the theory above yield a theoretical probability or an experimental probability?

71 What is the probability of correctly predicting right- or left-handedness when this test is performed on an individual? What is the probability that it is incorrect?

72 You can perform the test with yourself and your classmates. Do your results agree with the results given above?

FAMILY AND CONSUMER SCIENCE

You are surveying the prices for carpet padding for a client. You record the following prices from several different suppliers:

Carpet Padding Price ($ per sq. yd)

$2.09	$3.55
$2.75	$3.15
$2.99	$3.07
$2.95	

73 What is the range in carpet padding prices?

74 What is the mean carpet padding price?

75 What is the median carpet padding price?

You have recorded the temperatures from a refrigerated dairy case during a six-day period, twice each day.

Temperature (°F)		
Day 1	38	39
Day 2	38	40
Day 3	40	39
Day 4	40	38
Day 5	38	39
Day 6	38	40

76 What is the range of the temperatures?

77 What is the mean temperature?

78 What is the median temperature?

79 What is the mode of the temperature data?

Suppose that you have five different wall hangings and three locations on which to place them.

80 How many different ways can you arrange the five wall hangings?

81 If you found one more wall hanging to include in the arrangement, what would be the new number of choices available to you?

A large bowl is filled with a nut mix. Since your favorite type of nut is the almond, you decide to study the number of almonds in the bowl. You remove a few handfuls of nuts, then count and record the types.

Cashews	5	Peanuts	54
Brazil nuts	2	Almonds	7

82 Determine the probability of randomly removing each type of nut in the mix.

83 Suppose you remove one nut at random. You return the nut and remove another one. What is the probability of obtaining an almond on the second draw?

84 Suppose you randomly remove two nuts from the bowl. What is the probability of obtaining an almond?

INDUSTRIAL TECHNOLOGY

- -

You find the height of a stack of twenty pieces of plywood to be $7\frac{5}{8}$ inches.

85 What is the mean thickness of the pieces of plywood in this stack?

86 If you knew the boards were supposed to be either $\frac{1}{4}$, $\frac{3}{8}$, $\frac{1}{2}$, or $\frac{5}{8}$ inches thick, which thickness would you suspect the boards to be?

While overhauling an eight-cylinder engine, you forget to number the eight rod caps. When you reassemble the rods, you do not know which cap to match with each rod.

87 How many combinations of rods and caps are possible?

88 What is the probability of "accidentally" matching all eight caps with the correct rods on the first try?

A certain product is assembled in stages using three machines: Machine A, Machine B, and Machine C. During the previous six months, each of the machines has broken down, causing a slowdown in production while it was being repaired. You have kept a record of the last six months' history.

Machine	Days Operating	Days Down for Repair
A	116	16
B	106	26
C	122	10

89 For each machine, determine the experimental probability of the machine being down on any given day.

90 If all machines are down at the same time, production must cease. Assuming that each machine's status is independent of the others, determine the probability that all three machines will be down at the same time.

A factory assembles deodorant bottles. The bottles are randomly selected from the lot and assembled with randomly selected bottle caps. A quality engineering study estimates that about 3% of a lot of 1000 bottles were made with tops of too small a diameter. Unfortunately, these bottles were matched up with a lot of 1000 bottle caps, of which 1.8% were indicated to have too large a diameter. Whenever bottle tops that are too small are combined with caps that are too large, the bottles will leak.

91 What is the probability that a small bottle top in this lot is matched up with a large bottle cap?

92 How many of the lot of 1000 bottles should you expect to leak because of this mismatch of bottle tops and caps?

Your company has two safety vehicles; each is operable 95% of the time. Assume that the availability of each vehicle is independent of the other's condition.

93 What is the probability that both vehicles are operable at the same time?

94 What is the probability that only one of the two vehicles is operable at any time?

95 What is the probability that neither vehicle is operable at the same time?

A company produces a certain type of candy. Each bag contains small candies that can have one of five different color coatings: red, orange, brown, green, and yellow. Each bag, on the average, contains about 45 candies. The bags are filled from a bin that has a uniform mixture of each of these five colors.

96 What is the probability that the first piece of candy into a bag is red?

97 What is the probability that the first piece of candy into a bag is not red?

98 What is the probability that the second piece of candy into a bag is not red?

99 What is the probability that both the first *and* second candies into the bag are not red?

100 What is the probability that a bag gets filled with 45 pieces of candy from the bin, and not one of the pieces is red?

Applications

The highest annual commissions paid on sales by the Top-Flite Shoe Corporation were listed in its newsletter.

Region	Commission
1	$10,000
2	7,500
3	9,000
4	8,000
5	10,000
6	9,500

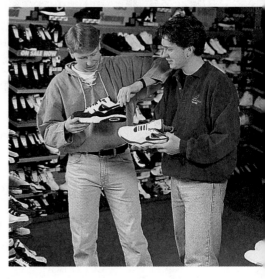

1. What is the mode of the commissions?

2. What is the median of the commissions?

3. What is the mean of the commissions?

4. Which measure of central tendency would you report to attract new salespeople interested in making a large commission?

5. Suppose the person in Region 3 is replaced by a salesperson who makes $6000 more in commissions. How is the mean affected?

The following final exam grades were posted in the Library:

98	94	94	74	87
100	80	89	96	68
88	88	98	92	79
67	97	94	84	76

6. Make a frequency table using the following classes:

61–70   71–80   81–90   91–100

7. What is the mean of the test scores?

8. Use the frequency distribution to create a histogram.

Use the following data:

$$(7, 7) \qquad (3, 5) \qquad (9, 6) \qquad (1, 2) \qquad (5, 3)$$

9. Make a scatter plot.

10. Describe the correlation exhibited by the scatter plot.

A six-sided, colored cube has two faces painted red, two faces painted blue, one face painted green, and one face painted yellow.

11. If the cube is rolled once, what is the probability it will come up red?

12. If the cube is rolled once, what is the probability it will come up blue or green?

13. If the cube is rolled twice, what is the probability the first roll will be green and the second roll yellow?

A license plate has seven positions. You can place any letter in the first position and any digit in the other positions.

14. How many arrangements of the letters and digits are possible?

15. What is the probability of randomly choosing a plate that begins with the letter A?

Math Labs

16. You hold a meterstick at the 100 centimeter end and drop it for Ann and John. Ann catches the stick at the 70 centimeter mark. John catches the stick at the 82 centimeter mark. Who has the quickest reaction time?

17. The histograms for the single-die throws were relatively flat. However, for the two-dice throws, the histograms have a higher frequency in the middle than at the ends. Explain why this is so.

18. Construct a tree diagram for left- or right-handed; dark brown, light brown, blue, green, or hazel eyes; and black, brown, blonde, or red hair. How many combinations are possible?

CHAPTER 8

WHY SHOULD I LEARN THIS?

In today's highly competitive world, businesses are constantly looking for solutions that save money or conserve scarce resources. Skillful problem solvers use such tools as systems of equations to find optimum solutions. You can take an early step toward career success by mastering the algebra skills found in this chapter.

Contents: Systems of Equations

SYSTEMS OF EQUATIONS

OBJECTIVES

1. Solve a system of two equations by graphing.
2. Solve a system of two equations by substitution.
3. Solve a system of two equations by addition or subtraction.
4. Solve a system of two equations by using determinants.
5. Use a system of equations to solve problems involving two unknowns.

In Chapters four through six, you used equations with one variable to model problems involving volume, speed, and money. However, some situations involve two variables and need two equations to model the problem. In these cases, you must set up **systems of equations** to solve the problems.

For instance, when a large aerospace company begins planning a new airplane, its engineers and salespeople use systems of equations to model production costs and projected sales revenue. This model can predict whether the new plane will earn a profit or cause the company to lose money.

In this chapter, you will examine several ways to solve a system of equations. These ways will include:

- *Using* graphing *to find points of intersection.*
- *Using* substitution *to reduce two equations with two unknowns to one equation with one unknown.*
- *Using* addition *or* subtraction *to reduce two equations with two unknowns to one equation with one unknown.*
- *Using* determinants *to find the unknowns in systems of equations.*

As you watch the video for this chapter, look for ways in which systems of equations are used in the workplace.

LESSON 8.1 SOLVING A LINEAR SYSTEM BY GRAPHING

Mr. Cassidy is trying to choose the best one-day rental plan from the A-1 Rental Agency.

The agency offers two rental plans:

Plan 1: $30.00 per day with unlimited mileage
Plan 2: $20.00 per day plus 10 cents per mile

To compare the two plans, you can model each with a function. Let C_1 represent the cost under Plan 1 and C_2 represent the cost under Plan 2. Let N represent the number of miles driven.

Plan 1: Cost = $30.00

$$C_1 = 30$$

Plan 2: Cost = ($0.10 per mile) ($N$ miles) + $20.00

$$C_2 = 0.10N + 20$$

Plan 1 is a constant function. Plan 2 is a linear function with slope 0.10 and y-intercept 20. When you graph the two functions on the same coordinate axes, you can see the solution.

N	0	50	100	150
C_1	30	30	30	30

N	0	50	100	150
C_2	20	25	30	35

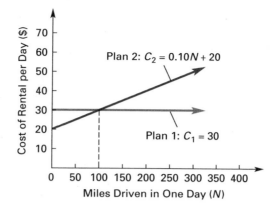

At the the point where the two graphs intersect, the plans cost the same. This is called the **break-even point.** The graph shows the break-even point occurs when the *x*-coordinate is 100. In other words,

- *Both plans cost the same if Mr. Cassidy drives exactly 100 miles.*
- *Plan 1 is less expensive if he drives over 100 miles.*
- *Plan 2 is less expensive if he drives under 100 miles.*

The two functions modeling the car-rental problem represent a **system of linear equations** in two variables. The following is another example of such a system:

$$y = 3x - 5$$
$$y = -x + 7$$

The solution to each linear equation or function is a set of ordered pairs. To solve the *system,* you must find all of the ordered pairs that solve *both* equations. These ordered pairs represent the points of intersection of the graphs of the system of equations

Examine the table of values for each equation.

Only the ordered pair (3,4) solves both equations. Thus, the ordered pair (3,4) is the solution to this system of equations.

$y = 3x - 5$		$y = -x + 7$	
x	y	x	y
−1	−8	−1	8
0	−5	0	7
1	−2	1	6
2	1	2	5
3	4	3	4
4	7	4	3
5	10	5	2
⋮	⋮	⋮	⋮

Sometimes you cannot find the solution by examining a table of values. However, you can graph the equations. The solution is the ordered pair representing the point where the graphs intersect.

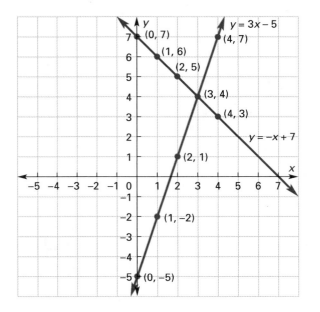

If you use a graphics calculator, you must rewrite the equations in slope-intercept form. When using graph paper, it is often easier to use the standard form of the equation and draw a line between the x-intercept and y-intercept. In either case, check your solution and make sure it is reasonable.

EXAMPLE 1 The Point of Intersection

Use graph paper or a graphics calculator to solve this system.

$$x - y = 3$$
$$2x + y = 3$$

SOLUTION

If you are using a graphics calculator, rewrite each equation in slope-intercept form, $y = mx + b$.

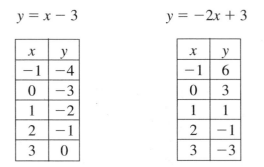

$y = x - 3$

x	y
-1	-4
0	-3
1	-2
2	-1
3	0

$y = -2x + 3$

x	y
-1	6
0	3
1	1
2	-1
3	-3

Graph the system and find the point of intersection.

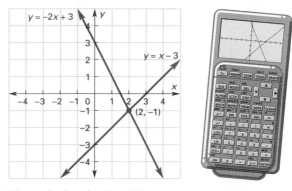

The solution is $(2, -1)$.

To check the solution, substitute 2 for x and -1 for y in the original equations.

$$x - y = 3 \qquad\qquad 2x + y = 3$$

$$2 - (-1) = 3 \qquad (2)(2) + (-1) = 3$$

$$3 = 3 \qquad\qquad 3 = 3$$

EXAMPLE 2 Solving a System Graphically

The band director must order 35 uniforms. There are usually five more boys than twice the number of girls in the band. How many uniforms of each type should the band director order?

SOLUTION

Write a system of equations to model the situation. Then solve the system using graph paper or a graphics calculator.

Let x represent the number of girls.

Let y represent the number of boys.

Since there are 35 band members, $x + y = 35$ or $y = 35 - x$.

The number of boys is five more than two times the number of girls. Thus,

$$y = 2x + 5$$

Now solve the system.

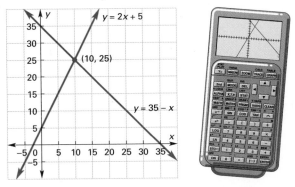

There are 10 girls and 25 boys. Check the solution in the original equations to make sure it is correct.

When you use a graphics calculator, make sure you use a friendly window. You may also need to use (TRACE) and (ZOOM) to find the point of intersection.

ACTIVITY   Using a Graphics Calculator to Solve a System

1 Graph $y = 4x - 5$ and $y = -x + 7$ on your graphics calculator.

2 Use (TRACE) to find the intersection of the graphs.

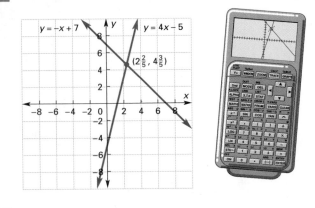

3 Write the coordinates of the intersection.

LESSON ASSESSMENT

Think and Discuss

1 How can you use the graphs of two equations to find the solution to a system of equations?

2 In what form must you write an equation to enter it in a graphics calculator?

3 If two straight lines intersect on a graph, how many points of intersection are there?

Practice and Problem Solving

Find the solution to each system graphically.

4. $y = 2x - 7$
$y = -3x + 3$

5. $y = -x + 7$
$y = 2x - 5$

6. $y = 2x - 4$
$y = 3x - 7$

7. $y = -2x + 3$
$y = 5x + 10$

8. $3x + y = 6$
$x + 2y = 2$

9. $3y - 2x = 12$
$2y + x = 8$

10. $2x + y = 9$
$x + y = 4$

11. $5x - 2y = 18$
$x + 2y = -6$

12. $4x + 2y = -2$
$3x + y = -2$

Write a system of equations to model each situation. Solve the system graphically.

13. In a basketball game, Juan made a total of twelve 2-point and 3-point baskets totaling 26 points. How many 2-point shots and how many 3-point shots did Juan make?

14. A class divides into two groups for a science experiment. The first group has five fewer students than the second group. If there are 27 students in the class, how many students are in each group?

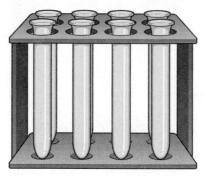

Let $f(x) = |x| + 3$.

15. Graph f.

16. What is the domain of f?

17. What is the range of f?

18. What is $f(-3)$?

Latisha recorded the number of hours she worked during a 2-week period in July.

$$6, 6, 7, 8, 7, 6, 4, 7, 5, 8$$

19. List the data in order from least to greatest.

20. Find the mode of the data.

21. Find the median of the data.

22. Find the mean of the data.

When you graph two equations on the same coordinate axes, one of three things can happen.

The lines may intersect. The system has one solution.

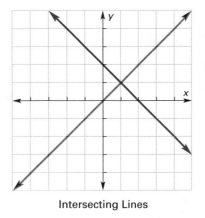

Intersecting Lines

The lines may be parallel. The system has *no* solution.

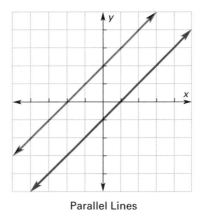

Parallel Lines

The lines may coincide. The system has infinite solutions.

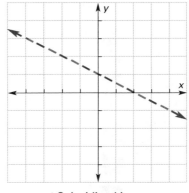

Coinciding Lines

In the worlds of agriculture, business, engineering, industry, and science, systems of linear equations model many problems. Most of these linear systems have single, or unique solutions. A system that has at least one solution is called **consistent**. If the system has *only* one solution, the system is **independent**.

ACTIVITY 1	**Modeling Profit in a Business**

Your younger sister wants to earn money by selling lemonade. The cost of starting the business is $1.20, and each cup of lemonade costs six cents to make.

Your sister's expenses in dollars (d) is equal to six cents times the number of cups sold (n) plus the $1.20 start-up cost.

$$d = 0.06n + 1.20$$

If the lemonade sells for ten cents a cup, the income (d) is equal to ten cents times the number(n) of cups sold.

$$d = 0.10n$$

1 How does this graph model your sister's profit or loss?

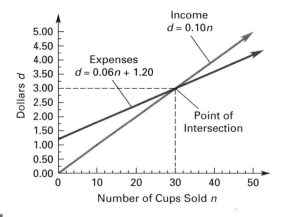

2 What does the point (0, 0) mean on the income graph?

3 What are some of the start-up costs?

4 If she sells 20 cups of lemonade, what is her profit or loss?

5 If she sells 40 cups, what is her profit or loss?

6 What are the coordinates for the break-even point?

7 How many cups must she sell to make a profit?

8 Is this system of equations consistent? Is it independent?

Lemons
$1.⁰⁰ / lb

Suppose the cost of making lemonade increases to 10 cents per cup. The equation that models the expenses becomes

$$d = 0.10n + 1.20$$

1 Graph the system $d = 0.10n + 1.20$
$$d = 0.10n$$

2 What is the slope of each line?

3 What is the y-intercept of each line?

4 How are the lines related?

5 What is her profit or loss if she sells 20 cups? 40 cups?

6 What is the break-even point?

7 Is this system consistent?

The lines in Activity 2 will never intersect. The expression $0.10n + 1.20$ will never equal $0.10n$. The two lines are **parallel.** If two lines are parallel, they have the same slopes but different y-intercepts.

Parallel Lines

Nonvertical lines are parallel if they have the same slope.

Vertical lines are also parallel even though their slopes are undefined.

Since the lines in Activity 2 never intersect, the system has *no solution.* A system with no solution is called **inconsistent.**

EXAMPLE   An Inconsistent System

Solve this system.

$$-2x + y = -1$$
$$-6x + 3y = 12$$

Use graph paper or a graphics calculator. First write each equation in slope-intercept form.

$$y = 2x - 1$$
$$y = 2x + 4$$

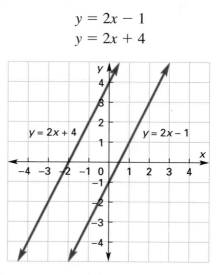

Notice that the slopes are equal, but the y-intercepts are different. Therefore, the lines are parallel. There is no solution to this system, and it is inconsistent.

Ongoing Assessment

Which of the systems is inconsistent?

a. $x + y = 5$ **b.** $x + y = 5$
     $x - y = 3$ $x + y = 4$

ACTIVITY 3   A Dependent System

1 Graph this system on the same coordinate axis.

$$4x + y = -2$$
$$8x + 2y = -4$$

2 What is the slope of each line?

3 What is the y-intercept of each line?

4 What is the intersection of the two lines?

5 Is this system consistent? Is it independent? How many solutions does the system have?

In Activity 3, the lines are the same. When each equation is written in slope-intercept form, the equations are identical. Every solution for one equation is also a solution for the other equation. The system is consistent because the system has at least one solution. In fact, the system has infinite solutions. The system is **dependent** because all the solutions are the same.

Critical Thinking Explain the difference between the following three systems:

a. a system that is consistent and dependent
b. a system that is consistent and independent
c. a system that is inconsistent

LESSON ASSESSMENT

Think and Discuss

1 Describe the graphs and slopes of a system of equations with no solution.

2 Describe the graphs, slopes, and y-intercepts of a system of equations with an infinite number of solutions.

3 Describe the graphs of a system of equations that has one solution.

4 What is the difference between consistent and inconsistent systems of equations?

5 What do you mean when you say a system of equations is dependent or independent?

Practice and Problem Solving

For Exercises 6-17, state whether the system is

a. consistent and independent
b. consistent and dependent
c. inconsistent

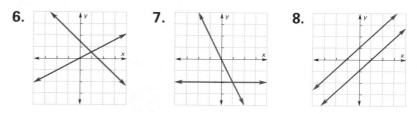

6. 7. 8.

9. 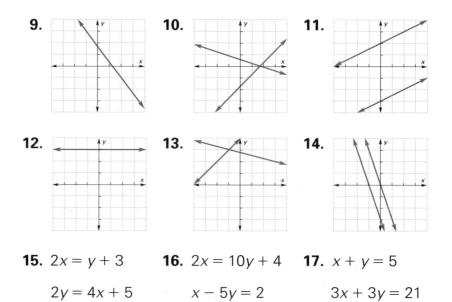 **10.** **11.**

12. **13.** **14.**

15. $2x = y + 3$ **16.** $2x = 10y + 4$ **17.** $x + y = 5$

$2y = 4x + 5$ $x - 5y = 2$ $3x + 3y = 21$

Write a system of linear equations to model the number of flights each flight attendant works over two consecutive weeks. Identify each system as independent, dependent, or inconsistent.

18. Keesha works a total of 22 flights over two weeks. Her average number of flights over the two weeks is 11.

19. Sal works a total of 22 flights over two weeks. The second week, he works four fewer flights than the first week.

20. Heather works a total of 22 flights over two weeks. The second week, she works eleven minus the number of flights she worked the first week.

The following table shows the number of miles a person of each occupation walks in one day.

A. City messenger   4.0   B. Dentist   0.8

C. Hospital nurse   5.3   D. Hotel employee   3.2

E. Secretary   2.2   F. Teacher   1.7

21. If two miles is the average distance walked in one day on the job for people of all occupations, use an integer to indicate the deviation from the average for each occupation listed in the table.

22. Draw a number line. Place the deviation from the average for each occupation (A–F) on the number line.

Let r represent the radius of the Earth. The formula for the circumference (C) of the Earth is $C = 2\pi r$. The formula for the surface area (S) of the Earth is $S = 4\pi r^2$. The average radius of the Earth is 6.37×10^6 kilometers at the equator.

23. If a road could be built around the Earth at the equator, how long would the road be?

24. If a road could be built on an elevated bridge around the equator just 1 meter above the Earth, what would be the length of the road?

25. What is the surface area of the Earth?

26. If approximately 70% of the Earth is covered by water, how many square kilometers of the Earth's area is water?

27. Suppose the radius of the Earth is doubled. By what factor is the circumference increased? By what factor is the surface area increased?

LESSON 8.3 SOLVING SYSTEMS BY SUBSTITUTION

Once again, recall Mr. Cassidy's car rental problem. You can use another method to choose the best plan.

Plan 1:   Cost = $30.00

$$C_1 = 30$$

Plan 2:   Cost = ($0.10 per mile) ($N$ miles) + $20.00

$$C_2 = 0.10\,N + 20$$

For what number of miles driven is the cost the same for both plans? In an earlier lesson, you used a graph to find that the break-even point occurs when N is 100 miles. Another way to find the break-even point uses an algebraic technique called **substitution**.

The break-even point is the value of N that makes C_1 and C_2 the same. Since $C_1 = C_2$, you can substitute the value of C_1 (30) for C_2 in the second equation and solve for N.

Thus, when $C_1 = C_2$, the following substitution is possible.

$$30 = 0.10N + 20$$

$$0.10N = 10$$

$$N = 100$$

You can use the method of substitution to solve any system. For example, consider this system.

$$3x - y = 1$$
$$2x + y = 0.2$$

To solve this system by substitution, follow these steps.

1. Solve the first equation, $3x - y = 1$, for y.

$$y = 3x - 1$$

2. Substitute the expression $3x - 1$ for y in the second equation.

$$2x + (3x - 1) = 9$$

3. Solve the equation in Step 2 for x.

$$5x - 1 = 9$$
$$5x = 10$$
$$x = 2$$

4. Substitute 2 for the value of x in the first equation and solve for y.

$$3x - y = 1$$
$$3(2) - y = 1$$
$$y = 6 - 1$$
$$y = 5$$

5. You have now found a solution $(2, 5)$ for the system. Check your solution in each of the original equations.

$$3x - y = 1 \qquad 2x + y = 9$$
$$3(2) - 5 = 1 \qquad 2(2) + 5 = 9$$
$$6 - 5 = 1 \qquad 4 + 5 = 9$$

The solution is correct. As a further check, use graph paper or a graphics calculator to graph the system.

ACTIVITY 1   **Special Systems**

1 Solve the following system by substitution.

$$-2x + y = -1$$
$$-6x + 3y = 12$$

2 Is the system consistent or inconsistent?

3 What happens to the variables when you solve an inconsistent system by substitution? Does the equation simplify to a true statement?

4 What are the solutions to the system?

5 Solve the following system by substitution.

$$4x + y = -2$$
$$8x + 2y = -4$$

6 Is the system dependent or independent?

7 What happens to the variables when you solve a dependent system by substitution? Does the equation simplify to a true statement?

8 What is the solution for the system?

EXAMPLE The Perimeter of a Rectangle

Suppose the perimeter of a rectangular picture frame is 56 inches. If the width is 6 inches longer than the length, what is the area enclosed by the picture frame?

SOLUTION

Let h represent the height of the frame. Let w represent the width of the frame.

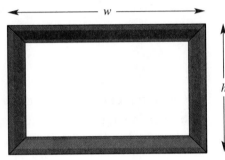

The perimeter is $2h + 2w = 56$.

The width is the height plus six, or $w = h + 6$.

Solve the system $2h + 2w = 56$ by substitution.

$$w = h + 6$$

Substitute $h + 6$ for w in the first equation and solve for h.

$$2h + 2(h + 6) = 56$$

$$2h + 2h + 12 = 56$$

$$4h = 44$$

$$h = 11$$

Substitute 11 for h in the equation $w = h + 6$ and solve for w.

$$w = 11 + 6 = 17$$

Check the solution (11, 17) in each of the original equations.

$$2h + 2w = 56 \qquad w = h + 6$$

$$2(11) + 2(17) = 56 \qquad 17 = 11 + 6$$

$$22 + 34 = 56$$

The height of the frame is 11 inches, and the width is 17 inches. Thus, the area enclosed by the frame is $(11)(17) = 187$ or 187 square inches.

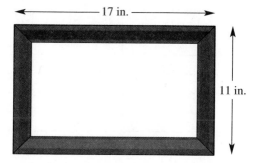

The first step in solving a system by substitution is to isolate one of the variables in one of the equations. You can always solve the system regardless of which variable you choose to isolate. But sometimes one of the variables is easier to isolate than the other.

ACTIVITY 2 Solving for Either Variable

Consider this system.

$$a - 4b = -8$$
$$3a + 5b = 27$$

1 Solve the first equation for a and then for b.

2 Does solving for a or b result in the simplest substitution?

3 Substitute the simplest expression into Equation 2 and solve the system.

Now rework the problem beginning with Equation 2.

4 Solve the second equation for a and then for b.

5 Does solving for a or for b result in the simplest substitution?

6 Substitute the simplest expression into the first equation and solve the system.

7 Check your solution using a graphics calculator.

WORKPLACE COMMUNICATION

1. Plot the data points for plasma source voltage as a function of average thickness of gold deposited for a target thickness of 8 microns. Use your graph to estimate the plasma source voltage for an 8 micron average deposition.

2. For a more accurate value, pick two data points for voltage (V) and two for thickness (t) for a target thickness of 8 microns. Using these data points, write two equations in the form $V = mt + b$. Solve these two equations for the slope and intercept of the line. Then calculate the plasma source voltage for an 8 micron deposition.

Date; Time of Production	Target Thickness of Gold Deposition (microns)	Average Thickness of Gold Deposited (microns)	Plasma Source Voltage (kilovolts)	Time of Exposure (hours)	Average Temperature (Celsius)
03/12;AM	8.0	15.1	20.1	4.67	192
03/12;PM	8.0	12.6	18.1	4.35	195
03/13;AM	8.0	4.2	9.1	4.14	189
03/13;PM	8.0	4.9	10.1	4.33	
03/14;AM	12.0	6.4			
03/14;PM	12.0				
03/15;AM	12.0				
03/15;PM	12.0				

Susan:

This is a portion of my team's data log from last week's solar cell production runs. As you can see, we are having a hard time controlling the thickness of gold in the plasma deposition process. Our specs called for target thicknesses of 8 and 12 microns, and we were off target every run.

Tomorrow we start another production run, and we need better control over the gold thickness. Can you adjust the plasma source voltage for an 8 micron deposition?

Thanks,
Anthony

LESSON ASSESSMENT

Think and Discuss

1 Describe the steps for solving a system of equations by substitution.

2 Given a system of equations that is inconsistent, what happens when you solve the system by substitution? Give an example.

3 In a system of equations that is consistent and dependent, what happens when you solve the system by substitution? Give an example.

Solve by substitution. Identify the system as inconsistent, dependent, or independent.

4. $x = -2y + 4$
  $3x - y = 26$

5. $2x + 4y = 12$
  $2y = 6 - x$

6. $x + y = 0$
  $y = x - 2$

7. $x - 5y = 6$
  $2x + 3y = 25$

8. $5x + 3y = 10$
  $2x + y = 3$

9. $x - y = 8$
  $3x + 3y = 24$

10. $6x + 2y = 5$
  $3x + y = 3$

11. $4x + 3y = 6$
  $2x + y = 4$

12. $2x + y = -1$
  $4x + 2y = -4$

Write a system of equations to model each situation. Solve the system by substitution. Check the solution.

13. At the beginning of school, Nancy bought one 3-subject notebook and four single-subject notebooks for $15. If the 3-subject notebook cost twice as much as a single-subject notebook, how much did each type of notebook cost?

14. Artier has a total of $7000 invested, part in a mutual fund and part in a certificate of deposit (CD). The CD pays 7% interest, and the mutual fund pays 9%. If he receives $550 per year in interest, how much does Artier have in each investment?

15. Together Nita and Joe do 95 haircuts a week. If Nita does 16 fewer than twice as many as Joe, how many haircuts does each person do?

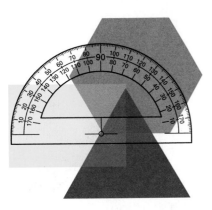

The sum (*S*) of the measures of the angles of a polygon with *n* sides is found by using the formula

$$S = 180(n - 2)$$

Find the total number of degrees in each polygon.

16. A triangle.   **17.** A quadrilateral.   **18.** A hexagon.

19. The measures of the angles in a triangle are represented by *a*, *a* + 20, and *a* + 25. Find the measure of each angle.

20. The measures of the angles in a parallelogram are represented by 2*m*, 2*m*, 3*m* − 10, and 3*m* − 10. Find the measure of each angle.

Suppose you have a hat containing 10 tickets. The tickets are numbered 1, 2, and so on through 10. You draw one ticket from the hat. What is the probability the number on the ticket will be

21. Even?   **22.** A prime number?

23. Greater than 3?   **24.** Odd and less than 8?

You remove one ticket and look at it. You put the ticket back in the hat and draw again. What is the probability that

25. The number on the first ticket is even, and the number on the second ticket is odd?

26. The number on the first ticket is odd, or the number on the second ticket is odd?

27. The sum of the numbers on the two tickets is 10?

28. The sum of the numbers on the two tickets is greater than 18?

LESSON 8.4 SOLVING SYSTEMS BY ADDITION OR SUBTRACTION

Nancy and Jim are team leaders at the Allied Computer Chip Manufacturing Company. The production supervisor needs to report the number of computer chips each team made on Friday. The supervisor knows the total number of chips produced by both teams is 130. Nancy's team made 10 more chips than Jim's team.

Let x represent the number of chips made by Nancy's team. Let y represent the number of chips made by Jim's team.

The following system models the information known by the supervisor.

$$x + y = 130$$
$$x - y = 10$$

Both the graphing and substitution methods will solve this system. However, you can also use another method based on an alternate form of the Addition Property of Equality.

> **Addition Property of Equality (Alternate Form)**
> For all numbers a, b, c, and d,
> if $a = b$ and $c = d$, then $a + c = b + d$.

This property allows you to form a new equation from two existing equations. For example, suppose you have the following system of equations:

$$a = b$$
$$c = d$$

If you add the left sides and then the right sides, a new equation results.

$$\begin{aligned} a &= b \\ +\ c &= d \\ \hline a + c &= b + d \end{aligned}$$

The reason for using the Addition Property of Equality to form a new equation is that in certain cases the new equation will contain variables that are opposites. Since opposites sum to zero, the new equation will contain only one variable, which you can solve using the methods described in Chapters 4–6. For instance, start with the system that models computer chip production.

$$x + y = 130$$
$$x - y = 10$$

Now use the Addition Property of Equality to add the left and right sides of these two equations.

$$(x + y) + (y + y) = 130 + 10$$

Notice that y and $-y$ are opposites, and their sum is zero. Thus the y variable is eliminated, leaving an equation with only the single variable x.

$$(x + x) + 0 = 130 + 10$$
$$2x = 140$$

The solution to this equation is $x = 70$. Substituting this value for x in either of the original equations results in $y = 60$. Since the ordered pair (70, 60) is the solution for the system, Nancy's team made 70 computer chips and Jim's team made 60 computer chips on Friday.

Chip Production

Nancy's Team	Jim's Team
70	60

Ongoing Assessment

Solve this system using the alternate form of the Addition Property of Equality.

$$x + y = 100$$
$$x - y = 20$$

You can also use an alternate form of the Subtraction Property of Equality to solve systems of linear equations.

Subtraction Property of Equality (Alternate Form)

For all numbers a, b, c, and d,

If $a = b$ and $c = d$, then $a - c = b - d$.

Subtraction can also solve the computer chip production equations.

$$x + y = 130$$
$$x - y = 10$$

This time subtract the expressions on each side of the equal signs. Notice in the following steps how the x-variable is eliminated.

$$x + y = 130 \qquad \Rightarrow \qquad x + y = 130$$
$$-(x - y) = -10 \qquad \Rightarrow \qquad -x + y = -10$$
$$\overline{} \qquad\qquad\qquad \overline{}$$
$$2y = 120$$
$$y = 60$$

The result of replacing y by 60 in either original equation gives $x = 70$.

EXAMPLE

Solve this system.

$$2x + 2y = 10$$
$$3x + 2y = 14$$

SOLUTION

Use the alternate form of the Subtraction Property of Equality to subtract the expressions on the left side and the expressions on the right side of the equal sign.

$$2x + 2y = 10 \qquad \Rightarrow \qquad 2x + 2y = 10$$
$$-(3x + 2y) = -14 \qquad \Rightarrow \qquad -3x - 2y = -14$$
$$\overline{} \qquad\qquad\qquad \overline{}$$
$$-x + 0y = -4$$
$$-x = -4$$
$$x = 4$$

To solve for y, substitute 4 for x in the first equation.

$$2(4) + 2y = 10$$
$$8 + 2y = 10$$
$$2y = 2$$
$$y = 1$$

The solution to the system is (4, 1). Check this solution in both equations.

Use the alternate form of the Subtraction Property of Equality to solve the following system:

$$2x + y = -2$$
$$2x + 3y = 0$$

ACTIVITY   **Solving a System by Subtracting**

On Monday, two computer chip manufacturing teams produced a total of 80 chips. Nancy's team made four fewer than twice the number produced by Jim's team.

Let x represent the number of items made by Nancy's team. Let y represent the number of items made by Jim's team.

1 What equation models the total number of chips made by the teams?

2 Write an equation that models the number of chips made by Nancy's team.

3 Write the equations from Steps 1 and 2 in the standard form $ax + bx = c$.

4 Solve this system using the alternate form of the Subtraction Property of Equality.

5 How many chips did each team produce on Monday?

You can use the addition or subtraction method to solve a linear system. Write each equation in standard form, $ax + by = c$. Add or subtract the terms on the same side of the equal signs of the two equations to eliminate one of the variables from the system. Then solve for the other variable.

Solve this system by the addition or subtraction method.

$$2x = 3y - 5$$
$$3y + 4x = 8$$

CULTURAL CONNECTION

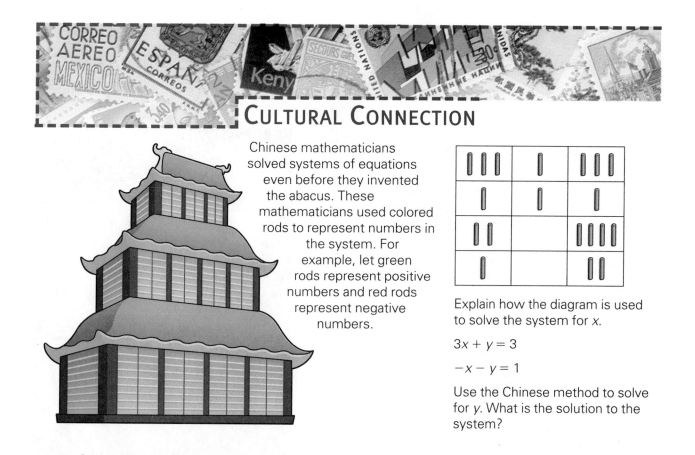

Chinese mathematicians solved systems of equations even before they invented the abacus. These mathematicians used colored rods to represent numbers in the system. For example, let green rods represent positive numbers and red rods represent negative numbers.

Explain how the diagram is used to solve the system for x.

$$3x + y = 3$$

$$-x - y = 1$$

Use the Chinese method to solve for y. What is the solution to the system?

LESSON ASSESSMENT

Think and Discuss

1 When is the addition method the best method for solving a system of equations in standard form? When is the subtraction method best?

2 How do you write an equation in standard form?

3 Why should the two equations be written in standard form before a system is solved?

Solve by addition or subtraction.

4. $2y - 2x = 3$
  $2y - 5x = 9$

5. $5y + 3x = 0$
  $3y - 3x = 0$

6. $6x - 4y = 4$
  $4y - 6x = 2$

7. $x - 3y = -6$
  $x + y = 6$

8. $x + y = 2$
  $2x - y = 1$

9. $3x + y = 6.5$
  $x + y = 2.5$

10. $2x + y = 2$
  $2x + 3y = 14$

11. $x - y = 2$
  $x + y = 4$

12. $x - y = -2$
  $2x - y = 1$

Write a system of equations for each situation. Solve and check your answers.

13. Because of limited storage space at the job site, a bricklayer must schedule two deliveries. The two deliveries must contain a total of 23 pallets of bricks. One delivery must have three more pallets than the other. How many pallets of bricks will be in each delivery?

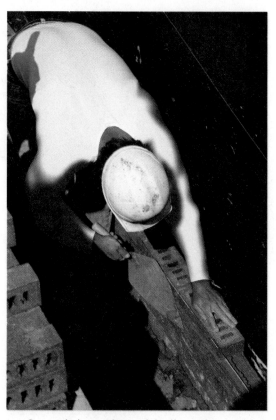

14. In a surveyor's drawing using a coordinate system, one side of an angle lies along the line $x + y = 3$, and the other side lies along the line $2x + y = -3$. Find the coordinates of the vertex of the angle.

15. To send a telegram, you must pay a fixed (or flat) rate for the first ten words and a fixed charge for each additional word. The cost to send a 17-word telegram is $14.55. The cost for a 21-word telegram is $17.15. What is the flat rate, and what is the fixed charge for each additional word?

The American Sand and Gravel Company charges $215 for one ton of sand. One of the American Sand and Gravel Company trucks can haul a load of sand in a bed that measures 8 feet by 4 feet by 2 feet. One cubic yard of sand weighs about 1.212 tons.

16. How many cubic yards of sand will one truck hold?

17. How many tons of sand will one truck hold?

18. If one driver makes six deliveries to a construction site, what is the cost of the sand delivered?

19. Write a formula to find the number of truckloads (N) that can be delivered for x dollars.

Let $g(x) = \dfrac{1}{x}$.

20. Find $g(5)$.

21. Find $g(0.5)$.

22. Graph g in intervals of 0.5 for the domain from 0.5 through 5.

23. What is the range for the domain given in Exercise 22?

24. What happens to $g(x)$ as x becomes very large?

25. What happens to $g(x)$ as x becomes very small?

LESSON 8.5 SOLVING SYSTEMS BY MULTIPLICATION

In the last lesson, you used addition and subtraction to solve systems of linear equations. This resulted in the elimination of one of the variables. However, most of the time the variables are not opposites, and another step is needed.

EXAMPLE 1   Use Multiplication First

Solve this system.

$$2x = 4y - 5$$
$$3y + 4x = 12$$

SOLUTION

First write the equations in standard form.

$$2x - 4y = -5$$
$$4x + 3y = 12$$

None of the terms are opposites. Thus, adding or subtracting will not eliminate a variable. However, if you use the Multiplication Property of Equality to multiply both sides of the first equation by -2, opposite terms will result. The sum of these opposite terms will equal zero, thus eliminating the x-variable.

$$-2(2x - 4y) = -2(-5) \quad \Rightarrow \qquad -4x + 8y = 10$$
$$4x + 3y = 12 \qquad\qquad \Rightarrow \quad + \underline{4x + 3y = 12}$$
$$0 + 11y = 22$$
$$y = 2$$

Substitute 2 for y in either of the original equations to find $x = 1.5$.

Check to make sure $(1.5, 2)$ is the solution to the system.

Ongoing Assessment

Solve this system of equations using multiplication and then addition.

$$3x - 5 = y$$
$$5 = x - 2y$$

Rachel works as a chemist's assistant. For an experiment, Rachel needs 25 gallons of a 24% alcohol solution. She has two solutions available: a 20% alcohol solution and a 30% alcohol solution. How much of each solution should Rachel mix to get a 24% alcohol solution?

You can model this problem using a system of linear equations.

Let x represent the number of gallons of 20% solution.

Let y represent the number of gallons of 30% solution.

The total number of gallons of the 24% solution is

$$x + y = 25$$

The alcohol in the 20% solution (20% of x gallons) plus the alcohol in the 30% solution (30% of y gallons) must equal the total alcohol in the 24% solution (24% of 25 gallons). That is,

$$20\%\ x + 30\%\ y = 24\%(25) \text{ or}$$
$$0.20x + 0.30y = 6$$

Rachel must solve this system.

$$x + y = 25$$
$$0.20x + 0.30y = 6$$

First, multiply each side of the second equation by 100 to eliminate the decimals.

$$x + y = 25$$
$$20x + 30y = 600$$

If you multiply each side of the first equation by 30, both equations will have $30y$ as a common term.

Multiplication

$x + y = 25$ $\Rightarrow$ $30x + 30y = 750$

$20x + 30y = 600$ $\Rightarrow$ $20x + 30y = 600$

Now you can use the subtraction method to eliminate $30y$.

$30x + 30y =\ \ \ 750$ $\Rightarrow$ $30x + 30y =\ \ \ 750$

$-(20x + 30y) = -650$ $\Rightarrow$ $\underline{-20x - 30y = -650}$

              $10x + 0 =\ \ \ 150$

The result is $10x = 150$, or $x = 15$. To find y, substitute 15 for x in the first equation. The result is $y = 10$.

Thus, Rachel needs to mix 15 gallons of the 20% solution and 10 gallons of the 30% solution to get 25 gallons of the 24% solution.

Critical Thinking To solve Rachel's system of equations by addition, what number should be used to multiply each side of the first equation?

EXAMPLE 2 Pick the Right Combination

Solve this system.

$$2x + 3y = 0$$

$$3x + 2y = -5$$

SOLUTION

One way to solve the system is to choose multipliers that will result in equal coefficients of x. Multiply the first equation by three and the second equation by two. Then eliminate a variable by subtraction.

$$\begin{aligned}
3(2x + 3y) = 3(0) &\Rightarrow 6x + 9y = 0 &\Rightarrow &\quad 6x + 9y = 0 \\
2(3x + 2y) = 2(-5) &\Rightarrow 6x + 4y = -10 &\Rightarrow &\underline{-(6x + 4y) = 10} \\
& & &\quad\quad 0 + 5y = 10 \\
& & &\quad\quad\quad\quad y = 2
\end{aligned}$$

Substitute 2 for y and solve for x in the first equation.

$$\begin{aligned}
2x + 3y &= 0 \\
2x + 3(2) &= 0 \\
2x &= -6 \\
x &= -3
\end{aligned}$$

Check $(-3, 2)$ in the original equations.

Ongoing Assessment

Solve the system in Example 2 by first eliminating the variable y.

Critical Thinking The addition, subtraction, and multiplication methods are sometimes referred to as the elimination method. Explain why.

LESSON ASSESSMENT

1 How do you use addition or subtraction to solve a system of equations in standard form if neither of the variables are eliminated when you add or subtract?

2 Given a system of equations that is consistent and dependent, what happens when you try to solve the system using multiplication?

3 Given a system of equations that is inconsistent, what happens when you try to solve the system using multiplication?

4 Explain how division can be used to solve a system of equations.

Practice and Problem Solving

Solve each system by multiplication. Check the solution. If the system cannot be solved, identify the system as either inconsistent or dependent.

5. $3x - 4y = -9$
$x - 3y = 7$

6. $5x - 2y = 5$
$x + 4y = 12$

7. $2x + 3y = 4$
$4x + 5y = 6$

8. $2x + 3y = 4$
$8x + 2y = 5$

9. $2x - y = 6$
$3x - 2y = 2$

10. $4x + 3y = -2$
$3x + 2y = 3$

11. $4x + 3y + 6 = 0$
$3x - 2y - 4 = 0$

12. $8x + 5y = 3$
$7x + 3y = -7$

13. $6x + 9y = -3$
$2x + 3y = -1$

Write a system of equations to model each situation. Solve and check your solutions.

14. To service her car before a trip, Ms. Stein bought motor oil for $1.25 per quart and engine coolant for $2.50 a quart. Altogether she bought 10 quarts and spent $15. How many quarts of motor oil and coolant did Ms. Stein buy?

15. A rental company charged Mr. Juarez $90 to rent a generator and sprayer for six hours. On his next job Mr. Juarez rented the same equipment, but used the sprayer for 4 hours and the generator for 8 hours at a total cost of $100. What was the hourly rate for each piece of equipment?

16. Rasheed spent $15 for three rolls of ASA400 film and one roll of ASA100 film for his camera. He decided to take some extra film on a vacation. So Rasheed went back to the same store and bought two more rolls of ASA400 film and three more rolls of ASA100 film. The extra film cost $17. What is the cost of each type of film?

17. An airline baggage truck can hold 95 pieces of baggage in two separate compartments. In one trip from the airplane to the terminal, the truck is filled to capacity. One compartment is holding 14 more than twice the other. How many pieces of baggage are in each compartment of the truck?

18. The Giant Bike Company builds dirt bikes and racing bikes. The cost to build a dirt bike is $55 for materials and $60 for labor. The cost to build a racing bike is $70 for materials and $90 for labor. For this quarter, the Giant Bike Company has a budget of $9000 for materials and $10,500 for labor. How many of each type of bike can the company build this quarter?

Mixed Review

Write each number in scientific notation.

19. The wavelength of an X-ray is about 0.000000095 centimeters.

20. The mass of the Earth is about 6,600,000,000,000,000 metric tons.

21. A TV signal travels about 30,000,000,000 centimeters per second.

22. A protective coating used on a metal is about 0.0005 inches thick.

23. A formula used to determine the volume of a log is

$$V = \frac{1}{2} L(B + b)$$

The volume (V) of a log is 40 cubic feet. The length (L) of the log is 20 feet. The area (B) of one end of the log is 2.5 square feet. Find the area (b) of the other end of the log.

24. The force on a beam is found using this formula:

$$M = \frac{L(wL + 2P)}{8}$$

Rearrange the formula to solve for P.

Write an equation in one unknown to model each situation. Solve the equation and check your solution.

25. The total weight of two machine parts is 8.6 pounds. One part weighs 5 pounds more than the other part. Find the weight of each machine part.

26. Jack needs 23 two-by-four boards of 16-foot and 20-foot lengths to build rafters for a house. He needs seven fewer 16-foot boards than 20-foot boards. How many of each size does Jack need?

LESSON 8.6 SOLVING SYSTEMS WITH DETERMINANTS

In Lesson 8.5 you solved the system

$$2x + 3y = 0$$
$$3x + 2y = -5$$

using multiplication and subtraction. You can also solve this system by substitution or graphing. The last method you will learn to use is similar to the multiplication method.

First, write the coefficients as a square array of two rows and two columns enclosed in brackets. This array of numbers is called a **matrix**. The numbers or variables are called the **elements** of the matix. A matrix is usually designated by a capital letter. The following matrix is referred to as matrix A.

$$A = \begin{bmatrix} 2 & 3 \\ 3 & 2 \end{bmatrix}$$

The **determinant** of matrix A is computed in the following way:

$$(2 \bullet 2) - (3 \bullet 3) = -5$$

A pair of vertical lines is used to designate the determinant of a matrix. The determinant of matrix A is written as det A.

$$\det A = \begin{vmatrix} 2 & 3 \\ 3 & 2 \end{vmatrix} = -5$$

In general, if a, b, c and d are elements of a square matrix, the determinant of the matrix is computed as follows:

$$\begin{vmatrix} a & b \\ c & d \end{vmatrix} = ad - bc$$

The rule for finding the determinant of a square matrix with four elements is:

Find the difference of the products of the diagonal elements.

EXAMPLE 1 Finding a Determinant

Evaluate the determinant of matrix A.

$$A = \begin{bmatrix} 3 & 0 \\ 2 & -5 \end{bmatrix}$$

Find the difference of the products of the diagonal elements of the determinant.

$$\det A = \begin{vmatrix} 3 & 0 \\ 2 & -5 \end{vmatrix}$$

The value of the determinant is $(3 \cdot -5) - (2 \cdot 0) = -15 - 0 = -15$.

Ongoing Assessment

Evaluate the determinant of matrix B.

$$B = \begin{bmatrix} 2 & 0 \\ 3 & -5 \end{bmatrix}$$

You can use determinants to solve for the values of x and y in a system of equations. First write the system with the equations in standard form.

$$ax + by = c$$
$$dx + ey = f$$

You can use the multiplication method to solve this system for x and y. The result is the following:

$$x = \frac{ce - bf}{ae - bd} \qquad y = \frac{af - cd}{ae - bd}$$

In the early 18th century a Swiss mathematician, Gabriel Cramer, devised a way to use determinants to solve a system of linear equations. To use Cramer's rule, you must first write the equations in standard form. The values for x and y can then be found by using determinants.

$$x = \frac{\begin{vmatrix} c & b \\ f & e \end{vmatrix}}{\begin{vmatrix} a & b \\ d & e \end{vmatrix}} \text{ and } y = \frac{\begin{vmatrix} a & c \\ d & f \end{vmatrix}}{\begin{vmatrix} a & b \\ d & e \end{vmatrix}}, \text{ where } \begin{vmatrix} a & b \\ d & e \end{vmatrix} \neq 0$$

Critical Thinking   Evaluate the determinants in the numerators and denominators of the equations for x and y. Compare the results with the equations found above from the multiplication method. Are the equations the same?

EXAMPLE 2 Using Determinants

Solve the following system using determinants.

$$x - 4y = 1$$
$$2x + 3y = 13$$

Compare the equations to the general form:

$$ax + by = c$$
$$dx + ey = f$$

Find the corresponding values for a, b, c, d, e and f.

$$a = 1 \qquad b = -4 \qquad c = 1$$
$$d = 2 \qquad e = 3 \qquad f = 13$$

Substitute these values into the formulas for x and y, expressed as ratios of two determinants, as follows:

$$x = \frac{\begin{vmatrix} c & b \\ f & e \end{vmatrix}}{\begin{vmatrix} a & b \\ d & e \end{vmatrix}} = \frac{\begin{vmatrix} 1 & -4 \\ 13 & 3 \end{vmatrix}}{\begin{vmatrix} 1 & -4 \\ 2 & 3 \end{vmatrix}}$$

$$y = \frac{\begin{vmatrix} a & c \\ d & f \end{vmatrix}}{\begin{vmatrix} a & b \\ d & e \end{vmatrix}} = \frac{\begin{vmatrix} 1 & 1 \\ 2 & 13 \end{vmatrix}}{\begin{vmatrix} 1 & -4 \\ 2 & 3 \end{vmatrix}}$$

Finally, use the definition of a determinant to solve for x and y.

$$x = \frac{\begin{vmatrix} 1 & -4 \\ 13 & 3 \end{vmatrix}}{\begin{vmatrix} 1 & -4 \\ 2 & 3 \end{vmatrix}} = \frac{(1)(3) - (13)(-4)}{(1)(3) - (2)(-4)} = \frac{3 + 52}{3 + 8} = \frac{55}{11} = 5$$

Thus, $x = 5$.

$$y = \frac{\begin{vmatrix} 1 & 1 \\ 2 & 13 \end{vmatrix}}{\begin{vmatrix} 1 & -4 \\ 2 & 3 \end{vmatrix}} = \frac{(1)(13) - (2)(1)}{(1)(3) - (2)(-4)} = \frac{13 - 2}{3 + 8} = \frac{11}{11} = 1$$

Thus, $y = 1$.

The solution to the system of equations is the ordered pair $(5, 1)$.

Solve the following problem by setting up a system of linear equations and using determinants. A car dealer nets $500 for each family car sold and $1000 for each sports car sold. The dealer is urged to sell 4 family cars for each sports car sold. To stay in business, the dealer must net at least $3000 per week to meet costs. How many family cars and how many sports cars must the dealer sell each week?

ACTIVITY | **Using Technology**

Most graphics calculators and some simple computer programs can evaluate determinants.

1 Explain how to find a determinant on the graphics calculator of your choice.

2 Find a computer program for solving a system using determinants. Program a computer to solve the system in Example 2.

WORKPLACE COMMUNICATION

The CELSAT Division has 7 technicians assigned. The Communications Electronics Division has 11. Write and solve a system of two linear equations to determine how many people each Division Chief can send to the course. What other factors might be considered in determining who attends the course?

At the weekly staff meeting, the Director says:

"I picked up this brochure at a training conference in Washington, DC last week. I want to send some of our top technical people from the CELSAT Division and the Communications Electronics Division to the course. The training budget will pay tuition and travel for nine people. The two Division Chiefs will decide who goes, but divide up the nine proportionately—depending on the number of technicians assigned to each Division."

> **The *Continuing Education Association* Announces a
> One-Week Short Course on
> Introduction to Modern Digital Communications**
>
> Time: This course is offered Monday through Friday, the first
> full week of every month.
> Place: Headquarters, Continuing Education Association,
> Fairfax, VA
> Instructors: Ms. Joyce Morrison-Bell
> Mr. Raymond D. Wong
> Course Outline: The Electromagnetic Spectrum
> Principles of Communication
> Wireless Communication
> Satellite Communications
> Digital Data Communications
> Examples of Modern Communication Systems
> Hands-On Laboratory Activities
> Who Will Benefit: This course is designed for newly assigned technicians or
> experienced technical personnel needing a refresher
> course in modern applications of digital communications.
> Prerequisites: Successful completion of High School Algebra and
> Geometry. Knowledge of Trigonometry will prove useful.
> Cost: $1499
> Registration Form and Hotel Information Are Attached
>
> CEA Announcement Number 32

LESSON ASSESSMENT

Think and Discuss

1 Explain how to use determinants to solve a system of equations.

2 What are the advantages and disadvantages of using determinants to solve a system of equations?

Practice and Problem Solving

Solve each system using determinants. If the system cannot be solved, identify the system as either dependent or inconsistent.

3. $y + 2x = -3$
$x - y = 1$

4. $3x + 5y = 2$
$2x - 3y = 5$

5. $2x - 3y = 12$
$5x + 2y = -8$

6. $7x + 6y = 5$
$5x + 4y = 4$

7. $2x - 5y = -5$
$15x - 6y = 15$

8. $3x + 8y = 0$
$6x + 7y = 5$

9. $2x - 4y = -14$
$3x = -7y + 5$

10. $3x + 4y = 4$
$9x + 12y = 16$

11. $6x + 3y = 1$
$3x - 2y = 4$

Write a system of equations to model each situation. Then solve the system using determinants and a calculator. Check the solution.

12. Home Security, Inc. monitors two separate radio signals from home security systems. The sum of the power levels of the signals is 4 milliwatts. The stronger signal has 10 times the power of the weaker signal. What is the power level of each signal?

13. A car travels 15 miles per hour faster than a truck. In the same amount of time that the car travels 195 miles, the truck travels 150 miles. Find the speed of each vehicle.

14. Lisa invests $10,000 in two different funds. One fund earns 5% annually, and the other earns 8% annually. At the end of one year, Lisa receives a check for $680 in earnings for her investment in the funds. How much did Lisa invest in each fund?

Mixed Review

Solve each equation for a.

15. $3a - 9 = -5a + 15$ **16.** $2(a + b) = c$ **17.** $\dfrac{a - 4}{9} = \dfrac{6}{15}$

Kim has recorded the number of video rentals for his store, Hello Hollywood, over a 14-day period.

$$125, 118, 126, 115, 221, 200, 101,$$
$$115, 135, 158, 140, 175, 218, 115$$

18. What is the mode of the data?

19. What is the median of the data?

20. Create a frequency distribution table where the first class is 101–125 videos rented and the remaining classes are in intervals of 25 videos rented.

The Smiths have four children, but you do not know the number of boys or girls.

21. Create a sample space to show the possible outcomes for boys and girls in the Smith family.

What is the probability that there are

22. Exactly three girls?   **23.** Exactly two boys?

24. At least two boys?   **25.** At least one boy and one girl?

MATH LAB

Activity 1: Solving a Linear Equation as a System

Equipment   Graphics calculator or graph paper

Problem Statement

You will solve linear equations by graphing the expressions on either side of the equal sign. This Activity assumes you are using a graphics calculator. Before you begin, be sure you know how to input an equation, determine the coordinates of a point, and use the ZOOM and TRACE features of your graphics calculator. If you do not have access to a graphics calculator, use graph paper to complete each step.

Procedure

a   Solve $3x - 7 = 11$ algebraically.

b   Graph $y = 3x - 7$ and $y = 11$ on the same coordinate axes. What are the coordinates of the point of intersection? How can you use these coordinates to find the solution to $3x - 7 = 11$?

c   Solve the equation $3x + 9 = 5x - 7$ algebraically.

d   Graph this system of equations.

$$y = 3x + 9$$
$$y = 5x - 7$$

What are the coordinates of the point of intersection? What is the solution to the system?

Consider this problem: Latoya has $30 and is saving additional money at the rate of $7 each week. Shawn has $50 and is saving at the rate of $3 each week. In how many weeks will each person have the same amount of money?

e   Set up a system of two equations to model the problem. Solve the system by graphing and check your results.

f You can solve the problem using one unknown. If x represents the number of weeks until each person has the same amount, you can model the problem with this equation:

$$30 + 7x = 50 + 3x$$

Solve this equation by graphing the expressions on each side of the equal sign.

g Compare the methods used in Steps **e** and **f**. How are they alike? How are they different?

h Solve the following problem using a graphics calculator. Then solve the problem algebraically using a system of equations. Which method do you prefer? Explain your answer.

The distance around a rectangular patio is 40 feet. The length is one and one-half times the width. Find the length and width of the patio.

Activity 2: Battleships and Mines

Equipment String or cord, approximately 20 feet long
Paper (different color for each lab group)
Scissors
Masking tape
Tape measure
Optional: Graphics calculator

Problem Statement

You are navigating a battleship during wargames. Your course will take you across several enemy shipping lanes. Your mission is to lay mines at the points where your course crosses the enemy lanes. The enemy shipping lanes are represented by the following equations:

Enemy Lane 1: $x - y = -4$
Enemy Lane 2: $3x - y = 10$
Enemy Lane 3: $x - 2y = -2$

Procedure

a Clear an area of the classroom floor for use as a coordinate system. If your floor has square tiles, use the lines between the tiles as a grid for the x- and y-coordinate plane. If not, measure a uniform grid with a tape measure.

b As a class, identify a length that will serve as "one unit." On a tile floor, the length of one tile is very convenient. For floors without tiles, use a length of about 10 inches or 20 centimeters.

c This Lab will focus on only Quadrant I of the xy-coordinate system. Identify two perpendicular reference lines to serve as the positive x- and y-axes. For example, you can select two perpendicular rows of tiles near a corner of the classroom.

d Measure the units along each axis with the tape measure (or count tiles on a tile floor). Using small pieces of masking tape, label the units along your "axes" from $x = 0$ to $x = 20$ and $y = 0$ to $y = 20$. For convenience you may label every other unit. Create the coordinate system as a class. Only one coordinate system is needed for all the lab groups.

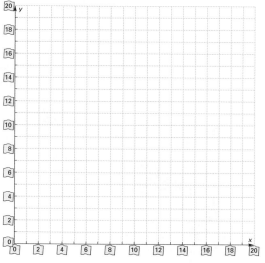

e Each group should use a different equation to represent the course of its battleship. Choose from the following equations.

Equation 1: $5x + 4y = 48$
Equation 2: $x + y = 18$
Equation 3: $3x + y = 10$
Equation 4: $15x - 200y = -1800$
Equation 5: $9x + 5y = 180$

f Determine the points where your equation (the path of your battleship) will intersect each of the enemy shipping lanes. Use any of the methods discussed in the text for solving a system of equations. Round your answers to the nearest 0.1 unit before they are recorded.

Points of Intersection	
Enemy Lane	xy-coordinates
1	
2	
3	

g Cut out three 1-inch squares of colored paper. Each square represents a "mine" dropped by your battleship. Each group should use a different color.

h For each point of intersection, locate the corresponding point on the classroom floor. Use masking tape to fasten a piece of colored paper to the floor at each point of intersection. This will indicate where your battleship has dropped a "mine."

i After all the lab groups have completed "dropping their mines," check each group's success. The equations for the shipping lanes are linear and represent lines. Find two ordered pairs on each of these lines. Use a string to connect the ordered pairs for each line. These three strings represent the three enemy shipping lanes. Did your "mines" lie within the shipping lanes?

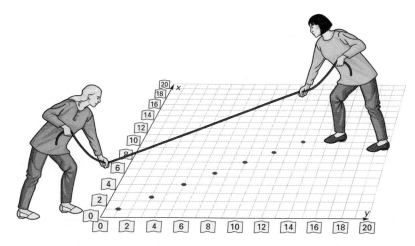

Activity 3: Unmixing a Mixture

Equipment   Box of small bolts
Box of small nuts
Paper cup, approximately 8 ounces
Balance scale with one gram resolution

Problem Statement

Suppose you have a mixture containing two different items. You know the total count and the total weight of the mixture. You also know the weight of each type of item. But you do not know the number of each item in the mixture. You will determine how many items of each type are in the mixture with a simulation using nuts and bolts.

Procedure

a First determine and record the weight of one nut (N) and the weight of one bolt (B). If your scale is not accurate enough to measure just one nut, count and weigh 20 nuts. Divide this weight by 20 to find the weight of a single nut. Similarly, use 20 bolts to determine the weight of a single bolt.

b Place a "random" number of nuts and bolts into a cup. Do not attempt to put an equal number of each into the cup. However, you need to know the *total* number of objects in the cup.

c Give your cup to another lab group. Tell this group only the total number of objects in the cup.

d At the same time, you will receive a cup of nuts and bolts from another group. They should tell you the total number of items in the cup. Record this total as T. Do *not* remove the nuts and bolts from the cup.

e Use the scale to determine the total weight of the cup with nuts and bolts. Record this total weight W. Set the cup aside and perform the calculations that follow.

f Let x represent the number of nuts and y the number of bolts in the cup. Write an equation for the total number of nuts and bolts in the cup. Write an equation for the total weight of the nuts and bolts. (Can you safely ignore the weight of the paper cup?)

g From the preceding step, you should have two equations with two unknowns: x (the number of nuts) and y (the number of bolts). Solve these two equations for x and y. Check your solution by substituting back into your equations.

h Examine your solution from Step **g**. Did your mathematical solution result in whole numbers? If not, what whole number values for x and y would you choose as the actual numbers of nuts and bolts in the cup?

i Now count the nuts and bolts in the cup. Compare this to your solution for x and y. If the actual number of nuts and bolts is different than you predicted, explain why. Will the actual values for x and y make your equations true (or almost true)?

The applications that follow are like the ones you will encounter in many workplaces. Use the mathematics you have learned in this chapter to solve the problems.

Wherever possible, use your calculator to solve the problems that require numerical answers.

A motorboat takes 40 minutes to travel 20 miles down a river with the help of the current. On the return trip, the boat is powered at the same level and moves against the current. The upstream trip (the same 20-mile distance) now takes 50 minutes.

1 Let x represent the speed of the boat in the water and y represent the speed of the water current. While traveling downstream, the boat's speed—as seen from the shore—is the *sum* of the speed of the current and the speed of the boat in the water. Write an equation that describes the downstream speed. This sum is equal to the speed of 20 miles per 40 minutes, or 20/40 miles per minute.

2 While traveling upstream, the boat's speed—as seen from the shore—is the *difference* between the speed of the current and the speed of the boat in the water. Write an equation that describes the upstream speed. This difference is equal to the speed of 20 miles per 50 minutes, or 20/50 miles per minute.

3 Solve the system of two equations to find the speed of the boat in the water (x) and the speed of the current (y).

4 What are the units for the speeds you determined: Miles per hour? Miles per minute? Feet per minute? Explain.

You and your friends are planning a kayak trip. You are in charge of deciding which kayak rental company will be the least expensive. Company A charges $5 per hour for the use of its kayaks. Company B charges an $8 usage fee plus $3 per hour. If c represents the cost and h represents the hours used, you can write the following equations for each company:

Company A: $c = 5h$
Company B: $c = 3h + 8$

5 Draw a graph for each company showing cost as a function of hours.

6 Use the graph to find the number of hours where the costs of the two plans are equal.

7 If you and your friends are going to take a 6-hour trip, which plan is least expensive?

Bill and Laura mow lawns during the summer to earn money. They each calculate their start-up expenses, operating expenses, and income per hour of mowing. Bill and Laura write the following equations for their total income (y) after mowing for x hours:

$$\text{Bill:} \quad y = 9.10x - 125$$
$$\text{Laura:} \quad y = 9.10x - 100$$

8 Graph each equation on the same coordinate plane. Use x-values from zero to 20 hours.

9 Before Bill and Laura mow their first lawn, what is their total income? Explain your answer.

10 After mowing for 20 hours, what is each of their total incomes?

11 If both Bill and Laura work the same number of hours, when will Bill have the same total income as Laura? Explain your answer.

AGRICULTURE & AGRIBUSINESS

A farmer plants two kinds of crops on his 2500 acres of land. The income from Crop A is $230 per acre. The income from Crop B is $280 per acre. The farmer's goal is to earn $625,000 from the sale of the crops. How many acres of each crop should the farmer plant?

12 Let x be the number of acres of Crop A and y be the number of acres of Crop B. Equate sales of x acres of Crop A plus y acres of Crop B to the desired total sales of $625,000.

13 If the farmer uses all 2500 acres of available land, write an equation for the total number of acres planted.

14 Solve this system of equations to find the number of acres of each type of crop the farmer should plant to obtain the desired total sales.

15 Check your result by substituting the x- and y-values into your equations.

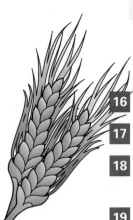

You have been offered a job marketing grain products. The company has two salary plans. Plan A pays a monthly salary of $700 plus a commission of $50 for each ton of grain products that you sell. Plan B pays a monthly salary of $500 plus a commission of $70 for each ton sold.

16 Write these two payment plans as a system of equations.

17 Draw a graph for each equation.

18 How many tons of grain must you sell for Plan A and Plan B to pay the same amount?

19 You estimate that you can sell about 25 tons per month. Which plan should you choose?

Roberta is selling two types of land. Type A land has access to water and sells for $900 per acre. Type B land does not have access to water and sells for $700 per acre. Roberta wants to earn $60,000 and will sell 75 acres.

20 Write an equation for the total earnings from the sale of land.

21 Write a second equation for the total number of acres Roberta will sell.

22 Solve this system of two equations by using determinants. How much of each type of land must Roberta sell to make $60,000?

23 Check your answers by substituting them into your original equations.

Mario is stocking a new pond for his fish farm. The pond is filled by a spring-fed creek at a rate that can support F fish:

$$F = 42m + 250$$

where m is the number of months the pond has been filling. Mario stocks the pond with an initial population of 110 fish (at month $m = 0$). He predicts a population growth rate of 8% per month. Using this information, Mario writes an equation for the population size P:

$$P = 110(1.08)^m$$

where P is the number of fish in the pond at month m, and m is the number of months since the fish were placed in the pond. Notice that this equation is *not linear*. This problem can be solved by graphing.

24 Graph both equations on a graphics calculator. Show values for m from zero months up to 50 months.

25 Mario predicts that the pond will take about 40 months to fill. He wants to begin harvesting the fish just before the population exceeds the limit that the pond can support. When will the predicted population of fish (*P*) equal the number of fish the pond can support (*F*)?

26 Should Mario wait till the pond fills to begin harvesting the fish, or should he start sooner?

BUSINESS & MARKETING

Chris is in charge of buying new suits for The Men's Shop. The purchasing budget is $20,000. Chris decides to stock the inventory with two different types of suits. One type is a designer label costing $400. The other is a nondesigner label costing $250. From previous sales records, Chris knows customers will buy nondesigner suits about 60% of the time and designer suits about 40% of the time.

27 Let *x* represent the number of nondesigner suits and *y* represent the number of designer suits. Equate the purchasing budget of $20,000 to the cost of purchasing *x* nondesigner suits and *y* designer suits.

28 To meet the sales demand, 60% of the purchases should be nondesigner suits and 40% should be designer suits. Write this as a proportion and solve the proportion for *y*.

29 Solve the system of equations from Exercises 27 and 28 by substitution.

30 How many of each type suit should Chris purchase to stay within her budget and meet the sales demand?

31 Check your result by substituting the *x*- and *y*-values into your equations.

You are starting a business that sells water filters. Your initial start-up cost is $1500, and each filter kit will cost $50. You plan to sell each unit for $75. From this information, you can write the following equations, where *n* is the number of filters sold:

$$\text{Cost} = 50n + 1500$$
$$\text{Income} = 75n$$

32 Write each equation as a function of *n*. Draw the graph of each function on the same coordinate plane.

33 From the graph, determine how many filters you need to sell to "break even." (That is, how many filters do you need to sell for income to equal cost?)

Susan's new company jet carries enough fuel for eight hours of flying when cruising at an airspeed of 220 miles per hour. Susan is heading west with a tailwind of 15 miles per hour. She needs to find the maximum distance west she can fly and still have enough fuel left to return home.

Westerly speed = 220 mph + 15 mph

Easterly speed = 220 mph – 15 mph

34 Consider the following variables: the time t_w for the westerly (outbound) flight and the time t_e for the easterly (return) flight. The total flight time is limited by the amount of fuel. Write an equation that shows the flight times, t_w and t_e, totaling eight hours.

35 Since the westerly flight is assisted by a tailwind, the airplane's speed is $220 + 15 = 235$ miles per hour. The distance traveled during the westerly flight is this speed times t_w. Write the product for the distance traveled during the westerly flight.

36 Since the easterly flight is into a headwind, the airplane's speed is $220 – 15 = 205$ miles per hour. The distance traveled during the easterly flight is this speed times t_e. Write the product for the distance traveled during the easterly flight.

37 Of course, Susan wants the outbound westerly flight distance to equal the easterly return flight distance. What happens if they are not equal? Equate the two products from Exercises 35 and 36 to obtain a second equation with t_w and t_e.

38 Solve the system of two equations. Check your results by substituting into your original equations.

39 Use the product from Exercise 35 to find how far west Susan can fly under these conditions and still have enough fuel to return home.

Your trucking company needs to move 21 tons of gravel. You have eight qualified drivers in the company and two types of trucks. One type of truck can haul 5 tons and the other type can haul 3 tons. An insurance requirement specifies that 5-ton trucks must have two drivers in the cab during operation. Three-ton trucks require only one driver. Let x be the number of 5-ton trucks you will use and y the number of 3-ton trucks.

40 Write two linear equations relating x and y that determine how many of each size truck is needed to move the gravel in one trip using all available drivers.

41 Solve the system using determinants.

42 How many of each size truck will you use?

You must determine the price schedule for concert seating at the local orchestra hall. You will sell tickets for two types of seats: one type will sell for $5 each, and the second type for $8 each. There are a total of 1500 seats, and you expect to sell tickets for all of them. Your ticket sales total $10,500. Let x be the number of seats that sell for $8 each and y be the number of seats that sell for $5 each.

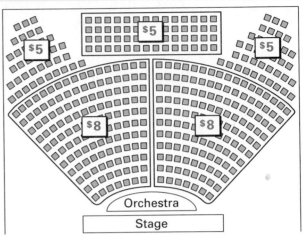

43 Write an equation for the total number of seats as the sum of the two different types of seats.

44 Write an equation for the income from the sale of tickets ($10,500).

45 Solve this system of equations.

Amad wants to earn $500 on a $6000 investment. He is going to split his investment between stocks that he predicts will have an annual yield of 10% and a mutual fund he predicts will have a 7% annual yield. Let x be the amount invested in stocks and y be the amount invested in the mutual fund.

46 Write the equation for the total amount invested as the sum of the amount invested in stocks (x) and the amount invested in the mutual fund (y).

47 Write an equation that equates the amount that Amad wants to earn with the sum of the yield from the stocks (10% of the amount invested in stock) and the yield from the mutual fund (7% of the amount invested in the mutual fund).

48 Graph both equations on the same coordinate axes.

49 From the graph, find the amount that Amad should invest in stocks and the amount he should invest in the mutual fund. about

50 Solve the two equations using the method of determinants and compare the answer to the results obtained using the graphical method.

A cellular telephone company offers two service plans. The economy plan charges a $20 monthly fee, allows 30 minutes of free call time, and charges 50¢ per minute thereafter. The deluxe plan charges a $50 monthly fee, 25¢ per minute for call time, but no free minutes.

51 The cost of the economy plan (*CE*) can be written in terms of the number of minutes (*m*) of call time.

$$CE = 0.50(m - 30) + 20.00, \text{ for } m \geq 30$$

Multiply the terms in this equation. Rewrite it in slope-intercept form.

52 Write a similar equation for the cost of the deluxe plan (*CD*) in terms of the number of minutes (*m*) of call time.

53 Graph these two equations. Use the number of minutes of call time (*m*) as the horizontal axis and cost (*C*) as the vertical axis. What is the cost for the economy plan when *m* is less than 30 minutes?

54 Interpret the graph to find the number of minutes for which the two plans cost the same.

55 Explain how you can use the graph to decide which plan is better for you.

FAMILY AND CONSUMER SCIENCE

You are in charge of catering a banquet. To keep the costs down, you will serve only two entrees. One is a chicken dish that costs $5; the other is a beef dish that costs $7. There will be 250 people at the banquet, and the total cost of the food is $1500.

56 Let *x* be the number of chicken dishes you will prepare. Let *y* be the number of beef dishes. Equate the total number of entrees to the total number of people.

57 Write an equation that equates the total cost of the food to the cost for all the chicken dishes plus the cost for all the beef dishes.

58 Find the number of each type of entree you should prepare. Check your answers.

You are redecorating your home. You have decided to put down a combination of carpet and vinyl floor covering in the family room. The carpet costs $2 per square foot, and the vinyl covering costs $1 per square foot. You can spend a total of $500 on the materials. The area you want to cover is 300 square feet.

59 Let x be the number of square feet of carpet you will use. Let y be the number of square feet of vinyl. Equate the total cost of the materials to the cost of the carpet and the vinyl.

60 Write an equation that equates the total floor area (300 square feet) to the sum of the area covered by carpet and the area covered by vinyl.

61 Solve this system of equations to find the number of square feet of carpet and vinyl that meet your requirements. Check your answers.

You are shopping for refrigerators. Brand A costs $600 and uses $68 per year in electricity. Brand B costs $900 but uses only $55 per year in electricity.

62 Write a system of equations to model the total cost of each refrigerator over t years.

63 Assume that the domain of each equation is from zero years to 15 years (why not 70 years?). Draw a graph of the system.

64 Over the lifetime (that is, the domain) of these refrigerators, will the total costs ever be the same? If so, when?

65 Which refrigerator has a lower total cost over a 15-year lifetime?

As an interior decorator, you are selecting the lighting for a client's new family room. You must decide between regular incandescent light bulbs and energy-saving fluorescent light bulbs. The incandescent bulbs cost 50¢ each and will last for 1000 hours. The fluorescent light bulbs cost $11 each and last 9000 hours. However, the incandescent bulb costs $0.006 per hour to operate, while the fluorescent bulb operates for only $0.0018 per hour. You want to show your client that you have chosen the more economical lighting system.

66 Write a system of equations for the total cost of each type of light bulb for equal time periods. (Hint: one fluorescent bulb lasts as long as nine incandescent bulbs.)

67 Graph the two equations for the total cost for each bulb over the span of 9000 hours.

68 Is there a solution to these two equations? That is, will the total cost of the incandescent bulbs ever equal the total cost of the fluorescent bulb? If so, when?

69 Based on cost alone, which bulb would you recommend to your client? Explain your reasoning.

HEALTH OCCUPATIONS

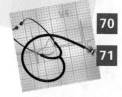

Shon needs three liters of an 8% saline solution. He has a 5% saline solution and a 9% solution in the lab stock room. Before he mixes the two solutions, he needs to calculate the right proportions. Let x be the number of liters of the 5% solution and y be the number of liters of 9% solution.

70 Equate the sum of x and y to the total number of liters needed.

71 To obtain a second equation, Shon starts with the word equation below. The equation uses the fact that the amount of *salt* in the saline solution is the concentration of the solution times the number of liters of solution. When Shon mixes solution A and solution B, the amount of salt in the mixture is the amount of salt from solution A plus the amount of salt from solution B.

(Concentration of A)(Liters of A) + (Concentration of B)(Liters of B) = (Concentration of mixture)(Liters of mixture)

Substitute the appropriate variables and numbers into the word equation and write Shon's second linear equation.

72 Use the method of substitution to solve the system of two equations. How many liters of 5% and 9% saline solution should Shon mix to obtain 3 liters of 8% saline solution? Check your answers.

INDUSTRIAL TECHNOLOGY

Fahrenheit and Celsius are two different scales of temperature measurement. Linear equations relate the two scales.

$$F = aC + b$$

F = the temperature in degrees Fahrenheit.
C = the temperature in degrees Celsius.
a and b = the constants to be determined.

The two scales are related by the fact that water freezes at 32°F or 0°C. Similarly, water boils at 212°F or 100°C.

73 Substitute the temperatures for freezing water into the equation above to obtain an equation with unknowns a and b.

74 Substitute the temperatures for boiling water into the equation above to obtain a second equation with unknowns a and b.

75 Solve this system of equations to determine the values of a and b that satisfy both equations.

76 Rewrite the equation $F = aC + b$ with the calculated values of a and b. Check the resulting equation for $C = 0$ and $C = 100$.

Laura needs 4000 pounds of a 17% copper alloy. To make the alloy, she will mix a 23% copper alloy and a 12% copper alloy. Let x be the number of pounds of the 23% copper alloy. Let y be the number of pounds of the 12% copper alloy.

77 Write the equation for the total number of pounds needed for the mixture as the sum of the pounds of 12% alloy and the pounds of 23% alloy.

78 To obtain a second equation, Laura starts with the word equation below. The equation uses the fact that the amount of copper in an alloy is the concentration of the alloy times the number of pounds of alloy. When Laura mixes the two alloys, the amount of copper in the mixture is the amount of copper from one alloy plus the amount of copper from the other alloy.

(Concentration of A)(Quantity of A) + (Concentration of B)(Quantity of B) = (Concentration of mixture)(Quantity of mixture)

Rewrite this word equation using the variables x and y. Remember to express the percentages as decimal values.

79 Use the method of substitution to solve the two equations. How many pounds of 23% and 12% alloy should Laura mix to obtain 4000 pounds of 17% alloy? Check your answers by substituting the results for x and y into your equations.

You are calibrating a new thermocouple that is used for measuring temperatures between 0°C and 100°C. To do this, you must be able to relate the thermocouple voltage to the temperature being measured. You know that the thermocouple's output voltage is roughly linear between 0°C and 100°C (that is, you can write a calibration equation in the form $y = mx + b$). Thus, for a temperature T, the thermocouple should produce a voltage V according to this calibration equation.

$$T = Vm + b$$

Two of the test measurements with your new thermocouple show that when $T = 0$°C, $V = -1.56$ millivolts, and when $T = 100$°C, $V = 4.76$ millivolts.

80 Substitute each pair of temperature and voltage measurements into the calibration equation to obtain a pair of equations with unknowns m and b.

81 Solve for m and b. Check the results by substituting your answers into the two equations.

82 Write the correct calibration equation with the values of m and b you determined.

Kirchhoff's loop rule analyzes the current (I) in an electronic circuit. This rule states that in any closed loop of a circuit, the sum of the voltage drops (across the resistors) must equal the sum of the voltages supplied. If you apply this rule to the circuit shown here, you obtain the following equations:

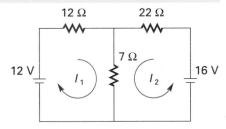

$$12 - 12I_1 - 7(I_1 + I_2) = 0$$

$$16 - 22I_2 - 7(I_1 + I_2) = 0$$

83 Simplify each equation to obtain a system of equations in standard form.

84 Solve this system of equations for I_1 and I_2. Check your solution.

Skills

Solve each system by substitution.

1. $x = -3y + 8$
    $2x - 5y = -17$

2. $x - 2y = -4$
    $5x - 3y = 1$

3. The sum of two integers is 33. One of the integers is 3 more than the other. What are the integers?

Solve each system by multiplication.

4. $2x - y = -7$
    $3x + y = -3$

5. $2x - y = 1$
    $4x - 3y = -5$

Applications

6. The perimeter of a rectangle is 48 feet. The width is 6 feet less than the length. Find the dimensions of the rectangle.

Use the following system of equations for Exercises 7-9.

$$x + y = 3$$
$$2x - 3y = 16$$

7. Write the ratio of determinants needed to solve for x.

8. Write the ratio of determinants needed to solve for y.

9. Solve the system using determinants.

Write a system of equations to model each situation. Solve the system by any method you choose.

10. The total cost for printing a book includes a setup fee and a charge for each book printed. The total cost to print 500 books is $1750. The total cost for printing 1000 books is $3250. What is the setup fee?

11. A car dealer makes a profit of $750 for each 4-door family car sold and $1250 for each 2-door sports car sold. The dealer needs to sell three times as many family cars as sports cars each week. How many of each car must the dealer sell to make a profit of $7000 each week?

12. A chemist needs a 5% sodium solution. The chemist has a 4% sodium solution and an 8% sodium solution available. How much of each solution should be mixed to make 400 milliliters of 5% sodium solution?

Math Lab

13. When solving a system of two equations graphically, what point represents the solution to the system?

14. The course of your battleship follows $3x + y = 10$. The enemy's shipping lanes follow two equations:

Enemy Lane 1: $3x - y = 10$
Enemy Lane 2: $6x + 2y = 10$

Can you lay mines to intercept both enemy lanes? Why or why not?

15. The weight of a bolt is measured at 125 gm. The weight of a nut is measured at 35 gm. A paper cup that contains a total of 29 nuts and bolts is measured at 2190 gm. How many nuts and how many bolts are in the cup?

CHAPTER 9

WHY SHOULD I LEARN THIS?

People in today's high tech world demand quality products. To meet these expectations, engineers design products to high standards using very tight tolerances. These tolerances are often expressed as inequalities. Learn the concepts in this chapter and you will be another step closer to a bright and profitable future.

INEQUALITIES

OBJECTIVES

1. Solve linear inequalities in one variable and graph their solutions.
2. Solve combined inequalities.
3. Solve inequalities involving absolute values and graph their solutions.
4. Graph linear inequalities in two variables.
5. Solve linear programming problems.

So far you have learned to translate work-related problems into equations and then solve the equations for unknowns. However, there are many workplace problems that involve unequal expressions or inequalities. For instance, in building the "Chunnel," technicians used precise tolerances to ensure the tunnel stayed on course as it was constructed under the English Channel. Engineers stated these tolerances to the technicians using such phrases as "greater than" or "less than." These phrases are also used to express inequalities.

For instance, suppose you need to know if a $\frac{3}{4}$-inch bolt is longer than a $\frac{5}{8}$-inch bolt. A number line can help you choose the longer bolt. Visualize a number line divided into eight parts.

You can quickly determine that $\frac{3}{4}$ is farther to the right on the number line than $\frac{5}{8}$. Thus, $\frac{3}{4}$ is greater than $\frac{5}{8}$, and the $\frac{3}{4}$-inch bolt is longer. In this chapter, you will use number lines to picture inequalities and explain their behavior.

As you watch the video for this chapter, notice how people use inequalities in the workplace.

LESSON 9.1 INEQUALITIES AND THE NUMBER LINE

The statement "five is greater than three" is an example of an **inequality**. You can write this statement using a *greater than* symbol.

$$5 > 3$$

You will use several symbols to describe inequalities.

greater than	$>$
greater than or equal to	$\geq$
less than	$<$
less than or equal to	$\leq$
not equal to	$\neq$

An inequality statement can be true, false, or conditional. The inequality $5 < 3$ is false. The inequality $-5 < 3$ is true. However, you cannot tell if the inequality $x < 3$ is true until you know what replacements to use for x. The inequality $x < 3$ is a *conditional inequality*.

You can specify an **interval** by combining conditional inequalities. For example, let n represent a member of the set of whole numbers in the interval between 5 and 10. This set is $\{6,7,8,9\}$. The set can be specified with the **compound inequality**

$$5 < n < 10$$

This inequality is read

$$5 < n \quad \text{and} \quad n < 10$$

and means that n is *between 5 and 10*.

The whole numbers (6, 7, 8, and 9) that make this inequality true are the **solutions** of the inequality. If the domain is the set of all real numbers, then there are an infinite number of solutions because there are an infinite number of real numbers between 5 and 10. In this case, you can use a graph to show the solution.

The graph of the inequality $5 < x < 10$ is called an **open interval**. An open interval does not include the endpoints. Notice that the endpoints of an open interval are represented by an open circle. If the solution *does* contain either endpoint, the corresponding circle is solid. If the solution contains both endpoints, the graph is called a **closed interval**.

Critical Thinking   Describe the graph of $2 \leq x \leq 3$.

EXAMPLE 1 Graphing Inequalities

Graph the solution for each inequality.

a. $x < 35$ **b.** $x \leq -10$ **c.** $x > 25$ **d.** $0 \leq x \leq 20$

SOLUTION

a. $x < 35$

b. $x \leq -10$

c. $x > 25$

d. $0 \leq x \leq 20$

Ongoing Assessment

Graph the solution to each inequality.

a. $x > -3$ **b.** $x < 5$ **c.** $-2 \leq x \leq 2$

EXAMPLE 2 Writing an Inequality

Write an inequality to describe each graph.

a.

b.

c.

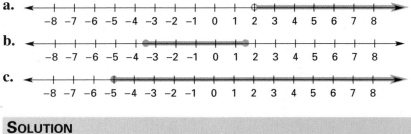

SOLUTION

a. $x > 2$ **b.** $-3.5 \leq x \leq 1.5$ **c.** $x \geq -5$

Many businesses' financial situations are modeled by inequalities.

EXAMPLE 3 Graphing an Interval

Every year since 1996, the Gorgeous Garment Company has made a profit of at least $20,000. It has never made a profit greater than $50,000. Graph an inequality to represent the Gorgeous Garment Company's profit since 1996.

SOLUTION

Let x represent the profit since 1996. The Gorgeous Garment Company's profit is described by the inequality

$$\$20,000 \leq x \leq \$50,000$$

The graph is as follows:

```
 ←——|——|——|——|——|——|——|——|——|——|——|——|——|——|——|——|——|——|→
   -80 -70 -60 -50 -40 -30 -20 -10  0  10  20  30  40  50  60  70  80
```
Scale in thousands of dollars

CULTURAL CONNECTION

Archimedes, one of the greatest of all Greek mathematicians, lived in Egypt more than 2000 years ago. Using only the Egyptian numerals available in his time, he performed complex computations with whole numbers and fractions. Archimedes was probably the first person to compute π correctly to two decimal places. He started with two regular hexagons, one drawn inside and one drawn outside a circle, and the knowledge that π is the ratio of the circumference of a circle to its diameter.

Archimedes computed the perimeters of the two hexagons. He then formed two ratios. The first ratio was the perimeter of the outer hexagon to the diameter of the circle.

The second ratio was the perimeter of the inner hexagon to the diameter of the circle. Archimedes knew the value of π must be between the values of these two ratios. Then he began doubling the number of the sides of the inside and outside polygons to get values closer and closer to π. He finally figured out that π is greater than $\frac{223}{71}$ and less than $\frac{22}{7}$.

1. Write an inequality to model Archimedes' approximation for π.

2. Change the fractions to decimals.

3. How do Archimedes' ratios compare to the value of π we use today?

LESSON ASSESSMENT

Think and Discuss

Let *a* and *b* represent two real numbers. Explain how to graph each inequality.

1 $x < a$

2 $x > b$

3 $a \le x \le b$

Practice and Problem Solving

Write an inequality to describe each graph.

4.

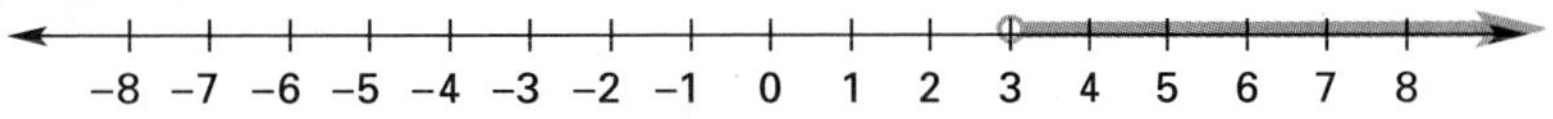

5.

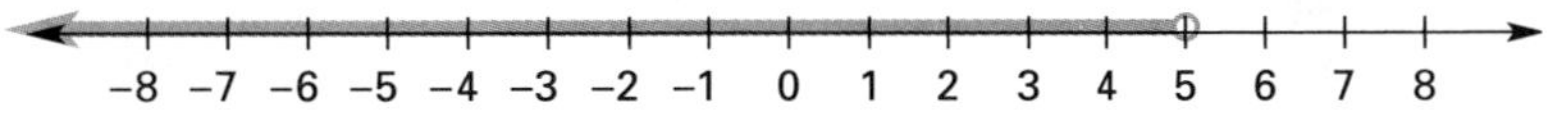

6.

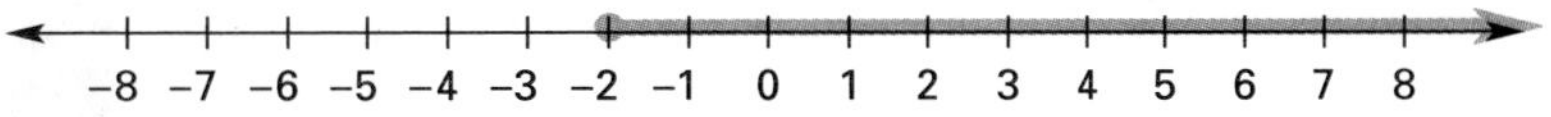

7.

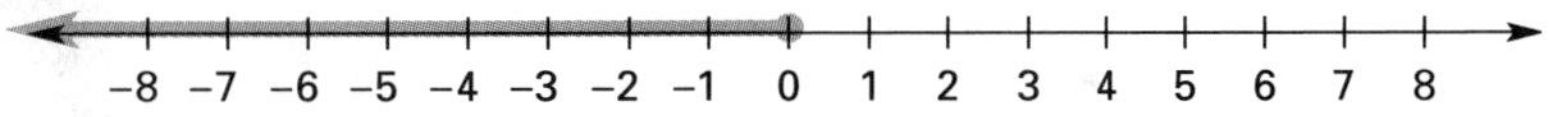

8.

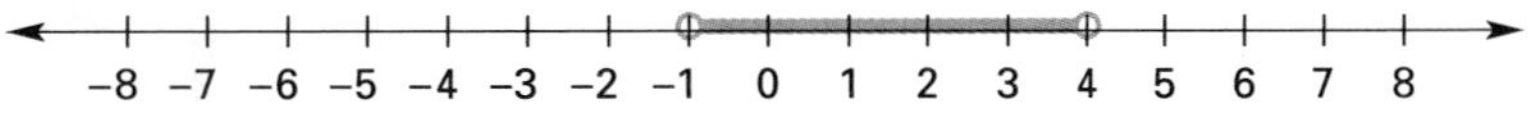

9.

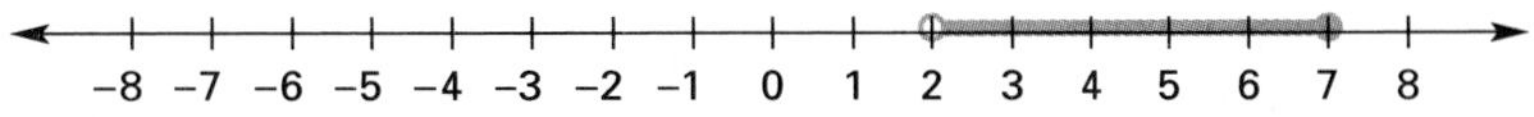

Graph each inequality.

10. $x < 5$

11. $x > 2$

12. $x < -4$

13. $x \ge -6$

14. $-2 \le x \le 2$

15. $4 < x < 7$

Write an inequality that models each situation.

16. Paul's monthly income is more than $1500.

17. Angie must drive less than 25.5 kilometers.

18. The winning time in the 100-meter dash was at least 10 seconds.

19. It takes between $1\frac{1}{2}$ and 2 hours to play a soccer game.

Mixed Review

An antique phonograph needle is approximately a smooth half sphere. Suppose one needle is 0.02 centimeters in diameter. A measurement of the tracking force on a phonograph yields 0.156 Newtons. Write your answers to Exercises 20–21 in scientific notation.

20. The surface area of a sphere is $4\pi r^2$. Convert the diameter of the phonograph needle half sphere to meters. What is the surface area of the needle tip in square meters?

21. Divide the tracking force by the area of the needle tip to find the pressure exerted by the needle on the phonograph record in Newtons per square meter.

The Air Care Furnace Company charges $50 for each service call plus $25 per hour for labor.

22. Write a function to model the cost of having Air Care repair a furnace at someone's home.

23. Graph the function.

24. What are the slope and y-intercept of the graph?

LESSON 9.2 THE ADDITION PROPERTY OF INEQUALITY

The cost of a matinee ticket is $3. After you pay for the ticket, you have less than $9 left. How much did you have before you bought the ticket?

Model this situation with an inequality. Let x represent your initial amount of money.

$$x - 3 < 9$$

How can you solve the inequality?

ACTIVITY   Exploring Inequalities

1 Draw a number line from −8 to 8.

2 Circle the point −4. Circle the point 2. An inequality describing the relationship between the coordinates you circled is −4 < 2.

3 Moving to the right on the number line represents addition. Move three points to the right of each point. Circle the new coordinates.

4 Add 3 to each side of the inequality in Step 2. What is the new inequality showing the addition of 3 to each side of the original inequality? Find the sum of each side of the inequality.

5 Moving to the left on the number line represents subtraction. Start again with the inequality in Step 2. This time move three points to the left. Circle the new coordinates.

6 Subtract 3 from each side of the inequality in Step 2. What is the new inequality showing the subtraction of 3 from each side of the original inequality? Find the difference on each side of the inequality.

7 Examine the new coordinates from Steps 3 and 5. Is the inequality still true when you add or subtract the same number from each side?

In the Activity, you modeled two properties of inequalities. These properties are the same as the properties you have used to solve equations.

Addition Property of Inequality
If the same number is added to each side of an inequality, the resulting inequality is still true.

Subtraction Property of Inequality
If the same number is subtracted from each side of an inequality, the resulting inequality is still true.

Now you can use the Addition Property of Inequality to solve the matinee ticket problem.

$x - 3 < 9$	Given
$x - 3 + 3 < 9 + 3$	Addition Property of Inequality
$x < 12$	Simplify

Thus, you started with less than $12.

EXAMPLE 1 Using the Subtraction Property of Inequality

Solve $a + 4 \geq -1$. Graph the solution.

SOLUTION

Use the Subtraction Property of Inequality to subtract 4 from each side of the inequality.

$a + 4 \geq -1$	Given
$a + 4 - 4 \geq -1 - 4$	Subtraction Property of Inequality
$a \geq -5$	Simplify

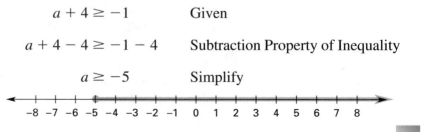

Solve and graph each inequality.

a. $8 + m \le 5$ **b.** $15 > s - 9$

EXAMPLE 2 Solving an Inequality

Juanita has at least $270 in her savings account after she makes a deposit of $25. How much did Juanita have in her savings before the deposit?

SOLUTION

Words like "at least" and "at most" lead to statements of inequality. Since Juanita has at least $270, she can have $270 or more in her account. Her savings is *greater than or equal to* $270. Model the problem with an inequality and solve. Let x represent the amount of money before the deposit.

$x + 25 \ge 270$	Given
$x + 25 - 25 \ge 270 - 25$	Subtraction Property of Inequality
$x \ge 245$	Simplify

Juanita had at least $245 in her savings before she made the deposit.

Critical Thinking   Suppose a, b, and c are real numbers. Use these variables to write the Addition Property of Inequality and the Subtraction Property of Inequality as algebraic statements.

LESSON ASSESSMENT

Think and Discuss

1 How is the Addition Property of Inequality used to solve inequalities such as $x - a < b$? Give an example.

2 How is the Subtraction Property of Inequality used to solve inequalities such as $x + a < b$? Give an example.

3 Replace the ? in $x \underline{\quad ? \quad} 10$ with an inequality sign to represent

a. at most   **b.** more than

c. less than   **d.** at least   **e.** no more than

Solve each inequality. Graph the solution.

4. $x + 5 < 7$ **5.** $x + 2 > -4$ **6.** $x - 3 \geq 4$

7. $x + 4 > -1$ **8.** $x + 3 < -5$ **9.** $x - 5 < 1$

10. $x + 2 \leq 2$ **11.** $x - 3 \geq -2$ **12.** $x - 1 \leq 4$

Write and solve an inequality that models each situation.
13. Tai has saved $35 toward the purchase of a stereo. She needs to save at least $150 by the end of the month to buy the stereo on sale. How much more does Tai need to save to purchase the stereo at the sale price?

14. You can take at most 100 pounds of luggage on an airplane. One of your bags weighs 55 pounds. How many more pounds of luggage can you take?

15. The difference between two numbers is less than 50. The smaller number is 15. What values can the larger number be?

16. State University accepts no more than 4500 students into its freshman class. So far 2000 students have been accepted. How many more students can State University accept?

17. The Home Town Gift Shop plans to spend no more than $350 on advertising in May. The shop has already spent $210. How much more can the shop spend on advertising?

18. At least 30,000 people attended a rally. This is 7500 more than attended the rally last year. How many people attended the rally last year?

Mixed Review

The daily production rates at an electronics assembly plant were recorded for 16 days:

42, 44, 38, 38, 43, 37, 44, 41
38, 46, 41, 37, 44, 42, 44, 37

19. Find the mode of the data.

20. Find the median of the data.

21. Find the mean of the data.

22. The president of the plant wants to know the average daily production rate. Which measure of central tendency would you use to answer this question?

A car lot has 20 sedans, 2 convertibles, 13 pickup trucks, and 11 sports cars for sale. Suppose each vehicle has an equally likely chance of being sold.

23. What is the probability that the first car sold is a convertible?

24. What is the probability that the first car sold is a sedan or a sports car?

25. Two cars were sold at once. What is the probability that both cars were convertibles?

26. A customer bought one of the vehicles. He brought it back and then bought one of the other vehicles. What is the probability that he bought a pickup first and a sports car second?

LESSON 9.3 THE MULTIPLICATION PROPERTY OF INEQUALITY

In Chapter 4, you used two methods to solve multiplication equations such as $2x = 12$. You found the solution by dividing each side by 2 or by multiplying each side by $\frac{1}{2}$. Either method results in the same solution, 6.

Can you use the same methods to solve the inequality $2x < 12$?

ACTIVITY 1 Multiplying and Dividing by a Positive Number

1 Draw a number line from −8 to 8.

2 Circle the points −4 and 2. Write an inequality describing the relationship between the coordinates you graphed.

3 Multiply each coordinate by 2. Plot the new coordinates on the number line.

4 Multiply each side of the inequality in Step 2 by 2. What is the new inequality showing the multiplication of 2 to each side of the original inequality. Now find the products on each side of the inequality.

5 Examine the new coordinates. Is the inequality still true when you multiply each side by the same positive number?

6 Repeat Steps 3–5. This time divide by 2.

7 What is the solution of $2x < 12$?

So far, it appears that solving a multiplication *inequality* is just like solving a multiplication *equality*. However, there is one case that is different. What happens when you solve $-2x < 12$?

ACTIVITY 2 Multiplying and Dividing by a Negative Number

1 Draw a number line from −8 to 8.

2 Circle the points −4 and 2. Write an inequality describing the coordinates you circled.

3 Multiply each coordinate by -2. Plot the new coordinates on the number line.

4 Rewrite the inequality in Step 2 to show each coordinate multiplied by -2. Find the products on each side of the inequality. Be careful to use the correct inequality sign.

5 Examine the new coordinates. What must you do to make the inequality true when you multiply each side of an inequality by the same negative number?

6 Repeat Steps 3–5. This time divide by -2.

7 What is the solution of $-2x < 12$?

The following chart summarizes the findings you made in the two Activities:

Given Inequality	Algebraic Operation Involved	Resulting Inequality	Observation
$-4 < 2$	multiply by 2	$-8 < 4$	Inequality is true with *same* symbol
$-4 < 2$	divide by 2	$-2 < 1$	Inequality is true with *same* symbol
$-4 < 2$	multiply by -2	$8 > -1$ (note: symbol has reversed)	Inequality is true only if $<$ *changes to* $>$
$-4 < 2$	divide by -2	$2 > -1$ (note: symbol has reversed)	Inequality is true only if $<$ *changes to* $>$

The inequality sign shows order. When you multiply or divide each side of an inequality by a *positive* number, the order remains the

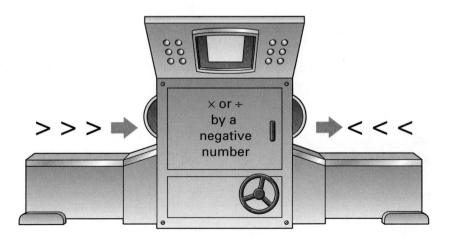

> > > ➡   × or ÷ by a negative number   ➡ < < <

same. When you multiply or divide each side of an inequality by a *negative* number, the order is reversed. These statements about inequality are combined as follows:

Multiplication and Division Properties of Inequality

For all numbers a, b, and c:

If $a < b$ and $c > 0$, then $ac < bc$ and $\frac{a}{c} < \frac{b}{c}$.

The order remains the same.

If $a < b$ and $c < 0$, then $ac > bc$ and $\frac{a}{c} > \frac{b}{c}$.

The order is reversed.

EXAMPLE 1 Solving an Inequality

Solve and graph $-3x + 1 \geq -2$.

SOLUTION

$$-3x + 1 \geq -2 \qquad \text{Given}$$

$$-3x + 1 - 1 \geq -2 - 1 \qquad \text{Subtraction Property of Inequality}$$

$$-3x \geq -3 \qquad \text{Simplify}$$

$$\frac{-3x}{-3} \leq \frac{-3}{-3} \qquad \text{Division Property of Inequality (reverse the inequality sign)}$$

$$x \leq 1 \qquad \text{Simplify}$$

```
◄——+——+——+——+——+——+——+——+——●——+——+——+——+——+——+——+——+——►
  -8  -7  -6  -5  -4  -3  -2  -1  0   1   2   3   4   5   6   7   8
```

Ongoing Assessment

Solve and graph each inequality.

a. $5n + 3 < 3n + 8$ **b.** $2(3 - 2n) \geq 12$

EXAMPLE 2 The Cost of Doing Business

To fix a pool heater, a pool service company estimates a charge of at least $138. This price includes a service charge of $40 and an hourly rate of $28. Write and solve an inequality to determine how many hours the pool service estimates the repair will take.

Let x represent the number of hours. Write the inequality to model the problem.

$$40 + 28x \geq 138$$

Solve the inequality.

$40 + 28x \geq 138$	Given
$28x \geq 98$	Subtraction Property
$x \geq 3.5$	Divide each side by 28

The pool service estimates the job will take at least three-and-a-half hours.

WORKPLACE COMMUNICATION

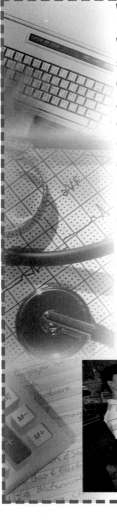

What exercise rates for the stationary bike and stair-climber do you think you could sustain? At these rates, how much time on each machine would meet or exceed the recommended amount of exercise? Some people prefer to use the bike for half the time of a workout and the stair climber for the other half. Under this condition, and at the same exercise rates as before, how much time on each machine would meet or exceed the recommended amount of exercise?

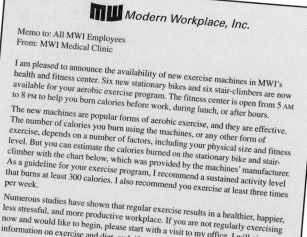

mw *Modern Workplace, Inc.*

Memo to: All MWI Employees
From: MWI Medical Clinic

I am pleased to announce the availability of new exercise machines in MWI's health and fitness center. Six new stationary bikes and six stair-climbers are now available for your aerobic exercise program. The fitness center is open from 5 AM to 8 PM to help you burn calories before work, during lunch, or after hours.

The new machines are popular forms of aerobic exercise, and they are effective. The number of calories you burn using the machines, or any other form of exercise, depends on a number of factors, including your physical size and fitness level. But you can estimate the calories burned on the stationary bike and stair-climber with the chart below, which was provided by the machines' manufacturer. As a guideline for your exercise program, I recommend a sustained activity level that burns at least 300 calories. I also recommend you exercise at least three times per week.

Numerous studies have shown that regular exercise results in a healthier, happier, less stressful, and more productive workplace. If you are not regularly exercising now and would like to begin, please start with a visit to my office. I will give you information on exercise and diet, and, if needed, I can do a simple cardiovascular screening.

If you need an appointment, please call Vickie at extension 6-7893. I look forward to seeing you at the fitness center.

S.C. McMahan

Sandra C. McMahan
Physician's Assis
Head, MWI Med

Stationary Bike		Stair-Climber	
Revolutions per Minute	Calories per Minute	Steps per Minute	Calories Burned per Minute
70	6	70	8
80	8	80	10
90	12	90	15
100	16	100	20
110	18	110	23

LESSON ASSESSMENT

Think and Discuss

 1 What happens to the order of the inequality when you multiply or divide each side of the inequality by a negative number?

2 What are the similarities and differences between solving inequalities and equalities?

Practice and Problem Solving

Solve each inequality. Graph the solution.

3. $3x < 6$

4. $4x < -8$

5. $\dfrac{x}{2} \geq -5$

6. $-2x < 7$

7. $-\dfrac{3}{4}x \geq 12$

8. $3x - 12 < 0$

9. $4x + 3 \geq -5$

10. $3x < -x - 5$

11. $5x - 5 > -9 + 3x$

Write and solve an inequality for each situation.

12. The perimeter of a rectangular patio is at most 68 feet. The length is 2 feet more than the width. What are the largest possible dimensions of the patio?

13. Margarita is choosing from among several jackets on sale for 30% off. All the jackets are originally over $75. What is the least amount Margarita can save on one jacket?

14. Lee's parents have agreed to give him an interest-free loan to buy a used car. Lee wants to borrow less than $5000. He plans to make 24 equal payments. What is the largest payment he will make?

15. Mr. Lewis is staying in a hotel for eight nights during a sales conference. He can spend no more than $1000 for lodging. What is the most Mr. Lewis can pay for each night's lodging?

16. Sarah charges $15 per hour to repair electronics equipment. Her goal is to earn at least $540 per week. So far this week, she has earned $450. What is the least number of hours Sarah must work during the remainder of this week to meet her goal?

17. Kelli needs at least 270 total points on three tests to earn an A. On her first two tests, she scored 86 and 88. What is the lowest score Kelli can make and still earn an A?

18. Wayne wrestles in the 148 pound division. He is training for a meet that is five days away. If Wayne's current weight is 154.5 pounds, what must be his least average weight loss (in pounds per day) for him to wrestle in his division?

19. John has $20 to buy gas for his truck. He needs to have at least $5 remaining after he pays the bill. If gas costs $1.39 per gallon, what is the greatest number of gallons of gas John should buy?

Mixed Review

A manufacturing company uses the linear function $y = 3x + 5$ to model the cost (y) of producing x gears.

20. Name the independent variable.

21. Find the slope of the graph.

22. Find the y-intercept of the graph.

23. Find the function that results when x is replaced by $x + 2$.

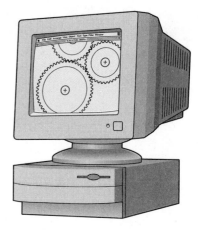

The same company uses the nonlinear function $y = \sqrt{x + 3} - 2$ as coordinate points in the computer-aided design of one of the gears.

24. What are the domain and range of the function?

25. What is the value of y when x is replaced by 1?

26. Graph the function.

LESSON 9.4 SOLVING COMPOUND INEQUALITIES

Inequalities are often combined into one statement. Consider these two inequalities:

$$2 < 3 \quad \text{and} \quad 4 < 5 \qquad 7 > 4 \quad \text{or} \quad 6 > 1$$

The first inequality states that "2 is less than 3 *and* 4 is less than 5." The second states that "7 is greater than 4 *or* 6 is greater than 1." Such inequalities are called **compound inequalities.** Compound inequalities are often used in the workplace.

Todd is a testing supervisor in a factory that produces and packages small metal parts. A packaging machine is set to drop 500 nails into each box. However, the nails vary slightly in weight. A 5-pound box of nails is approved for shipping if its weight is *between* 78.5 ounces and 81.5 ounces. To model the tolerance, Todd uses the inequality

$78.5 < x < 81.5$, where x is the weight of the nails and box

The inequality $78.5 < x < 81.5$ is an example of a compound inequality. It combines the two simple inequalities

$$78.5 < x$$
$$x < 81.5$$

connected by the word *and*. The inequality is read:

78.5 is less than x *and* x is less than 81.5

Conjunctions

A compound inequality that uses the word *and* to separate two simple inequalities is an example of a **conjunction**. The solution set of a conjunction is the set of solutions that make *both* simple inequalities true. This solution is the intersection of the solutions of the simple inequalities.

EXAMPLE 1 Solving a Conjunction

Solve and graph $-3 < (2x - 3) \le 5$.

SOLUTION

Write the simple linear inequalities separately. Solve each inequality.

$$
\begin{array}{ccc}
-3 < (2x - 3) & \text{and} & (2x - 3) \le 5 \\
-3 + 3 < (2x - 3) + 3 & \text{and} & (2x - 3) + 3 \le 5 + 3 \\
0 < 2x & \text{and} & 2x \le 8 \\
0 < x & \text{and} & x \le 4
\end{array}
$$

The solution is the set of all numbers greater than zero and less than or equal to four, which is

$$0 < x \le 4$$

To graph the conjunction, graph the common solution to the two simple inequalities on the same number line.

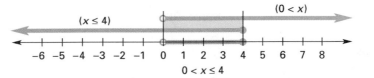

The solution set for the conjunction is the intersection of the solution sets of the two simple inequalities. This is shown as the shaded segment that is common to both graphs.

There is a simple method to check your solution.

1. Choose a value in the intersection and substitute it into each simple inequality. The value should make both of the original inequalities true.

2. Choose a value to the left of the interval. This value should make one of the simple inequalities true.

3. Choose a value to the right of the interval. This value should make the other simple inequality true.

Critical Thinking   How can you check the solution set of a conjunction by using the endpoint values of each simple inequality?

Ongoing Assessment

Solve and graph each conjunction. Check your solution set.

a. m is greater than -5 and less than 4   **b.** $-2 \le 3y + 1 \le 5$

The solution set of a conjunction is the **intersection** of two sets of numbers. The intersection of two sets consists of the numbers that are common to both sets; in other words, the numbers in one set that are also in the other set. The shaded area in the Venn diagram below represents the intersection of sets A and B. The intersection of the two sets is designated A ∩ B.

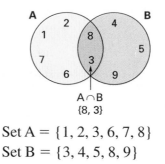

Set A = {1, 2, 3, 6, 7, 8}
Set B = {3, 4, 5, 8, 9}

Since only the elements 3 and 8 are in the shaded (overlap) region,

$$A \cap B = \{3, 8\}$$

Consider again the combined inequality in Example 1:

$$-3 < (2x - 3) \leq 5$$

For the left inequality, $-3 < 2x - 3$, the solution is the set of all values for which $x > 0$. Call these values Set A.

For the other inequality, $2x - 3 \leq 5$, the solution is the set of all values for which $x \leq 4$. Call these values Set B.

For the combined inequality, the solution is

$$0 < x \leq 4$$

The intersection of Sets A and B (or A ∩ B) is pictured below:

$$-3 < 2x - 3 \qquad 2x - 3 \leq 5 \qquad -3 < 2x - 3 \leq 5$$

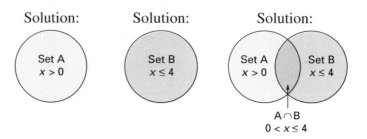

ACTIVITY 1 **The Empty Set**

1 Solve and graph $-3 > 2x - 3$.

2 Solve and graph $2x - 3 \geq 5$.

3 What is the solution to $-3 > 2x - 3$ and $2x - 3 \geq 5$?

4 Use a Venn diagram to picture the solution to the conjunction.

5 Explain why the solution is called the *empty set*.

Disjunctions

Wire stock used to make nails is tested on a machine that compresses the wire and then stretches it. An 8-inch length of wire fails the test if it is compressed to a length shorter than 7.95 inches or stretched to a length longer than 8.05 inches. A compound inequality models the wire length that fails the test:

$x > 8.05$ *or* $x < 7.95$, where x is the length of the nail wire

This compound inequality consists of two simple inequalities connected by the word *or*. It is an example of a **disjunction.** The solution set of the disjunction is the set of all solutions that make *either* of the simple inequalities true.

EXAMPLE 2 Graphing a Disjunction

Solve and graph $-3 > 2x - 3$ or $2x - 3 \geq 5$.

SOLUTION

Solve each simple linear inequality.

$$
\begin{array}{ccc}
-3 > (2x - 3) & or & (2x - 3) \geq 5 \\
-3 + 3 > (2x - 3) + 3 & or & (2x - 3) + 3 \geq 5 + 3 \\
0 > 2x & or & 2x \geq 8 \\
0 > x & or & x \geq 4
\end{array}
$$

Graph the two inequalities on a number line.

```
        (0 > x)                              (x ≥ 4)
 ←――+――+――+――+――+――+――+――○――+――+――+――+―●――+――+――+――+――+――→
   −7 −6 −5 −4 −3 −2 −1  0  1  2  3  4  5  6  7  8  9
```

All the points for $0 > x$ and $x \geq 4$ constitute the solution for the combined inequality connected by the word *or*.

To check the solution to the disjunction, check x-values less than zero and x-values greater than 4 to show that the values make the combined inequality true. Then check x-values greater than or equal to zero and less than 4 to show that these values do not satisfy either inequality.

Ongoing Assessment

Solve and graph $8y + 1 < 25$ or $-2y < -10$.

The solution set of a disjunction is the **union** of two sets of numbers. The union of two sets of numbers consists of all the numbers that are in either one of the sets. The shaded area in the Venn diagram below represents the union of sets A and B. The union is designated $A \cup B$.

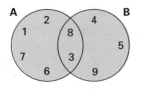

Set A = $\{1, 2, 3, 6, 7, 8\}$
Set B = $\{3, 4, 5, 8, 9\}$

Thus, the union of the two sets A and B, designated as $A \cup B$, is the set

$$A \cup B = \{1, 2, 3, 4, 5, 6, 7, 8, 9\}$$

ACTIVITY 2   Graphing a Subset

1 Solve and graph $5y - 2 < 8$.

2 Solve and graph $3y + 4 < 16$.

3 What is the solution to $5y - 2 < 8$ or $3y + 4 < 16$?

4 Use a Venn diagram to picture the solution to the disjunction.

5 Explain why the solution for $5y - 2 < 8$ is called a *subset* of the solution for $3y + 4 < 16$.

Critical Thinking   In the language of algebra, AND statements correspond to the intersection of two sets, and OR statements correspond to the union of two sets. How is this similar to the way the words are used in probability?

LESSON ASSESSMENT

Think and Discuss

1 What steps are the same when you solve a conjunction and a disjunction? What steps are different?

2 How is the word AND used to find the solution of a conjunction? How is the word OR used to find the solution to a disjunction?

3 How do you know if a compound inequality is a conjunction or a disjunction?

Practice and Problem Solving

Solve each compound inequality. Graph the solution.

4. $2x \leq 6$ or $x > 5$

5. $3x < 2.4$ and $x \geq -1.5$

6. $2x - 3 < 13$ and $x > 4$

7. $5x + 4 < -11$ or $3 - 4x < -1$

8. $3x + 1 > 10$ or $2x - 5 < 3$

9. $x + 3 < -1$ and $4x - 1 < 5$

10. $-5 < 7 - 2x < 5$

11. $3 < 4x + 7 < 15$

Write and solve a compound inequality for each situation.
12. Tad will spend at least $20 but no more than $50 on T-shirts for summer camp. The cost of a T-shirt is $8. How many T-shirts can Tad buy?

A thermostat will send an electrical signal if it senses a temperature below 65 degrees or above 80 degrees.

13. If the initial temperature increases by 8 degrees, the thermostat will send a signal. What are the possible initial temperatures?

14. If the initial temperature decreases by 8 degrees, the thermostat will send a signal. What are the possible initial temperatures?

15. A store owner adds $5 to twice the wholesale price in order to set the retail price of his sweaters. If the retail prices are between $145 and $155, what wholesale prices did the store owner pay for the sweaters?

Mixed Review

The heights in inches and weights in pounds of 12 students are recorded.

Ht. 68 71 70 69 72 70 69 71 65 65 68 72
Wt. 160 165 160 140 180 155 145 160 140 135 139 150

16. Find the mode of the heights.

17. Find the median of the weights.

18. Make a scatter plot from the data. Let height be the independent variable.

19. Describe the correlation between the two sets of data.

Solve each system of equations.
20. $3a + 2b = 0$ **21.** $4c - 5d = -8$
    $a - 3b = -7$ $6c + 2d = 7$

Write and solve a system of equations that models each situation.
22. The length of the molding around the ceiling of a rectangular room is 120 feet. The length of the room is twice its width. What are the dimensions of the ceiling?

23. Mario has saved $20 less than Nancy. Together they have saved $100. How much has each person saved?

24. A chemist needs 400 grams of a 5% salt solution. There is one available solution containing 4% salt and another available solution containing 8% salt. How many grams of each solution should the chemist mix to get the desired 5% solution?

25. Mal bought 15 gallons of 87 octane gas and 12 gallons of 92 octane gas for $34.20. Yuri bought 10 gallons of 87 octane gas and 7 gallons of 92 octane gas for $21.45. How much does a gallon of each type of gas cost?

LESSON 9.5 SOLVING ABSOLUTE VALUE INEQUALITIES

The construction plans for a new bridge specify steel reinforcement rods with a diameter of 1.00 centimeter. A rod is acceptable if its diameter is not more than 0.01 centimeter from the specification; that is, the tolerance is ±0.01 cm. An inequality containing an absolute value models the acceptable diameter of the rod.

$$|d - 1.00| \leq 0.01 \text{ where } d \text{ is the diameter of the rod in}$$
$$\text{centimeters}$$

This statement indicates that the difference between 1.00 centimeter and the actual diameter d cannot exceed 0.01 centimeter.

Recall that the definition of the absolute value of x is the distance between zero and x on a number line.

$$|x| = x, \text{ when } x \geq 0 \text{ and } |x| = -x, \text{ when } x < 0$$

By applying the definition of absolute value, you can write an absolute value equality as a disjunction. For example,

$$|x| = 2 \text{ is equivalent to the disjunction } x = 2 \text{ or } x = -2$$

The solution to an absolute value inequality such as $|x| > 2$ is a disjunction:

$$|x| > 2 \text{ is equivalent to the disjunction } x > 2 \text{ or } x < -2$$

The graph of $|x| > 2$ is shown below. Notice that all the points in the solution are a distance greater than 2 units from zero.

$$\begin{array}{ccccccccccc} & -5 & -4 & -3 & -2 & -1 & 0 & 1 & 2 & 3 & 4 & 5 \end{array}$$

On the other hand, the solution to an absolute value such as $|x| < 2$, is a *conjunction*.

$$|x| \leq 2 \text{ is equivalent to the conjunction } x \leq 2 \text{ and } x \geq -2$$

The graph of $|x| \leq 2$ is shown below. Notice that all the points in the solution are a distance less than or equal to 2 units from zero.

$$\begin{array}{ccccccccccc} & -5 & -4 & -3 & -2 & -1 & 0 & 1 & 2 & 3 & 4 & 5 \end{array}$$

You can also use disjunctions and conjunctions to represent absolute value inequalities that contain *linear functions*.

Solutions to Absolute Value Linear Functions
For all numbers a, b, and c, where $c \geq 0$,

If $|ax + b| \geq c$, then $ax + b \geq c$ or $ax + b \leq -c$

If $|ax + b| \leq c$, then $ax + b \leq c$ and $ax + b \geq -c$

Now look at the inequality for the diameter of the steel rod. Since it is an absolute value using *less than* or *equal to*, the solution must be the conjunction

$$(d - 1.00) \leq 0.01 \qquad \text{and} \qquad (d - 1.00) \geq -0.01$$

Thus,

$$(d - 1.00) + 1.00 \leq 0.01 + 1.00 \quad \text{and} \quad (d - 1.00) + 1.00 \geq -0.01 + 1.00$$
$$d \leq 1.01 \qquad \text{and} \qquad d \geq 0.99$$

Thus, an acceptable rod diameter is less than or equal to 1.01 centimeters and greater than or equal to 0.99 centimeters. In most industrial settings, this statement is written in the following way:

$$d = 1.00 \text{ cm} \pm 0.01 \text{ cm}$$

EXAMPLE Solving Absolute Value Inequalities

Solve $|2x - 1| > 5$.

SOLUTION

Since the absolute value is a linear inequality containing a *greater than* relation, the solution is a disjunction.

	Case 1		Case 2
	$2x - 1 > 5$	or	$2x - 1 < -5$
	$2x - 1 + 1 > 5 + 1$	or	$2x - 1 + 1 < -5 + 1$
	$2x > 6$	or	$2x < -4$
	$x > 3$	or	$x < -2$

The solution is the union of the two individual graphs.

Check the solution by choosing values from the three regions.

Solve and graph each absolute value inequality.

a. $|2r - 1| \geq 5$ **b.** $|5y + 2| < 3$ **c.** $|w - 4| \leq -2$

Critical Thinking   For which values of a are there no solutions for $|x - 4| < a$?

LESSON ASSESSMENT

Think and Discuss

1 Why must you be able to solve a compound inequality before you can solve an absolute value inequality?

2 What determines whether an absolute value inequality has a solution that is a conjunction or a disjunction? Give examples.

3 Explain why $|x + 2| \leq -4$ has no solution.

Practice and Problem Solving

Solve each absolute value inequality. Graph the solution.

4. $|x| < 5$ **5.** $|x| > 4$ **6.** $|x| \leq -1$

7. $\left|\dfrac{2}{3}x - 5\right| \geq 3$ **8.** $|x - 2| < \dfrac{1}{3}$ **9.** $|4x + 5| > 15$

10. $|x - 1| > 7$ **11.** $|5 - 3x| \geq 10$ **12.** $|3x - 4| \leq 8$

Write and solve an absolute value inequality that models each situation.

13. The arrival of an airline flight was predicted to be within 5 minutes of its scheduled flight time of 120 minutes. If the flight did not arrive within its predicted time, what was the airline's actual flight time?

14. A survey by a marketing research company shows that 56% of the customers polled prefer Super Cola over all other soft drinks. The survey has a margin of error of 13%. What percent of all soft drink customers prefer Super Cola?

15. The warranty for Captain Tires recommends a pressure of 30 pounds per square inch (psi). A tire is considered unsafe if its pressure is more than 3 psi from the recommended pressure. What is the unsafe pressure range for Captain Tires?

16. An aircraft design specifies a structural support beam as 1.2 centimeters thick, with an allowable error in thickness of 0.04 centimeters. If you are fabricating this beam, what is the allowable thickness of your final product?

Mixed Review

A warehouse needs exhaust fans. The warehouse is 500 feet wide and 700 feet long. It has a ceiling that is 25 feet high. Each fan moves 500 cubic feet of air per minute. The air in the warehouse must be replaced three times per hour.

17. What is the volume of the warehouse in cubic feet? Express your answer in scientific notation.

18. What volume of air is replaced in the warehouse each hour? Each minute? Express your answers in scientific notation.

19. How many exhaust fans are required?

20. The expression 16*a* represents the perimeter of a square. Write the expression that represents the area of the square.

Write an equation in one unknown to model each situation. Solve the equation.

21. Tickets to the theater are $9. If 25 people buy tickets as a group, there is a 25% discount. How much is one ticket at the discounted price?

22. You can buy six greeting cards for $10. How much will it cost to buy nine cards?

23. The sum of three consecutive numbers is 63. What are the numbers?

24. The Book Club sells books for the wholesale cost plus $3.50. The Book Club receives a check for $151 to purchase five books. What is the wholesale cost of one book?

LESSON 9.6 GRAPHING LINEAR INEQUALITIES IN TWO VARIABLES

You learned to solve and graph linear equations on a coordinate plane in Chapter 8. Solving and graphing inequalities with two variables is very similar. You can graph linear equations or inequalities on a coordinate plane with one of these methods:

- *Use a table to list ordered pairs.*
- *Use the slope-intercept form of the equation.*
- *Use a graphics calculator.*

Consider the following equation and inequality:

$$4x + 2y = 6 \qquad 4x + 2y \leq 6$$

Solve for y and write both the equation and the inequality in slope-intercept form ($y = mx + b$ and $y \leq mx + b$).

$$4x + 2y = 6 \qquad\qquad 4x + 2y \leq 6$$

$$2y = -4x + 6 \qquad\qquad 2y \leq -4x + 6$$

$$y = -2x + 3 \qquad\qquad y \leq -2x + 3$$

Now graph the equation and the inequality.

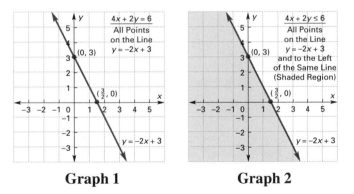

Graph 1   **Graph 2**

Look at Graph 1. If you pick a point on the line and substitute this ordered pair into $y = -2x + 3$, the result is a true statement.

The same is true for Graph 2. However, the graph of the inequality contains both the line $y = -2x + 3$ and the shaded area to the left of the line. The shaded area is the graph of $4x + 2y < 6$. Together, the line and the shaded area are the graph of $4x + 2y \leq 6$.

Critical Thinking   Describe the graph of $4x + 2y \leq 6$ as the union of two sets.

If you choose a point that is on the line or in the shaded area and substitute this ordered pair into $4x + 2y \leq 6$, the result will be a true statement.

Use the ordered pair $(4, -5)$, which is on the line.

$$4x + 2y = 6$$

$$4(4) + 2(-5) = 6$$

$$6 = 6$$

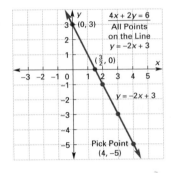

The ordered pair makes the statement true.

Now use $(-1, 2)$ in the shaded region and $(2, -1)$ on the line.

Shaded Region: $(-1, 2)$	On the Line: $(2, -1)$
$4x + 2y \leq 6$	$4x + 2y \leq 6$
$4(-1) + 2(2) \leq 6$	$4(2) + 2(-1) \leq 6$
$-4 + 4 \leq 6$	$8 - 2 \leq 6$
$0 \leq 6$	$6 \leq 6$

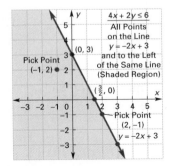

Each ordered pair results in a true statement.

To graph a linear inequality with two variables, first graph the line that represents the associated equation. The graph of the equation divides the coordinate plane into three parts:

- *the line itself*
- *the area to one side of the line*
- *the area to the other side of the line*

EXAMPLE 1 Describing Graphs of Inequalities

Write an equation or inequality to describe the following graphs. Each of these graphs compares the expression $(-2x + y)$ to 5 using an equality or an inequality.

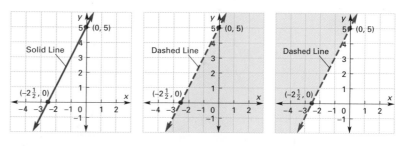

SOLUTION

a. $-2x + y = 5$

$y = 2x + 5$

b. $-2x + y < 5$

$y < 2x + 5$

c. $-2x + y > 5$

$y > 2x + 5$

Notice that in the graphs for inequalities, the lines are dashed. This means that the line itself is not included in the solution set. If the inequalities use $\leq$ or $\geq$, use a solid line to show the lines are included in the solution set.

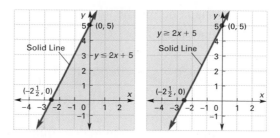

To determine which side of the line to shade, choose a *test point* above or below the line and substitute the corresponding ordered pair into the inequality. (Use the test point $(0, 0)$ whenever you can.) Check to see if the result is a true statement. If it is, you have chosen the side that represents the solution. Shade this side of the line.

EXAMPLE 2 Choosing the Area to Graph

Graph the inequality $-x + y \leq 2$.

SOLUTION

Use a table or change the inequality to slope-intercept form, $y \leq x + 2$. Then draw the graph.

First draw the graph of the line $-x + y = 2$. Then substitute test points on either side of the line to determine which side to shade.

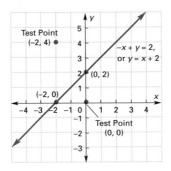

Above the Line: $(-2, 4)$ Below the Line: $(0, 0)$

$$-x + y \leq 2$$
$$-(-2) + 4 \leq 2$$
$$6 \leq 2$$

$$-x + y \leq 2$$
$$0 + 0 \leq 2$$
$$0 \leq 2$$

This is false. Thus, $(-2, 4)$ is not a solution. Do not shade above the line.

This is true. Thus, $(0, 0)$ is a solution. Shade below the line.

Look again at Example 2. If you change an inequality to slope-intercept form, you can tell by inspection whether you should shade above or below the line. If the inequality is $<$ (less than), shade below the line. If the inequality is $>$ (greater than), shade above the line.

Ongoing Assessment

Graph each inequality.

a. Use a test point to determine the region to shade.

   1. $(-4x + y) < -1$ 2. $(-4x + y) \geq -1$

b. Then change each inequality to slope-intercept form and use inspection to determine whether to shade above or below the line.

The following examples illustrate how inequalities help solve problems that arise in the workplace.

EXAMPLE 3 A Profit Model

Suppose Mountain Sales Bicycle Shop makes $100 on each Model X bike sold and $50 on each Model Y bike sold. The bike shop's overhead expenses are $1500 per month. At least how many of each model bike must be sold each month to avoid losing money?

SOLUTION

The bike shop must make at least $1500 each month. The condition "at least $1500" is represented by "≥ 1500." The "equal" part of ≥ means that if the owner makes exactly $1500, he meets expenses. The "greater than" part of ≥ means that if the owner makes over $1500, he meets expenses and makes a profit.

Let x and y represent the number of Model X and Model Y bikes sold per month. Since the owner must make at least $1500, an inequality models the situation.

$$100x + 50y \geq 1500$$

Solve for y to write the inequality in slope-intercept form.

$$100x + 50y \geq 1500$$

$$50y \geq -100x + 1500$$

$$y \geq -2x + 30$$

The line that defines the boundary of the shaded region is $y = -2x + 30$. Since the owner cannot sell less than zero of either model bike, the graph is confined to the first quadrant, where x- and y-values are both greater than or equal to zero.

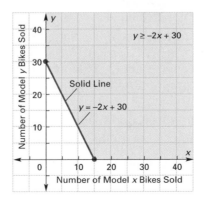

Conclusions

1. All points in the shaded region, on and above the line, are solutions.

2. All points below the line (unshaded triangle) are not solutions.

Ongoing Assessment

Determine whether the following points are in the solution set and, if so, how much money the bike shop makes.

$(7, 12)$; $(10, 10)$; $(15, 0)$; $(0, 29)$; $(20, 20)$

EXAMPLE 4 Graphing Distance Versus Time

A truck is traveling on a road with a posted speed limit of 55 miles per hour. An inequality that describes the distance that could be covered by a driver operating the truck legally is $d \leq 55t$. In this inequality, d is the distance the truck travels and t is the time it takes to travel that distance.

SOLUTION

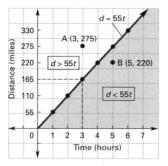

The lower shaded region represents the legal speed zone, where $d < 55t$; that is, d is less than $55t$ for any pair (t, d) in this region.

The upper shaded region represents the illegal speed zone, where $d > 55t$; that is, d is greater than $55t$ for any pair (t, d) in this region.

Since the point $(3, 165)$ lies on the line, the coordinates satisfy the equation $d = 55t$. What does the number 165 represent?

Point A has coordinates $(3, 275)$. Point A is "typical" of all points in the region above the line $d = 55t$. At what speed would a truck be traveling to satisfy coordinates $(3, 275)$? Is this a legal speed in the speed zone of Example 4?

Point B has coordinates (5, 220). Point B is "typical" of all points in the region below the line $d = 55t$. At what speed would a truck be traveling to satisfy coordinates (5, 220)? Is this a legal speed in the speed zone of Example 4?

EXAMPLE 5 Multiplying by a Negative Number

Graph this inequality.

$$-3x - 2y > -2$$

SOLUTION

First, write the inequality in slope-intercept form.

$$-3x - 2y > -2$$

$$-2y > 3x - 2$$

$$y < -\frac{3}{2}x + 1 \quad \text{(Note that} > \text{ changes to} < \text{ after division by } -2.\text{)}$$

Next, graph $y = -\frac{3}{2}x + 1$ as a dashed line. Then determine the shaded region.

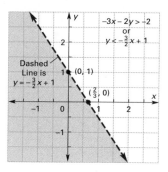

Since the inequality symbol in the slope-intercept form is $<$, the shaded area must be below the dashed line, as shown.

Choose a test point in the shaded region of the graph and substitute this ordered pair into both forms of the inequalities:

$$-3x - 2y > -2 \quad \text{and} \quad y < -\frac{3}{2}x + 1$$

Are the inequalities true?

$$-3 - 2y > -2 \qquad\qquad y < -\frac{3}{2}x + 1$$
$$\text{choose } (0,0) \qquad\qquad \text{choose } (0,0)$$
$$-3 > -2 \qquad\qquad\qquad 0 < 1$$

LESSON ASSESSMENT

 How do you determine if the line associated with an inequality is graphed as a dashed or a solid line?

 How do you determine which side of the line to shade when you graph an inequality?

3 Give examples of different methods for determining which side of the line graphed contains all the solutions to an inequality.

Practice and Problem Solving

Solve each inequality for y. Graph each inequality.

4. $x + y > 6$ **5.** $x + y > -2$

6. $x - y < 6$ **7.** $2x - y \geq 5$

8. $3y + 12 < 0$ **9.** $3x - 4y \leq 12$

10. $5x + 3y < 15$ **11.** $5y - 40 < -8x$

12. $6x - 3y + 12 < 0$

The perimeter of a rectangular garden is more than 24 meters.

13. Write an inequality to model the perimeter as a function of length and width.

14. Solve the inequality and graph the solution.

15. Give three whole number solutions to the inequality.

This semester, there are fewer than 30 students on the chess team.

16. Write an inequality to model the number of boys and girls on the team.

17. Solve the inequality and graph the solution.

18. If there are more girls than boys on the team, give three possible combinations of numbers of boys and girls on the team.

A manufacturing company uses two machines to cut patterns for wet suits. It costs $10 per wet suit to use the faster machine. It costs $6 per wet suit to use the slower machine. The daily cost of cutting out wet suits must be held to a maximum of $60.

19. Write an inequality to model the daily cost of cutting out wet suits as a function of the number of wet suits cut on each machine.

20. Solve the inequality and graph the solution.

21. Give three whole number solutions to the inequality.

22. A work order for one day requests 8 wet suits. How many wet suits can the company cut out on each machine and still meet the cost constraints for the day? Give two other possible answers.

Mixed Review

Suppose line AB passes through $A(0, 9)$ and $B(9, 0)$. Suppose line MN passes through $M(1, 4)$ and $N(-1, 2)$.

23. What is the slope of each line?

24. What is the equation of line AB?

25. What is the equation of line MN?

26. What is the y-intercept of each line?

27. Graph each line on the same coordinate axes.

28. At what point do the lines intersect?

29. Find the equation of a line CD that passes through the origin and is parallel to line AB.

30. At what point do lines CD and MN intersect?

LESSON 9.7 SOLVING SYSTEMS OF INEQUALITIES

Recall the car dealer from Chapter 8. Suppose the car dealer nets $500 for each family car (*F*) sold and $750 for each sports car (*S*) sold. The dealer must sell at least two family cars for every sports car and must earn at least $3500 per week. You can find *S* and *F* by graphing the two inequalities that meet these conditions.

Inequality	Meaning
$F \geq 2S$	sell at least 2 family cars for each sports car
$500F + 750S \geq 3500$	make at least $3,500 per week if net for each family car sold is $500 and net for each sports car sold is $750

The solution of this system of inequalities is the intersection of the solution regions of the two inequalities.

ACTIVITY   Intersecting Solution Sets

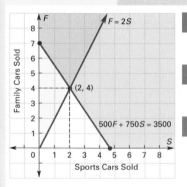

1 Graph the lines $F = 2S$ and $500F + 750S = 3500$ on the same axes.

2 Use test points to determine the solution region for each inequality.

3 Find the area where the two solution regions intersect. This is the solution region for the system of inequalities.

A solution region contains many solutions. In the car dealer problem, the dealer would want to know which solution maximizes profit. You can use linear programming to determine the *optimum* number of items to sell to *maximize* profits.

Recall the Mountain Sales Bicycle Shop problem. The shop made $100 on each Model X bike sold and $50 on each Model Y bike sold. The owner's objective is to maximize sales. You can model sales with the following equation:

$$\text{Sales} = 100x + 50y$$

In linear programming, this equation is called an **objective function**, because the objective is to maximize sales. In the objective function, x is the number of Model X bikes sold and y is the number of Model Y sold. Suppose that each month the owner can have in stock no more than 20 Model X bikes and no more than 30 Model Y bikes. These conditions are represented by inequalities.

$$x \leq 20 \quad \text{and} \quad y \leq 30$$

Furthermore, suppose the maximum number of bikes the shop can stock is 40. This condition is written as an inequality.

$$x + y \leq 40$$

Since the owner cannot order fewer than zero of either model, a final condition also exists.

$$x \geq 0 \quad \text{and} \quad y \geq 0$$

In the language of linear programming, each of the five inequalities listed above is called a **constraint**. Each inequality limits or constrains the set of values for x and y. This is illustrated in a graphical representation of the problem.

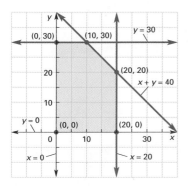

The shaded region is the intersection of all the constraining inequalities. The shaded region is a polygon bounded by the five lines that represent the constraints. All **feasible solutions** to the problem are contained in the shaded region. Note particularly the five corner points around the perimeter of the polygon: (0, 0); (20, 0); (20, 20); (10, 30); and (0, 30). To find the **optimum solution**, substitute each of these five coordinate pairs into the objective function (sales equation) and compare the sales.

Feasible Solution	Objective Function (Sales)
0, 0 ⟶	100(0) + 50(0) = 0
20, 0 ⟶	100(20) + 50(0) = 2000
20, 20 ⟶	100(20) + 50(20) = 3000
10, 30 ⟶	100(10) + 50(30) = 2500
0, 30 ⟶	100(0) + 50(30) = 1500

The table shows that the maximum sales for the bike shop is $3000. This occurs when the bike shop stocks and sells 20 of each model each month. Although any point in the shaded region of the graph of the system of inequalities satisfies all five inequalities (constraints), only the corner point (20, 20) maximizes the sales.

Check Mr. Cannon's math. Let x and y be the number of miles driven each day. Write an inequality modeling rule 5(a). Write two more inequalities involving x and y for rule 5(c). Graph the three inequalities to determine the combinations of x and y that satisfy both safety rules. If the distance between MWI and Carlisle is 775 miles, can the trip be made in two days within the safety rules?

Jackson Trucking Company
1-400-569-4398

Modern Workplace, Inc.
Attn: Ms. Anita Villarreal
    Chief, Environmental Control

Dear Ms. Villarreal:

I am writing in response to your request for clarification concerning our proposal to transport hazardous waste from the MWI facility to the Carlisle waste treatment plant.

Our cost proposal for this job is based on using a single driver for two days for each trip. Our internal safety rules, approved by state and federal transportation authorities, allow a single driver for hazardous cargo, subject to layover restrictions. The restrictions are described in the Jackson Trucking Safety Rules. I have attached a copy of the relevant page for your review.

I hope this clarifies our proposed use of a single driver. If you have any further questions, please do not hesitate to call me at 232-1600. Thank you for considering our proposal. I am sure Jackson Trucking can satisfy all MWI's hazardous material transportation needs.

Sincerely,
Graham Cannon
President

Attachment

JACKSON TRUCKING CO.
Hazardous Materials On Board

state regulations concerning transportation of hazardous cargo.

5. Hazardous cargo may be transported by a single driver; however, in such cases:
  (a) The average distance driven per day shall be less than 400 miles.
  (b) Trips greater than 400 miles require layovers. The driver shall rest at least eight hours during a layover.
  (c) No distance driven in one day shall be greater than 475 miles.

LESSON ASSESSMENT

Think and Discuss

1 What conditions must be true for a system of inequalities to have a common solution?

2 If the lines for two inequalities are parallel, can there be a common solution? Why or why not?

Graph each system of inequalities. Shade the area to show the solution.

3. $2x + y \geq 4$
$y - 2x \geq 4$

4. $x - y > -1$
$x - y \leq 2$

5. $2x + y \leq 4$
$y - 2x \leq 4$

6. $5x + 3y \leq 1$
$3x - 2y > 12$

7. $2x + 5y < 4$
$3x + 4y > -1$

8. $5x - 7y > 4$
$2x + 3y < 18$

9. $3x - 2y < 2$
$4y - 3x > 2$

10. $2y - 6x \leq 3$
$3y + 5x < -6$

11. $5y + 3x \leq 0$
$3y - 2x \leq 0$

A painting contractor estimates it will take 10 hours to paint a one-story house and 20 hours to paint a two-story house. The contractor submits a bid to paint 20 houses in less than 250 hours.

12. Write a system of equations to model the time to paint the houses and the number of houses to be painted.

13. Graph the system.

14. Give five whole-number solutions to the system. Explain what these solutions mean.

It costs 50 cents to make a bracelet and $1 to make a necklace. To make a profit, the total cost for bracelets and necklaces must be less than $10. The jeweler can make no more than 14 pieces of jewelry each day.

15. Write a system of inequalities to model the number of bracelets and necklaces to be made each day.

16. Graph the system.

17. The jeweler sells bracelets for $3 and necklaces for $4. Test the three corner points on your graph and determine how many bracelets and necklaces should be made to maximize profits.

Write a system of inequalities for each situation. Graph the solution.

18. A farmer needs to enclose a rectangular field with wire fencing. He has 800 meters of fencing available. The length of the field must be greater than 100 meters. What are the possible dimensions of the field? What is the maximum possible area of the field?

19. A manufacturer makes running shoes and basketball shoes. It costs $20 to make a pair of running shoes and $30 to make a pair of basketball shoes. The manufacturer can spend no more than $2000 making both types of shoes each day. The company needs to make no fewer than 50 pairs of running shoes each day. How many of each type can be produced each day?

20. A hardware distributor wants to sell at least 45 electric and gas mowers each week. The profit on an electric mower is $50 and the profit on a gas mower is $40. Which combinations will give the distributor a profit of up to $2500 per week?

Mixed Review

A jar contains 10 nickels, 15 dimes, and 25 quarters. One coin is removed at random.

21. What is the probability that the coin is a dime?

22. What is the probability that the coin is a nickel or a quarter?

A jar contains 10 nickels, 15 dimes, and 25 quarters. Two coins are removed at random.

23. What is the probability that both coins are dimes?

24. What is the probability that one coin is a nickel and the other is a quarter?

A jar contains 10 nickels, 15 dimes, and 25 quarters. One coin is removed at random. It is replaced, and another coin is removed at random.

25. What is the probability that both coins are the same?

26. What is the probability that both coins have a sum less than 25¢?

Solve using an equation in one unknown.

27. Tim and Shaneka have together invested $98 in an after-school business. Shaneka has invested $14 more than Tim. How much does each person have invested?

Solve using a system of linear equations.

28. Sherrill has $8000 invested in CD accounts. One account pays 6% simple interest, and the other pays 8% simple interest. This year Sherrill earned $580 interest on her two accounts. How much is invested in each account?

MATH LAB

Activity 1: Distribution of Student Heights

Equipment Calculator
Tape measure with metric scale

Problem Statement

You will plot the heights of the students in your class on a number line. Then you will make inequality comparisons with the height measurements of the group.

Procedure

a Use the tape measure to determine the height of each member of your group. Record the measurements to the nearest centimeter.

b Create a number line that includes the height measurements for each member of your group. Plot the height measurements on your number line. Place each member's name above the point representing his or her height. An example is shown below.

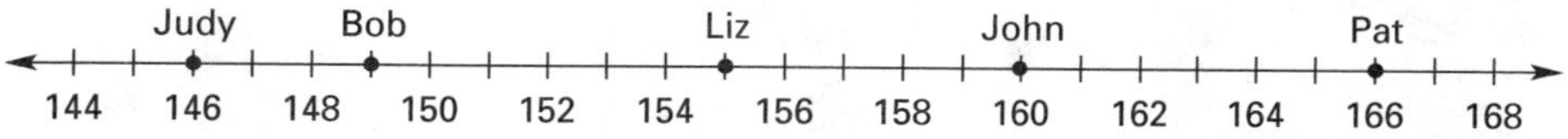

If two members have the same height, "stack" the points and their names above each other.

c Use the three relationships (<, =, and >) to write an expression relating various pairs of people. Use each group member at least once. For example, using the sample number line data above, you can write "Liz > Judy."

d Select three people from your number line and write a compound inequality. For example, using John, Pat, and Bob from the sample number line, you can write "Bob < John < Pat."

e Record your data on a master number line that includes the results for your entire class.

f Determine how many students from your class would represent 20% of all the members in the class. Round your answer to the nearest whole number. Use this number to identify the students who are in the highest 20% and lowest 20% of the class according to height. The remaining students are in the middle 60%. Write a compound inequality that describes the range of heights for the middle 60% of your class. For example, let H represent the height of a student in the middle 60%. Then write a compound inequality such as the following:

$$152 \text{ cm} < H < 166 \text{ cm}$$

g Can you use your inequality to predict the heights of other students in the same age group in your school? Explain your answer.

Activity 2: Sampling the Quality of a Product

Equipment   Calculator with statistical functions
20 large machine bolts and nuts
Vernier caliper
Micrometer caliper

Problem Statement

Most products include tolerances in their specifications. In this laboratory, you will measure the dimensions of a sample of machine bolts. Then, based on your measurements, you will determine how well the machine bolts meet a given set of tolerances.

Procedure

a Measure the length of each of the 20 bolts with the vernier caliper, and measure the head width of each bolt with the micrometer caliper. Record the results in a table similar to the one shown below:

Head Width	Overall Length
0.789 in.	5.11 cm
·	·
·	·
·	·
Mean: _____	Mean: _____

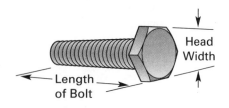

Head Width

Length of Bolt

b Write six inequality statements that describe the maximum and minimum measured values. The six statements will include:

> Two *greater than* relationships. For example, head width > 0.760 in., length > 5.03 cm

> Two *less than* relationships. For example, head width <0.792 in., length < 5.16 cm

> Two *compound inequality* relationships. For example, 0.760 in. < head width < 0.792 in.; 5.03 cm < length < 5.16 cm

c Create a histogram of the head width measurements and also of the length measurements. Adjust the width of each cell so that the data are spread over six to eight cells. For example, suppose your range of measurements for head width was 0.032 inches—that is, from 0.760 inches to 0.792 inches—Then the width of each cell could be $\frac{0.032}{8} = 0.004$, and your histogram might look like the sample.

d Use your calculator to determine the mean of the head width measurements and the mean of the length measurements. Use the statistics keys or the formula for the mean. Record the mean for each set of data in your table, under the appropriate column.

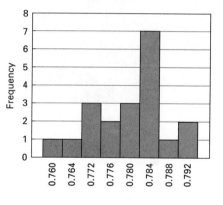

e If you work in quality assurance for a company, you may help establish tolerances for a production process. Quality assurance personnel sample the product under ideal production conditions and recommend a tolerance for the particular machine or process. For this activity, use the following tolerance for the head width.

$$\overline{w} \text{ inches} \pm 0.003 \text{ inches}$$

where $\overline{w}$ is the mean head width from Step **d**. Use the following tolerence for the length.

$$\overline{L} \text{ cm} \pm 0.01 \text{ cm}$$

where $\overline{L}$ is the mean length from Step **d**.

Write the tolerances in Step **e** as compound inequalities. Use the horizontal axes of your histograms as number lines. Graph your inequalities on the appropriate number line.

g Do any of your bolt measurements fall outside the tolerance you wrote in Step **f**?

h Compare your inequalities with those of the other groups in your class. Do any of their bolt measurements fall outside your group's tolerance interval?

Activity 3: Linear Programming

Equipment   Calculator
Paper
Pencil
Scissors
Stopwatch

Problem Statement

Production line schedules are determined in advance to maximize a manufacturer's profits. Among the many factors that determine production schedules are: profit per item, time to produce the item, the total number of items to produce, and available worker hours.

In this Activity, you will measure the average time it takes to produce two different production items. You will then use this data with given information to graph inequalities. Finally, you will use a form of linear programming to determine how many of each item should be produced to maximize profits.

Use the following information to complete the Activity:

x is the total number of Widget 1s.
y is the total number of Widget 2s.
The total time available to construct all of the widgets cannot exceed 30 minutes (1800 seconds).
The total number of widgets constructed will not exceed 35.
The profit from the sale of one Widget 1 is $0.75.
The profit from the sale of one Widget 2 is $1.50.

Procedure

a Have one student in the group trace five outlines of Widget 1, shown below. Trace all lines and the center dot. Have a second student use a stopwatch to time the first student in completing the tracing of the five outlines of Widget 1. Start the stopwatch when the pencil first touches the paper. Stop the watch when the pencil is lifted from the paper after the fifth Widget 1 is outlined. Record the time in seconds under the column "Total Tracing Time" in the data table shown below.

Object	Total Tracing Time (sec)		Average Tracing Time (sec)	Total Cutting Time (sec)		Average Cutting Time (sec)
Widget 1		÷5			÷5	
Widget 2		÷5			÷5	

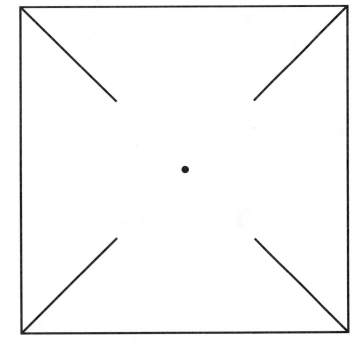

Widget 1

b Have a third member of the group cut out the five outlines of Widget 1. Make certain to cut the length of each heavy line toward the center. The second student will time the cutting of the five outlines of Widget 1. Start the stopwatch when the scissors first touch the paper. Stop the watch when the cutting of the fifth Widget 1 is completed. Record the time in seconds under the column "Total Cutting Time" in the Data Table.

c Repeat Steps **a** and **b** for Widget 2 shown below.

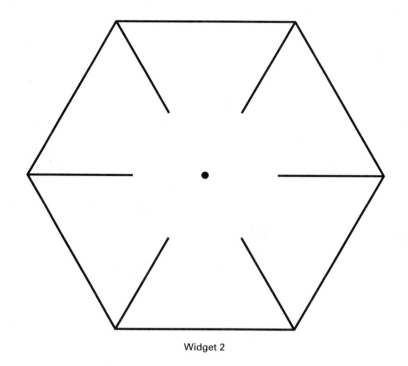

Widget 2

d Calculate the average time to trace one Widget 1. Round your answer *up* to the next whole second. Record this time in the Data Table under the column "Average Tracing Time." Repeat the process for Widget 2.

e Calculate the average time to complete the cutting of one Widget 1. Round your answer *up* to the next whole second. Record this time in the Data Table under the column "Average Cutting Time." Repeat the process for Widget 2.

f Use the information in the Problem Statement to write an equation for the profit *P* in selling *x* Widget 1s and *y* Widget 2s.

g Write inequalities that show the following:

- You cannot make fewer than zero of either type of widget. (Note: There will be one inequality for *x* and one for *y*.)

- The total time to trace and cut out *x* Widget 1s cannot exceed 30 minutes. Use the average times for tracing and cutting found in Steps **d** and **e** to write this inequality.

- The total time to trace and cut out *y* Widget 2s cannot exceed 30 minutes (1800 seconds). Use the average times for tracing and cutting found in Steps **d** and **e** to write this inequality.

- The total number of widgets constructed cannot exceed 35.

These five inequalities are the constraints that limit the profit from widget production.

h Graph the five inequalities from Step **g** on a common set of axes.

Calculations

1. Determine the coordinates of the points of intersection of the lines that represent the constraints. Record the coordinates in the table below.

2. Evaluate the equation in Step **f** using the coordinates above to determine the profit for this combination of sales. Record the profit in the table.

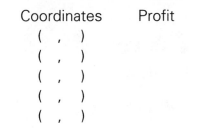

Coordinates   Profit
(,)
(,)
(,)
(,)
(,)

3. How many of each type of widget should be made to produce the maximum profit?

MATH APPLICATIONS

The applications that follow are like the ones you will encounter in many workplaces. Use the mathematics you have learned in this chapter to solve the problems.

Wherever possible, use your calculator to solve the problems that require numerical answers.

Rewrite each sentence as an inequality. For example, "Valarie received a lower score on the algebra test than Vicky" can be written as

Valarie's score on the algebra test < Vicky's score on the algebra test.

1 Bill is taller than Jack but not as tall as Sam.

2 This car will cost at least $10,500.

3 The most I can spend for a new suit is $160.

4 Phil is the same age as Janet, but he is older than Richard.

5 The water level does not reach the top of the dam.

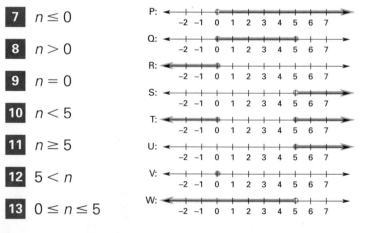

6 The trip will be canceled unless at least 12 persons are scheduled to go.

Match each inequality with the correct number line in the list at the right.

7 $n \leq 0$

8 $n > 0$

9 $n = 0$

10 $n < 5$

11 $n \geq 5$

12 $5 < n$

13 $0 \leq n \leq 5$

P: $\quad$ -2 -1 0 1 2 3 4 5 6 7

Q: $\quad$ -2 -1 0 1 2 3 4 5 6 7

R: $\quad$ -2 -1 0 1 2 3 4 5 6 7

S: $\quad$ -2 -1 0 1 2 3 4 5 6 7

T: $\quad$ -2 -1 0 1 2 3 4 5 6 7

U: $\quad$ -2 -1 0 1 2 3 4 5 6 7

V: $\quad$ -2 -1 0 1 2 3 4 5 6 7

W: $\quad$ -2 -1 0 1 2 3 4 5 6 7

The lengths (in inches) of several fish drawn from a lake on two separate occasions are listed as Set A and Set B.

Set A = {8, 9, 10, 11, 12}
Set B = {10, 11, 12, 13, 14}

14 Write the union of Sets A and B: A ∪ B.

15 Write the intersection of Sets A and B: A ∩ B.

16 If the length of a fish in Set A is represented by *f*, write an inequality that describes the interval of Set A.

17 Let *g* represent the length of a fish in Set B. Write an inequality for the interval of Set B.

18 Let *h* represent a number of the intersection of Set A and Set B. Write an inequality for the interval of the intersection.

In Exercises 14 and 15, you found the union and the intersection of two sets, A and B. Suppose you obtain a third sample of fish lengths, Set C.

Set A = {8, 9, 10, 11, 12}

Set B = {10, 11, 12, 13, 14}

Set C = {8, 11, 12, 7, 15}

19 Use the results of Exercise 14 to find the union of Sets A and B with Set C: (A ∪ B) ∪ C.

20 Find the union of Sets B and C: B ∪ C. Use the results to find the union of Set A with the union of Sets B and C: A ∪ (B ∪ C). Compare your answer to the results of Exercise 19. Do you think the following equality is true for any Set A, B, and C?

(A ∪ B) ∪ C = A ∪ (B ∪ C)

21 Use Sets A, B, and C to see if the following equality is true.

(A ∩ B) ∩ C = A ∩ (B ∩ C)

You need a semester average of at least 80 in Algebra to stay on the track team. You have taken six tests but you must still take the semester exam. The semester exam counts as two test grades when computing the semester average. Your six test grades are 88, 77, 75, 80, 80, and 70.

22 Write a formula to determine your semester test average *T*, if *E* is your score on the semester exam.

23 Write an inequality for the semester test average that describes an average of at least 80.

24 Combine the results of Exercises 22 and 23 to write an inequality that contains one variable, *E.*

25 Solve the inequality to find the score that you must make on the semester exam to stay on the track team.

AGRICULTURE & AGRIBUSINESS

Your landscaping company has been hired to mow a field with an area of 150,000 square yards. You have two mowers. To finish this job in 8 hours, you must rent additional mowers. You must rent no more mowers than needed. Each mower can cut at a rate of 2500 square yards per hour.

26 Determine how many square yards each mower can cut in 8 hours.

27 Write an expression for the area that can be mowed by your two mowers and *n* additional mowers (in 8 hours).

28 Write an inequality that relates the expression from Exercise 27 to the total mowing area.

29 Solve your inequality for *n* to find the fewest number of mowers you should rent to finish the job in eight hours.

Mr. Jackson owns 2800 acres of farmland. He must decide how much of two different types of crops to plant. When Mr. Jackson sells Crop A, he expects to make $240 per acre. When he sells Crop B, he expects to make $270 per acre. Federal regulations limit Mr. Jackson's planting to no more than 2000 acres of Crop A and no more than 1200 acres of Crop B.

30 Mr. Jackson plants *a* acres of Crop A and *b* acres of Crop B. Write an inequality for the total number of acres Mr. Jackson can plant.

31 Write inequalities for the total number of acres that can be planted for each type of crop.

32 Construct a graph for *a* versus *b.* Graph all the inequalities and shade the area that contains the points which satisfy all the conditions of your inequalities; that is, shade the region that contains feasible solutions for Mr. Jackson.

33 Mr. Jackson wants to maximize income from the sales of his crops. Write an equation for the amount of income (*I*) from the sale of Crop A and Crop B.

34 Using the techniques of linear programming, find the number of acres of Crop A and Crop B that satisfies the constraints and results in the maximum income.

A small accounting firm wants to subscribe to a nationwide database service. The service charges a monthly fee of $50 plus a $15 per hour access fee. The firm does not want to spend more than $500 per month for the database service.

35 Write an inequality that describes the relationship between the maximum amount the firm wants to spend and the total cost per month.

36 Solve the inequality to find the number of hours the accounting firm can be connected to the database and stay below budget.

As a travel agent, you have been hired to schedule a trip for a school's French club. The club has raised $5500 for the trip. Your agency charges a setup fee of $250. The cost per person will be $500. You must keep the cost below $5500.

37 Write an inequality that shows the relationship between the travel agency charges and the amount the club has raised.

38 Solve the inequality to find the number of club members that can go on the trip within the cost constraints.

Your ice cream parlor is running a promotional sale. For every regularly priced ice cream cone, a customer can buy a sundae on sale. On every sundae you sell at the sale price, you lose $0.19. You can recover your losses with the $0.24 profit you make by selling the regularly priced cones.

39 If "breaking even" is the worst you are willing to accept from the promotion, you will want the losses from the sales of sundaes to be less than or equal to the profits from the sale of cones. Write an inequality for this situation, letting s represent the number of sundaes sold and c the number of cones sold.

40 Graph the inequality, plotting the number of sundaes sold on the horizontal axis and the number of cones sold on the vertical axis. Indicate the region where your store will profit from the promotion.

Your company produces electronic games. To meet increased demand, you must purchase a number of new circuit board assembly machines. Brand X machines can assemble 280 games per month and cost $4000 each. Brand Y machines can assemble 320 games per month, and cost $7000 each. The company must maintain a production rate of at least 4100 games per month, and your budget for purchasing the new machines cannot exceed $75,000.

41 Write an inequality that relates the total cost of buying x Brand X machines and y Brand Y machines to your maximum purchase price.

42 Write an inequality that relates the total number of electronic games that you can produce with the new machines to the minimum acceptable production rate.

43 Graph each inequality and find the numbers of each brand of machine that you could order to meet the production requirement and stay within your budget. Remember that you must order whole machines, not fractional parts of a machine.

Your computer store sells two types of computers. The profit on the sale of a Model X is $250, while the profit from the sale of the more powerful Model Y is $350.

44 Write an expression that shows the income from selling x Model X computers and y Model Y computers during a given month.

45 You want the income from the sales of these two models to be at least $2000. Write an inequality for this condition.

46 Write the inequality in slope-intercept form and draw a graph of the inequality.

47 Use your graph to list three combinations of sales of Model X and Model Y computers that would allow you to meet your income goal.

48 Use your graph to list three combinations of sales of Model X and Model Y computers that would *not* allow you to meet your income goal.

You can spend $2400 on a vacation at a health and fitness resort. The round-trip air fare is $900. Each day at the resort costs $198. You want to determine the maximum number of days you can stay at the resort.

49 Write an expression for the total cost of airfare plus staying d days at the resort.

50 Use the result of Exercise 49 to write an inequality relating the total cost of the vacation to the spending limit.

51 Solve the inequality for d to find the maximum number of days you can stay at the resort.

Robert works in the Health Physics department at a nuclear waste disposal site. His job is to monitor the radiation exposure of the employees. The National Council on Radiation Protection (NCRP) recommends that a worker receive no more than 500 millirem of radiation exposure in a one-year period. At this disposal site, Robert has measured an average dose of 1.85 millirem per day for the workers. (assuming 8-hour work days)

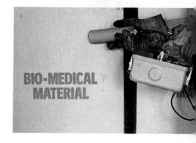

BIO-MEDICAL MATERIAL

52 Write an expression for the total amount of radiation a worker receives (on the average) after working d days in a year.

53 Write an inequality relating the NCRP recommendation to the total radiation received by an employee who works d days in a year.

54 Solve the inequality to determine the maximum number of days employees should work in a year.

55 Write a sentence summarizing the result from Exercise 54.

A clothing store manager wants to restock the men's department with two types of suits. A Type X suit costs $250. A Type Y suit costs $325. The store manager needs to stock at least $7,000 worth of suits to be competitive with other stores, but the store's purchasing budget cannot exceed $10,000.

56 Write an expression for the purchase cost of Type X suits and Type Y suits.

57 Write an inequality relating the purchase cost and the minimum value of suits in stock.

58 Write another inequality relating purchase cost and the maximum purchasing budget.

59 Graph the inequalities from Exercises 57 and 58 on the same axes.

60 Find the region that satisfies both inequalities and shade this region. Name one combination of purchases that will not exceed the maximum budget, yet will provide more than the minimum inventory.

Martin manages a beauty salon that offers a facial with manicure and a haircut with wash and styling. The salon staff can do as many as 10 facials a day, or they can give up to 24 haircuts a day. If the staff does both procedures at the same time, they can use the waiting time during the procedures more efficiently. In such a case, they can perform up to 29 facials and haircuts combined. The income from the facial procedure is $55, while the haircut brings in $35.

61 Let y represent the number of facials given in a day. Write an inequality for the number of facials the salon staff can give in one day.

62 Let x represent the number of haircuts given in a day. Write an inequality for the number of haircuts that can be given in one day by the salon staff.

63 Write an inequality for the total number of haircuts and facials the staff can do when they give both at the same time.

64 Draw a graph of these three inequalities. Shade the region where all three conditions are satisfied.

65 Write a formula for the total income from x haircuts and y facials given in a day.

66 Use the techniques of linear programming to determine the optimum number of haircuts and facials the salon should give to maximize income.

A chemical formula calls for a solution of pH level 6.3 with a tolerance of ±0.4.

INDUSTRIAL TECHNOLOGY

67 Let x represent the pH level of the solution. Use an absolute value expression to write an inequality for the acceptable values of pH.

68 Convert the absolute value inequality to two inequalities.

69 List two values of pH that will satisfy both inequalities.

A lab technician must classify rock samples according to their ore content. Group I samples have less than 2.5% ore content. Group II samples have from 2.5% to 7.0% ore content. And Group III samples have greater than 7.0% ore content. The table below of the lab's test results shows the ore contents of several rock samples.

Lab Test Results	
Sample ID	Ore Content
1	0.3%
2	4.8%
3	2.4%
4	10.0%
5	12.1%
6	5.9%
7	15.7%
8	7.4%
9	2.5%
10	7.0%

70 Write inequalities that describe the percent ore content for each of the three groups.

71 Identify which samples in the table belong to Group I, which belong to Group II, and which belong to Group III.

The specification for the diameter *D* of a metal rod is

$$2.775 \text{ inches} \pm 0.025 \text{ inches}$$

72 Write a compound inequality that represents the range of values for the rod's diameter *D*.

73 Use your combined inequality to find the range of dimensions of the radius *R* of the metal rod.

A technician working for an automobile manufacturer is testing the brake system of new car designs. She uses a formula to estimate braking distance on dry asphalt: $b = 0.039v^2$, where *b* is the braking distance in feet, and *v* is the car's speed in miles per hour. The braking test is conducted at three different speeds. A car fails the test if its braking distance exceeds the distance predicted by the formula for any of the three speeds.

74 Write an inequality that describes the maximum braking distance for a car that passes the braking test at a given speed.

75 The braking test is conducted at speeds of 30 miles per hour, 45 miles per hour, and 60 miles per hour. Write a statement that specifies the requirements for passing the entire braking test. (*Hint:* Use the word *and*.)

76 Draw a graph of the inequality in Exercise 74 for speeds up to 60 miles per hour. Show the region that indicates an acceptable braking performance.

Digital computers use logic circuits. Two such circuits are the AND gate and the OR gate. These circuits each have two input lines and one output line. To determine the output, the AND gate "looks" at the two input lines. The output line of an AND gate will be ON only if input line 1 is ON and input line 2 is ON. Similarly, the OR gate determines its output by "looking" at its two input lines. It will turn ON the output line if either line 1 is ON or line 2 is ON. The illustration shows how the AND gate and the OR gate are drawn in circuit diagrams.

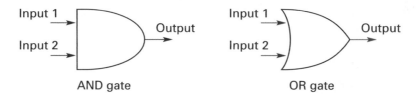

A certain circuit controls the two input lines. Input line 1 is ON only when a voltage $V_1 \geq 5.0$ volts. Input line 2 is ON only when the voltage V_2 satisfies the inequality $1.5 \leq V_2 \leq 3.0$ volts. Copy and complete the table below. For each V_1 and V_2, indicate whether line 1 and line 2 will be ON or OFF. Then, for each case, indicate whether the output of the AND gate and the output of the OR gate will be ON or OFF.

	V_1	V_2	Input: Line1	Input: Line2	Output: AND gate	Output: OR gate
77	6.0	6.0				
78	3.0	2.7				
79	4.0	1.2				
80	5.0	2.0				
81	2.0	1.9				
82	7.0	4.0				

A certain assembly requires three identical parts. Each part weighs x grams. After the three parts are connected, five grams are trimmed from the completed assembly. The final assembly must weigh more than 144.1 grams and less than or equal to 161.0 grams.

83 Write an expression for the mass of the final assembly in terms of x, the mass of one part.

84 Write a compound inequality using the result of Exercise 83 and the weight limits of the final assembly.

85 Write your compound inequality as two separate inequalities and solve each one for x.

86 Graph each solution on a number line. Indicate the region that satisfies your combined inequality.

87 Interpret the graph of your compound inequality. What is the range of mass for the parts in the assembly?

You have a box of resistors rated at $\frac{1}{2}$ watt; that is, the resistors can dissipate up to $\frac{1}{2}$ watt of power without being damaged. The resistance values vary from $R = 100$ ohms to $R = 2200$ ohms of resistance. For a resistor of resistance R (in ohms) and carrying a current I (in amperes), the power dissipated is equal to $I^2 R$.

88 Write an inequality stating that the power dissipated by a resistor is less than $\frac{1}{2}$ watt of power.

89 Solve the inequality for the current I for the minimum and maximum resistance values R in the box.

90 Draw a graph of the current I versus the resistance R for the inequality. Shade the region of allowable currents for these resistors.

Susan is an electrician. The wire she is using for a certain construction job comes on 100-foot spools. She uses only $14\frac{1}{2}$-foot lengths and 12-foot lengths of wire on this job.

91 Write an expression for the length of wire used to create a number x of $14\frac{1}{2}$-foot lengths of wire.

92 Write an expression for the length of wire used to create a number y of 12-foot lengths of wire.

93 Write an inequality indicating that the total number of $14\frac{1}{2}$-foot lengths of wire and 12-foot lengths of wire must not exceed the 100 feet of wire on each spool.

Skills

Graph each inequality.

1. $x \geq -1$ **2.** $-2 < x \leq 2$

Solve each inequality.

3. $2x + 3 \leq 8$ **4.** $-5x - 4 > 6$

5. $3x < 12$ or $x - 3 > 6$ **6.** $|x - 4| \leq 2$

Graph each inequality.

7. $x + 2y < 4$ **8.** $2x - 3y \geq -6$

9. Solve each system. Shade the solution.

$$y - 2x < 5$$
$$2y + x \geq 8$$

Applications

Write an inequality or a system of inequalities to model each situation. Graph the solution.

10. A mechanic has estimated that Carla's repair bill will be less than $250. Carla knows the garage charges $60 for a service call and $45 per hour for the repair work. How many hours did the mechanic estimate the work will take?

11. Clariss needs to bowl an average score of at least 150 to make the finals of the school bowling tournament. She bowled 128 and 160 in her first two games. What score does Clariss need to bowl in her third game to make the finals?

12. Virgil needs to rent a truck and a front-end loader for at least 4 hours to complete a construction project. He can rent a truck for $30 per hour and a front-end loader for $60 per hour. Virgil must keep his expenses under $180. The rental company charges by the whole hour. What combinations of rental times can Virgil use for a truck and a front-end loader in order to satisfy these constraints?

13. The following number line shows the height in centimeters of Janet, Jim, and Martha. Write an inequality to express the relationship of their heights.

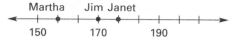

14. The specification for a bolt head requires that it be 0.75 ± 0.0015 inches across. Which of the following seven bolts do not meet this specification?

a. 0.7513
b. 0.7401
c. 0.7483
d. 0.7514
e. 0.7519
f. 0.7491
g. 0.7507

15. Write three inequalities that express the following constraints:

- The number of widget x's produced in 30 minutes cannot be less than zero.
- The number of widget y's produced in 30 minutes cannot be less than zero.
- The sum of widget x's and widget y's produced in 30 minutes cannot exceed 45.

Graph the constraints and show the region containing feasible solutions.

CHAPTER 10

WHY SHOULD I LEARN THIS?

Problem solving is an essential skill in the workplace. The key to solving a problem is dividing it into manageable parts. In this chapter, you will learn to solve complex mathematical expressions by breaking them into simpler parts called factors. If you practice the techniques presented in this chapter, you will develop effective problem-solving strategies.

Contents: Polynomials and Factors

POLYNOMIALS AND FACTORS

OBJECTIVES

1. Find the product of any two binomials.
2. Factor any trinomial.
3. Solve problems that involve factoring.

The word *factor* is often used in everyday conversations. Consider a few examples. In basketball, you know that height is an important factor in a player's ability to score or block a shot. In a jury trial, a certain piece of evidence may be an important factor in reaching a just verdict. When you plan for a day on the beach, the weather may be a big factor in your decision. The gauges and instruments in a nuclear power plant monitor many important factors related to safety and electrical production. In each of these cases, *factor* means a "part" of something—"part" of an athlete's ability, "part" of the evidence, "part" of a decision-making process or "part" of a power plant's operation.

In mathematics, the word *factor* has a similar meaning. A **factor** is "part" of a mathematical expression. When you factor a mathematical expression, you are writing it in a simpler form. One example of factoring is to use the Distributive Property to rewrite the expression $3x^2 + 5x^2$ as $(3 + 5)x^2$ or $8x^2$. The expression $8x^2$ is simpler than $3x^2 + 5x^2$. The Distributive Property lets you factor out the x^2 term from the original expression.

In this chapter you will see how factoring helps to simplify several different kinds of **polynomials**. In the next chapter you will use the factored form of a polynomial to model many real world situations.

As you watch the video for this chapter, look for examples of how polynomials are factored using geometric models.

LESSON 10.1 POLYNOMIALS

A contractor is buying paint to cover the exterior of two cubical storage tanks. Is there an algebraic expression for finding the amount of paint needed to cover both objects?

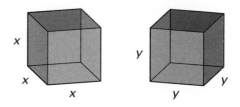

The first storage tank has sides with a length of x feet. The area of one of the faces of this tank is x^2 square feet. Since the tank has six faces, the total exterior area is $6x^2$ square feet. The second storage tank has sides of length y feet and an exterior area of $6y^2$ square feet. The total exterior area of the two tanks is described by the expression

$$6x^2 + 6y^2$$

Monomials, Binomials, and Trinomials

The expression $6x^2 + 6y^2$ contains two monomial terms. A **monomial** is a constant, a variable, or the product of a constant and one or more variables. The sum of one or more monomials is a **polynomial**. For example, $3x^2 + 4x + 2$ is a polynomial in one variable. The monomial terms are $3x^2$, $4x$, and 2.

In this chapter, you will discover patterns that can be represented by three special polynomials.

Name	Number of Terms	Example
Monomial	1	$3x^2$
Binomial	2	$3x^2 + 4x$
Trinomial	3	$3x^2 + 3x + 2$

The expression $3x^2$ is a *mono*mial because it has one term. The expression $3x^2 + 4x$ is a *bi*nomial because it has *two* terms. The expression $3x^2 + 3x + 2$ is a *tri*nomial because it has *three* terms.

Algeblocks can model monomials, binomials, and trinomials. In Chapter 4, you used Algeblocks to model an expression such as $3x + 2$.

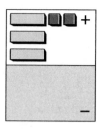

To model x^2, a new Algeblock is needed. Use a block that is x units on each side. The area of the block is represented by x^2.

Area = $1 \cdot x$
= x

Area = $x \cdot x$
= x^2

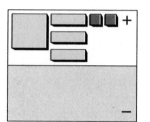

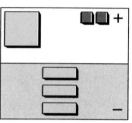

Algeblock model for
$x^2 + 3x + 2$

Algeblock model for
$x^2 - 3x + 2$

Adding and Subtracting Polynomials

ACTIVITY **Adding Trinomials**

1 Use Algeblocks to model $3x^2 + 4x + 1$.

2 Use other Algeblocks to model $2x^2 - 3x + 2$.

3 Combine the Algeblocks.

4 Record your result.

5 What is the sum of $(3x^2 + 4x + 1)$ and $(2x^2 - 3x + 2)$?

6 Complete the equation.

$$(3x^2 + 4x + 1) + (2x^2 - 3x + 2) = ?$$

Recall that monomials that have the same variable factors are called **like terms**. For example, $3x^2$ and $2x^2$ are like terms because each term has x^2 as a common factor. You can add like terms using the Distributive Property.

$$3x^2 + 2x^2 = (3 + 2)x^2 = 5x^2$$

If you add the coefficients of the like terms in the expression

$$(2x^2 + 3x + 4) + (x^2 - 3x + 2)$$

the result is $3x^2 + 6$.

Critical Thinking   Explain how to simplify

$$(2x^2 + 3x + 1) + (x^2 - 3x + 2)$$

EXAMPLE 1 Subtracting Polynomials

Subtract. $(3x^2 - 5) - (x^2 + 2x - 3)$

SOLUTION

You can find the difference by using the basic properties you learned in earlier chapters.

Horizontal Method

$$(3x^2 - 5) - (x^2 + 2x - 3) = (3x^2 - 5) + (-1)(x^2 + 2x - 3)$$
Definition of subtraction

$$= 3x^2 - 5 + (-x^2 - 2x + 3)$$
Distributive Property

$$= 3x^2 - x^2 - 2x - 5 + 3$$
Rearrange terms

$$= 2x^2 - 2x - 2$$
Add like terms

Vertical Method

$$3x^2 + 0x - 5$$ Use zero as coefficient of missing term

$$-(x^2 + 2x - 3)$$ Subtract like terms

$$2x^2 - 2x - 2$$

Use any method to simplify $-4x^2 + x - (2x^2 - 3x + 5)$.

EXAMPLE 2 The Perimeter of a Triangle

Find the simplest expression for the perimeter of triangle ABC.

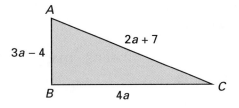

SOLUTION

The perimeter is the total distance around the triangle. To find the perimeter, add the lengths of the three sides; that is, simplify

$$(3a - 4) + (2a + 7) + 4a$$

If you rearrange the terms, the result is $3a + 2a + 4a - 4 + 7$. Now combine like terms. The perimeter is $9a + 3$.

One way to check the solution is to substitute a number for a in both the original expression and in the simplified expression. In this case, substitute 10 for a to see if the results are equal.

$$3a - 4 \quad + \quad 2a + 7 \quad + 4a \quad \overset{?}{=} \quad 9a + 3$$

$$[3(10) - 4] + [2(10) + 7] + 4(10) \quad \overset{?}{=} \quad 9(10) + 3$$

$$26 \quad + 27 + 40 \quad \overset{?}{=} \quad 90 + 3$$

$$93 \quad = \quad 93$$

Since the results are equal, the simplification is correct.

LESSON ASSESSMENT

Think and Discuss

1 Explain what is meant by *like terms* of an expression.

2 Explain the meaning of *monomial*, *binomial*, and *trinomial*. Give an example of each.

3 If *polynomial* means "many names," explain what *polygon* means.

4 How is the Distributive Property used to combine like terms?

Practice and Problem Solving

Classify each of the following expressions as a monomial, binomial, or trinomial.

5. $4x^2 + 3x$ **6.** $15x^4$ **7.** $x - 5$

8. -2 **9.** $7x^2 - 3x + 1$ **10.** $3x$

Use Algeblocks to find each sum.

11. $(3x + 5) + (2x + 3)$

12. $(4x^2 + 3x - 7) + (2x^2 + 5x - 3)$

13. $(3x^2 - 5x + 2) + (4x - 5)$ **14.** $(7x - 5) + (3x - 2)$

Add.

15. $(2x - 7) + (x^2 + 3x - 5)$ **16.** $(3x) + (2x^2 - 3x + 4)$

17. $(3x^2 - 7x + 4) + (x^2 + 7x - 4)$

18. $(4x^2 + 5x - 7) + (3x^2 + 7)$

Subtract.

19. $(2x + 5) - (x + 2)$ **20.** $(5x + 4) - (2x + 7)$

21. $(3x^2 + 8x + 5) - (2x^2 + x + 1)$

22. $(6x^2 + 5x + 7) - (3x - 3)$

23. $(4x^2 - 3x + 4) - (5x + 4)$

24. $(x^2 + 5x - 3) - (3x^2 + 7x + 4)$

25. The sides of a triangle are represented by the expressions $2x^2 + 3$, $4x^2 - 7x$, and $5x + 4$. Write the simplest expression for the perimeter of the triangle.

26. The length of a rectangle is represented by the expression $3x^2 - 2x + 1$. Its width is represented by the expression $x^2 + 2x - 3$. Write the simplest expression for the perimeter of the rectangle.

Mixed Review

27. Solve the distance formula $d = rt$ for the rate r.

28. Find the area of a circular garden with radius 5 meters.

Solve each equation.
29. $2x - 8 = 10$ **30.** $13 = 45 + 8x$ **31.** $3x + 14 = -4x$

32. If the perimeter of a square is 40 cm, what is the length of a diagonal across the square?

33. Barbara bought several CDs for $12 each and paid $5 for a poster. If she spent a total of $41, how many CDs did she buy?

Factor Pairs

When two or more numbers and/or variables are multiplied together, the numbers or variables are **factors**, and the result is the **product**.

$$a \quad \bullet \quad b \quad = \quad c$$

factor   factor   product

Since $3 \bullet 4 = 12$, 3 and 4 are factors of 12. Although $3 \bullet 4$ and $4 \bullet 3$ are two different multiplication facts, they are considered the same **factor pair**. This text lists factor pairs with the smaller number written first. Thus, you can write the factor pairs of 12 as

$$1 \bullet 12, \ 2 \bullet 6, \text{ and } 3 \bullet 4$$

The area of a rectangle is the product of two factors (the length and the width). Thus, you can use an Algeblock model or an area diagram to represent a factor pair.

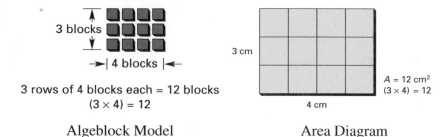

3 blocks

→| 4 blocks |←

3 rows of 4 blocks each = 12 blocks
$(3 \times 4) = 12$

3 cm

4 cm

$A = 12 \text{ cm}^2$
$(3 \times 4) = 12$

Algeblock Model   **Area Diagram**

ACTIVITY | **Building Rectangles with Factor Pairs**

1 Use Algeblocks or area diagrams to model the factor pairs for each whole number between 1 and 13. Record the results in a table like the one shown on the opposite page.

Number	Factor Pairs	Number of Factor Pairs
2	?	?
3	?	?
4	?	?

and so on until you reach the following:

12	1 • 12; 2 • 6; 3 • 4	3

2 Which numbers have exactly one factor pair? These numbers are **prime numbers**.

Prime Numbers
If a whole number, p, greater than 1 has exactly two unique whole-number factors, 1 and the number itself, then p is a prime number.

3 Which numbers have a factor pair in which both numbers are the same? These numbers are called **perfect square numbers**.

4 Which numbers have two or more factor pairs? These numbers are called **composite numbers**.

5 Continue the chart to find all of the prime numbers less than 30.

You can use factor pairs to write the factors of a number. For example, the factor pairs of 12 are

$$1 • 12, 2 • 6, \text{ and } 3 • 4$$

Thus, the factors of 12 are

$$1, 2, 3, 4, 6, \text{ and } 12$$

Ongoing Assessment

What are the factor pairs for 18? List the factors of 18.

Prime Factorization

A number is **factored** when it is written as a product of two of its factors. A number is *completely factored* when it is written as a product of prime factors.

EXAMPLE 1 Listing Prime Factors

List the prime factors of 24.

SOLUTION

Start with 24 and use a factor tree to write the factor pairs. Begin by writing down any factor pair of 24.

24

$4 \cdot 6$

Continue the tree.

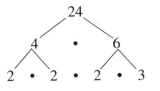

The prime factors of 24 are 2 and 3.

When you factor 24 into $2 \cdot 2 \cdot 2 \cdot 3$, you are writing the **prime factorization** of 24. You can also write the prime factorization of 24 using exponents. Thus,

$$2 \cdot 2 \cdot 2 \cdot 3 = 2^3 \cdot 3$$

EXAMPLE 2 Prime Factorization

Write the prime factorization of 36.

SOLUTION

Start with 36 and build a factor tree.

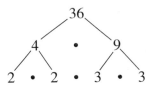

Thus, $36 = 2 \cdot 2 \cdot 3 \cdot 3$

$\qquad\qquad = 2^2 \cdot 3^2$

Since 2 and 3 are prime numbers, $2^2 \cdot 3^2$ is the prime factorization of 36.

Find the prime factorization of 36 by starting with a factor pair that is different than the one used in the example.

Greatest Common Factors

The numbers 24 and 36 have several **common factors**.

Factors of 24 $\qquad$ **1, 2, 3, 4, 6, 12**, 24

Factors of 36 $\qquad$ **1, 2, 3, 4, 6,** 9, **12,** 18, 36

The common factors of 24 and 36 are **1, 2, 3, 4, 6,** and **12**. The **greatest common factor** of 24 and 36 is 12.

> **Greatest Common Factor (GCF)**
> The greatest common factor of two or more integers is the largest factor that all the integers share.

You can use prime factorization to find the greatest common factor of a set of numbers. The GCF is the product of the common prime factors of the numbers in the set.

$$24 = 2 \cdot \mathbf{2} \cdot \mathbf{2} \cdot \mathbf{3}$$

$$36 = \qquad \mathbf{2} \cdot \mathbf{2} \cdot \mathbf{3} \cdot 3$$

Thus, the GCF is $\mathbf{2} \cdot \mathbf{2} \cdot \mathbf{3}$, or 12.

In the next section, the idea of GCF will be extended to monomials and polynomials.

CULTURAL CONNECTION

Many people have tried to find a formula for generating the set of all prime numbers. One such formula, $n^2 - n + 41$, will generate primes by substituting whole numbers less than 41 for n. Can you explain why the formula breaks down at $n = 41$?

An early Greek mathematician, Eratosthenes, used a method called a sieve to find prime numbers.

List the numbers 1–10 in the first row of a table. Then, continue listing the numbers 11–100 in rows of 10. Start at 2 and cross out every second number except 2. Start at 3 and cross out every third number except 3. Continue through all the rows of the table until no more numbers can be crossed out. Describe the set of numbers remaining.

Prime numbers such as three and five are called **twin primes** because their difference is two. A famous unproved theorem states that there is an infinite number of twin primes. Use the Sieve of Eratosthenes to list the twin primes less than 100.

LESSON ASSESSMENT

Think and Discuss

1 What are factor pairs?

2 How are factor pairs used in prime factorization?

3 What is the difference between a prime number and a composite number?

4 What is the difference between the prime factors of a number and the prime factorization of a number?

5 What is the greatest common factor of a set of numbers?

6 Can a set of numbers have more than one common factor?

7 Give an example.

What are the factor pairs for each number?

7. 45   **8.** 17   **9.** 32

10. 27   **11.** 26   **12.** 33

Write the prime factorization of each number.

13. 42   **14.** 35   **15.** 18

16. 84   **17.** 28   **18.** 51

Use prime factorization to find the greatest common factor of each pair of numbers:

19. 6 and 12   **20.** 5 and 7   **21.** 36 and 60

22. 50 and 52   **23.** 39 and 52   **24.** 72 and 90

25. Find the greatest common factor of 18, 36, and 60.

Mixed Review

Solve for x and y.

26. $3x - y = 7$ **27.** $3x - 4y = 9$ **28.** $2x - 3y = 12$
       $2x + 3y = 12$ $2x - 3y = 7$ $5x + 2y + 8 = 0$

29. An iron beam 68 inches long is divided into two parts. One piece is three inches longer than the other. What are the two lengths?

30. Find the angles of a triangle if one angle is three times a second angle and six times the third angle.

LESSON 10.3 MONOMIAL FACTORS

The Distributive Property

An area diagram is a geometric model of the Distributive Property.

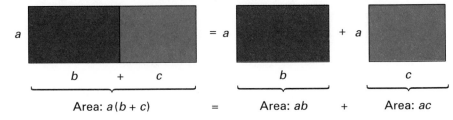

Area: $3(4 + 5)$ =   Area: 3×4 +   Area: 3×5
27 squares   =   12 squares   +   15 squares

In general, you can multiply a monomial a and a binomial $(b + c)$ using the Distributive Property. The model looks like this.

Area: $a(b + c)$ =   Area: ab +   Area: ac

Likewise, you can use the Distributive Property with expressions that contain exponents.

$$3x^2(x^3 + 4) = 3x^2\,(x^3) + 3x^2\,(4)$$

To simplify the expression $3x^2(x^3)$, you can use the properties of exponents you learned in Chapter 2.

EXAMPLE 1 Adding Exponents

Simplify $x^2 \bullet x^3$

SOLUTION

$$x^2 \bullet x^3 = (x \bullet x) \bullet (x \bullet x \bullet x)$$

$$= x \bullet x \bullet x \bullet x \bullet x$$

$$= x^5$$

Notice that the exponent for the product is the sum of the exponents of the factors; that is, $x^2 \bullet x^3 = x^{2 + 3} = x^5$. This property is the Product of Powers Property.

Now the simplification can be completed.

$$3x^2(x^3 + 4) = 3x^2\,(x^3) + 3x^2\,(4) = 3x^5 + 12x^2$$

Ongoing Assessment

The electrical power of a solar car is found using the equation $P_E = I(V - IR)$. Use the Distributive Property to multiply the terms on the right side of the equality.

Some polynomials have a common monomial factor. For example, the binomial $20H^2 + 50H$ estimates part of a company's cost of renting advertising space on a billboard. Notice that each term in the binomial contains the common factor $10H$.

Use the Distributive Property to factor out the common monomial.

$$20H^2 + 50H = (10H \cdot 2H) + (10H \cdot 5)$$
$$= 10H(2H + 5)$$

To *factor* $20H^2 + 50H$ means to write it as $10H(2H + 5)$. When you factor a polynomial, be sure to identify *all* the common factors.

EXAMPLE 2 Removing Common Monomial Factors

The U.S. Forest Service uses aerial photography in a survey to estimate the diameter of trees. If H is the height of a tree and V is the visible crown diameter from the photograph, then the tree's diameter is given by the polynomial $2HV - 2HV^2 - 6HV^3$. Factor this polynomial.

SOLUTION

To identify the common monomial factors, factor each term of the polynomial.

$$(2 \cdot H \cdot V) - (2 \cdot H \cdot V \cdot V) - (2 \cdot 3 \cdot H \cdot V \cdot V \cdot V)$$

The greatest common monomial factor is $2HV$. Factor out $2HV$ and write

$$2HV - 2HV^2 - 6HV^3 = 2HV(1 - V - 3V^2)$$

Factor the expression $4x^3 + 8x^2 + 12x$.

Dividing by a Monomial

The Distributive Property is true for all numbers, including the rational numbers. Thus, you can use the Distributive Property to simplify an expression such as $\frac{1}{x^2}(x^5 + x^3)$ where $x \neq 0$.

$$\frac{1}{x^2}(x^5 + x^3) = \frac{1}{x^2}(x^5) + \frac{1}{x^2}(x^3)$$

$$= \frac{x^5}{x^2} + \frac{x^3}{x^2}$$

To simplify the right side of the equation, you can use another property of exponents from Chapter 2.

EXAMPLE 3 Subtracting Exponents

Simplify $\dfrac{x^5}{x^2}$.

SOLUTION

$$\frac{x^5}{x^2} = \frac{x \cdot x \cdot x \cdot \cancel{x} \cdot \cancel{x}}{\cancel{x} \cdot \cancel{x}}$$

$$= x^3$$

What do you notice about the difference between the exponents in the original expression and the exponent in the answer?

Quotient of Powers Property
For all non-zero numbers x and integers m and n,

$$\frac{x^m}{x^n} = x^{m-n}$$

Now you can simplify the right side of the equation above.

$$\frac{x^5}{x^2} + \frac{x^3}{x^2} = x^{5-2} + x^{3-2} = x^3 + x$$

Simplifying $\frac{1}{x^2}(x^5 + x^3)$ where $x \neq 0$, is the same as dividing the expression $(x^5 + x^3)$ by the monomial x^2. Remember, when you divide an expression by a monomial, divide each of the terms by the monomial.

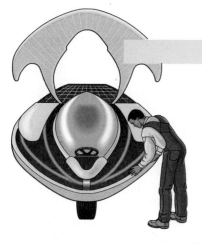

Ongoing Assessment

An electronics technician analyzes a simple circuit and verifies that the equation below models the voltage V and current I in the circuit, where the resistance R and k are constant.

$$VI = I^2R_1 + I^2R_2$$

a. Factor the expression on the right side of the equation.

b. Solve the equation for V.

LESSON ASSESSMENT

Think and Discuss

1 How is the Distributive Property used to multiply a monomial by a polynomial?

2 How is the Distributive Property used in factoring?

Practice and Problem Solving

Multiply.

3. $a(b + 3)$ **4.** $x(y - 5)$ **5.** $7x(x + 5y)$

6. $3x^2(4x^2 + 7)$ **7.** $2x(4x^3 + 7x - 5)$ **8.** $x^4(x^2 - 2x + y)$

Divide.

9. $4x^2 + 8y^2$ by 4 **10.** $2x^3 + 4x$ by $2x$

11. $3x^4 - 6x^2y^4$ by $3xy$

12. An automated assembly line is programmed to produce a variety of sizes of rectangular toolboxes. The volume of a toolbox produced on this line is given by $6ab^2$. If the width is $3b$ and the depth is a, what is the height of this toolbox?

The volume of a cylinder of radius r and height h is $V = \pi r^2 h$.

13. Write an equation in simplified form for the volume of air in a scuba diver's cylindrical tank if the height of the tank is three times its radius.

14. Rewrite this equation for the diameter (d) of the tank and simplify.

Factor out the greatest monomial factor.

15. $3xy - 5y^2$ **16.** $-3ax + cx - x^2$ **17.** $cx^2 + 2cx^3$

Mixed Review

A jar contains 7 pennies, 8 nickels, and 5 dimes.

18. If one coin is taken at random from the jar, what is the probability it will be a penny?

19. If one coin is taken at random from the jar, what is the probability it will be a nickel or a dime?

20. If two coins are taken at random from the jar, what is the probability that both coins are dimes?

Describe the set of whole numbers that satisfies each of the following inequalities:

21. $3 + n < 10$

22. $n + 2 > 8$

23. $3 + n < 9$ *and* $2 + n < 6$

Write and solve an equation for each situation.

24. The sum of three consecutive integers is 39. What is the middle integer?

25. A bank charges $3 for every $500 worth of certified checks it issues. If the charge for Mary's certified check is $12, what is the value of the check issued by the bank?

Zero and Negative Exponents

The following Example will help you recall two of the properties of exponents used in Chapter 2.

EXAMPLE 1 Zero Exponents

How can you convince someone that $x^0 = 1$?

SOLUTION

Show that the expression $x^6 \div x^6$ can be written as either 1 or x^0.

a. Factor the numerator and denominator.

$$\frac{x^6}{x^6} = \frac{\cancel{x} \cdot \cancel{x} \cdot \cancel{x} \cdot \cancel{x} \cdot \cancel{x} \cdot \cancel{x}}{\cancel{x} \cdot \cancel{x} \cdot \cancel{x} \cdot \cancel{x} \cdot \cancel{x} \cdot \cancel{x}} = \frac{1}{1} = 1$$

b. Use the Quotient of Powers Property.

$$\frac{x^6}{x^6} = x^{6-6} = x^0$$

Thus, $x^0 = 1$.

This example leads to the following property.

> **Zero Property of Exponents**
> For all numbers x (except zero), $x^0 = 1$

EXAMPLE 2 Negative Exponents

How can you convince someone that $\dfrac{1}{x^2} = x^{-2}$?

SOLUTION

Start with $\dfrac{x^3}{x^5}$.

a. Factor the numerator and denominator.

$$\frac{x^3}{x^5} = \frac{\cancel{x} \cdot \cancel{x} \cdot \cancel{x}}{\cancel{x} \cdot \cancel{x} \cdot \cancel{x} \cdot x \cdot x} = \frac{1}{x \cdot x} = \frac{1}{x^2}$$

b. Use the Quotient of Powers Property.

$$\frac{x^3}{x^5} = x^{3-5} = x^{-2}$$

Thus, $x^{-2} = \dfrac{1}{x^2}$.

This example leads to the following property.

Property of Negative Exponents

For all numbers x (except zero) and integers n,

$$x^{-n} = \frac{1}{x^n}$$

Critical Thinking   Explain why x^{-1} and x are reciprocals.

Powers of Products and Quotients

In some problems, it is necessary to raise a product or a fraction to a power before you can find a solution.

EXAMPLE 3 Raising a Monomial to a Power

Simplify $(5x)^3$.

SOLUTION

$$(5x)^3 = 5x \cdot 5x \cdot 5x$$
$$= (5 \cdot 5 \cdot 5)(x \cdot x \cdot x)$$
$$= 5^3 \cdot x^3$$
$$= 125x^3$$

Notice that when the product is raised to a power, each factor is raised to that same power.

Power of a Product

For all numbers x and y, and integers n,

$$(xy)^n = x^n y^n$$

EXAMPLE 4 Power Dissipation

An electric component dissipates power (in watts) at a rate given by the following formula:

$$P = I^2R$$

In this formula, I is the current passing through the component (in amperes) and R is the resistance of the component (in ohms). An electronics technician needs to predict the heat that will be generated by a resistor on a control board. He measures a current of 0.1 amperes passing through the resistor. The resistor has a resistance of 300 ohms. The power dissipated as heat is

$$P = I^2R$$
$$= (0.10^2)(300)$$
$$= 3 \text{ watts}$$

What power will the same resistor dissipate if the current doubles?

SOLUTION

Let I and R be the same numerical values as before. This time, the current is $2I$. Substitute $2I$ for I and simplify.

$$P = (2I)^2R$$
$$= (2)^2(I)^2R$$
$$= 4I^2R$$

If the current doubles, the power quadruples.

What happens when a fraction is raised to a power?

EXAMPLE 5 The Power of a Quotient

Simplify $\left(\dfrac{x}{3}\right)^3$.

SOLUTION

$$\left(\frac{x}{3}\right)^3 = \frac{x}{3} \cdot \frac{x}{3} \cdot \frac{x}{3}$$
$$= \frac{x \cdot x \cdot x}{3 \cdot 3 \cdot 3}$$
$$= \frac{x^3}{27}$$

When a fraction is raised to a power, the numerator and the denominator are each raised to the same power.

Power of a Quotient

For all numbers x and y (except zero) and integers n,

$$\left(\frac{x}{y}\right)^n = \frac{x^n}{y^n}$$

Ongoing Assessment

The resistor in Example 4 dissipated 3 watts of power with a current of 0.1 amperes. Use the Power of a Quotient Property to find the power dissipated when the current is halved.

Power of a Power

EXAMPLE 6 Raising a Power to a Power

Simplify $(x^3)^2$.

SOLUTION

$$(x^3)^2 = x^3 \cdot x^3$$

$$= x^{3+3}$$

$$= x^6$$

Notice that the product of the original two exponents $2 \cdot 3$ also equals 6. Thus,

$$(x^2)^3 = x^{2 \cdot 3} = x^6$$

When a power is raised to a power, the two powers are multiplied.

Power of a Power

For all numbers x and integers n and m,

$$(x^n)^m = x^{nm}$$

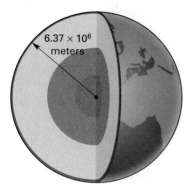

6.37 × 10⁶ meters

EXAMPLE 7 The Volume of the Earth

The radius of the Earth is 6.37×10^6 meters. What is its volume? Write the answer in scientific notation.

SOLUTION

The Earth is approximately a sphere. The volume of a sphere is found by using a formula.

$$V = \frac{4}{3}\pi r^3$$

$$= \frac{4}{3}\pi(6.37 \times 10^6)^3$$

$$= \frac{4}{3}\pi(6.37)^3(10^6)^3$$

$$= \frac{4}{3}\pi(6.37)^3 \times 10^{18}$$

$$\approx 1083 \times 10^{18}$$

$$\approx 1.08 \times 10^{21}$$

The volume of the Earth is approximately 1.08×10^{21} cubic meters.

WORKPLACE COMMUNICATION

What electrical current should the maintenance technician expect from the radiometer when measuring the furnace temperature at $\frac{1}{2}T_{ave}$ and $\frac{3}{2}T_{ave}$?

INSTRUMENTATION MAINTENANCE REQUEST FORM

SUBMITTED BY: Melanie Montoya

EXTENSION: 6-0318

OFFICE: Semiconductor Dev Team

EMAIL ADDRESS: melmont@mwi.org

DATE/TIME SUBMITTED: Oct 22/9:30 AM

DATE/TIME NEEDED: Oct 23/5:00 PM

COMPLETE DESCRIPTION OF INSTRUMENTATION MAINTENANCE REQUIRED:

We have built an instrument for measuring the temperature of the furnace in the semiconductor production area. This instrument is a *Stefan-Boltzmann* radiometer. The radiometer detects heat energy radiated from the furnace and converts it into an electrical current. This conversion is described by the formula

$I = CT^4$

where I is the instrument current, T is the furnace temperature, and C is a constant.

In a recent test, we found that the radiometer produces 1 milliampere of current when the furnace is at the average temperature (T_{ave}).

Please calibrate the radiometer so that it will accurately read temperatures from a low of $\frac{1}{2}T_{ave}$ to a high of $\frac{3}{2}T_{ave}$.

TO BE COMPLETED BY MAINTENANCE TECHNICIAN
DATE/TIME RECEIVED: Oct 22/11:00 AM

SIGNATURE:

DATE/TIME COMPLETED:

LESSON ASSESSMENT

Think and Discuss

1 Explain why $(2x)^3$ is not the same as $2x^3$.

2 Suppose the side of a square is doubled. Explain why the area of the square is four times as large as the original square.

3 Suppose the edge of a cube is doubled. What is the effect on the volume of the cube? Explain your answer.

4 Explain why $\left(\dfrac{a}{b}\right)^{-n}$ is the same as $\left(\dfrac{b}{a}\right)^{n}$.

Practice and Problem Solving

Simplify each expression.

5. $a^2(a^5)$

6. $(-3x)^3$

7. $2x^3(5x^4)$

8. $(3x^2)^4$

9. $\left(\dfrac{2x}{3}\right)^3$

10. $(5y^{-2})^{-3}$

11. $2x^3(3x^2 + 4)$

12. $5xy(3x^2 + 2xy - 4y^3)$

13. $\dfrac{16s^3t^2}{(2st)^2}$

Mixed Review

Solve each problem. Write an equation to model each situation.

14. A circular pipe has a radius of 3 inches. The pipe is placed inside a square box that has an inside dimension of 6 inches. What is the cross-sectional area outside the pipe but inside the box?

15. If 48 pounds of steel cost D dollars, write a formula that determines the cost of x pounds of steel.

16. A group of 36 people is divided into two groups. The first group is four-fifths the size of the second group. How many people are in each group?

17. The perimeter of a rectangle is 36 inches. The length is 4 inches longer than the width. Find the area of the rectangle.

18. The sum of two angles is 180 degrees. The ratio of the angle measures are two to one. What are the measures of the angles?

19. Solve the system $x > 2$ and $x + y \le 5$ by graphing.

LESSON 10.5 ADDING RATIONAL EXPRESSIONS

Tammy can fill a tank in 10 hours by opening Pipe A. In one hour, $\frac{1}{10}$ of the tank is filled. In 3 hours, the tank is $3 \cdot \frac{1}{10}$ or $\frac{3}{10}$ full, and in x hours, $x \cdot \frac{1}{10}$ or $\frac{x}{10}$ of the tank is full. In 10 hours, $\left(10 \cdot \frac{1}{10} = 1\right)$ the tank is completely full. Pipe B can fill the tank by itself in 6 hours. How can Tammy find the time it will take to fill the tank if she opens both pipes? Let x represent the time it takes both pipes to fill the tank. In x hours, Pipe A will fill $\frac{x}{10}$ of the tank, and Pipe B will fill $\frac{x}{6}$ of the tank. To find the time to fill the tank, add the contributions of both pipes.

$$\frac{x}{10} + \frac{x}{6} = 1$$

Adding rational expressions is like adding rational numbers. If necessary, first rewrite the terms with a common denominator. Then use the following rule:

Adding Rational Expressions

If $\frac{a}{b}$ and $\frac{c}{b}$ are rational expressions ($b \neq 0$), then

$$\frac{a}{b} + \frac{c}{b} = \frac{a + c}{b}$$

To solve Tammy's problem, start by writing the rational expressions with common denominators. Thus,

$$\frac{3}{3} \cdot \frac{x}{10} + \frac{5}{5} \cdot \frac{x}{6} = 1$$

Now add the rational expressions and solve for x.

$$\frac{3x}{30} + \frac{5x}{30} = 1$$
$$\frac{8x}{30} = 1$$
$$8x = 30$$
$$x = \frac{30}{8} \quad \text{or} \quad 3\frac{3}{4}$$

It will take $3\frac{3}{4}$ hours to fill the tank with both pipes open.

EXAMPLE 1 Completing a Job Alone

Ron and Al are preparing an estimate to repair a car. They estimate it will take 6 hours to repair the car if they work on it together. Ron can complete the repair by himself in 10 hours. How long will it take Al to do the job alone?

SOLUTION

Let x represent the amount of time it takes Al to do the job alone.

Then $\frac{1}{x}$ represents the part of the job that Al can complete in 1 hour. Remember that the denominator cannot be zero. In 6 hours, Al can complete $\frac{1}{x} \cdot 6$, or $\frac{6}{x}$ of the total job.

Ron can complete $\frac{1}{10}$ of the job in 1 hour. In 6 hours, he can complete $\frac{6}{10}$, or $\frac{3}{5}$ of the job.

The total job, represented by 1, is equal to the sum of Al's part and Ron's part.

The equation that models this situation is

$$\frac{6}{x} + \frac{3}{5} = 1$$

Write the rational expressions with common denominators. Then add the expressions and solve for x.

$$\frac{5}{5} \cdot \frac{6}{x} + \frac{x}{x} \cdot \frac{3}{5} = 1$$

$$\frac{30}{5x} + \frac{3x}{5x} = 1$$

$$\frac{30 + 3x}{5x} = 1$$

$$30 + 3x = 5x$$

$$30 = 2x$$

$$x = 15$$

It will take Al 15 hours to repair the car by himself. Check to see if this is the correct answer.

Ongoing Assessment

It takes two people 4 hours to clean an apartment if they work together. One person can clean the apartment in 6 hours. How long will it take the other person to clean the apartment working alone?

EXAMPLE 2 Adding Rational Expressions

Simplify $\dfrac{3}{x} + \dfrac{4}{x^2}$, $x \neq 0$.

SOLUTION

First, rewrite the expressions with a common denominator. Then add the expressions.

$$\frac{x}{x} \cdot \frac{3}{x} + \frac{4}{x^2} = \frac{3x}{x^2} + \frac{4}{x^2}$$

$$= \frac{3x + 4}{x^2}$$

The final expression $\dfrac{3x + 4}{x^2}$ carries the restriction that $x \neq 0$.

When you solve a rational equation, this restriction is very important.

Ongoing Assessment

Simplify $\dfrac{a}{x^2} + \dfrac{1}{x^3}$, $x \neq 0$.

You can also subtract rational expressions. Use the addition rule with the addition sign changed to a subtraction sign.

EXAMPLE 3 Subtracting Rational Expressions

a. Simplify $\dfrac{3}{2x} - \dfrac{1}{5x}$, $x \neq 0$.

b. Solve $\dfrac{3x}{x - 2} - \dfrac{5}{2 - x} = 1$, $x \neq 2$.

SOLUTION

a. The least common denominator is $10x$. Thus,

$$\frac{3}{2x} - \frac{1}{5x} = \frac{5}{5} \cdot \frac{3}{2x} - \frac{2}{2} \cdot \frac{1}{5x}$$

$$= \frac{15}{10x} - \frac{2}{10x}$$

$$= \frac{13}{10x}, x \neq 0$$

b. The denominators are opposites. Multiply the first rational expression by $\frac{-1}{-1}$ so that the rational expressions will have common denominators. The result is

$$\frac{-1}{-1} \cdot \frac{3x}{x-2} = \frac{-3x}{-(x-2)} = \frac{-3x}{2-x}$$

Now subtract and solve the equation for x.

$$\frac{-3x}{2-x} - \frac{5}{2-x} = 1$$

$$\frac{-3x-5}{2-x} = 1$$

$$-3x - 5 = 2 - x$$

$$-2x = 7$$

$$x = -3.5$$

Check to see if -3.5 is a solution to the original equation.

LESSON ASSESSMENT

Think and Discuss

1 What is a rational expression?

2 How do you add rational expressions?

3 How do you subtract rational expressions?

4 Why is it necessary to use restrictions when you add and subtract rational expressions?

Practice and Problem Solving

Add or subtract.

5. $\dfrac{x}{2} + \dfrac{3x}{4}$

6. $\dfrac{x+1}{3} - \dfrac{2x-1}{4}$

7. $\dfrac{3}{2x} - \dfrac{4}{5}$

8. $\dfrac{x+5}{2x} + \dfrac{3}{x}$

9. $\dfrac{x}{2} + \dfrac{1}{x}$

10. $\dfrac{x-1}{4x} + \dfrac{3-x}{6x}$

11. $\dfrac{x+3}{x^2} - \dfrac{2x-5}{x^2}$

12. $\dfrac{7}{x} + \dfrac{3x+1}{x^2}$

13. $\dfrac{x^2+2}{x^2} - \dfrac{x-1}{-2x}$

14. A farmer owns a tractor that can cultivate a field in 8 hours. The farmer is trying to decide whether to buy another tractor that can cultivate the same field in 6 hours. How long would it take to cultivate the field using both tractors?

15. Miquel and Rosa are painting contractors. They spent 5 hours painting a room together. Last year Miquel painted the same room in 10 hours. How long would it take Rosa to paint the room by herself?

16. A company is preparing a mass mailing to advertise its new stereo system. One employee can complete the address labels for the mailing in 4 hours. Another employee can do the job in 3 hours. If both employees work together, how long will it take to prepare the address labels?

17. Two electricians wired a new house in 30 hours. On a similar house, one of the electricians spent 50 hours to install all the wiring. How long would it take the other electrician to wire the house if he worked alone?

Mixed Review

Solve each problem.

18. Althea paid $250 for her utility bills in May and June. In June, the bill was $25 plus twice her bill for May. How much was Althea's utility bill for each month?

19. There are 91 students in three classes. One class has 9 more students than the first class. The other class has 5 fewer students than the first class. How many students are there in each class?

20. After Ronald deposited $125.50 into his savings account, he still had less than $200 in his account. How much did Ronald have in his account before he made the deposit?

21. The cost of three shirts and two pairs of shorts is $149. The cost of two shirts and three pairs of shorts is $136. Find the cost of the shirts and the shorts.

22. A 25% soap solution is mixed with a 40% soap solution to obtain 16 gallons of a 30% solution. How many gallons of each soap solution are needed?

A small patio is 4 feet by 6 feet. You want to expand the patio by the same length on each side. What formula expresses the new area?

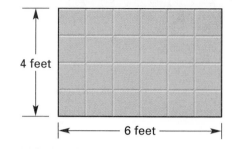

4 feet

6 feet

Examine the diagram of the patio. Let x represent the amount added to each side of the patio. Then one side will be represented by $x + 4$ and the other side by $x + 6$. The expression for the new area of the patio will be the product of the binomials.

$$(x + 4)(x + 6)$$

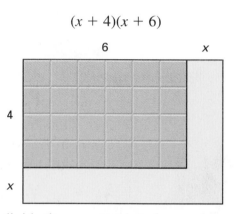

6

x

4

x

Notice you can divide the new area into four regions with the following areas:

$$x^2 \quad 6x \quad 4x \quad 24$$

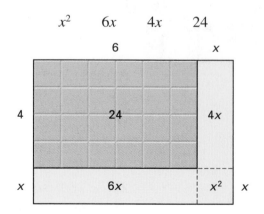

6

x

4

24

4x

x

6x

x^2

x

The total area is the sum of the areas:

$$x^2 + 6x + 4x + 24 \quad \text{or} \quad x^2 + 10x + 24$$

Suppose you need to expand the patio by 3 feet on each side. Substitute 3 for x to check if $(x + 4)(x + 6) = x^2 + 10x + 24$.

$$(3 + 4)(3 + 6) = 3^2 + 10(3) + 24$$

$$(7)(9) = 9 + 30 + 24$$

$$63 = 63$$

EXAMPLE 1 Multiplying Two Binomials

Multiply $(x + 2)(x + 1)$.

SOLUTION

The product $(x + 2)(x + 1)$ can represent a rectangle with length $(x + 2)$ and width $(x + 1)$. The area of this rectangle is

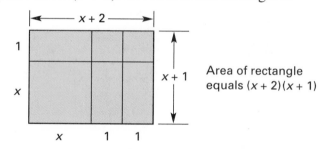

Area of rectangle equals $(x + 2)(x + 1)$

If you label the area for each of the small rectangles, you can determine the total area of the overall rectangle. This area is the same as the product of $(x + 2)(x + 1)$.

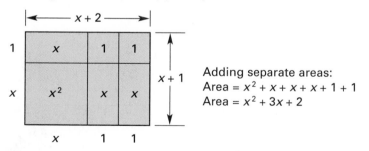

Adding separate areas:
Area $= x^2 + x + x + x + 1 + 1$
Area $= x^2 + 3x + 2$

You have just used a geometric representation to multiply two binomials, $x + 2$ and $x + 1$. The result is the trinomial $x^2 + 3x + 2$.

Thus, $(x + 2)(x + 1) = x^2 + 3x + 2$.

The geometric model can be used to derive an algebraic pattern for multiplying two binomials. Consider the product $(x + 3)(2x + 1)$.

Let $(x + 3)$ be the length of a rectangle and $(2x + 1)$ be the width.

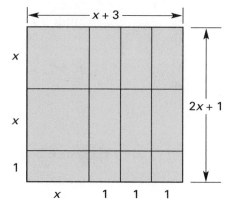

The development of the algebraic product $(x + 3)(2x + 1)$, term by term, is shown in a series of four drawings. The geometric equivalent of each term-by-term product is highlighted by the shaded portions in the rectangle.

Step 1: $(x + 3)(2x + 1)$
$(x)(2x) = 2x^2$

Step 2: $(x + 3)(2x + 1)$
$(x)(1) = x$

Step 3: $(x + 3)(2x + 1)$
$(3)(2x) = 6x$

Step 4: $(x + 3)(2x + 1)$
$(3)(1) = 3$

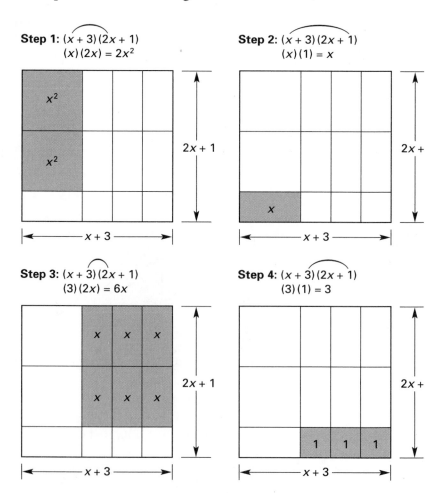

Add together all the highlighted areas in the four rectangles to find the total area of the rectangle of size $(x + 3)$ by $(2x + 1)$.

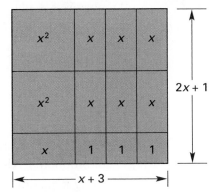

$(x + 3)(2x + 1) = x^2 + x^2 + x + x + x + x + x + x + x + 1 + 1 + 1$

$$= 2x^2 + 7x + 3$$

When the four steps in the multiplication process are done in sequence, the result is the following:

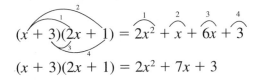

$$(x + 3)(2x + 1) = 2x^2 + 7x + 3$$

A simple memory device will help you keep track of which terms are being multiplied. This device is known as the FOIL technique.

To find the product of $(x + 3)(2x + 1)$, proceed as follows:

First terms are multiplied $(\boldsymbol{x} + 3)(\boldsymbol{2x} + 1)$ $x \cdot 2x = 2x^2$

Outer terms are multiplied $(\boldsymbol{x} + 3)(2x + \boldsymbol{1})$ $x \cdot 1 = x$

Inner terms are multiplied $(x + \boldsymbol{3})(\boldsymbol{2x} + 1)$ $3 \cdot 2x = 6x$

Last terms are multiplied $(x + \boldsymbol{3})(2x + \boldsymbol{1})$ $3 \cdot 1 = 3$

Add and then simplify by combining similar terms. The result is

$$(x + 3)(2x + 1) = 2x^2 + x + 6x + 3$$

$$= 2x^2 + 7x + 3$$

The FOIL method is a shortcut for using the Distributive Property. It helps you multiply two binomials in an orderly manner. Practice will help make the process automatic.

When one or both binomials in a product contains a negative term, the geometric model becomes complicated to use. For example, if you are asked to find the product $(x - 2)(x + 1)$, it is hard to imagine how to represent -2.

However, the FOIL method readily handles the negative terms.

$$\overset{\text{F}}{} \quad \overset{\text{O}}{} \quad \overset{\text{I}}{} \quad \overset{\text{L}}{}$$
$$(x - 2)(x + 1) = (x)(x) + (x)(1) + (-2)(x) + (-2)(1) = x^2 - x - 2$$

Algeblocks are another way to use areas to model the multiplication of binomials. Algeblocks solve the problem of negative areas by using a quadrant system to identify areas as either positive or negative.

EXAMPLE 2 Using Algeblocks to Multiply Binomials

Multiply $(x - 2)(x + 1)$ using Algeblocks as a model.

SOLUTION

1. Select the blocks that model $x - 2$ and $x + 1$.

2. Arrange these blocks along the axes on the Algeblock mat as shown in the diagram. Note that the 2 unit blocks that model -2 are placed along the negative portion of the horizontal axis.

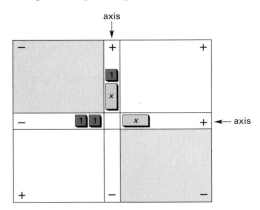

3. Now use additional Algeblocks to form a rectangle using the blocks on the axes to model the length and width of the rectangle. Note that the *areas* in the second quadrant are labeled negative. List the six *areas* that make up the rectangle you formed.

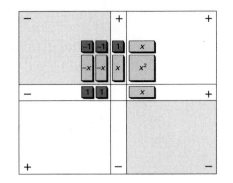

4. Add the *areas* of the rectangle formed by the Algeblocks. Do not include in the sum the *areas* of the Algeblocks on the axes. What is the sum of the *areas*? What is the product of $(x - 2)$ and $(x + 1)$?

Critical Thinking   Explain how the Distributive Property is used to find the following product:

$$(x + 3)(2x + 1) = (x + 3)(2x) + (x + 3)(1)$$

$$= (x)(2x) + (3)(2x) + (x)(1) + (3)(1)$$

$$= 2x^2 + 6x + x + 3$$

$$= 2x^2 + (6 + 1)x + 3$$

$$= 2x^2 + 7x + 3$$

LESSON ASSESSMENT

Think and Discuss

1 How does the geometric model illustrate the product of two binomials? Show an example.

2 How do you use the FOIL method to find the product of two binomials? Show an example.

3 Try doing the multiplication by placing one binomial under the other and multiplying as if they were whole numbers. How is this method similar to numerical multiplication?

4 Use the FOIL method to find the product $(10 + 4)(10 + 6)$. Explain how the FOIL method can be used to find the product of two numbers such as $(24)(35)$.

Practice and Problem Solving

Find the products using a geometric model.

5. $(x + 4)(x + 3)$ **6.** $(x + 5)(x + 1)$ **7.** $(2x + 1)(x + 2)$

Each pair of factors represents the length and width of a rectangle. Draw an area model and find the area of the rectangle.

8. $(x + 3)(x + 3)$ **9.** $(x + 5)(x + 2)$ **10.** $(3x + 4)(2x + 3)$

Find each product using the FOIL method.

11. $(3x + 1)(x + 3)$ **12.** $(2x + 3)(4x + 1)$ **13.** $(5x + 1)(2x - 3)$

14. $(3x - 5)(2x + 1)$ **15.** $(5x - 2)(3x - 1)$ **16.** $(2x - 1)(3x - 1)$

Mixed Review

Solve each equation for x.

17. $5x + 9 = -18 + 2x$

18. $\dfrac{15}{x} = \dfrac{9}{20}$

19. $x + 30\%x = \$65$

Write and solve an equation for each situation.

20. Taj and Ramen invested $360 in stock. Taj invested $120 more than Ramen. How much did Taj invest?

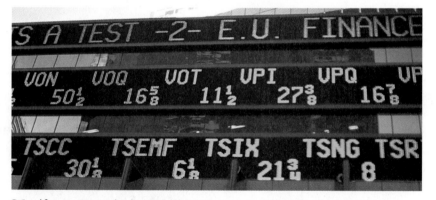

21. If a rectangle has a length that is 3 cm less than four times its width, and its perimeter is 19 cm, what are its dimensions?

22. John has $15,000 to invest. He invests part of the money in an investment that pays 12% simple interest and the rest in another that pays 8% simple interest. If his annual income from the two investments is $1456, how much did he invest at 12%?

LESSON 10.7 FACTORING TRINOMIALS GEOMETRICALLY

When multiplying two binomials, you are given two factors, and you must find the product.

$$(x + 2)(x + 1) = x^2 + 3x + 2$$

Binomial factors   Trinomial product

When factoring, you do the *opposite*. For example, if given the trinomial $x^2 + 3x + 2$, you can find its factors.

$$x^2 + 3x + 2 = (x + 2)(x + 1)$$

Given trinomial   Binomial factors

Factoring $x^2 + bx + c$

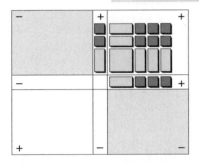

Algeblocks or area diagrams are useful models for factoring trinomials that are written in the form $x^2 + bx + c$. In an area diagram, $x^2 + bx + c$ represents the area of a rectangle. The factors represent the length and width of the rectangle. For example, to find the factors of $x^2 + 5x + 6$, arrange the blocks that represent $x^2 + 5x + 6$ into a rectangle.

The sides of the rectangle, $x + 2$ and $x + 3$, are the factors of $x^2 + 5x + 6$.

| ACTIVITY 1 | **Using Algeblocks to Factor a Trinomial** |

1   Select the blocks that model the trinomial $x^2 + 4x + 3$.

2   Build a rectangle from your blocks.

3   What are the dimensions of your rectangle?

4   What are the factors of $x^2 + 4x + 3$?

5   Multiply your factors to check your work.

Factoring $ax^2 + bx + c$

You can represent a trinomial such as $2x^2 + 9x + 4$ with a combination of area diagrams.

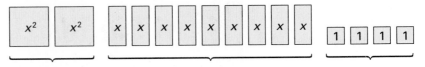

Area: $2x^2$ Area: $9x$ Area: 4

$$2x^2 + 9x + 4$$

The 15 individual areas form a rectangle.

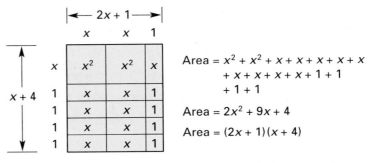

$$\text{Area} = x^2 + x^2 + x + x + x + x + x$$
$$+ x + x + x + x + 1 + 1$$
$$+ 1 + 1$$

$$\text{Area} = 2x^2 + 9x + 4$$

$$\text{Area} = (2x + 1)(x + 4)$$

The length $2x + 1$ and the width $x + 4$ of the rectangle represent the factors of $2x^2 + 9x + 4$.

The model for $2x^2 + 9x + 4$ formed a rectangle. However, some trinomial models such as $x^2 + x + 1$ cannot be arranged in a rectangle. The trinomial $x^2 + x + 1$ is *prime* and *cannot* be factored.

Ongoing Assessment

Use an area diagram to factor $2x^2 + 5x + 3$.

Critical Thinking Trinomials often contain negative terms. To model the trinomial, some of the areas must be drawn in the second or fourth quadrants. Explain why this is so.

ACTIVITY 2   Factoring Trinomials with Negative Terms

Algeblocks Quadrant Mat

1 Use Algeblocks to model the trinomial $x^2 - 5x + 6$.

2 Place the x^2 block in the first quadrant.

3 Place the six unit blocks in the third quadrant.

4 Place the five x-blocks in the second and fourth quadrants.

5 Arrange the blocks to form a rectangle.

6 What are the dimensions of the rectangle?

7 What are the factors of $x^2 - 5x + 6$?

8 Multiply the factors to check your answer.

LESSON ASSESSMENT

Think and Discuss

1 Can you rearrange your Algeblocks to form different rectangles from the same trinomial? Why or why not?

2 How does using a quadrant map allow you to represent negative values?

Practice and Problem Solving

Use Algeblocks or area diagrams to factor these trinomials.

3. $a^2 + 8a + 7$ **4.** $x^2 - 4x + 3$ **5.** $x^2 - 2x + 1$

6. $x^2 - 8x + 7$ **7.** $a^2 + a - 6$ **8.** $c^2 - 4c + 3$

9. $8 - 6w + w^2$ **10.** $7x^2 + 12x + 5$ **11.** $6x^2 + x - 15$

Mixed Review

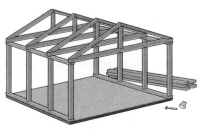

As seen from the front, the rafters of the roof of a storage shed form a triangle. The base of the triangle is 12 feet. The perpendicular distance from the peak of the roof to the base is 3 feet.

12. Draw a diagram to model the triangle.

13. If the triangular area formed by the rafters is to be covered with plywood, how many square feet of plywood are needed?

14. If two-thirds of the plywood is painted red, how many square feet of the plywood are painted?

Solve each equation for x.

15. $5x^2 = 125$ **16.** $2x^2 = 3$ **17.** $3x + 6 = x - 18$

From a group of five men and ten women, a committee of four is chosen by drawing names. Find the probability that

18. Four women are chosen.

19. Four men are chosen.

20. Two women are chosen, and then two men are chosen.

LESSON 10.8 FACTORING TRINOMIALS

In Lesson 10.7, you used a geometric model to factor $x^2 + 3x + 2$ into the product of two binomials.

$$x^2 + 3x + 2 = (x + 2)(x + 1)$$

As you know, the factors can be checked by using the Distributive Property or the FOIL method. You can also use the Distributive Property to factor a trinomial.

Factoring $x^2 + bx + c, c > 0$

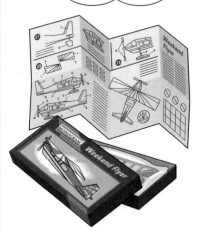

Factoring trinomials involves a step-by-step process and is just like building a model airplane

Factoring $x^2 + 5x + 6$ requires you to find two binomials of the form $(x + m)(x + n)$, whose product is $x^2 + 5x + 6$.

If the Distributive Property is used to multiply $(x + m)(x + n)$, the product is

$$x^2 + mx + nx + mn \quad \text{or} \quad x^2 + (m + n)x + mn$$

Thus, the two numbers m and n must have a sum of 5 and a product of 6. There are four different pairs of integers whose product is 6.

$$1 \cdot 6 \qquad -1 \cdot -6 \qquad 2 \cdot 3 \qquad -2 \cdot -3$$

However, only one pair (2 and 3) has a sum of 5. Consequently, you replace m by 2 and n by 3.

$$x^2 + 5x + 6 = (x + m)(x + n)$$

$$= (x + 2)(x + 3)$$

EXAMPLE 1 Factoring $x^2 - 7x + 12$

Factor $x^2 - 7x + 12$.

SOLUTION

This time, you must find a pair of integers whose product is 12 and whose sum is -7. First, list the integer pairs whose products are 12.

$$
\begin{array}{ll}
1 \cdot 12 & -1 \cdot -12 \\
2 \cdot 6 & -2 \cdot -6 \\
3 \cdot 4 & -3 \cdot -4
\end{array}
$$

The sum of -3 and -4 is -7. Thus,

$$x^2 - 7x + 12 = (x - 3)(x - 4)$$

Always use the FOIL method to check your factors. Do the factors check in this example?

Ongoing Assessment

Factor each trinomial. Check your answer using the FOIL method.

a. $x^2 + 12x + 20$ **b.** $x^2 - 9x + 20$

Factoring $x^2 + bx + c$, $c < 0$

ACTIVITY

Factoring $x^2 + bx + c$, $c < 0$

1 Explain why you can write the trinomial $x^2 + 2x - 15$ in the form $x^2 + 2x + (-15)$.

2 List the factor pairs for -15.

3 Which pair has a sum of 2?

4 Use the pair you found in Step 3 to make this statement true.

$$x^2 + 2x - 15 = (x + ?)(x + ?)$$

5 Rewrite the statement in the following form:

$$x^2 + 2x - 15 = (x + ?)(x - ?)$$

6 Why must one of the signs in final factored form be negative? How did you choose the correct integer to place after the negative sign?

EXAMPLE 2 Factoring $x^2 - x - 12$

Factor $x^2 - x - 12$.

SOLUTION

When the product is negative, you write one of the factors in the form $(x + ?)$ and the other in the form $(x - ?)$. Since the last term of the trinomial is -12, the factored form of the trinomial is

$$(x + ?)(x - ?)$$

You need to find two integers whose product is -12 and whose sum is -1. First, list the integer pairs whose products are -12.

$$1 \bullet -12 \qquad -1 \bullet 12$$
$$2 \bullet -6 \qquad -2 \bullet 6$$
$$3 \bullet -4 \qquad -3 \bullet 4$$

The sum of 3 and -4 is -1. Thus,

$$x^2 - x - 12 = (x + 3)(x - 4)$$

Ongoing Assessment

Factor each trinomial. Check your answer using the FOIL method.

a. $x^2 + 6x - 16$ **b.** $x^2 - 2x - 3$

Factoring $ax^2 + bx + c$

Factoring a trinomial is a guess and check process. Once you have listed the possible factors, the signs in the trinomial determine which pair to choose. After you write the factors, use the FOIL method to check your answer.

Using guess and check on expressions such as $2x^2 + 13x + 15$ may take considerable time. However, the FOIL method does tell us some things about the possible factors.

If the expression $2x^2 + 13x + 15$ can be factored, the first terms of the binomial factors must be x and $2x$. Since the product of the last two terms is 15 and the middle term is positive, the integer pairs must be

$$1 \bullet 15 \quad \text{or} \quad 3 \bullet 5$$

Here are the four possibilities:

1. $(2x + 1)(x + 15)$ **2.** $(2x + 15)(x + 1)$

3. $(2x + 3)(x + 5)$ **4.** $(2x + 5)(x + 3)$

Since the middle term is +13, the sum of the inner and outer products must be 13. Only possibility **3** works.

$$2x^2 + 13x + 15 = (2x + 3)(x + 5)$$

Factoring expressions such as $2x^2 + 13x + 15$ is relatively easy because 2 is prime. When the numbers in the first and last positions are composite numbers, the task can become very tedious. In Chapter 11, you will learn a formula to help you factor any trinomial expression. Remember, factoring a trinomial is like building an airplane. If you take a step-by-step approach, you will be successful.

LESSON ASSESSMENT

Think and Discuss

In the trinomial $x^2 + bx + c$, what do you know about the factors of c if

1 c is positive and b is negative?

2 c is positive and b is positive?

3 c is negative and b is positive?

4 c is negative and b is negative?

Practice and Problem Solving

Factor each trinomial. Check your answer by using the FOIL method.

5. $x^2 + 5x + 6$ **6.** $x^2 + 7x + 6$ **7.** $x^2 - 5x + 6$

8. $x^2 - x + 6$ **9.** $x^2 + 5x - 6$ **10.** $x^2 - 7x + 6$

11. $5x^2 + 7x + 2$ **12.** $2x^2 + 5x - 12$ **13.** $3x^2 + 14x + 15$

14. $4x^2 + 8x + 3$ **15.** $6x^2 + 7x - 20$ **16.** $6x^2 - 19x + 15$

Mixed Review

Write and solve an equation in one variable for each situation.

17. Reece charges $18 to cut and trim a yard. How much does Reece charge for trimming if he charges $15 to cut the yard?

18. Mary, Pat, and Armond have each earned money for basketball camp. Armond has earned twice as much as Pat, and Mary has earned three times as much as Pat. If the total earned is $210, how much has each person earned?

19. The length of a rectangle is equal to the base of a triangle. The altitude of the triangle is 20 feet, and the height of the rectangle is 25 feet. If the combined areas of the triangle and the rectangle are 280 square feet, what is the length of the rectangle?

Solve each system of equations.

20. $3x + y = 11$ **21.** $2x + 5y = -3$
$\ 2x - y = 10$ $\ x - 3y = -7$

Write and solve a system of equations for this problem.
22. A store owner made $2950 during a shirt and pants sale. The shirts sold for $40, and the pants sold for $65. There were 5 more pants sold than shirts. How many shirts were sold?

Lesson 10.9 Special Products and Factors

The Sum and Difference of Two Squares

The binomial $a + b$ is the sum of two terms. The binomial $a - b$ is the difference of the same two terms. The product of the sum and difference is represented as

$$(a + b)(a - b)$$
$$\text{sum} \cdot \text{difference}$$

Simplify $(9 + 1)(9 - 1)$ using the FOIL method.

$$(9 + 1)(9 - 1) = 81 - 9 + 9 - 1$$
$$= 81 + 0 - 1$$
$$= 9^2 - 1^2$$

Notice that:
- the sum of the inner product and the outer product is zero.
- the final product is the difference of two squares.

The multiplication $(9 + 1)(9 - 1)$ results in a **special product**:

$$9^2 - 1^2$$

Special products enable you to use a shortcut to write the answer for certain multiplication and factoring problems.

Examine the procedure for multiplying $(a + b)(a - b)$. Again, you see that the inner product (ab) and outer product $(-ab)$ cancel because they add to zero. The product of the sum and difference of two numbers is the difference of their squares.

$$(a + b)(a - b) = a^2 - ab + ab - b^2$$
$$= a^2 + 0 - b^2$$
$$(a + b)(a - b) = a^2 - b^2$$

Activity 1   Factoring the Difference of Two Squares

1 Factor $x^2 - 4$ using Algeblocks.

2 Check your answer using multiplication.

3 Factor $x^2 - 25$ using Algeblocks.

4 Check your answer using multiplication.

5 The terms x^2 and 25 are perfect squares. How can you use square roots to factor the expression $x^2 - 25$?

6 Explain how to factor the difference of two perfect squares.

7 Factor $a^2 - 36$.

In Activity 1, you found a pattern for factoring the difference of two squares. For example, $x^2 - 100 = x^2 - 10^2 = (x + 10)(x - 10)$.

Factoring the Difference of Two Squares

Since $(a + b)(a - b) = a^2 - b^2$,

$$a^2 - b^2 = (a + b)(a - b)$$

EXAMPLE 1 The Product of a Sum and Difference

a. Multiply $(2x + 3y)(2x - 3y)$.

b. Factor $16s^2 - 49t^2$.

SOLUTION

a. Use the shortcut.

Square the first term $2x$, $(2x)^2 = 4x^2$

Square the second term $3y$, $(3y)^2 = 9y^2$

Write the difference of the squares, $4x^2 - 9y^2$

Thus,

$$(2x + 3y)(2x - 3y) = 4x^2 - 9y^2$$

b. Use the shortcut. The first term $16s^2$ is a perfect square, $[(4s)^2]$. The second term $49t^2$ is also a perfect square $[(7t)^2]$. Thus,

$$16s^2 - 49t^2 = (4s + 7t)(4s - 7t)$$

Ongoing Assessment

a. Multiply $(m - 3n)(m + 3n)$.   **b.** Factor $16 - 81w^2$.

Critical Thinking   Why is $x^2 + 9$ a prime expression?

Perfect Square Trinomials

When you square a number, you use the number as a factor two times. Thus, $3^2 = 3 \cdot 3$, or 9. In general, the square of x is x^2. What does it mean to square a binomial?

ACTIVITY 2 Squaring a Binomial

1 Find each product using Algeblocks or an area model.

   a. $(x + 3)(x + 3)$ **b.** $(x - 3)(x - 3)$

   c. $(x - 4)(x - 4)$ **d.** $(x + 4)(x + 4)$

2 Find each product using the FOIL method.

   a. $(2 + m)(2 + m)$ **b.** $(2 - m)(2 - m)$

   c. $(m - n)(m - n)$ **d.** $(m + n)(m + n)$

3 Complete the equation.

$$(a + b)^2 = (\underline{\;?\;} + \underline{\;?\;})(\underline{\;?\;} + \underline{\;?\;}) = \underline{\;?\;} + \underline{\;?\;} + \underline{\;?\;}$$

4 What pattern do you observe when you square a binomial?

A **perfect square trinomial** is a special trinomial that can be factored into two identical binomials called a **binomial square**.

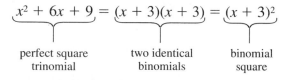

$$\underbrace{x^2 + 6x + 9}_{\substack{\text{perfect square}\\\text{trinomial}}} = \underbrace{(x + 3)(x + 3)}_{\substack{\text{two identical}\\\text{binomials}}} = \underbrace{(x + 3)^2}_{\substack{\text{binomial}\\\text{square}}}$$

EXAMPLE 2 Factoring $4x^2 + 4x + 1$

Is the polynomial $4x^2 + 4x + 1$ a perfect square trinomial?

SOLUTION

If $4x^2 + 4x + 1$ is a perfect square trinomial, then it factors into a binomial square.

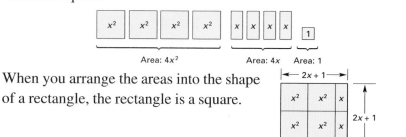

When you arrange the areas into the shape of a rectangle, the rectangle is a square.

In Activity 2, you observed a pattern that is used to square a binomial or factor a perfect square trinomial such as $4x^2 + 4x + 1$. From this pattern, note these three questions, which will help determine if a trinomial is a perfect square trinomial.

- Are the first and last terms positive? In the trinomial $4x^2 + 4x + 1$, both $4x^2$ and 1 are positive.

- Are the first and last terms perfect squares? $4x^2 = (2x)^2$ and $1 = 1^2$.

- Find the square roots of the first and last terms. Is twice the product of the square roots equal to the middle term?

$$2(2x)(1) = 4x$$

If a trinomial satisfies these three conditions, it is a perfect square.

When a binomial is squared, the result is a perfect square trinomial. When a perfect square trinomial is factored, the result is a binomial square.

Perfect Square Trinomials
$$(a + b)^2 = a^2 + 2ab + b^2$$
$$(a - b)^2 = a^2 - 2ab + b^2$$

Ongoing Assessment

Which of the following expressions are perfect square trinomials? Explain your answer.

a. $x^2 - 14x + 49$ **b.** $x^2 + 5x + 25$ **c.** $x^2 + 12x - 36$

EXAMPLE 3 Factoring a Perfect Square Trinomial

Factor $x^2 - 10x + 25$.

SOLUTION

The first term x^2 is a perfect square.

The last term is a perfect square $(-5)^2 = 25$.

The middle term is $2(x)(-5)$.

Thus, $x^2 - 10x + 25 = (x - 5)(x - 5) = (x - 5)^2$.

a. Multiply $(2t + r)^2$.   **b.** Factor $4x^2 + 12xy + 9y^2$.

A word of caution about squaring a binomial: *Do not forget the middle term.* A common error is to think that $(a + b)^2$ equals $a^2 + b^2$.

Substitute 1 for a and 2 for b, and you will see that this is not the case.

WORKPLACE COMMUNICATION

Show how to obtain the approximation: (a) expand the square of the binomial in the equation for the exact volume, and then (b) multiply by π using the Distributive Property. What assumption is required? Compare the exact and approximate volumes for $r = 10$ mm, $a = 0.1$ mm, and $L = 1m$.

File Edit View Label Special

mwi electronic mail

URGENT MESSAGE FROM: jengrove@mwi.org
TO: walsh@mwi.org
SUBJECT: Delay in Computer Control for HAZGAS

MESSAGE:

The newly installed computer control for the hazardous gas treatment system cannot go online as scheduled this week. The controller manufacturer has warned us that any squared terms (such as r²) in the control software will cause an unstable feedback response. This instability could cause our entire treatment system to fail under certain conditions. This creates a problem. Our software uses the following equation for the volume of gas in the heat exchanger tubes of the treatment system:

$V = L\pi(r + a)^2$

In this equation, L is the length of the tube, r is the original (unexpanded) tube radius, and a is the amount of tube expansion caused by high temperature. We can eliminate the square term in this equation by substituting a linear function for the cross-sectional area of the tube:

$V = L(A + Ca)$

In this equation, A is the original, unexpanded cross-sectional area ($A = \pi r^2$) and C is the original, unexpanded circumference ($C = 2\pi r$) of the tube. This function is a very close approximation to the exact volume as long as a is small compared to r. We cannot make the computer control operational until we make this change in the control software and thoroughly test the system. This will take about three more days.

LESSON ASSESSMENT

Think and Discuss

1 What happens to the middle term when you multiply the sum and difference of the same two numbers or terms?

2 How do you determine the middle term when you square a binomial?

3 What is the general rule or shortcut for finding the product of the sum and difference of the same terms?

4 What is the general rule or shortcut for squaring a binomial?

5 How can you use the general rules for multiplying in these special cases to find the factors of special trinomials?

Multiply.

6. $(x + 1)(x - 1)$ **7.** $(x - 3)(x + 3)$ **8.** $(2x - 5)(2x + 5)$

9. $(x + y)(x + y)$ **10.** $(2x - 1)(2x - 1)$ **11.** $(x - 3)(x + 3)$

12. $(a - 4)^2$ **13.** $(2y + 3)^2$ **14.** $(4x - 3y)(4x + 3y)$

Factor.

15. $x^2 - 25$ **16.** $4y^2 - 9$ **17.** $9 - 16y^2$

18. $x^2 - 6x + 9$ **19.** $y^2 - 18y + 81$ **20.** $9x^2 + 24x + 16$

21. $y^2 - 2y + 1$ **22.** $y^2 - 1$ **23.** $x^2 - 2ax + a^2$

Mixed Review

24. Solve $F = \dfrac{9}{5}C + 32$ for C.   **25.** Solve $P = 2a + 2b$ for b.

A cartographer has located three points on a map. The points are designated as $A(-3,0)$, $B(2,1)$, and $C(4,-1)$.

26. Connect the points in order from A to B to C. What geometric figure is formed?

27. Find the equation of the line that passes through points A and B.

28. Find the equation of the line that passes through Point C which is parallel to the x-axis.

29. Find the equation of the line that passes through point A which is perpendicular to the x-axis.

Solve each system of equations.

30. $x - y = 1$ **31.** $2x - 5y = 9$ **32.** $\dfrac{1}{2}x - \dfrac{1}{3}y = 1$

     $2x + y = 0$ $2x - 2y = 3$ $\dfrac{1}{3}x + \dfrac{1}{2}y = 5$

MATH LAB

Equipment Cardboard, 7.5 inches by 4 inches
Scissors

Problem Statement

Use the template to create a set of math tiles.

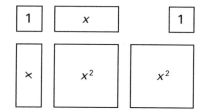

Mark the large squares with an x^2. Mark the long rectangles with an x. Mark the small squares with a 1. After you cut the tiles apart, you will have four x^2-tiles, eight x-tiles, and twenty-one 1-tiles.

| 1 | x | | 1 |

| x | x^2 | x^2 |

You will use the math tiles to model trinomial factorization. You will start with the tiles that represent each term of the trinomial and use those tiles to try and form a rectangle.

Procedure

a The polynomial $x^2 + 5x + 6$ is modeled by one x^2-tile, five x-tiles, and six 1-tiles.

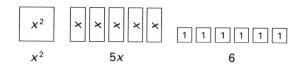

Start with the x^2-tile and add on the areas until you can make a rectangle. Always line up the long sides of the x-tiles against each other.

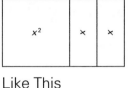 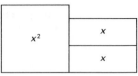

Like This   Not Like This

b Now complete the rectangle.

c When you have made a rectangular area with the tiles, draw a picture of your rectangle and record the dimensions along its base and side. These dimensions represent the factors of the polynomial. What are the dimensions of the rectangle you have just made? What are the factors of $x^2 + 5x + 6$?

d As you work with more examples, it will become easier to form the rectangles. If it is not possible to form a rectangle with the tiles, the trinomial is prime and cannot be factored. Explain why $x^2 + x + 2$ is prime.

e Use tiles to find the factors for each trinomial.

1. $x^2 + 5x + 6$

2. $x^2 + 7x + 6$

3. $x^2 + 7x + 12$

4. $x^2 + 8x + 12$

5. $x^2 + 6x + 9$

6. $x^2 + 4$

f Based on your observations in Step **e**, explain how you would factor $x^2 + bx + 8$.

g Use tiles to find the factors of each trinomial.

7. $2x^2 + 7x + 3$ **8.** $2x^2 + 5x + 3$

h How can you make a set of tiles to factor $x^2 + 2xy + y^2$? Explain how to use your tiles to show that $x^2 + 2xy + y^2 = (x + y)(x + y)$.

Activity 2: Listing Prime Numbers

Equipment Set of 30 small objects, such as pennies, beans, or math tiles

Problem Statement

You will discover if a number is *prime* by arranging objects into rectangular arrays.

Procedure

a For each of the numbers n greater than 1 and less than 30, count out n objects.

b Find all the possible rectangular arrangements, called **arrays**, of n objects. Here is an example for the number 6:

1 by 6	6 by 1	2 by 3	3 by 2
# # # # # #	#	# # #	# #
	#	# # #	# #
	#		# #
	#		
	#		
	#		

c For each array of n objects, record the array with the number that represents the length and the width next to the array.

d List the numbers from 2 through 29.

e Circle any number n on your list that can only be represented by an array of 1 by n and n by 1. The numbers that you have circled are the prime numbers less than 30.

f Compare your list of prime numbers with others in your class. Did all of you get the same results?

g Write a definition of a prime number.

MATH APPLICATIONS

The applications that follow are like the ones you will encounter in many workplaces. Use the mathematics you have learned in this chapter to solve the problems.

Wherever possible, use your calculator to solve the problems that require numerical answers.

A formula to find the sum of the first n whole numbers $1 + 2 + 3 + ... + n$ is

$$\text{Sum} = \frac{n^2 + n}{2}$$

Thus, for $n = 6$, the formula should give the sum of the first six whole numbers, or $1 + 2 + 3 + 4 + 5 + 6 = 21$.

1 With a calculator, it might be easier to evaluate this formula without having to compute the n^2 term. Identify the common monomial term in the numerator of the formula and factor out this term to obtain a different version that does not have the squared term.

2 Test the formula, using a relatively small value for n, to show that it works.

3 What is the sum of the first 200 whole numbers?

4 What is the sum of the first 1000 whole numbers?

Multiply each pair of binomials below using the FOIL method.

5 $(x + 5)(x + 2)$ **6** $(x - 3)(3x + 8)$

7 $(4x + 1)(2x - 5)$ **8** $(7x - 2)(x - 4)$

Match each binomial product with the correct "area" on the right that geometrically represents it. Multiply the products for each and verify that the areas match the polynomial.

9 $(2x + 1)(x + 2)$ **a.**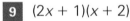

10 $(x + 3)(x + 1)$ **b.**

11 $(x + 1)(2x + 1)$ **c.**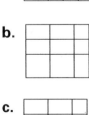

Factor each of the monomials below into its prime factors.

12 12   **13** 94   **14** $243x^2$ **15** $180x^5$

Factor the GCF from each polynomial.

16 $3x^2 + 18x$ **17** $12x^3 - 30x^2$ **18** $105x^4 + 21x^3 - 63x^2$

Use the Distributive Property to find each product.

19 $(x + 7)(x^2 + 6x - 9)$ **20** $(2x - 3)(3x^2 - 1)$

21 $(2x + 5)(9x^2)$ **22** $(x - 5)(x^2 - x)$

To factor a trinomial, one of the skills needed is the ability to identify pairs of numbers that add to give a certain sum and multiply to yield a given product. For example, 3 and 2 add to 5, and when multiplied give 6. Copy the table and complete it by providing a pair of numbers whose sum is the number in the first column and whose product is the number in the second column. (The first row is completed for you.)

	Sum ($a + b$)	Product (ab)	a	b
	5	6	3	2
23	2	1	?	?
24	8	15	?	?
25	11	10	?	?
26	14	48	?	?
27	−4	4	?	?
28	−6	−7	?	?
29	2	−8	?	?
30	9	−52	?	?
31	0	−9	?	?

Factor each trinomial. If the trinomial cannot be factored, then indicate that it is *prime*.

32 $x^2 + 6x + 5$ **33** $x^2 + 14x + 45$

34 $x^2 - x - 20$ **35** $x^2 + x + 16$

36 $x^2 + x - 2$ **37** $x^2 - 5x - 6$

Factor each trinomial below or state that it is *prime*.

38 $2x^2 + 11x + 5$

39 $3x^2 - 11x - 4$

40 $4x^2 - 15x - 4$

41 $6x^2 - 5x - 4$

Identify which of the following trinomials are considered "special products" and factor them. (You need not factor those that are not "special products.")

42 $x^2 - 4$

43 $x^2 + 6x + 9$

44 $4x^2 - 2x + 1$

45 $4x^2 + 1$

46 $x^2 - 49$

47 $x^2 - 8x + 16$

AGRICULTURE & AGRIBUSINESS

You are constructing a large rack to hold small pots while the seeds in them germinate. You plan to have approximately 500 pots in the rack. You would also like the rack to be as close to "square" as possible. The rack should have an equal number of pots in each row and an equal number of pots in each "column" (although possibly different from the number in each row). What are some possible numbers of pots per row and pots per column that would give you a "nearly square" layout for the following total number of pots?

48 480

49 500

50 520

51 Is there some total number of pots near 500 that you can use to create a perfectly square layout? If so, what is it?

BUSINESS & MARKETING

You are considering packaging small objects into large rectangular shaped cartons for shipping. The cartons should hold no more than 6 objects per row (length), be no more than 6 objects high (height), and be no more than six objects deep (depth). You would like to package no less than 100 objects per carton.

52 What are all your choices of total number of objects per carton? (Empty spaces in the carton are not allowed.)

FAMILY AND CONSUMER SCIENCE

A quilt is to be made of patchwork squares. The quilt has a length of L squares and a width of W squares. Although many different sizes are possible, you would like the quilt's length-to-width ratio to be 5 to 3. Thus, for a quilt of "size" S, if the quilt has a length of L squares equal to $5S$, the acceptable width would be W squares equal to $3S$.

53 Write an equation that gives the total number of squares T in the quilt as a function of the size S.

54 The quilt can be made slightly larger by adding a row of squares along the length and a row along the width. Modify your equation to show the total number of squares if the quilt is $L + 1$ squares long and $W + 1$ squares wide. Write your expression as two factors.

55 Multiply the two factors using the FOIL technique to obtain a polynomial expression.

INDUSTRIAL TECHNOLOGY

The cost to produce round metal washers is directly related to the amount of metal used. And the amount of metal used is directly related to the area of the washer, as shown in the shaded region.

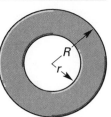

56 What is the area of the large disc of radius R before the middle hole is stamped out?

57 What is the area of the middle hole of radius r?

58 Write a formula for the area of the washer that has an outer radius R and a hole of radius r, using the formulas from Exercises 56 and 57. Simplify the formula by factoring out all common monomial terms.

59 You should recognize one of the factors in your formula as a "special product." Show your formula for the area of a washer in yet another form by rewriting the "special product" as two factors.

In the previous exercises, you found a formula that gives the area of metal used in a washer. When making this washer, a certain manufacturer will start with a square piece of metal and make the two circular stamps.

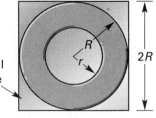

The length of the square's side will be equal to the diameter of the washer, or *2R*. Thus, the material outside the diameter of the washer as well as the hole from the middle becomes "waste."

Metal Plate

60 Write a formula for the area of the square piece of metal that will be used to stamp out a washer with a diameter of *2R*.

61 Create a formula for the area of metal waste from the production of a single washer of outer radius *R* and inner radius *r*. You can do this by subtracting the area of the washer (see Exercise 60) from the area of the square piece of metal. Simplify the formula by factoring out the GCF.

62 Use your formula from Exercise 58 to determine how many square inches of metal are used in creating a washer that has an outer radius of $\frac{3}{4}$ inch and an inner radius of $\frac{1}{2}$ inch.

63 How many square inches of metal would be wasted in creating such a washer (using your formula from Exercise 61)?

64 Comparing the metal wasted to the area of the initial metal square, what percent of metal is "wasted" in creating one of these metal washers?

A rectangular piece of metal has a length that is 5 cm greater than its width *W* in centimeters. As part of an assembly, the plate must have four square sections removed from it, each having a width of $\frac{1}{4}$ *W*.

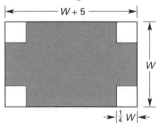

65 Write a formula for the area of the metal plate that remains after the corner pieces are removed.

66 Simplify the formula from Exercise 65 by factoring out the GCF.

67 Use your formula to find the area of the metal plate that remains when the width is chosen as *W* = 24 cm.

CHAPTER 10 ASSESSMENT

Skills

1. Find the sum. $(5x^2 + 3x - 9) + (2x^2 - 2x + 3)$

2. Find the difference. $(3x^2 - 5x + 2) - (4x^2 + 3x - 4)$

3. What is the greatest common factor of 36 and 90?

4. What expression represents the shaded area between the circle and the square? Write the expression in factored form.

Find each product.

5. $3x^2(x^2 - 3x + 8)$ **6.** $(x + 4)(x - 1)$ **7.** $(x - 3)^2$

Simplify each expression.

8. $(5x^2)^3$ **9.** $(3x^2)(2x^5)$ **10.** $(4.1 \times 10^2)^3$

Factor each expression.

11. $x^2 + 2x - 15$ **12.** $4x^2 - 9$ **13.** $4x^2 - 4x + 1$

Applications

14. The length of a rectangular floor is 3 meters longer than its width. Write an expression that represents the area of the rectangle.

15. A square playground's area is represented by the trinomial $x^2 - 10x + 25$. Write an expression using the variable x that represents the perimeter of the playground.

CHAPTER 11

WHY SHOULD I LEARN THIS?

There are many ways to solve a problem. Some are more efficient than others. To design a supersonic aircraft, an engineer must select a design method that completes the job quickly and efficiently. In this chapter, you will learn several methods to solve quadratic equations. Master these methods and learn to select the one that most efficiently solves the problem at hand.

Contents: Quadratic Functions

QUADRATIC FUNCTIONS

OBJECTIVES

1. Solve quadratic equations by graphing.
2. Solve quadratic equations by factoring.
3. Solve quadratic equations with the quadratic formula.
4. Use graphing, factoring, and the quadratic formula to solve practical problems involving quadratic equations.

In Chapter 6 you graphed several types of nonlinear equations. Now you will revisit one particular type of nonlinear equation. In this chapter, you will examine how the quadratic equation and its graph, the parabola, are used in science and industry.

The parabola has many applications in the modern world of technology. Supersonic airplanes are parabolic in shape to reduce air resistance. Flashlights and satellite dish antennas have reflectors shaped like a parabola. The parabolic reflector in a flashlight reflects the light from a tiny bulb, sending out a beam of nearly parallel rays of light. The parabolic reflector of a satellite dish reflects incoming radio signals and focuses them on a receiver unit. Compare the shape of the parabola drawn on the x- and y-axes with the drawings of the flashlight and the satellite dish below. Can you find the parabolic reflectors?

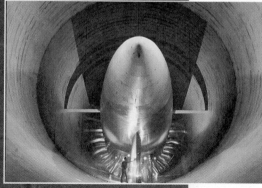
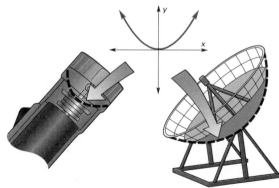

As you watch the video for this chapter, look for graphs that represent parabolas.

LESSON 11.1 GRAPHING QUADRATIC EQUATIONS

Suppose an arrow is fired vertically into the air with an initial speed of 112 feet per second. The height (h) of the arrow in feet is a function of the time (t) in seconds. You can model the relationship between the height and the time of the arrow's flight with the following **quadratic equation**.

$$h = -16t^2 + 112t$$

To graph $h = -16t^2 + 112t$, make a table of ordered pairs (t, h). Plot the points and connect them with a smooth curve. The parabola shows how the arrow rises until it reaches its maximum height. Then it falls back to Earth.

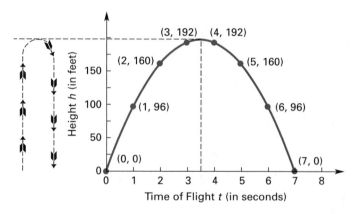

ACTIVITY 1   Maximum Height

The vertical dashed line is drawn at the maximum height of the arrow, which is the vertex of the parabola. The maximum height occurs when t is 3.5 seconds.

1 What is the total time of flight of the arrow?

2 Use the graph to estimate the maximum height of the arrow.

3 Use the equation $h = -16t^2 + 112t$ to calculate the actual value for the maximum height of the arrow.

Juanita is a pitcher for a softball team. Her coach measured the speed of her fastball at 64 feet per second. If Juanita throws the ball upward with an initial speed of 64 feet per second, the following quadratic equation models the relationship between the height (h) of the ball and the time (t) of travel.

$$h = -16t^2 + 64t$$

Critical Thinking Which coefficients in the last two equations represent the initial speeds of the arrow and the softball?

The height of the softball, like the height of the arrow, is a function of time. Thus, you can graph the softball's height on the same set of axes as the arrow's height.

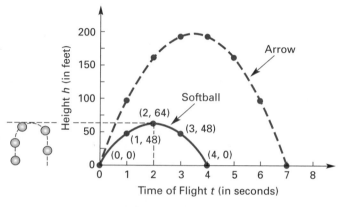

ACTIVITY 2 Time in the Air

Examine the graph of $h = -16t^2 + 64t$.

1 What is the maximum height the softball reaches?

2 How many seconds is the ball in the air?

In Chapter 6, you saw how the coefficients a and c in $y = ax^2 + c$ altered the graph of the parent quadratic function $y = x^2$.

Changing the sign of a reflects the graph across the x-axis.

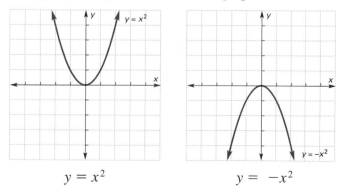

$y = x^2$ $y = -x^2$

Increasing *a* shrinks the graph. Decreasing *a* stretches the graph.

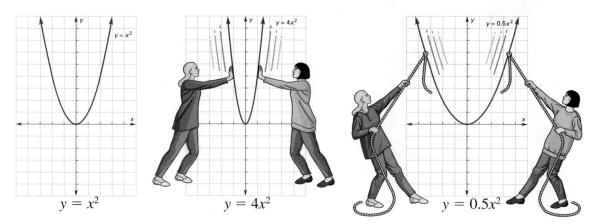

$y = x^2$ $\qquad\qquad$ $y = 4x^2$ $\qquad\qquad$ $y = 0.5x^2$

Adding *c* to the parent function translates the graph vertically.

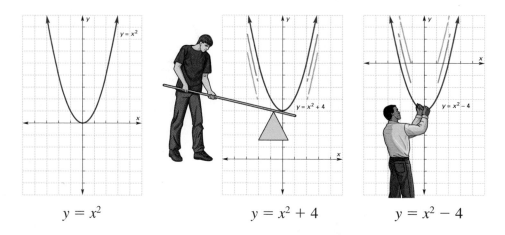

$y = x^2$ $\qquad\qquad$ $y = x^2 + 4$ $\qquad\qquad$ $y = x^2 - 4$

If a constant is added or subtracted from the variable before it is squared, the graph is translated horizontally.

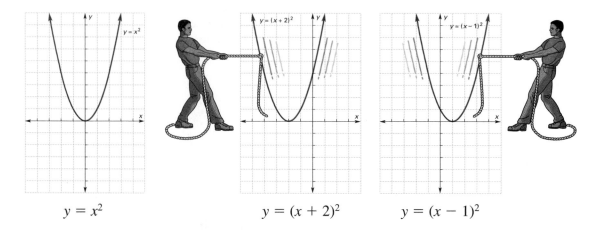

$y = x^2$ $\qquad\qquad$ $y = (x + 2)^2$ $\qquad\qquad$ $y = (x - 1)^2$

Reflections, stretching and shrinking (also called dilations), and vertical and horizontal translations are called **transformations.**

> ### Transformation
> A *transformation* is a rule that tells how a function is related to its parent function.

EXAMPLE Transforming a Parabola

Explain how transformations are used to graph $y = (x - 2)^2 + 1$.

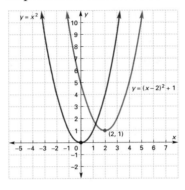

SOLUTION

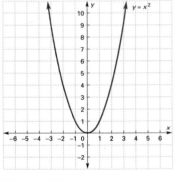

Start with the parent function.

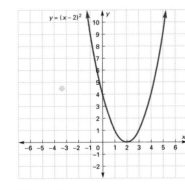

Move the vertex two units to the right.

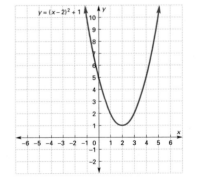

Move the vertex one unit up.

Critical Thinking Explain how to find the coordinates of the new vertex when $y = x^2$ is moved to $y = (x - m)^2 + n$?

You know that changing the coefficients of the terms in a quadratic equation has an effect on the graph of the equation. In the following lessons, these coefficients will play a key role in defining the characteristics of a quadratic function.

LESSON ASSESSMENT

Let the equation $h = -16t^2 + 250t$ represent the flight of a model rocket.

1 Explain how to use the graph of the equation to find the maximum height of the rocket.

2 Explain how to find the total time the rocket is in the air.

Start with the parent function $y = x^2$. How would you write the new equation for each of the following transformations? Give the coordinates of the new vertex.

3 The parent graph is translated three units to the left.

4 The parent graph is translated two units up.

5 The parent graph is translated one unit to the right and 1 unit down.

Practice and Problem Solving

Graph each function. Describe the transformation of the parent graph. Give the coordinates of the vertex of the transformed graph.

6. $y = -2x^2$ **7.** $y = 3x^2 + 5$ **8.** $y = -(x + 3)^2$

9. $y = (x - 2)^2 + 1$ **10.** $y = \frac{1}{2}x^2 - 3$ **11.** $y = -\frac{1}{2}(x + 1)^2 + 2$

Mixed Review

The formula for finding compound interest is $A = p(1 + r)^n$.

A is the amount of the new balance.
r is the rate of interest each period.
p is the principal.
n is the number of interest periods.

Shondra

Alan

8%   6%

12. Alan has $800 invested in a savings account. The account pays 6% interest compounded yearly. How much will the balance be after four years?

13. Shondra has $800 invested in a savings account. The account pays 8% interest compounded yearly. How much more interest will Shondra earn in four years than Alan?

14. State Bank compounds interest monthly. City Bank compounds interest semiannually. Each bank pays an annual rate of 6%. How much more interest will you earn if you invest $1000 in State Bank for one year? (*Hint:* The interest rate for each period is the annual rate divided by the number of interest periods.)

Write and solve an equation for each situation.

15. The side of a large square is 2 inches longer than the side of a small square. The sum of the perimeters of the two squares is 40 inches. What is the length of each side of the small square?

16. Raul receives a bonus of 5% of his salary when he meets his sales goal. One week Raul made his goal and earned a total of $256. How much is his bonus?

17. You have three test scores of 85, 90, and 93. What is the lowest score you can make on your fourth test and still have a test average of 90?

Two cars are located on a grid at the coordinates $M(-3, -2)$ and $N(1, 4)$.

18. What is the slope of line MN?

19. What is the equation of line MN?

20. Another car is located at point $P(2, 2)$. What is the equation of the line passing through point P and parallel to line MN?

LESSON 11.2 THE QUADRATIC FUNCTION

In Lesson 11.1, you graphed the flight of a softball thrown upward by Juanita. The softball had an initial speed of 64 feet per second, and you used the quadratic equation $h = -16t^2 + 64t$ to model the flight. Since at time $t = 0$, the height is $h = 0$, this equation models a softball starting its flight at ground level.

Suppose a person stands on a 75-foot cliff and throws a rock with the same upward speed as Juanita's fastball. You need a new equation to show the effect of moving the initial starting height from the ground to the 75-foot cliff. This new quadratic equation is

$$h = -16t^2 + 64t + 75$$

You can use the same coordinate axes to graph the equations for the height of the arrow, the softball, and the rock.

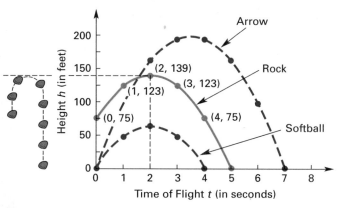

Critical Thinking   What is the maximum height reached by the rock? What is the approximate flight time of the rock from release to impact with the ground?

The graphs representing the softball and the rock are different because at any given moment, the rock parabola is "higher" than the softball parabola. If you slide the rock parabola vertically downward 75 units on the y-axis, it fits right over the softball parabola, point for point. Two figures that match point for point are **congruent**.

The softball and the rock equations are represented by congruent parabolas. They are congruent because the coefficients of t^2 and t in each equation are identical.

Critical Thinking Use the language of transformations to describe the graphs for both the rock and the softball in terms of the parent function $y = t^2$.

An equation written in the form

$$y = ax^2 + bx + c$$

is a *quadratic equation written in standard form.* The heights of the arrow, the softball, and the rock are described by quadratic equations.

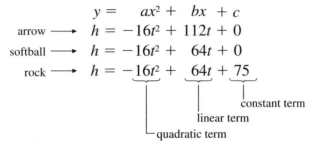

$$
\begin{array}{rccc}
y = & ax^2 + & bx & + c \\
\text{arrow} \longrightarrow \quad h = & -16t^2 + & 112t & + 0 \\
\text{softball} \longrightarrow \quad h = & -16t^2 + & 64t & + 0 \\
\text{rock} \longrightarrow \quad h = & -16t^2 + & 64t & + 75
\end{array}
$$

constant term
linear term
quadratic term

A quadratic equation contains three terms. The quadratic term is degree two, the linear term is degree one, and the constant term is a number with no variable.

EXAMPLE 1 Graphing a Quadratic Function

Graph the quadratic equation $y = x^2 + x - 12$.

SOLUTION

The equation $y = x^2 + x - 12$ defines a relationship between x and y. Thus, y is a function of x. The equation written in *functional notation* is

$$f(x) = x^2 + x - 12$$

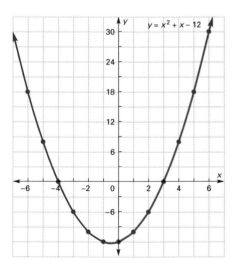

x	y
−6	18
−5	8
−4	0
−3	−6
−2	−10
−1	−12
0	−12
1	−10
2	−6
3	0
4	8
5	18
6	30

Substitute integer values for x from −6 to 6 ($-6 \leq x \leq 6$) to determine the corresponding values of y. The values of x from −6 to 6 are real numbers in the domain of f. The corresponding values of y are in the range of f. Complete a table of ordered pairs, graph the points, and draw the graph. The graph of a quadratic function is *always* a parabola.

Graph the quadratic function $h(m) = m^2 - m - 2$.

Look at the table and graph in Example 1. Where does the graph cross the x-axis? What are the approximate coordinates of the minimum point on the graph? When using a graphics calculator, use the ZOOM key to find a more accurate answer.

The graph crosses the x-axis at $(-4, 0)$ and $(3, 0)$. Since the x-intercepts result when $y = 0$, you have solved the equation

$$x^2 + x - 12 = 0$$

The solutions, also called the **roots,** of the equation $x^2 + x - 12$ are -4 and 3. One way to check your graphing results is to substitute -4 and 3 for x in the function

$$f(x) = x^2 + x - 12$$

$$f(-4) = (-4)^2 + (-4) - 12$$
$$= 0$$

$$f(3) = (3)^2 + (3) - 12$$
$$= 0$$

The Zeros of a Quadratic Function

Since $f(-4) = 0$ and $f(3) = 0$, -4 and 3 are also called the **zeros** or **roots of the function** $f(x) = x^2 + x - 12$.

> ## Zeros or Roots of a Function
> If f is a function and $f(a) = 0$,
>
> then a is a zero or root of f.

Look again at the graph of $f(x) = x^2 + x - 12$. The graph indicates that the vertex, in this case the minimum point, is at $(-0.5, -12.25)$. The vertex is on a vertical line midway between the x-intercepts of the parabola. Since this is true for all parabolas, you can use an algebraic technique to find the minimum or maximum value.

EXAMPLE 2 Finding a Minimum Value

Find the minimum value of $f(x) = x^2 + x - 12$ and the coordinates of the vertex.

SOLUTION

The minimum value of f is the y-value of the vertex. Since the vertex is midway between $(-4,0)$ and $(3,0)$, it must have a first coordinate of $\frac{-4+3}{2}$, or -0.5. Its y-coordinate, $f(-0.5)$, is the minimum value.

$$f(-0.5) = (-0.5)^2 + (-0.5) - 12 = 0.25 - (0.5) - 12 = -12.25$$

The vertex of the parabola for f is at $(-0.5, -12.25)$. The minimum value is -12.25.

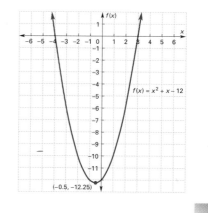

The line that passes through the vertex and is parallel to the y-axis is called the **axis of symmetry.** The axis of symmetry for the parabola in Example 2 is given by the equation $x = -0.5$.

Find the axis of symmetry, the vertex, and the minimum value for the function $f(x) = x^2 - 5x + 6$.

The graphs of some quadratic functions touch the x-axis at only one point. In this case, there is only one root. Consider the equation

$f(x) = x^2 - 6x + 9$

$f(3) = 3^2 - 6(3) + 9$

$f(3) = 0$

The root of the equation is 3.

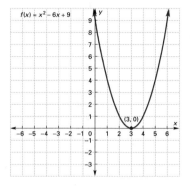

In some instances, the graph does not cross the x-axis at all.

Consider the function $y = f(x) = x^2 + 2x + 6$

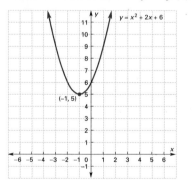

There is no real number that is a solution of this equation when $y = 0$. Thus, there are no real roots.

Approximating Zeros

You can use a graphics calculator to find the zeros of a quadratic function.

EXAMPLE 3 Using a Graphics Calculator

Use a graphics calculator to find the zeros of $f(x) = x^2 + 3x + 1$.

SOLUTION

Graph $y = x^2 + 3x + 1$ on your calculator. Use the ⟨ZOOM⟩ and ⟨TRACE⟩ keys to find the coordinates where the graph intersects the x-axis. These x-coordinates are the zeros. Since the zeros are not integers, you can approximate the zeros by rounding to the nearest thousandth.

$$x \approx -2.618 \text{ and } x \approx -0.382$$

The zeros of the function are about -2.618 and -0.382.

LESSON ASSESSMENT

Think and Discuss

Use $f(x) = x^2 - x - 6$ for 1–5.

1 Explain how to graph f.

2 Explain how to find the zeros of f.

3 Explain how to find $f(3)$.

4 How do you find the maximum or minimum of f?

5 How do you find the vertex and the axis of symmetry for the graph of f?

Practice and Problem Solving

Use a graph to find the x-intercepts and the vertex.

6. $y = x^2 - 8x + 15$ **7.** $y = x^2 + 6x + 8$

8. $y = x^2 + x - 2$ **9.** $y = x^2 + 2x + 1$

10. $y = x^2 + 5x + 6$ **11.** $y = x^2 + 3x - 10$

Use a graphics calculator to find the roots of each function.

12. $y = x^2 - 2x + 1$ **13.** $y = x^2 - 4x - 5$

14. $y = x^2 - 1$ **15.** $y = x^2 + 8x + 3$

16. $y = x^2 + 4x - 6$ **17.** $y = x^2 - 3.8x - 0.8$

The profit *P* for Nancy's Cookie Company is determined by subtracting overhead *O* from income *I*.

18. Write a formula to find the profit for Nancy's company.

19. Find the profit when the income is $2500 and the overhead is $1600.

20. Find the income when the profit is $500 and the overhead is $1200.

21. A rectangular box has a length of 20 centimeters, a width of 15 centimeters, and a depth of 8 centimeters. Suppose the length is increased by 2 centimeters and the depth is decreased by 2 centimeters. By how much is the volume of the new box increased or decreased?

Solve each equation.
22. $7a - 9 = 2a + 11$ **23.** $\dfrac{2x}{8} = \dfrac{12}{24}$ **24.** $80\% \, m + m = 126$

25. Find the slope and *y*-intercept of $x + y = 6$.

26. Find the slope and *y*-intercept of $y - 3x = 6$.

27. Solve this system of equations: $x + y = 6$
$$y - 3x = 6$$

Solve using a system of equations.
28. Roberto sells caps for $20 and T-shirts for $40. On Saturday Roberto sold a total of 12 caps and T-shirts. He received $320. How many caps did Roberto sell?

Examine the graph of $y = x^2 - 3x - 10$.

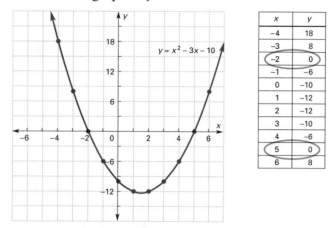

x	y
−4	18
−3	8
−2	0
−1	−6
0	−10
1	−12
2	−12
3	−10
4	−6
5	0
6	8

The roots of the quadratic equation $y = x^2 - 3x - 10$ are the values of x when the value of y is zero. Thus, you find the roots when you solve the quadratic equation

$$x^2 - 3x - 10 = 0$$

In Chapter 10, you learned how to rewrite a quadratic expression in factored form:

$$x^2 - 3x - 10 = (x - 5)(x + 2)$$

Thus, you can substitute the factored form of $x^2 - 3x - 10$ in the quadratic equation.

$$x^2 - 3x - 10 = 0$$

$$(x - 5)(x + 2) = 0$$

The substitution results in an equation that equates the product of two binomials to zero. If the product of two numbers is zero, what do you know about the numbers?

Zero Product Property

Let a and b be any real numbers.

If $ab = 0$, then $a = 0$ or $b = 0$.

To solve $(x - 5)(x + 2) = 0$, use the Zero Product Property. Set each factor equal to zero and solve the resulting linear equations.

$$\text{If } (x - 5)(x + 2) = 0, \text{ then}$$

$$x - 5 = 0 \qquad \text{or} \qquad x + 2 = 0$$

$$x = 5 \qquad \text{or} \qquad x = -2$$

You have solved $x^2 - 3x - 10 = 0$ by factoring. The solutions are 5 and -2. The solutions are the same x-intercepts shown on the graph.

EXAMPLE 1 The Roots of a Quadratic Equation

Find the roots of $y = 8x^2 - 30x + 25$.

SOLUTION

To find the roots of a quadratic equation, substitute zero for y and solve the resulting equation.

$8x^2 - 30x + 25 = 0$	Set original equation equal to zero
$(2x - 5)(4x - 5) = 0$	Factor the trinomial
$2x - 5 = 0 \quad \text{or} \quad 4x - 5 = 0$	Zero Product Property
$2x = 5 \quad \text{or} \quad 4x = 5$	Solve each linear equation
$x = \dfrac{5}{2} \quad \text{or} \quad x = \dfrac{5}{4}$	The solutions $\dfrac{5}{2}$ and $\dfrac{5}{4}$ are the roots

Substitute the solutions into the original equation $y = 8x^2 - 30 + 25$. Did you get $y = 0$?

Ongoing Assessment

Find the roots of $y = 3x^2 + 7x + 2$. Check the solution by substitution and by graphing.

EXAMPLE 2 Solving a Quadratic Equation

Solve $x^2 = 5x - 6$.

SOLUTION

Use the Addition Property of Equality to rewrite the equation in the form $x^2 + bx + c = 0$. Then factor and use the Zero Product Property.

$$x^2 = 5x - 6 \qquad \text{Given}$$

$$x^2 - 5x + 6 = 0 \qquad \text{Addition Property of Equality}$$

$$(x - 3)(x - 2) = 0 \qquad \text{Factor the trinomial}$$

$$x - 3 = 0 \ \text{ or } \ x - 2 = 0 \qquad \text{Zero Product Property}$$

$$x = 3 \ \text{ or } \ x = 2 \qquad \text{Addition Property of Equality}$$

The solutions are 3 and 2. Check the solutions in the original equation.

Critical Thinking Explain how to solve $x^2 - 5x + 6 = 0$ by graphing $y = x^2$ and $y = 5x - 6$ on a graphics calculator.

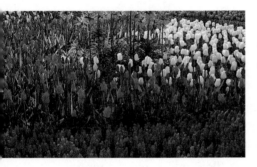

EXAMPLE 3 Solving a Practical Problem

The Chambers family has a garden that measures 15 feet by 30 feet. They plan to increase the area of the garden by 250 square feet. However, they must increase the length and the width by the same number of feet. What are the dimensions of the garden after the increase?

SOLUTION

Draw a picture of the original garden.

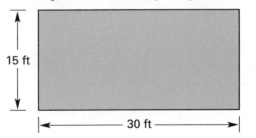

Draw a picture of the new garden.

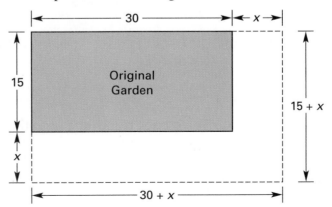

The area of the new garden is 450 + 250, or 700 square feet. Thus, the area of the new garden is modeled by this equation.

$$(30 + x)(15 + x) = 700$$

However, you cannot use the Zero Product Property with the equation written in this form. To use the Zero Product Property, first multiply the binomials. Then combine the constant terms and rewrite the equation with zero on the right side of the equal sign.

$(30 + x)(15 + x) = 700$	Given
$450 + 45x + x^2 = 700$	Multiply the binomials
$x^2 + 45x - 250 = 0$	Set trinomial equal to zero
$(x - 5)(x + 50) = 0$	Factor the trinomial
$x = 5 \text{ or } x = -50$	Zero Product Property

There are two possible solutions. The Chambers family can increase the garden by 5 feet on each side, or they can decrease it by 50 feet on a side. You can see from the drawing that a decrease of 50 feet is impossible. The solution to an equation does not always make sense in a real world problem. Always check the domain and the range before you give the answer. Does a solution of 5 make the equation true? Look at the drawing. Does the solution make sense?

LESSON ASSESSMENT

Think and Discuss

1 Explain how to find the roots of a quadratic equation by factoring.

2 How is the Zero Product Property used to solve a quadratic equation?

Practice and Problem Solving

Solve each quadratic equation by factoring. Check your answer.

3. $x^2 + 5x + 6 = 0$ **4.** $x^2 + 5x + 4 = 0$ **5.** $x^2 + 2x = 3$

6. $x^2 - 2x = 3$ **7.** $x^2 + 3x + 2 = 0$ **8.** $x^2 + 3x = 10$

9. $2x^2 + 5x = 12$ **10.** $3x^2 = 8x + 3$ **11.** $x^2 = 16 - 6x$

12. $4x^2 + 2x = 2$ **13.** $6x^2 + x - 12 = 0$ **14.** $4x^2 = 4x - 1$

Write and solve an equation for each situation.

15. The number of square meters in the area of a square deck is 12 greater than the number of meters around the deck's perimeter. What is the length of each side of the deck?

16. Shemekia is cutting material for a triangular sail. The height of the sail must be 2 feet longer than the base. The area of the sail is 40 square feet. What dimensions should Shemekia cut for the sail?

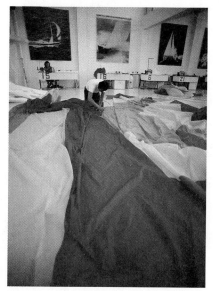

The amount of matter in a body is called its mass. Scientific notation is used to approximate very large and very small masses. For example,

the mass of a human cell is about 10^{-18} kg
the mass of a paper clip is about 10^{-3} kg
the mass of the moon is about 6×10^{22} kg

17. How many times greater is the mass of the moon than the mass of a paper clip? Write a ratio in simplified form comparing the two masses.

18. How many times less is the mass of a human cell than the mass of a paper clip? Write a ratio in simplified form comparing the two masses.

The length of a rectangle is one foot longer than its width. The perimeter is 180 inches.

19. Use one unknown to write an equation for the perimeter.

20. Solve the equation in Exercise 19.

21. Use two unknowns to write a system of equations for the perimeter.

22. Solve the system in Exercise 21 by substitution.

23. How are the solutions in Exercise 20 and 22 similar?

24. Which method of solution seems easier? Explain your answer.

The function $f(t) = -16t^2 + 400$ represents the height of a freely falling ballast bag that starts from rest on a balloon 400 feet above the ground.

25. Find $f(t)$ when t is 5.

26. What does this solution mean with regard to the height of the object?

In Lesson 10.7, you used geometric models to factor quadratics such as $x^2 + 4x + 4$.

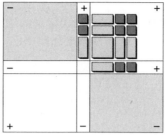

In this lesson you will learn a technique that can be used to solve any quadratic equation. This technique is called **completing the square**.

Start with the model for $x^2 + 6x$. How many 1-blocks must you add to one x^2-block and six x-blocks to build a square?

x^2 $6x$

ACTIVITY 1   Completing a Square

Draw an area diagram for each step.

1 Start with one x^2-block and six x-blocks.

2 Divide the six x-blocks into two equal sets.

3 How many blocks are in each set?

4 Square the number found in Step 3. Add this number of 1-blocks to the area diagram.

5 Form a square using all the blocks. What are the dimensions of the square?

6 Complete the Activity again. This time start with

    a. $x^2 + 8x$ **b.** $x^2 + 10x$

In Activity 1, you used a geometric method to complete a square. There is also an algebraic method for completing the square of a quadratic equation in the form $y = x^2 + bx$. First, divide the coefficient of the x-term by 2. The result is $\frac{b}{2}$. Next, square the result of the division. The product is $\left(\frac{b}{2}\right)^2$. Finally, add $\left(\frac{b}{2}\right)^2$ to both sides of the equation.

$$y + \left(\frac{b}{2}\right)^2 = x^2 + bx + \left(\frac{b}{2}\right)^2$$

The right side of the equation is now a perfect square trinomial.

$$x^2 + bx + \left(\frac{b}{2}\right)^2 = \left(x + \frac{b}{2}\right)^2$$

Ongoing Assessment

Complete the square for $x^2 + 100x$.

ACTIVITY 2 The Vertex of a Parabola

1 Use a table or graphics calculator to graph $y = x^2 - 6x$. What are the coordinates of the vertex?

2 What number should be added to make a perfect square trinomial from $x^2 - 6x$? Add this number to both sides of the equation. What property lets you do this without changing the equality?

3 Factor the perfect square trinomial.

4 Solve the equation for y. How can you use this equation to write the vertex of the parabola?

In Activity 2, you completed the square to write the equation

$$y = x^2 - 6x \text{ as } y = (x - 3)^2 - 9$$

which is in the form $y = (x - h)^2 + k$. Using your knowledge of transformations, you can see that the vertex of a parabola is the point (h, k). Thus, the vertex is $(3, -9)$.

When a quadratic equation is written in the form

$$y = a(x - h)^2 + k$$

the values of a, h, and k provide useful information for sketching the resulting parabola.

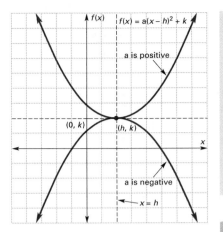

f(x)

f(x) = a(x − h)² + k

a is positive

(0, k)

(h, k)

a is negative

x = h

x

The Parabola Form of a Quadratic Function

The parabola form of a quadratic equation is

$$f(x) = a(x - h)^2 + k$$

The graph of f is a parabola with vertex (h, k). When a is positive, the *minimum* value of f is k. When a is negative, the *maximum* value of f is k. The axis of symmetry is the line $x = h$.

EXAMPLE 1 Completing the Square to Find the Vertex

Find the vertex of $y = x^2 + x - 12$ by completing the square.

SOLUTION

First, isolate the $x^2 + bx$ terms. Then, complete the square. Finally, write the equation in parabola form.

$$y = x^2 + x - 12 \qquad \text{Given}$$

$$y + 12 = x^2 + x \qquad \text{Addition Property of Equality}$$

$$y + 12 + \left(\frac{1}{2}\right)^2 = x^2 + x + \left(\frac{1}{2}\right)^2 \qquad \text{Complete the square}$$

$$y + 12.25 = (x + 0.5)^2 \qquad \text{Factor a perfect square}$$

$$y = (x + 0.5)^2 - 12.25 \qquad \text{Subtraction Property of Equality}$$

Compare the final equation to the parabola form of a quadratic. In this form, h is -0.5 and k is -12.25. The vertex is $(-0.5, -12.25)$.

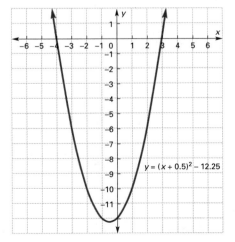

$y = (x + 0.5)^2 - 12.25$

Find the vertex of $y = x^2 - x - 12$ by completing the square.

EXAMPLE 2 Finding the Minimum of a Quadratic

Find the minimum of $f(x) = x^2 + 3x$.

SOLUTION

Solve the equation $y = x^2 + 3x$ by completing the square.

$$y = x^2 + 3x \qquad\qquad \text{Given}$$

$$y + \left(\frac{3}{2}\right)^2 = x^2 + 3x + \left(\frac{3}{2}\right)^2 \qquad \text{Complete the square}$$

$$y + \frac{9}{4} = \left(x + \frac{3}{2}\right)^2 \qquad\qquad \text{Factor a perfect square}$$

$$y = \left(x + \frac{3}{2}\right)^2 - \frac{9}{4} \qquad\qquad \text{Subtraction Property of Equality}$$

The quadratic is now written in parabola form. The vertex (h, k) is at $\left(-\frac{3}{2}, -\frac{9}{4}\right)$. Since a is positive, the vertex is the minimum value for the function. Thus, the minimum of the function occurs at $f\left(-\frac{3}{2}\right)$ or $-\frac{9}{4}$.

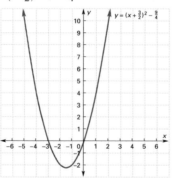

EXAMPLE 3 Finding the Maximum of a Quadratic

Find the maximum of $f(x) = -x^2 + 6x - 4$.

SOLUTION

Solve the equation $y = -x^2 + 6x - 4$ by completing the square.

$$y = -x^2 + 6x - 4 \qquad \text{Given}$$

$$y + 4 = -(x^2 - 6x) \qquad\qquad \text{Addition Property of Equality and Distributive Property}$$

$$y + 4 - 9 = -(x^2 - 6x + 9) \qquad \text{Complete the square and Addition Property of Equality}$$

$$y - 5 = -(x - 3)^2 \qquad \text{Factor perfect square trinomial}$$

$$y = -(x - 3)^2 + 5 \qquad \text{Addition Property of Equality}$$

Compare this equation to the parabola form of the quadratic equation. The vertex (h, k) is at $(3, 5)$. Since a is negative, the vertex is the maximum point on the graph. Thus, the maximum of the function occurs for $f(3)$ or 5.

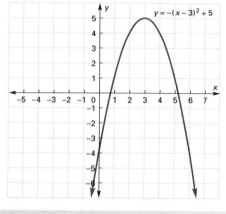

Ongoing Assessment

Find the minimum or maximum of $f(x) = -x^2 - 8x + 1$.

WORKPLACE COMMUNICATION

Divide both sides of the equation for total area by 4π. Assume a value for the height, such as $h = 4$ meters. Then, complete the square of the quadratic equation. Where is the vertex? Can you minimize A, as stated by Mr Bakorsi? How will you respond to Mr. Bakorsi? What other factors might influence your choice for r and h?

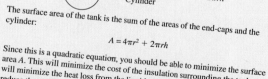

Modern Workplace, Inc.
Interoffice Memo

To: Ruby Mahan, Hazardous Material Handling Team Leader
From: Jack Bakorsi, Director of Plans and Programs
Subject: Optimum Design for Liquid Nitrogen Storage Tanks

I understand that you are planning to redesign the current liquid nitrogen storage tanks. Before you begin construction of the replacements, please design the most cost effective, efficient tank. I think this is the tank with minimum surface area.

To fit the constraints of the storage facility, the shape of the tanks must be right-circular cylinders with hemispherical end-caps. You cannot control the shape, but you can control the height and the radius and, therefore, the total surface area of the tanks.

Surface Area of Spherical End-Caps $= 4\pi r^2$

Surface Area of Cylinder $= 2\pi rh$

The surface area of the tank is the sum of the areas of the end-caps and the cylinder:

$$A = 4\pi r^2 + 2\pi rh$$

Since this is a quadratic equation, you should be able to minimize the surface area A. This will minimize the cost of the insulation surrounding the tank and will minimize the heat loss from the liquid nitrogen. This should greatly reduce the number of times we have to refill the tank each year.

In the long run, we can save the company a lot of money by using the optimum values of r and h for the new liquid nitrogen storage tanks. Please let me know what values you plan to use.

LESSON ASSESSMENT

Think and Discuss

Pick a value for b and substitute it into the equation $y = x^2 + bx$. Use this function to answer 1-2.

1 Explain how to use a geometric model to complete the square of your function.

2 Explain how to use the method of completing the square to find the vertex of the graph of your function.

3 Explain how to use the method of completing the square to find the vertex of the parabola formed by the graph of $y = x^2 + 5x + 6$.

Practice and Problem Solving

For each quadratic equation,

a. write the equation in parabola form $y = a(x - h)^2 + k$.

b. find the minimum or maximum value of the function.

c. find the coordinates of the vertex.

d. write the equation for the line of symmetry.

4. $y = x^2 + 4x$

5. $y = -x^2 + 8x$

6. $y = x^2 - 10x$

7. $y = -x^2 + 3x + 2$

8. $y = x^2 + 6x + 8$

9. $y = -x^2 - 7x + 10$

10. $y = x^2 - 3x - 4$

11. $y = -x^2 - 4x - 12$

12. $y = x^2 - 5x + 6$

13. Suppose you earn $160 in simple interest during one year on a $2000 investment. What is the rate of interest?

14. The diameter of a baseball is about 7.6 centimeters. The diameter of a basketball is about 24 centimeters. Find the difference between the volume of a baseball and a basketball.

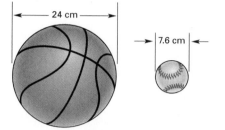

24 cm

7.6 cm

Write and solve a system of linear equations to model each situation.

15. The Right Company manufactures hammers for $4 each and wrenches for $2 each. On Tuesday the Right Company manufactured a total of 960 hammers and wrenches at a cost of $2880. How many hammers did the Right Company manufacture on Tuesday?

16. The Great Coffee Shop's bestselling coffees are hazelnut and Brazilian. A package containing 2 pounds of hazelnut and 3 pounds of Brazilian costs $34.50. A package containing 3 pounds of hazelnut and 2 pounds of Brazilian costs $35.50. How much does one pound of each coffee cost?

LESSON 11.5 SOLVING QUADRATIC EQUATIONS

In previous lessons, you used graphing and factoring to find the zeros of a quadratic function such as $f(x) = x^2 - 2x - 3$. You can also find the zeros by completing the square.

EXAMPLE 1 Finding the Zeros of a Function

Find the zeros of $f(x) = x^2 - 2x - 3$ by completing the square.

SOLUTION

The zeros of the function occur when $f(x) = 0$.

$x^2 - 2x - 3 = 0$	Given
$x^2 - 2x = 3$	Addition Property of Equality
$x^2 - 2x + 1 = 3 + 1$	Completing the Square and Addition Property of Equality
$(x - 1)^2 = 4$	Factoring
$x - 1 = 2 \quad \text{or} \quad x - 1 = -2$	If $x^2 = c$, $x = \pm \sqrt{c}$
$x = 3 \quad \text{or} \quad x = -1$	Addition Property of Equality

The zeros are 3 and -1.

Check:

$f(3) = (3)^2 - 2(3) - 3 = 9 - 6 - 3 = 0$

$f(-1) = (-1)^2 - 2(-1) - 3 = 1 + 2 - 3 = 0$

Consider the function $f(x) = x^2 - 5x - 14$. How can you find the value or values of x such that $f(x) = -20$?

You can rewrite the equation $x^2 - 5x - 14 = -20$ as $x^2 - 5x = -6$. Then solve $x^2 - 5x = -6$ by any one of the methods you have learned in this chapter.

Solving a Quadratic Equation

1 Let $f(x) = -20$. Rewrite $f(x) = x^2 - 5x - 14$ as the quadratic equation $x^2 - 5x - 14 = -20$.

2 What property allows you to rewrite $x^2 - 5x - 14 = -20$ as $x^2 - 5x = -6$?

3 Solve $x^2 - 5x = -6$ by completing the square.

4 Solve $x^2 - 5x = -6$ by factoring. (*Hint:* add 6 to both sides.)

5 Solve $x^2 - 5x = -6$ by graphing. (Use a graphics calculator if available.)

6 Which method do you prefer? Explain why.

EXAMPLE 2 Finding the Dimensions of a Rectangle

Taj wants to construct a rectangular cabin floor that has a perimeter of 54 feet and an area of 180 square feet. What dimensions should Taj use to construct the cabin?

SOLUTION

Let x represent the length of the cabin floor. Let y represent the width of the cabin floor.

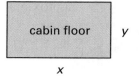

cabin floor y

x

The perimeter (54) of the cabin floor is $p = 2x + 2y$. Thus,

$$2x + 2y = 54 \quad \text{or} \quad x = 27 - y$$

The area (180) of the cabin floor is $A = xy$. Thus,

$$A = (27 - y)y \quad \text{or} \quad 180 = 27y - y^2$$

The solution will be the roots of this quadratic equation.

$$y^2 - 27y + 180 = 0$$

You can use any of the three methods to solve this quadratic equation. Since the roots are 15 and 12, the dimensions of the cabin floor should be 15 feet by 12 feet.

In an auditorium, the number of chairs in each row is 8 fewer than the number of rows. The auditorium holds 609 people. How many rows are in the auditorium?

Sometimes the solution to a quadratic equation is an irrational number, and you must approximate the roots.

EXAMPLE 3 Irrational Roots

Solve the quadratic $x^2 - 4x = 2$.

SOLUTION

$x^2 - 4x = 2$	Given
$x^2 - 4x + 4 = 2 + 4$	Completing the Square and Addition Property of Equality
$(x - 2)^2 = 6$	Factoring
$x - 2 = \sqrt{6}$ or $x - 2 = -\sqrt{6}$	If $x^2 = c$, $x = \pm\sqrt{c}$
$x = 2 + \sqrt{6}$ or $x = 2 - \sqrt{6}$	Addition Property of Equality

The roots are irrational because they cannot be written as the ratio of two integers. Use your calculator to find that the roots are approximately 4.45 and −0.45.

LESSON ASSESSMENT

Think and Discuss

Explain how to solve the quadratic equation $y = x^2 - 6x + 8$ by each of the following methods.

1 Graphing.

2 Factoring.

3 Completing the square.

Practice and Problem Solving

Solve each equation by completing the square. Round any irrational roots to the nearest hundredth.

4. $x^2 - 2x - 15 = 0$ **5.** $x^2 = 4x + 12$ **6.** $x^2 = 6x - 5$

7. $x^2 - 4x = 7$ **8.** $x^2 = 6x - 9$ **9.** $x^2 = 7 - 8x$

10. $x^2 + 12x = 0$ **11.** $x^2 = x + 3$ **12.** $x^2 + 3x = 3$

13. $x^2 - 3 = 5x$ **14.** $x^2 = x$ **15.** $x^2 = 1 - 4x$

Let $f(x) = x^2 - 8x + 10$. Find the value of x when $f(x)$ is the given value.

16. $f(x) = -5$ **17.** $f(x) = 10$ **18.** $f(x) = -2$

Solve using quadratic equations.

19. The length of a playing field is 2 meters longer than its width. If the area of the playing field is 1200 square meters, find the dimensions of the playing field.

20. The sum of two whole numbers is 21. Their product is 108. Find the numbers.

Mixed Review

The formula $P = R - B - C$ is used by insurance companies to find their profit (P). R is the revenue received as premiums. B is the benefits paid out. C is the overhead cost.

21. The Universal Insurance Company receives $400 in premiums for a certain life insurance policy. The overhead cost for the policy is $25 per policy. During one year, the company sold 20,000 policies and paid out $6.45 million in benefits. What is the profit for this policy?

22. Solve the formula for the overhead cost.

A dart is tossed and hits the dartboard in random fashion.

23. What is the probability the dart lands in section A?

24. What is the probability the dart lands in section B?

25. What is the probability the dart lands in section C?

26. What is the probability the dart does *not* land in section D?

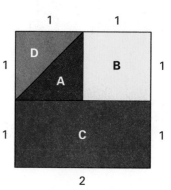

LESSON 11.6 THE QUADRATIC FORMULA

When a quadratic equation is written in the form

$$y = ax^2 + bx + c$$

you can find the roots by graphing, factoring, or completing the square. There is one more method you can use to find the roots. This method uses the **quadratic formula**. This formula can be derived by solving $ax^2 + bx + c = 0$ for x.

ACTIVITY 1   Deriving the Quadratic Formula

Give the reason for each step.

$$ax^2 + bx + c = 0$$

$$ax^2 + bx = -c$$

$$x^2 + \frac{b}{a}x = -\frac{c}{a}$$

$$x^2 + \frac{b}{a}x + \frac{b^2}{4a^2} = -\frac{c}{a} + \frac{b^2}{4a^2}$$

$$\left(x + \frac{b}{2a}\right)^2 = -\frac{c}{a} + \frac{b^2}{4a^2}$$

$$\left(x + \frac{b}{2a}\right)^2 = \frac{4ac}{4a^2} + \frac{b^2}{4a^2}$$

$$\left(x + \frac{b}{2a}\right)^2 = \frac{b^2 - 4ac}{4a^2}$$

$$x + \frac{b}{2a} = \pm\sqrt{\frac{b^2 - 4ac}{4a^2}}$$

$$x = \frac{\pm\sqrt{b^2 - 4ac}}{2a} - \frac{b}{2a}$$

$$x = \frac{-b \pm \sqrt{b^2 - 4ac}}{2a}$$

You can use the quadratic formula to find the roots of any quadratic equation written in the form $ax^2 + bx + c = 0$.

The Quadratic Formula

Let a, b, and c be real numbers with $a \neq 0$.

If a quadratic equation is written in the form

$ax^2 + bx + c = 0$, then the roots of the equation are

$$x = \frac{-b \pm \sqrt{b^2 - 4ac}}{2a} \qquad \text{where } a \neq 0$$

EXAMPLE 1 Applying the Quadratic Formula

Find the x-intercepts of $y = 8x^2 - 30x + 25$.

SOLUTION

To find the x-intercepts, first substitute zero for y. Then use the quadratic formula to solve $8x^2 - 30x + 25 = 0$. In the formula, a is 8, b is -30, and c is 25.

$y = 8x^2 - 30x + 25$

$y = ax^2 - bx \ + c$

$$x = \frac{-b \pm \sqrt{b^2 - 4ac}}{2a}$$

$$x = \frac{-(-30) \pm \sqrt{(-30)^2 - (4)(8)(25)}}{(2)(8)}$$

$$x = \frac{30 \pm \sqrt{900 - 800}}{16}$$

$$x = \frac{30 \pm \sqrt{100}}{16}$$

$$x = \frac{30 \pm 10}{16}$$

$x = 2.5$ (when $+ 10$ is used)

$x = 1.25$ (when $- 10$ is used)

The table shows the pattern that occurs when a number of people talk once to each other on the telephone.

Number of People	Number of Phone Calls	Geometric Representation of Phone Calls
1	0	•
2	1	•———•
3	3	△
4	6	▷◁
5	10	▷◁
etc.	etc.	etc.

You can use a formula to express the relationship between the number of phone calls completed and the number of people talking to each other.

Let c represent the number of phone calls.
Let p represent the number of people making calls.

The following equation models the relationship in the table:

$$c = \frac{p(p-1)}{2}$$

1 Let c represent 45 phone calls. Substitute 45 for c in the equation

$$c = \frac{p(p-1)}{}$$

2 Show how to write the new equation as $p^2 - p - 90 = 0$.

3 Solve the quadratic equation using the quadratic formula.

4 Check each root in the original equation.

5 If 45 phone calls are completed, how many people talked to each other?

WARNING

This home is protected by

SPARTAN SECURITY SYSTEMS

EXAMPLE 2 The Profit Function

Spartan Security installs security systems. To track monthly revenues for the number of systems installed (x), they use the function

$$R(x) = 10x^2 + 10x + 100$$

Spartan Security projects their costs using the function

$$C(x) = 8x^2 + 50x + 350$$

Find a function to project monthly profit and use it to find the break-even point.

SOLUTION

To find the profit function, $P(x)$, subtract the cost function from the revenue function.

$$P(x) = R(x) - C(x)$$

$$P(x) = (10x^2 + 10x + 100) - (8x^2 + 50x + 350)$$

$$= 2x^2 - 40x - 250$$

The break-even point occurs when revenues equal cost—in other words, when $P(x) = 0$.

Thus, the roots of $2x^2 - 40x - 250 = 0$ will give the break-even point. If you use the quadratic formula with $a = 2$, $b = -40$, and $c = -250$, the result is $x = 25$ or $x = -5$. Since there cannot be a negative number of security systems, the only root of the quadratic that makes sense is 25. The company will break even after it installs 25 systems.

The quadratic formula gives more than just the roots of quadratic equations. The portion of the quadratic formula under the radical, $b^2 - 4ac$, is called the **discriminant**. In Example 1, the discriminant is 100. When the discriminant is greater than zero, the equation has two unequal roots.

$$x = \frac{-b \pm \sqrt{b^2 - 4ac}}{2a}$$

Critical Thinking   Why does a positive discriminant mean the graph of a quadratic equation intersects the x-axis twice?

Consider the quadratic $y = x^2 - 8x + 16$.

When you use the quadratic formula with $a = 1$, $b = -8$, and $c = 16$, the following result is obtained.

$$x = \frac{-b \pm \sqrt{b^2 - 4ac}}{2a}$$

$$x = \frac{-(-8) \pm \sqrt{(-8)^2 - (4)(1)(16)}}{(2)(1)}$$

$$x = \frac{8 \pm \sqrt{64 - 64}}{2} = \frac{8 \pm 0}{2}$$

$$x = 4$$

The graph of $y = x^2 - 8x + 16$ intersects the x-axis at exactly one point: $(4, 0)$. When the discriminant is equal to zero, the equation has exactly one root.

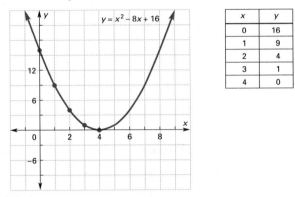

x	y
0	16
1	9
2	4
3	1
4	0

Now examine the graph of the equation $y = 3x^2 + 4x + 5$.

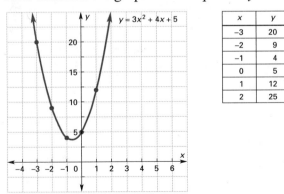

x	y
-3	20
-2	9
-1	4
0	5
1	12
2	25

Notice that the graph does not intersect the x-axis. There are no real numbers that make the equation true when y is equal to zero. Also notice that the discriminant is $(4)^2 - (4 \cdot 3 \cdot 5)$, or -44. If the discriminant is less than zero, there are no real roots.

EXAMPLE 3 Using the Discriminant

Find the discriminant of the equation $y = 2x^2 + 5x + 15$. Use the discriminant to describe the roots.

SOLUTION

For the discriminant, a is 2, b is 5, and c is 15. The discriminant $b^2 - 4ac$ is $25 - 120$, or -95. Since the discriminant is less than zero, the equation $2x^2 + 5x + 15 = 0$ has no real roots. Thus, the expression $2x^2 + 5x + 15$ is prime.

CULTURAL CONNECTION

The Great Pyramid at Giza utilizes a special ratio between the altitude of a triangular face and one-half the length of its base. This ratio is known as the *golden ratio* and has been used by artists and architects to create many works of art over the centuries.

To find the golden ratio, start with segment AC and point B that divides segment AC in the following way:

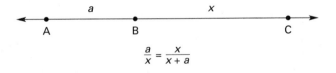

$$\frac{a}{x} = \frac{x}{x + a}$$

The ratio of the shorter segment AB to the longer segment BC is equal to the ratio of the longer segment BC to the total length AC.

1. Let $a = 1$. Solve the quadratic equation that results when you cross-multiply. The positive root is the golden ratio.

Research the relationship between the golden ratio and

2. the Parthenon.

3. a chambered nautilus.

4. the Fibinocci sequence.

5. the Pentagon in Washington, D.C.

LESSON ASSESSMENT

Think and Discuss

1 How is the quadratic formula used to find the roots of a quadratic equation?

2 How does the discriminant describe the roots of a quadratic function?

Identify a, b, and c for each quadratic equation. Find the value of the discriminant.

3. $3x^2 = 5x - 1$ **4.** $4x^2 - 16 = 0$ **5.** $-x^2 = 2x - 1$

Use the quadratic formula to solve each equation. Use your calculator to approximate roots to the nearest hundredth when necessary. Indicate if there are no real roots.

6. $x^2 - 2x - 3 = 0$ **7.** $x^2 - 4x = 21$ **8.** $3x^2 - 2x = 1$

9. $x^2 = x - 6$ **10.** $-x^2 + 6x = 9$ **11.** $8x^2 = 2 - 7x$

12. $2x^2 = 2x + 5$ **13.** $3x^2 = x - 1$ **14.** $0.5x^2 = 0.2x + 0.1$

Solve using a quadratic equation.
15. The length of a rectangular garden is 5 meters longer than its width. If the area of the garden is 84 square meters, find the dimensions of the garden.

16. A rectangular garden has an area of 76 square feet. If the perimeter of the garden is 80 feet, what are the dimensions of the garden?

17. A piece of copper tubing 60 inches long is cut into two unequal pieces. Each piece is formed into a square frame. If the total area of the two frames is 117 square inches, what is the length of each piece of tubing?

Marguerite Gomez owns a company that makes sensors for air bags. She uses the function $P(x) = x^2 - 280x - 14,625$ to estimate her yearly profit. $P(x)$ represents the profit, and x represents the number of sensors she sells.

18. How many sensors must Ms. Gomez sell to break even?

19. How many sensors must Ms. Gomez sell to make a profit of $1125?

20. Ms. Gomez's company has the capacity to produce 500 sensors per year. If the company operates at full capacity, what is the expected profit?

The distance (*d*) in feet an object falls in *t* seconds is
$d = 16t^2 + 5t$.

21. About how long will it take for an object to fall 1000 feet?

22. About how long will it take for an object to fall 4000 feet?

Mixed Review

The Lakeside Running Club recorded the following distances in miles jogged by its members in one week:

14 18 18 16 16 23 21 15 20 25
15 16 22 24 21 16 25 23 17 15

23. Find the range of the data.

24. Find the mode of the data.

25. Find the median of the data.

26. Make a frequency distribution table using intervals of three units.

27. Find the mean of the data.

28. Use the frequency distribution table to draw a histogram.

MATH LAB

Equipment   Flashlight reflector, preferably from a 6 volt
                      flashlight
                  Vernier caliper
                  Ruler with centimeter scale
                  Masking tape

Problem Statement

Reflectors of light and radio waves often have the shape of a
parabola. Parabolic reflectors are useful because incoming
radio waves reflect to a common point, called the *focus*. In
the opposite fashion, if a bright light source is located right at
the focus, its light rays will reflect away from the reflector as
parallel rays. The illustrations below show a cross-sectional
view of a reflector and a collector.

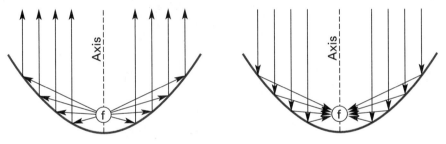

Reflector   Collector

In the case of the reflector, a light bulb is positioned at a point
called the *focal point*, *f*. Light from the focal point is reflected
out in rays parallel to the axis of the parabola. When used as
a collector, the parabolic mirror takes incoming light rays that
are parallel to the axis and reflects them to the focal point, *f*.

You will measure the coordinates of a flashlight reflector to
determine if its shape is really parabolic. You will locate the
focus of the reflector and determine if the bulb's filament is
at the focal point of the reflector.

Procedure

a Disassemble the flashlight so that the reflector is exposed.
    (Do not remove the bulb.)

b Lay the ruler flat across the top lip of the reflector so that the edge of the centimeter scale is directly above the center of the bulb. Use masking tape to hold the ruler in place

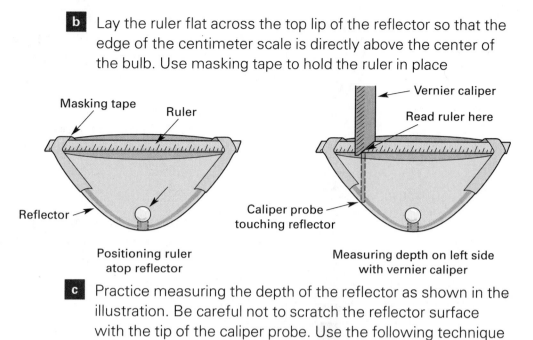

Positioning ruler
atop reflector

Measuring depth on left side
with vernier caliper

c Practice measuring the depth of the reflector as shown in the illustration. Be careful not to scratch the reflector surface with the tip of the caliper probe. Use the following technique to take at least three measurements on each side of the bulb:

1. Open the caliper to 1.0 cm.

2. Carefully place the probe into the reflector while resting the edge of the caliper on the ruler (see illustration).

3. Slide the caliper along the ruler until the tip just touches the reflector. If the probe has an uneven tip, be sure the tip of the longest edge touches the reflector surface.

4. With the caliper probe held in a vertical position (that is, perpendicular to the ruler's edge), read the measurement on the ruler (to the nearest 0.1 cm) where the probe contacts the ruler.

5. Record the measurement from the ruler as the x-value and the preset vernier caliper reading as the y-value.

Ruler Position x (cm)	Vernier Caliper Setting y (cm)
	—

6. With the same setting, move the vernier to the opposite side of the bulb and obtain a measurement from the ruler. Record these *x*- and *y*-values in the Data Table. (You should have two *x*-measurements for the same *y*-value.)

7. Repeat Steps 1–6, each time increasing the opening of the caliper about 1.0 cm. You should have at least three measurements on each side of the bulb.

d Before removing the ruler, use the vernier caliper probe to estimate the location of the filament. Estimate the *x*-value by looking at the ruler position directly above the bulb. Find the *y*- value by positioning the vernier caliper probe beside the bulb, with the tip at the same height as the filament. Record both the *x*-value (the ruler position) and the *y*-value (the vernier caliper reading).

e Remove the masking tape and the ruler and reassemble the flashlight.

f Graph the *x*- and *y*-values you recorded for the reflector.

g Plot the point you measured for the bulb filament. This point should be close to the focus of the parabola. Complete the following steps to plot the first point on a line called the *directrix*.

1. Draw a line from the focus to a point you plotted in **f**.

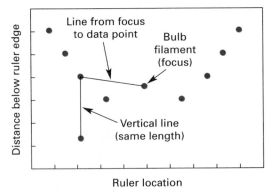

2. Measure the length of this line in centimeters.

3. From the plotted point, draw a vertical line segment of the same length. Darken the endpoint at the bottom of this line segment.

h Repeat Steps 1–3 of **g** for each of the points you plotted. Draw a horizontal line that best fits the dark points. This line will estimate the position of the directrix. Your graph should look similar to the one shown here.

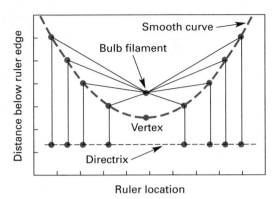

i Draw a final point halfway between the focus and the directrix. This is the vertex of the parabola.

j Draw a smooth curve through your data points and through the vertex. Does your curve resemble a parabola?

k Answer the following questions.

1. If the bulb filament is really *not* at the focus but a little high (that is, too far from the vertex), will the dark points form a straight line? Explain.

2. What do you think would be the effect on the flashlight beam if the bulb were too high?

3. Suppose you have a satellite dish antenna. You know that it is shaped like a parabola. Describe how you would locate the focus so that you could accurately position the receiver element.

Equipment Heavy construction paper
Ruler

Problem Statement

You need to construct a rain gutter from a rectangular piece of sheet metal (simulated in this Activity by a piece of paper) that is 12 inches wide.

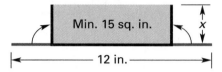

To form the gutter, you will make a 90° bend on each of the two sides of the sheet metal to form a U-shaped gutter. For good drainage, the gutter must have a cross-sectional area of at least 15 square inches. You will determine the length of the gutter sides to meet this area requirement.

Procedure

a Use your ruler to place marks at 1-inch intervals along both 12-inch sides of your construction paper. Draw a line between each pair of marks.

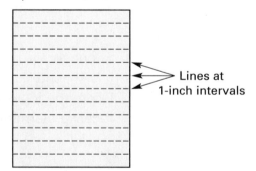

b Make a 90° fold along the outer pair of lines to "make a gutter" with 1-inch sides. Make a sketch of the cross section of the gutter. Record the height (*h*) of the sides, the width (*w*) of the base, and the cross-sectional area. Include proper units in your data.

Side (*h*)	Width (*w*)	Area (*A*)

c Smooth out the folds made in the previous step. Repeat Step **b** for each of the other pairs of lines to make "gutters" with sides of 2 inches, 3 inches, and so on.

d Examine your data. Which of the folds would make a gutter that has at least 15 square inches of cross-sectional area?

e Draw a graph of your measurements. Let the horizontal axis represent the size of the fold. Let the vertical axis represent the cross-sectional area.

f Identify the portion of the graph that will produce a gutter with a cross-sectional area of at least 15 square inches.

g Write an equation for the width (*w*) of the gutter in terms of the height (*h*). Write a second equation that expresses the gutter's cross-sectional area (*A*) in terms of *w* and *h*. Substitute the expression for the width from the first equation into the second equation. Rewrite the resulting equation in standard form for a quadratic equation.

h Find the values of *h* that will yield an area *A* = 15. You can use the quadratic formula or the method of completing the square to determine the values of *h* that will make the equation true when you substitute *A* = 15 in your quadratic equation.

i What value of *h* makes the cross-sectional area a maximum? (*Hint:* Use your graph to help answer this question.)

j Would your graph be any different if you had chosen the width (*w*) to be the horizontal axis instead of the height (*h*)? Explain.

Activity 3: Using BASIC

Equipment   Calculator
Computer with BASIC installed
Tape measure

Problem Statement

In this Activity, you will find the roots of a quadratic equation using a computer program. The quadratic equation will model the area of your teacher's desktop. The solution or roots to this equation will provide information for equally expanding the dimensions of the desktop to double its area.

Procedure

a Measure the length and the width of the top of your teacher's desk. Record the measurements in the Data Table below.

Object	Length (feet or meters)	Width (feet or meters)	Area (Length × Width) (square feet or square meters)
Desktop			

b Calculate the area of the desktop. Write this value in the Data Table.

c Write an expression for increasing the length of the desktop by an amount x. Write a similar expression for increasing the width of the desktop by the same amount x.

d Write an equation for the expanded area of the desktop using the expressions from Step **c**. Let the variable A represent the expanded area.

e Double the value of the area of the desktop calculated in Step **b**. Form a new equation by replacing A in the equation from Step **d**. with the value of the doubled area. Write the quadratic equation in the form of $ax^2 + bx + c = 0$.

f Write a BASIC computer program to find the roots of the equation in Step **e**. Use the program on the opposite page as a guideline.

```
10 REM PROGRAM TO PRINT VALUES OF F(X)

20 PRINT "(X, Y)"

30 PRINT

40 FOR X = d₁ TO d₂
```

(*Note:* d_1 and d_2 are selected upper and lower limits of the domain of the function that allow you to determine if there are any values of x which will yield a zero value for y.)

```
50 Y=(Enter your equation from Step e, with 0 replaced
by Y.)
```

```
60 PRINT"(";X;",",";Y;")"
```

```
70 NEXT X
```

```
80 END
```

g RUN the program.

h Between what two x-values do the y-values change from positive to negative? From negative to positive?

i Change line 40 of the computer program as follows:

```
40 FOR X = d₁ TO d₂ STEP .1
```

Note: d_1 and d_2 are the positive x-values found in Step **h** that caused the y-values to change from negative to positive.)

j RUN the new program.

k Repeat Steps **g** through **j** until you determine the positive x-value (to the nearest tenth) that results in a y-value of zero. (*Note:* the y-values will only change from negative to positive.)

l Explain the meaning of the x-value found in Step **k**.

m In Step **h**, you found two x-values between which the y-values changed from positive to negative. These x-values were not used to find the second root of the equation. Use either the quadratic formula or the BASIC program above to find the second root of the equation. Does this root have any physical meaning in the problem? Explain why or why not.

n Compare the area of the expanded desktop to that of the original desktop. Is the area doubled using the x-value found in Step **k**?

MATH APPLICATIONS

The applications that follow are like the ones you will encounter in many workplaces. Use the mathematics you have learned in this chapter to solve the problems.

Wherever possible, use your calculator to solve the problems that require numerical answers and use your graphics calculator to draw graphs.

Identify which of the following equations are quadratic functions. If the function is quadratic, sketch a graph or use your graphics calculator to find the maximum or minimum and the roots of the function.

1. $y = 12x - 7$ **2.** $y = x^2 - 8x + 15$

3. $y = x^3 + 5x + 6$ **4.** $y^2 = 2x^2$

5. $y = x^2 + 1$ **6.** $y = 2x^2 + 5x - 3$

The function $d = 9t$ represents the distance (d) traveled by a sprinter in time (t) when running at a constant speed of 9 meters per second. The function $d = 0.9t^2$ represents the distance traveled if the sprinter starts at rest and accelerates at a constant rate of 1.8 meters per second.

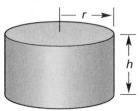

7 Describe the graph of each function.

8 Graph each function on the same coordinate axes.

9 At what point do the graphs intersect? What is the physical meaning of this point?

AGRICULTURE & AGRIBUSINESS

You want to construct a cylindrical tank for storing grain. You know that the formula for the surface area of a right-circular cylinder with no bottom is $A = \pi r^2 + 2\pi rh$, where A is the surface area, r is the radius of the cylinder, and h is the height of the cylinder.

10 The material for constructing the tank is a rectangular piece of sheet metal 12 feet wide and 40 feet long. What is the total area of available sheet metal?

11 Since the sheet metal is 12 feet wide, you must restrict yourself to a cylinder that is 12 feet high (that is, $h = 12$ feet). Equate the area of a cylinder to the area of sheet metal available. Write a quadratic equation in the form $ax^2 + bx + c = 0$, where x represents the unknown radius of the cylindrical tank.

12 Use the quadratic formula to determine the radius of a tank that will use all of the sheet metal.

13 What is "physically" wrong with the answer to Exercise 12? How might you overcome this problem in your equation? (*Hint:* Think of cutting a circular top from the rectangular sheet. Draw a diagram.)

A farmer normally plows a rectangular field 80 meters by 120 meters. This year the government allows the farmer to expand his area under cultivation by 20%. The farmer will extend the width and the length of the plowed area by equal amounts, x.

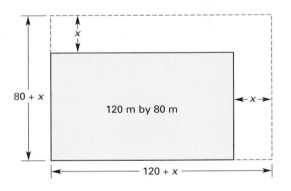

14 Write an expression for the area of the new field in terms of x.

15 Equate this expression to the area of the new field; that is, the original area plus 20% of the original area. Combine and simplify terms to obtain a quadratic equation in the form $ax^2 + bx + c = 0$.

16 Solve the quadratic equation for x. How much wider and longer should the farmer plow the field to obtain a 20% increase in area?

An airplane drops feed bags to horses wintering in a mountain pasture. When dropped from a plane at 1000 feet above the horses, the approximate height h of the feed bags after t seconds is modeled by the quadratic equation $h = -16t^2 + 1000$.

17 How long will it take the feed bags to reach the ground? Give the time to the nearest tenth of a second.

18 Suppose the plane takes a second pass at 1500 feet. Write the equation to model the height h and the time t of the feed bags dropped from this altitude. How long will it take the feed bags to reach the ground?

BUSINESS & MARKETING

You have been instructed to get quotes on the cost for a billboard advertisement. One company determines the total cost from a formula based on an initial setup fee and the height (h) and width (w) of the sign. The company's specifications assume the width of the sign is 5 feet plus twice the height.

Total cost =
    Setup fee + (Cost per square foot) (Area of sign)

19 For the type of sign you need, the setup fee is $150 and the cost per square foot is $10. Substitute these values and the width specifications into the cost equation to obtain a quadratic equation in h, the height.

20 Your budget allows a total cost of no more than $4000 for the billboard. Set the Total cost equation equal to $4000 and solve the quadratic equation for the height (h) of the sign you can afford.

A manufacturing company uses the function $y = 81x - 81x^2$ to model the profit on one of its products. In the model, x represents the number of items in thousands.

21 What is the maximum profit the company can make on this product?

22 How many items must the company produce to make this profit?

23 What are the break-even points for production?

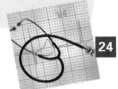

HEALTH OCCUPATIONS

The recommended dosage of a certain type of medicine is determined by the patient's body weight. The formula to determine the dosage is

$$D = 0.1w^2 + 5w$$

where D is the dosage in milligrams (mg) and w is the patient's body weight in kilograms.

24 What is the maximum weight of a patient that you can treat with 1800 mg of this medicine?

A report on river contamination downstream from a waste treatment plant contains a graph, as shown below. The graph shows the amount of contaminant in parts per million (ppm), measured at various distances downstream from the plant. The distances are in units of thousands of feet. (Thus, 5 represents 5000 feet, 10 represents 10,000 feet, and so on.)

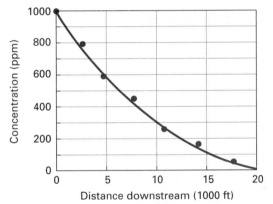

25 Which of the following quadratic equations best describes the trend shown by the data? In the equations, d represents the downstream distance in units of 1000 feet and c represents the concentration in parts per million.

a. $c = 2d^2 - 90d + 1000$
b. $c = -1000d^2 + 1000$
c. $c = d^2 + 20d - 1000$
d. $c = d^2 + 1000$

26 Use the equation that you selected to predict how far downstream the concentration drops to 10 ppm.

As a safety engineer for a chemical manufacturing company, you are designing a chemical waste holding area. It is located on a rectangular lot that is 200 yards long and 75 yards wide. The holding area must be 10,000 square yards. There must be a "safety zone" of uniform width around the perimeter of the holding area.

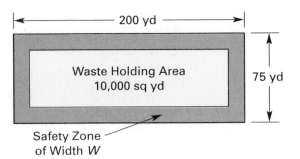

200 yd

Waste Holding Area
10,000 sq yd

75 yd

Safety Zone
of Width *W*

27 Let *W* represent the width of the safety zone. Write expressions for the length and the width of the interior holding area in terms of *W*.

28 Write an equation for the area of the interior holding region, using the specified area of 10,000 square yards and the product of the length and the width from Exercise 27.

29 Multiply the length and width terms and write a quadratic equation in the form $ax^2 + bx + c = 0$.

30 Use the quadratic formula to determine the width of the safety zone that makes your equation true.

FAMILY AND CONSUMER SCIENCE

The boiling point of water is affected by altitude. If you are boiling water at an altitude of *h* feet above sea level, it will not boil at exactly 212°F, but a few degrees lower. Since the water is not quite as hot, you may have to boil vegetables longer to completely cook them. The following equation approximates the number of degrees Fahrenheit below 212°F (the drop *D*) at which water will boil at various altitudes above sea level:

$$D^2 + 520\,D = h$$

31 Substitute 10,000 feet for *h* and rewrite the equation above in the standard form for a quadratic equation. For an altitude of 10,000 feet above sea level, about how much will the boiling point of water drop below 212°F?

32 What is the approximate boiling point at the altitude of your school?

You have just won a contract to select a wall hanging for the lobby of a new theater. The wall is 32 feet long, and the ceiling is 20 feet high. The theater owners want the wall hanging no closer than 2 feet from the ceiling, the floor, and each side of the wall. They also want the hanging to have the length and height dimensions of the ratio of the Golden Rectangle.

The Golden Rectangle has dimensions of length (L) and height (H) that satisfy the following formula:

$$\frac{L + H}{L} = \frac{L}{H}$$

33 Cross-multiply the terms in the proportion to obtain a quadratic equation. Substitute the values for either the maximum vertical distance available or the maximum horizontal distance available. Organize the resulting equation into the form $ax^2 + bx + c = 0$, where x represents the unknown dimension of the wall hanging.

34 Solve the quadratic equation for the unknown dimension. Determine whether there is enough room for a wall hanging that satisfies the ratio of the Golden Rectangle.

You are designing the layout for a square quilt: a patchwork of 3-inch squares with a solid piece as a backing. You need to determine how many squares will fit onto the backing material, which is 60 inches by 60 inches, or 3600 square inches.

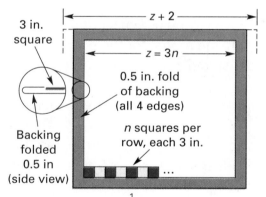

3 in. square

$z + 2$

$z = 3n$

0.5 in. fold of backing (all 4 edges)

n squares per row, each 3 in.

Backing folded 0.5 in (side view)

The backing will be folded to make a $\frac{1}{2}$-inch border on each of the four edges of the quilt (see the enlarged side view in the illustration). This will use 1 inch of material for each edge. To obtain a quilt with rows of squares that total z inches long, you must allow for a backing that is $z + 2$ inches long (see the illustration). Each row of the quilt is made of 3-inch squares so that the length of a row is $3n$ inches, where n is the number of squares. Thus, $z = 3n$.

35 The area of the backing equals its length times its width. Write an equation for the area of the backing in terms of *z*, using the information and the drawing. (*Hint:* Remember that the final quilt will be square.)

36 The length *z* of a finished row of squares is a multiple of the 3-inch size of each square; that is, $z = 3n$. Substitute this expression for *z* into your equation from Exercise 35.

37 Write your equation in the standard form for a quadratic equation. Solve for *n*.

38 Based on the amount of backing material you have, what is the maximum number of 3-inch squares per row that you can use? (*Hint:* Remember that *n* must be a whole number.)

39 What is the length and the width of the backing material you will use?

INDUSTRIAL TECHNOLOGY

A BASIC computer program is used to help locate the roots of a quadratic equation. The program calculates *y*-values for a range of *x*-values that is input by the user. The program prints the values as ordered pairs. Use the computer output shown to answer each of the following.

40 What range of *x*-values was input by the user?

41 How would you use this output to identify the root(s) for the equation under study?

42 Estimate the root(s) of the equation.

43 How could you modify the program to increase the accuracy of your estimate?

44 Since the equation is quadratic, it has a minimum or a maximum *y*-value. Which one does the output show: a maximum or a minimum? Estimate the maximum (or minimum).

X		Y	X		Y
(−2	,	1.331)	(0	,	−1.969)
(−1.9	,	.976)	(.1	,	−1.924)
(−1.8	,	.641)	(.2	,	−1.859)
(−1.7	,	.326)	(.3	,	−1.774)
(−1.6	,	.031)	(.4	,	−1.669)
(−1.5	,	−.244)	(.5	,	−1.544)
(−1.4	,	−.499)	(.6	,	−1.399)
(−1.3	,	−.734)	(.7	,	−1.234)
(−1.2	,	−.949)	(.8	,	−1.049)
(−1.1	,	−1.144)	(.9	,	−.844)
(−1.	,	−1.319)	(1	,	−.619)
(−.9	,	−1.474)	(1.1	,	−.374)
(−.8	,	−1.609)	(1.2	,	−.109)
(−.7	,	−1.724)	(1.3	,	.176)
(−.6	,	−1.819)	(1.4	,	.481)
(−.5	,	−1.894)	(1.5	,	.806)
(−.4	,	−1.949)	(1.6	,	1.151)
(−.3	,	−1.984)	(1.7	,	1.516)
(−.2	,	−1.999)	(1.8	,	1.901)
(−.1	,	−1.994)	(1.9	,	2.306)
			(2	,	2.731)

In an electric circuit, the power dissipated in a load is given by the following equation.

$$P = VI - I^2R$$

P is the power used by the load in watts.
V is the voltage supplied in volts.
R is the resistance of the circuit external to the load in ohms.
I is the circuit current in amperes.

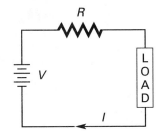

45 In a certain circuit, the voltage supplied is 110 volts, and the resistance of the circuit external to the load is 30 ohms. What is the current if the load uses 60 watts of power?

You need to construct a rain gutter from a rectangular piece of sheet metal. The sheet metal is 20 inches wide. To form the gutter, you will fold up two sides of equal width to form right angles, as shown. For proper drainage, the gutter needs to have a cross-sectional area of 38 in².

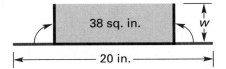

38 sq. in.   w
20 in.

46 Write an expression for the width of the gutter, in terms of the height x of the gutter's sides.

47 Write an equation for the cross-sectional area of the gutter, in terms of the width from Exercise 46 and the height x. Your equation should satisfy the requirements for a cross-sectional area of 38 square inches. Rewrite the equation in the standard form for a quadratic equation.

48 Use the quadratic formula or the method of completing the square to determine the values of the height of the gutter.

49 Sketch a scale drawing that shows the two solutions obtained in Exercise 48. Which solution would you choose for the gutter height?

You are designing an archway that has a radius of 25 feet and spans a distance of 40 feet. The formula below gives the relationship between the center height (h) of the arch, the radius of curvature (R) of the arch, and the distance (s) spanned by the arch.

$$s^2 = 8Rh - 4h^2$$

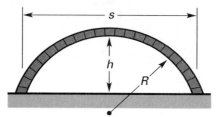

50 Substitute the given values to obtain a quadratic equation for the center height h.

51 Solve the quadratic equation. Which value of h would you use for the center height of your arch? Why did you choose this value?

You are selecting a piece of angle iron for a construction project. The weight of the angle iron will exceed specifications if it has a cross-sectional area greater than 60 square centimeters. You need a vertical height of 16 cm and a horizontal width of 10 cm. Find the thickness of the stock x that satisfies these conditions.

52 The dotted line divides the angle iron's cross section into two rectangles, each with a width of x. Write an expression for the length of each rectangle. Multiply length and width to write equations for the area of each rectangle.

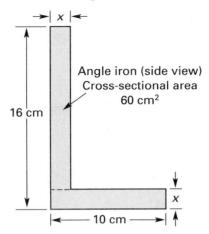

Angle iron (side view)
Cross-sectional area
60 cm²

16 cm

10 cm

53 Add the two areas and write an equation for the total cross-sectional area of the angle iron. Set this equation equal to the maximum area of 60 square centimeters.

54 Rewrite the equation in standard form and solve for x. Which value of x would you choose for the thickness of the angle iron?

A cylindrical tank is placed along a wall, as shown in the side-view drawing below. The tank has a diameter ($2R$) of 84 inches. You need to run a pipe along the wall between the wall and the tank. You need to determine the radius r of the largest pipe that can fit in the space left by the tank.

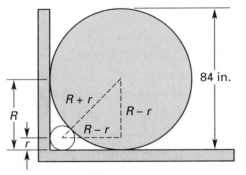

84 in.

$R + r$

$R - r$

$R - r$

R

r

55 A right triangle is shown by dotted lines. The hypotenuse of the triangle is formed by the line that joins the centers of the two circular cross sections and has a dimension of $R + r$. Each side of the triangle is equal to the radius of the tank (R) less the radius of the pipe (r), or $R - r$. Use the Pythagorean Theorem ($c^2 = a^2 + b^2$) with these dimensions to obtain a quadratic equation for r and R. Substitute for R and write the equation in standard form.

56 Solve the quadratic equation for r to find the largest pipe that will fit between the tank and the wall.

You want to increase the weight of an iron counterweight by 20% by bolting two iron plates (of the same type of iron) along the top and the side. The weight of the iron stock is proportional to its cross-sectional area. Find the thickness x of iron plate stock needed to increase the overall cross-sectional area by 20%.

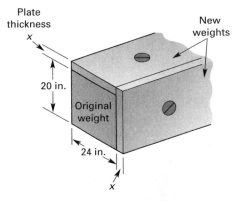

Plate thickness x

New weights

20 in.

Original weight

24 in.

x

57 The original counterweight is 20 inches by 24 inches and its cross-sectional area is 480 square inches. If you increase this area by 20%, what is the new cross-sectional area?

58 The new cross-sectional area will be the product of the new height and the new width. Refer to the drawing. Write an expression for the new area in terms of the plate thickness x. Equate this expression to the larger area from Exercise 57.

59 Rewrite the equation for the new area in standard form and solve for the plate thickness x.

Most three-way light bulbs have only two filaments. For example, one filament may be 50 watts, and the second filament may be 100 watts. The bulb produces three levels of brightness by using the first filament (50 watts), the second filament (100 watts), or both filaments together (150 watts). However, your eye will not see equal ratios of brightness for these three power levels. Switching from the first to the second level yields a brightness ratio of 100:50, or 2:1. However, going from the second to the third level yields a ratio of 150:100, or 3:2.

Suppose you are asked to design a three-way bulb that has a 50-watt element as its first filament and yields equal ratios for 2nd level:1st level and 3rd level:2nd level.

Let x be the wattage of the second filament. The wattage of the third level will be $x + 50$. You want the following relationship:

$$\frac{\text{2nd level}}{\text{1st level}} = \frac{\text{3rd level}}{\text{2nd level}}$$

or

$$\frac{x}{50} = \frac{x + 50}{x}$$

60 Find a quadratic equation for x by cross multiplying the terms in this proportion.

61 Use the quadratic formula to determine the wattage x for the second filament.

62 Check the results from Exercise 61. What is the ratio of the second-to-first-filament wattages? What is the ratio of the third-to-second-filament wattages? Are the ratios equal?

63 Suppose the first filament is 40 watts. Find the wattage of the second filament.

64 Now let the first filament be any power level W. Show that as long as the ratios of 2nd level:1st level and 3rd level:2nd level are equal, the ratios are equal to 1.62.

CHAPTER 11 ASSESSMENT

Skills

Graph each function. Give the vertex of the graph.

1. $y = x^2 + 3$

2. $y = (x - 1)^2 + 3$

Let $f(x) = x^2 - 3x + 2.$

3. Find the zeros of the function by factoring.

4. Write the function in parabola form.

5. Give the coordinates of the vertex.

6. Write the equation for the axis of symmetry.

Solve each equation using the quadratic formula.

7. $6x^2 + 16x + 9 = 0$

8. $-20x^2 + 9x - 1 = 0$

9. Use the discriminant to find the number of real roots of $y = x^2 + 5x + 8.$

Applications

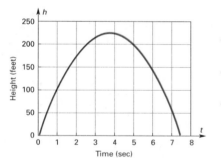

The parabola describes a ball's height versus time of flight.

10. Estimate the maximum height of the ball.

11. Estimate the time the ball stays in the air.

12. The sum of the first n whole numbers, $1 + 2 + 3 + \cdots + n,$ is determined by the formula

$$\text{Sum} = \frac{n^2 + n}{2}$$

Suppose the sum is 2628. Find the value of n.

Your garden measures 16 feet by 48 feet. You want to reduce its area of 768 square feet to 585 square feet. However, you need to reduce the length and the width by the same number of feet.

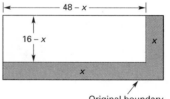

Original boundary of garden

13. Write a quadratic equation to model the situation.

14. By how many feet will you reduce the length and the width of the original garden?

15. The formula that relates the center height (*h*) of an arch, the radius of curvature (*R*), and the distance spanned (*s*) is

$$s^2 = 8Rh - 4h^2$$

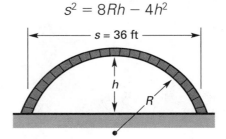

You are designing an archway to span a 36-foot distance. The arch must have a radius of curvature of 24 feet. Find the center height to the nearest tenth of a foot.

Math Lab

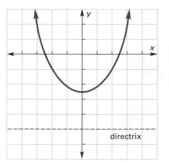

16. What are the coordinates of the focus of the parabola shown at the left?

17. Write an equation for the cross-sectional area of a gutter constructed from a piece of sheet metal 12 inches wide with sides that are equal to each other. (*Hint:* Sketch the gutter cross section.)

18. You are using a computer program to solve a quadratic in the form $y = ax^2 + bx + c$. You input upper and lower values for *x*, and the program calculates values for *y*. When you run the program, it prints the following:

x	*y*
5.0	−2.33
5.1	−1.55
5.2	−0.27
5.3	0.42
5.4	1.84
5.5	2.96

Estimate one of the roots of the quadratic.

CHAPTER 12

WHY SHOULD I LEARN THIS?

Describing geometric figures with algebra is an essential workplace skill. Architects use algebra to design the shape of a sloped roof or determine the spacing between rafters. Carpenters transform architectural designs into buildings by using several tools including algebra. If designing or building ever become part of your career goals, you will use the concepts in this chapter.

RIGHT TRIANGLE RELATIONSHIPS

OBJECTIVES

1. Name the parts of a right triangle.
2. Use the Pythagorean Theorem to find a side of a right triangle.
3. Use the characteristics of 3-4-5, 45°-45°, and 30°-60° right triangles to solve problems.
4. Use the ratios for the sine, cosine, and tangent of an angle to solve problems that involve right triangles.
5. Add, subtract, and simplify expressions containing radicals.

Look around you. How many triangles can you see? Triangles are everywhere. Architects use triangles to brace corners, add strength to buildings, and support signs. Triangles are used to construct homes, offices, dams, and towers. They are also used in navigation to help airplanes and ships find their exact positions.

People have known for a long time that triangles are important and useful. Long before books were printed or motors invented, people knew about the mathematics of triangles. The Greeks, more than 2500 years ago, worked out most of the mathematics that use properties of triangles to solve problems. You will use these properties to solve many workplace problems.

In this chapter, you will study several properties of right triangles. As you watch the video, notice how people use right triangles on the job.

Before you begin working with triangles, review this list of general properties of triangles. They will be used throughout this chapter.

General Properties of Triangles

a. Triangles have 3 sides and 3 angles.

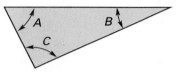

b. $m\angle A + m\angle B + m\angle C = 180°$

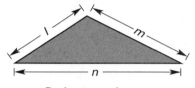

c. Perimeter = $l + m + n$

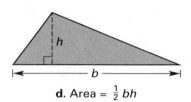

d. Area = $\frac{1}{2}bh$

• All triangles have three sides and three angles. The three sides lie in the same plane.

• The sum of the measures of the three angles of any triangle is equal to 180°.

• The perimeter of a triangle is equal to the sum of the lengths of the three sides.

• The area of a triangle is equal to the product of one-half the base times the height.

Triangles are labeled in a special way. Examine the information below to refresh your memory.

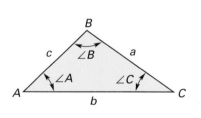

• The three vertices of a triangle are often labeled with capital letters, such as A, B, and C.
• The triangle is referred to as $\triangle ABC$.
• The sides opposite the angles are often labeled with lowercase letters, such as a, b, and c. Note that a = length of BC, b = length of AC, and c = length of AB.
• The three angles of the triangle are often labeled with the appropriate vertex letter, using the symbol $\angle$ for angle, such as $\angle A$, $\angle B$, and $\angle C$.

A right triangle has one **right angle** with a measure of 90°. The sides that form the right angle are perpendicular to each other.

A right triangle has one angle equal to 90°.

You can divide any triangle into two right triangles. For example, Triangle *ABC*, written Δ*ABC*, is *not* a right triangle. But by drawing a perpendicular line segment from vertex *B* to point *M* on side *AC*, you can form two right triangles, Δ*ABM* and Δ*CBM*.

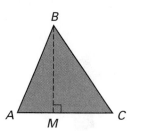

This illustration labels the parts of a right triangle.

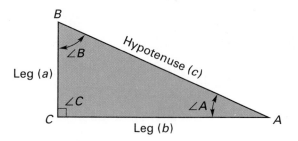

The following list summarizes the important facts about a right triangle.

1. A right triangle has one right angle. Angle *C* is the right angle in triangle *ACB*. The measure of angle *C* is 90 degrees. This is written

$$m\angle C = 90°$$

2. The side opposite the right angle is the longest side and is called the **hypotenuse.** The hypotenuse is labeled as *c*.

3. The two sides that meet at right angles are called the **legs** of the right triangle. The length of the leg opposite *A* is labeled *a*, and the length of the leg opposite *B* is labeled *b*.

Similar Right Triangles

ACTIVITY | **Similar Right Triangles**

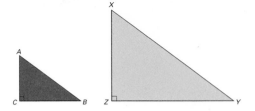

1 Use a protractor to measure each angle to the nearest whole degree in the two right triangles. Complete the following information.

m∠C = 90°	m∠Z = 90°
m∠A = ?	m∠X = ?
m∠B = ?	m∠Y = ?

2 What is the relationship between m∠A and m∠X? m∠B and m∠Y?

3 Use a ruler to measure the length of each side of the two triangles. Copy and complete this information.

length of AB = ?	length of XY = ?
length of AC = ?	length of XZ = ?
length of BC = ?	length of YZ = ?

4 Write each ratio in decimal form.

$$\frac{AB}{XY} \qquad\qquad \frac{AC}{XZ} \qquad\qquad \frac{BC}{YZ}$$

5 Summarize your findings.

You found in Step 1 that the three angles of $\triangle ABC$ are equal respectively to the three angles of $\triangle XYZ$. Therefore, $\triangle ABC$ is *similar* to $\triangle XYZ$. The three pairs of equal angles are called **corresponding angles.** The sides opposite the corresponding angles are called **corresponding sides.** You found in Step 4 that the ratios of corresponding sides of similar triangles are equal.

Similar Triangles

Two triangles are similar if they have corresponding angles with the same measure.

If two triangles are similar, then the lengths of their corresponding sides are proportional.

Similar triangles have the same shape but not necessarily the same size. Model builders use the properties of similar triangles to build scale models.

EXAMPLE 1 Finding a Missing Side

$\triangle MNO$ is similar to $\triangle QRS$.

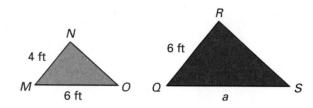

Find the length of side QS.

SOLUTION

Since the two triangles are similar, the lengths of the corresponding sides are proportional. Write a proportion and solve.

$$\frac{MN}{QR} = \frac{MO}{QS} \qquad \text{Similar Triangle Proportion}$$

$$\frac{4}{6} = \frac{6}{a} \qquad \text{Substitution}$$

$$4a = 36 \qquad \text{Proportion Property}$$

$$a = 9 \qquad \text{Division Property of Equality}$$

The length of QS is 9 feet.

Ongoing Assessment

For $\triangle MNO$ and $\triangle QRS$, change the lengths to the following: Let the length of RS equal 15 inches, the length of MO equal 6 inches, and the length of NO equal 9 inches. Find the length of QS.

EXAMPLE 2 Indirect Measurement

How can a surveyor find the distance across the lake from point *A* to point *D*?

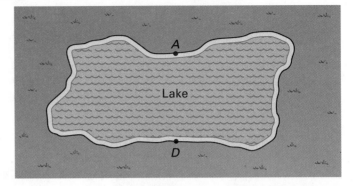

Lake

A

D

SOLUTION

The surveyor can draw two right triangles, △*ABC* and △*AED*. The triangles share a common angle, ∠*A*. Both ∠*C* and ∠*D* are 90°. Thus, m∠*B* = m∠*E*. (Can you show why?) Since △*ABC* and △*AED* have three equal angles, they are similar triangles.

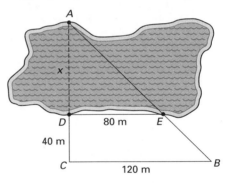

A

x

D   80 m   E

40 m

C   120 m   B

The surveyor can measure the distances *CB, DE,* and *CD* on land. Then the surveyor can use the similar triangle property to write the following ratios:

$$\frac{AD}{AC} = \frac{DE}{CB}$$ Similar Triangle Property

If the distance *AD* across the lake is represented by *x*, the ratios become

$$\frac{x}{x + 40} = \frac{80}{120}$$ Substitution

$$120x = 80x + 3200$$ Proportion Property; Distributive Property

$$40x = 3200$$ Subtraction Property

$$x = 80$$ Division Property

The distance across the lake is 80 meters.

LESSON ASSESSMENT

Think and Discuss

Use the diagram for Exercises 1 and 2.

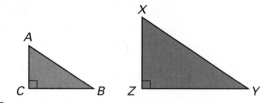

1 Suppose that angle *A* has the same measure as angle *X*. Explain why the two triangles are similar.

2 Suppose the length of *AC* is 6 inches, the length of *CB* is 9 inches, and the length of *XZ* is 10 inches. Explain how to find the length of *ZY*.

Practice and Problem Solving

Each pair of triangles is similar. Find the length of each missing side.

3.

4.

5.

6.

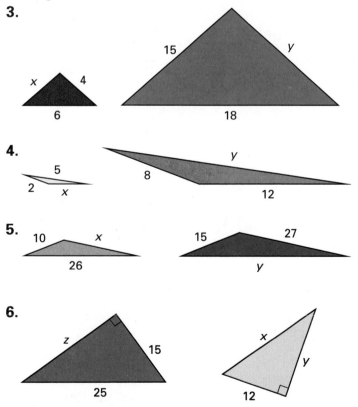

Solve each problem. Draw a diagram when needed.

7. The length of *DE* is 40 yards, the length of *CB* is 180 yards, and the length of *BE* is 50 yards. How far is it across the lake from point *C* to point *A*?

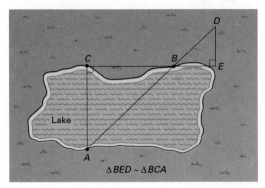

△*BED* ~ △*BCA*

8. Suppose the length of *DE* is 4 meters, the length of *AD* is 8 meters, and the length of *BC* is 6 meters. What is the length of *AC*?

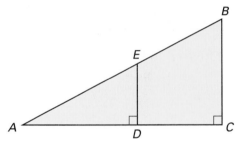

△*AED* ~ △*ABC*

9. The lengths of the three sides of a triangle are 10 inches, 15 inches, and 20 inches. The longest side of a similar triangle is 25 inches. Find the lengths of the other two sides.

10. A tree casts a 25-foot shadow at the same time that an upright yardstick casts a 10-foot shadow. Use similar triangles to find the height of the tree.

11. The top of a 30-foot ladder touches the side of a building at a point 25 feet above the ground. A 12-foot ladder is placed against the building at the same angle. How far from the ground is the top of the 12-foot ladder?

For Exercises 12–15, use the following ordered pairs:
$A\,(-3,1)$ $B\,(1,1)$ $C\,(1,-2)$ $D\,(-3,-2)$

12. Graph the four points on the same coordinate plane.

13. Find the slope of a line containing AD and a line containing BC.

14. Write the equation of a line containing AB.

15. Write the equation of a line containing CD.

Graph the solution to each inequality.
16. $4x + 3 > -15$ and $2x < 6$

17. $2x - 1 \leq 5$ or $-x > -3$

18. $|2x + 1| \leq 9$

Write and solve an inequality for each statement.
19. The length of a missile is ten times the length of its nose cone. The difference in length between the missile and its nose cone is at least 16 feet. What is the length of the missile?

20. The sum of two acute angles in a triangle is less than or equal to 90°. One angle is twice the measure of the other. What is the greatest possible measure of the smaller angle?

Recall the special relationship between the sides of a right triangle that you learned in Chapter 6.

In a right triangle, the square of the length of the hypotenuse equals the sum of the squares of the lengths of the two legs. Although many civilizations knew and used this fact, it is now known as the **Pythagorean Theorem.** The Pythagorean Theorem states that for any right triangle:

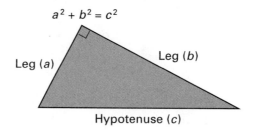

$$a^2 + b^2 = c^2$$

Leg (a)   Leg (b)

Hypotenuse (c)

ACTIVITY 1 Pythagorean Triples

A set of three positive integers that satisfy $a^2 + b^2 = c^2$ is called a **Pythagorean triple.** For example, the set 3, 4, and 5 is a Pythagorean triple because

$$3^2 + 4^2 = 5^2$$

That is, $9 + 16 = 25$.

Complete each set of integers to form a Pythagorean triple.

		a	b	c
1	Triangle One	7	24	?
2	Triangle Two	8	15	?
3	Triangle Three	?	63	65
4	Triangle Four	28	?	53

The Sides of a Right Triangle

In Chapter 6, the Pythagorean Theorem provided a way to find the length of a leg or the hypotenuse of a right triangle. The Pythagorean Theorem is also a tool for determining if a triangle is a right triangle. If you measure the three sides of a triangle and find that the sum of the squares of two sides is equal to the square of the longest side, then the triangle is a right triangle.

EXAMPLE 1 Checking for a Right Triangle

A 12-foot antenna pole is held in place by two guy wires of unequal length. One end of each guy wire is connected to the top of the pole. The shorter guy wire is 15 feet long. It is fastened to the ground 9 feet from the base of the pole. Does the pole form a right angle with the ground?

SOLUTION

First, sketch the situation to see the position of the pole and the wires.

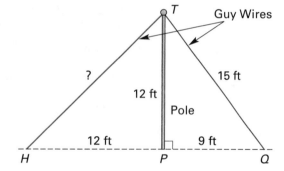

Label the triangle and its sides. Substitute the values in the formula $a^2 + b^2 = c^2$. Does 15^2 equal $12^2 + 9^2$?

Since $225 = 144 + 81$, $\triangle TPQ$ is a right triangle, and the pole forms a right angle with the ground.

Ongoing Assessment

Look at Example 1 again. Suppose you fasten the second guy wire to the ground at point H, 12 feet from the base of the pole. Find the length of the guy wire to the nearest foot.

An electrical circuit wired in series contains an alternating voltage source, a resistor, and a coil. The voltage across the resistor (V_R) and the voltage across the coil (V_L) are related to the voltage source (V_S) by the following formula:

$$(V_R)^2 + (V_L)^2 = (V_S)^2$$

Notice that this relationship is the same as the relationship between the sides of a right triangle. Because of this, these voltages are often represented in a voltage triangle.

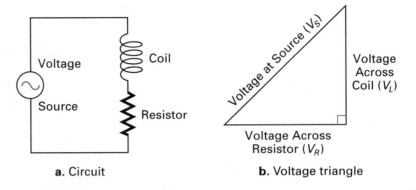

a. Circuit   b. Voltage triangle

EXAMPLE 2 An Electrical Circuit in Series

In an electrical circuit such as the one above, the voltage source is 185 volts. The voltage across the resistor is 76 volts. Find the voltage across the coil to the nearest tenth.

SOLUTION

First, sketch a voltage triangle to help you see the situation.

Next, label the hypotenuse and legs of the triangle.

Substitute the values in the Pythagorean Theorem and solve for the unknown voltage.

$$(V_R)^2 + (V_L)^2 = (V_S)^2$$

$$(76)^2 + (V_L)^2 = (185)^2$$

$$(V_L)^2 = (185)^2 - (76)^2$$

$$(V_L)^2 = 34,225 - 5776$$

$$(V_L)^2 = 28,449$$

$$(V_L) = \sqrt{28,449}$$

$$(V_L) \approx 168.7$$

The voltage across the coil is about 168.7 volts.

Ongoing Assessment

Firefighters in training practice a rescue operation from a building. Their ladder is 20 feet long, and they place the foot of the ladder 6 feet from the base of the building. How high up the wall will the top of the ladder reach?

The Distance Formula

ACTIVITY 2   Distance on the Coordinate Plane

1 In a coordinate plane, draw a line segment from A (3,2) to B (6,4).

2 Graph point C (6, 2). Draw $\triangle ACB$. Explain why $\triangle ACB$ is a right triangle.

3 Find the length of AC and BC.

4 Use the Pythagorean Theorem to find AB to the nearest tenth.

5 Explain how to use the Pythagorean Theorem to find the distance between M (2, 3) and N (5,7).

This Activity demonstrates that you can use the Pythagorean Theorem to find the distance between any two points on the coordinate plane.

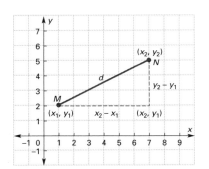

The Distance Formula

Let M (x_1, y_1) and N (x_2, y_2) be any two points on the Cartesian coordinate plane. The distance (d) between the two points M and N is

$$d = \sqrt{(x_2 - x_1)^2 + (y_2 - y_1)^2}$$

Critical Thinking Explain why you can choose either point M or N for (x_1, y_1) in the distance formula.

EXAMPLE 3 Applying the Distance Formula

Two ships are several miles apart. A ship's navigator plots their locations using coordinates. One ship is located at P (30,50). The other ship is located at Q (105,150). What is the distance between the ships?

SOLUTION

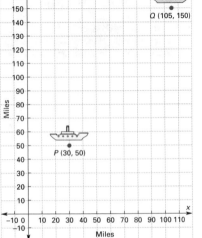

Draw a diagram. Use the distance formula.

$$d = \sqrt{(105 - 30)^2 + (150 - 50)^2}$$

$$d = \sqrt{(75)^2 + (100)^2}$$

$$= \sqrt{15,625}$$

$$= 125$$

The ships are 125 miles apart.

LESSON ASSESSMENT

Think and Discuss

1 The lengths of the three sides of a triangle are represented by a, b, and c. Under what condition is the triangle a right triangle?

2 Explain how to find the distance between two points on a coordinate plane.

Practice and Problem Solving

Each set of integers represents the lengths of the sides of a triangle. Use the Pythagorean Theorem to determine whether the triangle is a right triangle. Answer *yes* or *no*.

3. 5, 12, 15   **4.** 24, 7, 25   **5.** 17, 15, 8

6. 12, 37, 35   **7.** 81, 82, 18   **8.** 30, 16, 34

Use a diagram to help solve each problem. When necessary, round the answer to the nearest tenth.

9. A rectangular swimming pool is 9 yards wide and 20 yards long. A person swims diagonally across the pool from one corner to the other. How far does the person swim?

10. A lawn is in the shape of a rectangle. A water pipe 45 feet long is buried diagonally across the lawn. The length of the lawn is 25 feet. How wide is the lawn?

11. Two telephone poles are 50 feet apart. The height of one pole is 50 feet. The height of the other pole is 55 feet. What is the minimum length of wire that can be strung between the tops of the two poles?

12. A carpenter is building a rectangular gate 4 feet by 6 feet. She needs to cut a metal rod to brace the gate diagonally. How long should she cut the brace?

Find the distance between each pair of points.

13. A $(4, -1)$ and B $(7,3)$ **14.** C $(2, -3)$ and D $(-1,3)$

15. M $(-5,9)$ and N $(3, -6)$ **16.** X $(0,1)$ and Y $(-1,2)$

Three ships are located at the following points: A $(4, -1)$; B $(4,11)$; and C $(8, -1)$.

17. Prove the ships are at the corners of a right triangle.

18. Find the perimeter of the triangle.

Mixed Review

A flooring contractor is tiling a kitchen floor that is 9 feet by 12 feet. The tile costs $2 per square foot. The tile cement costs $4.50 for a gallon can and covers 10 square yards.

19. What is the cost of the kitchen floor tile?

20. How many square yards of cement are needed?

21. How many gallon cans of cement must be purchased? What is the cost of the cement?

22. Write a formula that can be used to find the total cost of the tile and cement for a floor x feet by y feet.

LESSON 12.3 IRRATIONAL NUMBERS AND RADICALS

Solving problems with right triangles requires skill in working with **radicals.** Often, when you evaluate radicals, the result is an **irrational number.** For example, $\sqrt{7}$ is an irrational number. You cannot write $\sqrt{7}$ in the form $\frac{a}{b}$ where $b \neq 0$ and a and b are integers.

Recall that the symbol $\sqrt{\ }$ is a radical sign. The expression under the radical sign is called the **radicand.** The expression $\sqrt{7}$ is read "radical seven."

Simplifying Radicals

All Pythagorean triples are integers. However, not all numbers that make the Pythagorean Theorem true are integers. For example, the right triangle in this illustration has two legs that measure 2 centimeters on a side. What is the length of the hypotenuse?

If you use the Pythagorean Theorem, you will find the following:

$$c^2 = (2)^2 + (2)^2$$
$$c^2 = 8$$
$$c = \pm \sqrt{8}$$

Since length is always a positive number, the hypotenuse is $\sqrt{8}$ centimeters.

The number $\sqrt{8}$ is an irrational number. You cannot write $\sqrt{8}$ as a decimal or a fraction. However, you can use your calculator to approximate $\sqrt{8}$.

Rounded to the nearest thousandth, $\sqrt{8}$ is 2.828.

You have seen that a Pythagorean triple cannot represent a right triangle with legs 2 centimeters in length. In fact, the lengths of the sides of most triangles cannot be described by a combination of three integers. At least one of the sides is usually an irrational number. Thus, most of the work you do with right triangles will involve irrational numbers written as radicals.

Use your calculator to approximate each expression to the nearest thousandth.

	A	B	C	D
1	$\sqrt{64}$	$\sqrt{16 \cdot 4}$	$\sqrt{16} \cdot \sqrt{4}$	$4 \cdot 2$
2	$\sqrt{32}$	$\sqrt{16 \cdot 2}$	$\sqrt{16} \cdot \sqrt{2}$	$4\sqrt{2}$
3	$\sqrt{125}$	$\sqrt{25 \cdot 5}$	$\sqrt{25} \cdot \sqrt{5}$	$5\sqrt{5}$
4	$\sqrt{27}$	$\sqrt{9 \cdot 3}$	$\sqrt{9} \cdot \sqrt{3}$	$3\sqrt{3}$

5 Examine the relationship between column B and column D. What is the relationship between the square root of a product of two numbers and the product of their square roots?

This activity demonstrates two important properties of radicals that simplify your work with irrational numbers.

Calculating with Radicals

If a and b are any two numbers,

$$\sqrt{a} \cdot \sqrt{b} = \sqrt{ab}$$

If a and b are any two numbers with $b \neq 0$,

$$\sqrt{\frac{a}{b}} = \frac{\sqrt{a}}{\sqrt{b}}$$

A radical is in **simplest form** when the radicand does not contain a factor (other than one) whose square root is a whole number. For example, $\sqrt{32}$ is not in simplest form because one factor of 32 is 16 and the square root of 16 is 4. However, $\sqrt{32}$ can be written as $4\sqrt{2}$. The expression $4\sqrt{2}$ is in simplest form because $\sqrt{2}$ does not contain a factor whose square root is a whole number.

EXAMPLE 1 Multiplying Radicals

Write each of the following in simplest form.

a. $\sqrt{6} \cdot \sqrt{8}$ **b.** $\sqrt{8} \div \sqrt{2}$

SOLUTION

a. $\sqrt{6} \cdot \sqrt{8} = \sqrt{48}$ **b.** $\sqrt{8} \div \sqrt{2} = \sqrt{4}$
$$= \sqrt{16} \cdot \sqrt{3} \qquad\qquad\qquad = 2$$
$$= 4\sqrt{3}$$

Write each of the following in simplest form.

a. $\sqrt{12} \cdot \sqrt{6}$ **b.** $\sqrt{50} \div \sqrt{2}$

When dividing radicals, it is customary to write the denominator without a radical. For example, to eliminate the radical in the denominator of $\frac{3}{\sqrt{6}}$, multiply $\frac{3}{\sqrt{6}}$ by 1, in the form $\frac{\sqrt{6}}{\sqrt{6}}$.

The result is

$$\frac{3}{\sqrt{6}} \cdot \frac{\sqrt{6}}{\sqrt{6}} = \frac{3\sqrt{6}}{\sqrt{36}}$$

$$= \frac{3\sqrt{6}}{6}$$

$$= \frac{\sqrt{6}}{2}$$

This process is called **rationalizing** the denominator.

CULTURAL CONNECTION

The mathematicians living in Alexandria, Egypt, about 75 BCE discovered a way to find the area (A) of any triangle given only the length of each of the sides. Represent the lengths of the sides of a triangle by a, b, and c. Let s equal $\frac{1}{2}$ the perimeter of the triangle; that is, $s = \frac{1}{2}(a + b + c)$.

The following area formula is named for Heron of Alexandria.

$$A = \sqrt{s(s - a)(s - b)(s - c)}$$

To find the area of a triangle with sides of 6 inches, 9 inches, and 9 inches, first find s.

$$s = \frac{1}{2}(6 + 9 + 9) = 12$$

Substitute the numerical values into Heron's Formula.

$$A = \sqrt{12(12 - 6)(12 - 9)(12 - 9)}$$

$$= \sqrt{(12)(6)(3)(3)}$$

$$= \sqrt{36 \cdot 9 \cdot 2}$$

$$= \sqrt{36} \cdot \sqrt{9} \cdot \sqrt{2}$$

$$= 6 \cdot 3 \cdot \sqrt{2}$$

$$= 18\sqrt{2}$$

The area of the triangle is $18\sqrt{2}$, or about 25.46 square inches.

1. Use Heron's Formula to find the area of a triangle with sides that measure 8 meters, 12 meters, and 10 meters.

2. Find the area of a triangle with sides that measure 8 inches, 12 inches, and 4 inches. Describe the shape of this triangle. Draw the triangle to check your thinking.

LESSON ASSESSMENT

Think and Discuss

1 How do you know when a radical is in its simplest form?

2 How do you rationalize a fraction that contains a radical in the denominator?

Practice and Problem Solving

Write in simplest form.

3. $\sqrt{12}$ **4.** $\sqrt{75}$ **5.** $\sqrt{200}$

6. $\sqrt{72}$ **7.** $\dfrac{4}{\sqrt{24}}$ **8.** $\sqrt{6} \cdot \sqrt{15}$

9. $\sqrt{90} \div \sqrt{5}$ **10.** $\sqrt{6}(\sqrt{6} + \sqrt{18})$

The formula for the area (A) of a circle with radius r is $A = \pi r^2$.

11. Solve the formula for r.

12. Write a radical expression for r in simplest form for a circle with an area of 72 square feet.

You can approximate the velocity (in feet per second) of a tidal wave in water of depth d (in feet) with the formula $V = \sqrt{15d}$.

13. Write a radical expression in simplest form for the velocity of a tidal wave in 400 feet of water.

14. Find the approximate velocity of the tidal wave.

Graph the following functions on the same coordinate plane.

15. $y = \sqrt{x}$ **16.** $y = \sqrt{2x}$ **17.** $y = \sqrt{\left(\frac{1}{2}\right)x}$

18. Draw the graph of $y = \sqrt{2x} + 2$. What affect does adding a positive number to $\sqrt{2x}$ have on the graph of $y = \sqrt{2x}$?

Write an equation in one unknown to model each situation. Solve the equation and check the solution.

19. The width of a rectangular swimming pool is 25 feet less than its length. The distance around the pool is 250 feet. Find the dimensions of the swimming pool.

20. Two guy wires are attached to a TV transmitter. The wires form a triangle with the ground as its base. The base angles are equal, and each base angle is four times the third angle. What is the measure of each base angle?

Write a system of equations to model each situation. Solve the system and check the solution.

21. Althea is installing an underground sprinkler system. She needs to cut a 24-foot water pipe into two sections. The longer pipe must be 6 feet longer than twice the length of the shorter one. How long are the two sections?

22. Keiko ordered $427.50 worth of books in July. She ordered $382.50 worth in August. Keiko ordered the same number of fiction books and the same number of nonfiction books each month, but the number of fiction books was not the same as the number of nonfiction books. How many of each did Keiko order?

| | July and August Book Prices for | |
	Fiction	Nonfiction
July	$8.50	$6.50
Aug	$7.50	$6.00

LESSON 12.4 SPECIAL RIGHT TRIANGLES

The 3-4-5 Ratio

Sometimes you can find the length of a side of a right triangle without directly using the Pythagorean Theorem. The ratios of the sides of three special right triangles can simplify your calculations.

ACTIVITY 1   **The 3-4-5 Right Triangle**

Four triangles with sides *a*, *b*, and *c* have the following lengths:

	a	*b*	*c*
Triangle 1	3	4	5
Triangle 2	6	8	10
Triangle 3	15	20	25
Triangle 4	18	24	30

1 Use the Pythagorean Theorem to show that each triangle is a right triangle.

2 Use the definition of similar triangles to show that each pair of triangles is similar.

3 For each triangle, find the greatest common factor *m* of *a*, *b*, and *c*.

4 Write the ratio of the lengths of each triangle as

$$a:b:c = x \cdot m : y \cdot m : z \cdot m$$

For example, since the greatest common factor for Triangle 1 is 1. The ratio is written:

$$3:4:5 = 3 \cdot 1 : 4 \cdot 1 : 5 \cdot 1$$

5 What do you notice about *x*, *y*, and *z* for each of these similar right triangles?

The triangles in Activity 1 are called **3-4-5 right triangles**.

3-4-5 Right Triangle

Every 3-4-5 right triangle has sides whose ratios are 3 to 4 to 5. Any triangle with sides of these ratios is a right triangle.

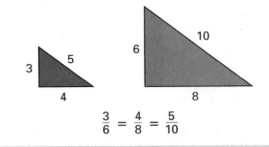

$$\frac{3}{6} = \frac{4}{8} = \frac{5}{10}$$

EXAMPLE 1 Using a 3-4-5 Right Triangle

A ship gathering specimens for marine biologists starts from an island and sails 100 miles due north. Then the ship turns and sails 75 miles due east. At that point, what is the shortest distance from the ship to the island?

SOLUTION

Draw a sketch to identify the right angle and the hypotenuse.

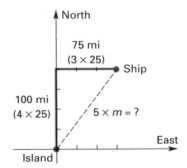

Notice that 100 and 75 have a greatest common factor of 25. Thus, you can write the ratio of the two legs as

$$\frac{75}{100} = \frac{3(25)}{4(25)}$$

Since the two legs of the right triangle have the ratio of 3 to 4, the three sides of the right triangle have the ratio of 3 to 4 to 5. To find the length of the hypotenuse, use the ratio and the greatest common factor, 25.

$$3{:}4{:}5 = 3 \bullet 25{:}4 \bullet 25{:}5 \bullet 25$$

The shortest distance back to the island is 125 miles.

To determine if a right triangle is a 3-4-5 right triangle, find the greatest common factor of the lengths of the sides. Then see if you can write the known sides as a ratio of

$$3 \cdot m{:}4 \cdot m{:}5 \cdot m$$

where m is the greatest common factor.

The side written as $5 \cdot m$ is always the longest side or hypotenuse.

The 45° - 45° Right Triangle

A triangle with two equal sides is called an **isosceles triangle.** In an isosceles triangle, the angles opposite the equal sides are also equal. If an isosceles triangle contains a right angle, it is called a **45°- 45° right triangle.**

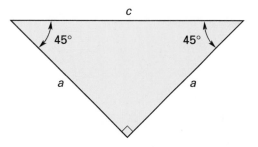

What happens when you apply the Pythagorean Theorem to the 45°- 45° right triangle? Since the two legs are equal in length,

$$a^2 + a^2 = c^2$$

$$2a^2 = c^2 \qquad \text{Combine like terms}$$

Solve the second equation for c.

$$c^2 = 2a^2$$

$$c = \sqrt{2}a \quad \text{or} \quad a\sqrt{2}$$

These steps show one unique characteristic of 45°- 45° right triangles.

45°- 45° Right Triangle

For every 45°- 45° right triangle, the length of the hypotenuse is $\sqrt{2}$ times the length of a leg.

This drawing shows the main characteristics of a 45°- 45° right triangle.

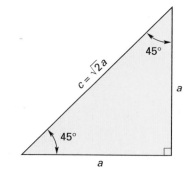

You can use a 45°- 45° right triangle to find the diagonal of a square.

Since the sides of a square are equal, the diagonal is the hypotenuse of a right triangle with two legs of equal length. Thus, the two right triangles formed by the diagonal are 45°- 45° right triangles.

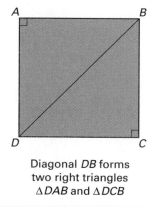

Diagonal *DB* forms
two right triangles
△*DAB* and △*DCB*

EXAMPLE 2 Finding the Diagonal of a Square

How long is the diagonal brace connecting the opposite corners on the back of a square frame that is 6 feet on a side?

SOLUTION

Make a sketch. Mark the right angle and the hypotenuse.

Since the length of each side of the 45° - 45° right triangles formed by the diagonal is 6 feet, the diagonal will be

$$6\sqrt{2} \approx 8.5$$

The length of the diagonal brace is approximately 8.5 feet.

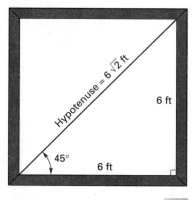

The measure of one angle of a right triangle is 45°. The length of one leg is 7.3 inches. How long is the other leg? How long is the hypotenuse?

The 30° - 60° Right Triangle

Another special right triangle has one angle equal to 30°. If one angle of a right triangle is 30°, how many degrees are in the other angle?

ACTIVITY 2   **The Sides of a 30° - 60° Right Triangle**

1   Use your protractor to draw three right triangles of different sizes, each having one angle of 30°. Label the 30° angle in each right triangle.

2   Label the shorter leg in each drawing with the letter *a*. Label the hypotenuse with the letter *c*. Notice that in each case the shorter leg of the triangle is opposite the 30° angle.

3   Complete the chart by carefully measuring the shorter leg *a* and the hypotenuse *c* in each of the triangles.

	a (side opposite the 30° angle)	*c* (hypotenuse)
Triangle 1	?	?
Triangle 2	?	?
Triangle 3	?	?

4   What is the ratio of *c* to *a* for each of the three triangles?

In the Activity, you found an important relationship. In a 30°- 60° right triangle, the hypotenuse is always twice the length of the shorter leg; that is,

$$c = 2a$$

To find the longer leg of a 30°- 60° right triangle, use the Pythagorean Theorem and the relationship $c = 2a$.

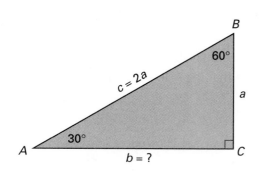

Use a sketch to label the sides and the angles. Label the side opposite the 30° angle a. Label the hypotenuse $2a$. Apply the Pythagorean Theorem.

$$a^2 + b^2 = c^2$$

Substitute $2a$ for c and solve for b.

$$a^2 + b^2 = (2a)^2$$

$$b^2 = 4a^2 - a^2$$

$$b^2 = 3a^2$$

Solving for b, the result is

$$b = \sqrt{3}a \text{ or } a\sqrt{3}$$

The leg AC (opposite the 60° angle) is equal to $\sqrt{3}$ times the length of the shorter leg. This is true for all 30°- 60° right triangles.

30°- 60° Right Triangle

For every 30°- 60° right triangle,

a. the hypotenuse is twice the length of the shorter leg.

b. the longer leg is $\sqrt{3}$ times the length of the shorter leg.

Ongoing Assessment

A 10-foot board is used as a ramp to slide crates from a warehouse floor to a loading platform. The board makes a 30° angle with the floor. What is the height (H) of the loading platform? What is the length of the horizontal distance (D) along the floor?

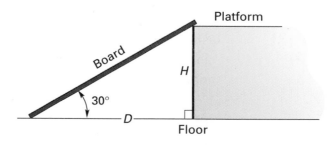

Important facts relating to the two special right triangles are summarized below.

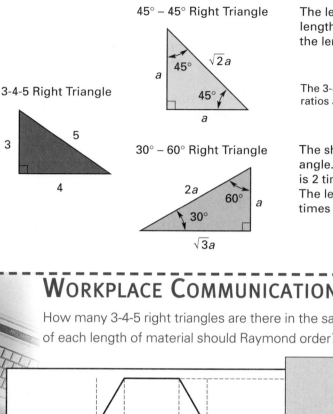

45° – 45° Right Triangle

The legs have equal length. The length of the hypotenuse is $\sqrt{2}$ times the length of either leg.

3-4-5 Right Triangle

The 3-4-5 right triangle has sides whose ratios are 3:4:5.

30° – 60° Right Triangle

The shorter leg is opposite the 30° angle. The length of the hypotenuse is 2 times the length of the short leg. The length of the longer leg is $\sqrt{3}$ times the length of the short leg.

WORKPLACE COMMUNICATION

How many 3-4-5 right triangles are there in the satellite framework? How many pieces of each length of material should Raymond order?

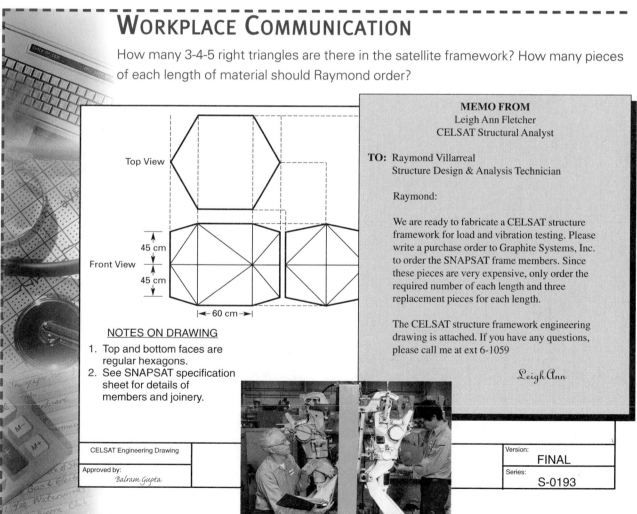

Top View

Front View

45 cm
45 cm

|← 60 cm →|

NOTES ON DRAWING

1. Top and bottom faces are regular hexagons.
2. See SNAPSAT specification sheet for details of members and joinery.

CELSAT Engineering Drawing
Approved by:
Balram Gupta

MEMO FROM
Leigh Ann Fletcher
CELSAT Structural Analyst

TO: Raymond Villarreal
Structure Design & Analysis Technician

Raymond:

We are ready to fabricate a CELSAT structure framework for load and vibration testing. Please write a purchase order to Graphite Systems, Inc. to order the SNAPSAT frame members. Since these pieces are very expensive, only order the required number of each length and three replacement pieces for each length.

The CELSAT structure framework engineering drawing is attached. If you have any questions, please call me at ext 6-1059

Leigh Ann

Version:
FINAL
Series:
S-0193

LESSON ASSESSMENT

Think and Discuss

1 If a right triangle *ABC* is similar to a 3 - 4 - 5 right triangle, the hypotenuse of triangle *ABC* is a multiple of what number?

2 Why are 45°- 45° right triangles and 30°- 60° right triangles considered special right triangles?

3 How do you find the hypotenuse of a 45°- 45° right triangle when you know the length of a leg?

4 How do you find the hypotenuse of a 30°- 60° right triangle when you know the length of the shorter leg?

Practice and Problem Solving

5. Which of the following are right triangles?

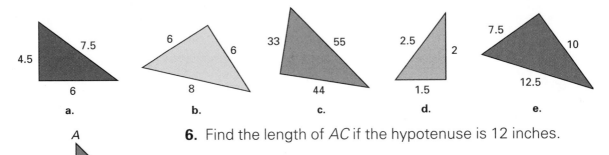

a. b. c. d. e.

6. Find the length of *AC* if the hypotenuse is 12 inches.

7. Find the length of *BC* if the hypotenuse is 20 feet.

8. Find the length of *AB* if each leg is 8 centimeters.

9. The length of the hypotenuse of a 45°- 45° right triangle is 16 inches. Find the length of each leg.

M ∠*A* = 45°

10. Find the length of *XY* and the length of *XZ* if the length of *YZ* is 12 centimeters.

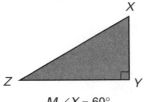

M ∠*X* = 60°

11. Find the length of *XZ* and the length of *YZ* if the length of *XY* is 6 feet.

12. Triangle *ABC* is an equilateral triangle. If the length of the altitude *AD* is $4\sqrt{3}$ inches, find the length of *BC*.

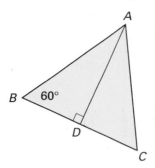

13. The length of the hypotenuse of a 30°- 60° right triangle is 20 inches. Find the length of each leg.

14. A bridge joins two points *A* and *B* across a river (see diagram). A surveyor determines that ∠*ABC* is 90° and ∠*ACB* is 60°. If the length of *BC* is 200 yards, how long is the bridge?

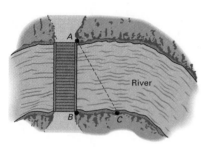

15. A carpenter is building trusses to support the roof of a new house. The trusses are each in the shape of a 3 - 4 - 5 right triangle. One leg of the triangle is 108 inches. The other leg is 144 inches. How long should the carpenter cut the lumber for the hypotenuse?

16. A 15-foot ladder is leaning against a building. The ladder makes a 30° angle with the building. How far from the building is the foot of the ladder?

17. Isaias is a catcher on his school's baseball team. He wants to practice throwing from home plate to second base with his dad. He knows the distance is the same as the diagonal of a square that has sides 90 feet long. How far apart should Isaias and his dad stand for Isaias to practice his throws?

18. A crane with a 120-foot boom can safely lift a maximum weight of 10,000 pounds when the boom angle (∠*A*) is 60° or more. What is the maximum distance between the base of the crane and the weight when the crane lifts the maximum weight?

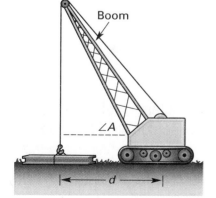

The following formula is used by a truck conversion business to find the number of four-wheelers (N) needed to break even. C is the start-up cost of production, S is the selling price of the four wheelers, and P is the cost of converting the four-wheelers.

$$N = \frac{C}{S - P}$$

19. Use the formula to find the number of four-wheelers that must be sold to break even if the start-up cost is $12,000, the selling price is $4000 per truck, and the production cost is $1000 per truck.

20. Solve the formula for the selling price (S).

Solve each equation for c.
21. 15% of $c = 180$ **22.** $c^2 + c = 20$

Write a system of equations to model each situation. Solve the system and check the solution.
23. Todd needs a total of 30 two-inch and three-inch bolts to assemble a piece of equipment. He needs six more two-inch bolts than three-inch bolts. How many two-inch bolts does Todd need?

24. Alicia needs 8 liters of a 55% saltwater solution for an experiment. She has one solution that is 40% saltwater and another that is 60% saltwater. How many liters of each solution should Alicia mix together to get a 55% saltwater solution?

Write a quadratic equation to model each situation. Solve the equation and check the solution.
25. The sum of the squares of two consecutive, even integers is 155 more than the odd integer between them. What are the two even integers?

26. A 9- by 12-foot rug is placed in a rectangular family room so that it leaves a uniform strip of flooring on all sides. The area of the exposed flooring is 100 square feet. What are the dimensions of the family room?

LESSON 12.5 TRIGONOMETRIC RATIOS

The hypotenuse of a right triangle is its longest side. It is also the side *opposite* the right angle. The other two sides of the right triangle are its legs.

In the drawing, the two legs are described in reference to angle *A*.

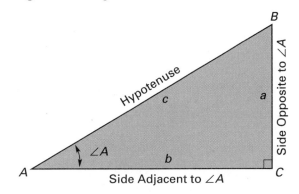

The Tangent Ratio

The tangent of an angle is an especially useful trigonometric ratio.

> **Tangent of an Angle**
> In a right triangle, the tangent of an acute angle is the ratio
> $$\frac{\text{Length of opposite leg}}{\text{Length of adjacent leg}}$$

The tangent of angle *A* is usually abbreviated

$$\tan A = \frac{\text{opp}}{\text{adj}}$$

Notice that the word *tangent* is shortened to *tan*. You can read the abbreviation as "tangent of angle A" or as "tan A." Using the labels in the drawing above,

$$\tan A = \frac{a}{b}$$

You can also write the tangent ratio for angle *B*

$$\tan B = \frac{b}{a}$$

Use the sketch of the 45°- 45° right triangle to answer these questions.

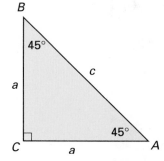

1 What is the tangent of ∠A and of ∠B?

2 What is the value of the tangent of ∠A if the legs of the 45°- 45° right triangle are both increased by the same amount? Explain your answer.

3 If an angle of a right triangle has a tangent of one, what can you conclude about the angles in the triangle?

You can find the slope of a line using the tangent ratio. Look at the drawing below. The slope is defined as the ratio of the rise to the run. The rise is the length of segment BC. The run is the length of segment AC.

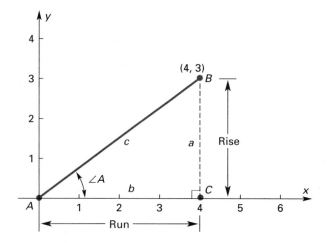

Thus, the slope of AB is

$$\text{Slope} = \frac{BC}{AC} = \frac{a}{b}$$

The ratio $\frac{a}{b}$ is also equal to the tangent of $\angle A$.

Slope and Tangent

The slope of a line is the same as the tangent of the angle formed by the line and the *x*-axis.

Critical Thinking   For a line that passes through the origin of the Cartesian coordinate system, there are two quadrants where the tangent of the angle formed by the line and the *x*-axis can be negative. What are the two quadrants?

EXAMPLE 1 Using the Tangent to Find the Slope

Find the slope of line segment *AB*.

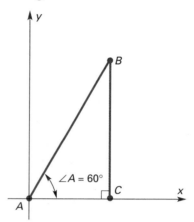

SOLUTION

Notice that $\angle A$ is opposite the leg of the right triangle that represents the rise of the line.

Now use a calculator to find tan *A*.

- Make sure your calculator is in the *degree mode*.
- Enter the measure of angle *A*.
- Press the TAN key.
- Read the value for the tangent of angle *A*.

Since m$\angle A$ is 60°, the tangent of $\angle A$ is about 1.7. The slope of the line is equal to the tangent of $\angle A$, which in this case is 1.7.

To check your result, measure the rise and run of the line and compute the slope using the slope formula. Is this ratio the same as the tangent of 60°?

The Sine and Cosine

The tangent of an angle is a useful ratio to use when you are solving problems involving right triangles. There are two other ratios that are also very useful.

Sine and Cosine of an Angle

In a right triangle, the sine of an acute angle is the ratio

$$\frac{\text{Length of opposite leg}}{\text{Length of hypotenuse}}$$

and the cosine of an acute angle is the ratio

$$\frac{\text{Length of adjacent leg}}{\text{Length of hypotenuse}}$$

The sine and cosine of $\angle A$ are abbreviated like this:

$$\sin A = \frac{\text{opp}}{\text{hyp}} \qquad \cos A = \frac{\text{adj}}{\text{hyp}}$$

ACTIVITY 2 — Trigonometric Ratios

1 Measure the lengths of the hypotenuse and legs of the right triangle.

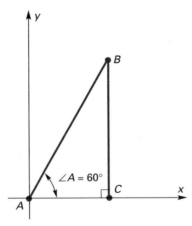

2 Use the lengths to calculate $\sin A$ and $\cos A$.

3 Use your calculator to find $\sin B$ and $\cos B$. Be sure your calculator is in degree mode before you enter the measure of $\angle A$.

4 How do the results from Step 2 and Step 3 compare?

Use the data from Activity 2 to show that the sine of an angle divided by its cosine is equal to its tangent.

EXAMPLE 2 Finding Trigonometric Ratios

The tangent of an angle in a right triangle is $\frac{5}{12}$. Find the sine of the angle.

SOLUTION

Draw a sketch.

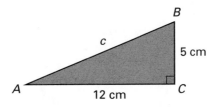

The side opposite $\angle A$ equals 5. The side adjacent to $\angle A$ equals 12. The hypotenuse is c.

Use the Pythagorean Theorem or Pythagorean triple 5, 12, 13 to find that the hypotenuse c is equal to 13.

Thus, the sine of $\angle A$ is

$$\frac{\text{opp}}{\text{hyp}} \quad \text{or} \quad \frac{5}{13}$$

Ongoing Assessment

The cosine of an angle of a right triangle is $\frac{8}{17}$. Find the sine and tangent of the angle.

The branch of mathematics that uses the sine, cosine, and tangent ratios is called **trigonometry**. The word itself comes from *tri* (meaning "three") and *metry* (meaning "measure"). The sine, cosine, and tangent are called **trigonometric ratios**. Most early civilizations used trigonometry to measure distances indirectly. For example, people used the tangent ratio to find the angle between the horizontal and a line of sight toward a point above the horizontal. This angle is called the **angle of elevation**.

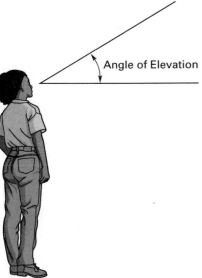

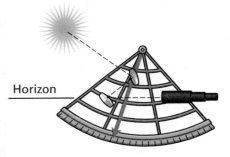

EXAMPLE 3 Using Indirect Measurement

The angle of elevation from a point on the ground to the top of a 35-foot cliff is 40°. How far is this point from the base of the cliff?

SOLUTION

To find the distance, draw a diagram and use the tangent ratio to write a trigonometric equation.

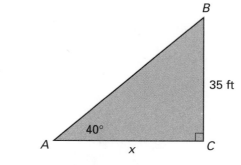

Solve for x.

$$x = \frac{35}{\tan 40}$$

$$x \approx \frac{35}{0.839} \text{ (Use your calculator to find } \tan 40° \approx 0.839)$$

$$x = 41.7$$

The distance to the base of the cliff is 41.7 feet (to the nearest tenth of a foot).

Ongoing Assessment

The angle of elevation from a point 300 feet from the base of a microwave relay tower to its top is 26°. What is the height of the tower?

LESSON ASSESSMENT

1 Compare the tangent ratio to the sine and cosine ratios. How are they alike? How are they different?

2 If you know the sine of one acute angle in a right triangle, how can you find the cosine of the other acute angle?

3 Explain how to use a calculator to find the sine of 60°.

4 Explain how to find the angle of elevation.

Practice and Problem Solving

Find the sine, cosine, and tangent for angle A and angle B. Write each ratio as a fraction in simplest form.

5.

6.

7.

Use a calculator to find the following. Round each answer to the nearest thousandth.

8. Sin 25°

9. Cos 45°

10. Tan 75°

Write the trigonometric equation that models each situation. Solve the equation, rounding to the nearest tenth.

11. You must cut a wire to connect the top of a 40-foot telephone pole with the ground. The wire must make a 35° angle with the ground. What length of wire should be cut?

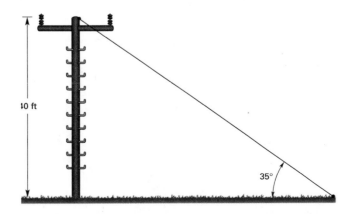

12. From a point on the ground 80 feet from the base of a building to the top of the building, the angle of elevation is 30°. Find the height of the building.

13. What is the length of a shadow cast by a 40-foot flagpole when the angle of elevation of the sun is 40°?

The angle of depression is the angle between the horizontal and a line of sight below the horizontal. Write the trigonometric equation that models each situation. Solve each equation, rounding to the nearest tenth.

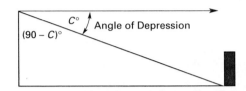

14. From a point on top of a 100-foot cliff, the angle of depression to a cabin below the cliff is 25°. How far is the cabin from the base of the cliff?

15. A lookout on a ship spots a school of whales. The lookout is standing 55 feet above the ocean surface and measures an angle of depression to the whales of 3°. How far from the ship is the school of whales?

16. A person standing on a 235-foot cliff spots a boat in a lake. The eye level of the person is 5 feet. The angle of depression to the boat is 30°. How far from the base of the cliff is the boat?

Mixed Review

Solve each equation for y.

17. $10(y + 3) = 8 + 6y$

18. $(y + 2)(y - 5) = 0$

Graph the system of inequalities. Shade the solution area.

19. $x + y \le 1$
$2x - y > 2$

20. $x \ge 1$
$y \ge x$
$x + y < 5$

The function $P(a) = a^2 - 2a - 3$ is used to project profit for the Lite Shade Company.

21. Find the minimum value of P.

22. Find $P(0)$.

23. The center of a circle is located at $A(2, -3)$. One of the points on the circle is $B(5, 1)$. Use the distance formula to find the radius of the circle.

LESSON 12.6 SOLVING EQUATIONS CONTAINING RADICALS

Adding and Subtracting Radicals

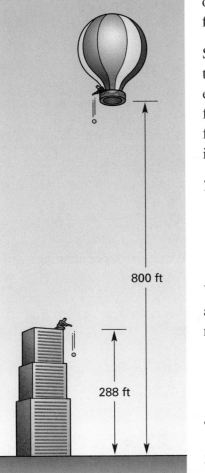

A radical expression is used to find the time (in seconds) it takes an object to fall a given distance (in feet) when air resistance is not a factor.

Suppose a ball is dropped from a height of 800 feet. At the same time, another ball is dropped from 288 feet. How much time will elapse between each ball's impact with the ground? You can use the formula $t = \frac{1}{4}\sqrt{h}$ to find how many seconds it will take a ball to fall in h feet. The elapsed time (T) will be the difference in the time it takes each ball to reach the ground.

$$T = \frac{1}{4}\sqrt{800} - \frac{1}{4}\sqrt{288} \qquad \text{Given}$$

$$= \frac{1}{4}(20)\sqrt{2} - \frac{1}{4}(12)\sqrt{2} \qquad \text{Simplify radicals}$$

$$= 5\sqrt{2} - 3\sqrt{2} \qquad \text{Simplify fractions}$$

You can further simplify the difference $5\sqrt{2} - 3\sqrt{2}$. When an addition or subtraction expression contains terms with the same radicands, you can apply the Distributive Property.

$$= (5 - 3)\sqrt{2} \qquad \text{Distributive Property}$$

$$= 2\sqrt{2}$$

The time difference is about 2.8 seconds.

$5\sqrt{2}$ and $3\sqrt{2}$ are *like radicals* because they have the same radicands. You can add and subtract two like radicals in the same way you add and subtract other like terms. Thus,

$$5\sqrt{2} - 3\sqrt{2} = 2\sqrt{2}$$

Addition and Subtraction of Radicals
Expressions containing like radicals can be added and subtracted in the same way that like terms can be added and subtracted.

EXAMPLE 1 Adding Radical Expressions

Simplify $2\sqrt{3} + 2\sqrt{8} + 3\sqrt{12}$.

SOLUTION

Simplify the second and third radicals.

$$2\sqrt{3} + 2\sqrt{8} + 3\sqrt{12} = 2\sqrt{3} + 2\sqrt{4}\sqrt{2} + 3\sqrt{4}\sqrt{3}$$
$$= 2\sqrt{3} + 4\sqrt{2} + 6\sqrt{3}$$

Then, combine like radicals.

$$= 8\sqrt{3} + 4\sqrt{2}$$

Critical Thinking   How can you use your calculator to show the simplification is correct?

Ongoing Assessment

Simplify $\sqrt{80} - \sqrt{125} + \sqrt{20}$.

Equations with Radicals

The following property is used to solve equations containing radicals.

Solving $\sqrt{x} = a$
  If $\sqrt{x} = a$ and $x \geq 0$, then $x = a^2$.
  That is, $(\sqrt{x})^2 = x$.

EXAMPLE 2 The Free Fall Formula

To calculate the energy water will produce in a hydroelectric plant, a technician needs to know the distance the water falls from the inlet of the dam to the turbine. The technician knows it takes the water 4 seconds to fall this distance. How can the technician find the distance?

SOLUTION

Solve the formula $t = \frac{1}{4}\sqrt{h}$ for h.

$$t = \frac{1}{4}\sqrt{h} \qquad \text{Given}$$

$$(t)^2 = \left(\frac{1}{4}\sqrt{h}\right)^2 \qquad \text{Square both sides}$$

$$t^2 = \frac{1}{16}h \qquad \left(\frac{1}{4}\right)^2 = \frac{1}{16} \text{ and } (\sqrt{h})^2 = h$$

$$h = 16t^2 \qquad \text{Multiplication Property of Equality}$$

When the time is 4 seconds, the distance the water falls is $h = 16(4)^2$, or 256 feet.

Critical Thinking   Explain why the following is true for all real numbers a and b: $(\sqrt{a} + \sqrt{b})(\sqrt{a} - \sqrt{b}) = a + b$

Sometimes an equation contains variables inside and outside the radical. To solve for a variable inside the radical:

• Isolate the radical.
• Square both sides of the equation.
• Solve the resulting equation.
• Check your solutions.

EXAMPLE 3 Radicals and Quadratic Equations

Solve the equation $\sqrt{3x - 2} = x$.

SOLUTION

$$\sqrt{3x - 2} = x \qquad \text{Given (The radicand is isolated.)}$$

$$(\sqrt{3x - 2})^2 = x^2 \qquad \text{Square both sides}$$

$$3x - 2 = x^2 \qquad (\sqrt{a})^2 = a$$

$$x^2 - 3x + 2 = 0 \qquad \text{Addition and Subtraction Properties of Equality}$$

$$(x - 2)(x - 1) = 0 \qquad \text{Factor the expression}$$

$$x = 2 \text{ or } x = 1 \qquad \text{Zero Product Property}$$

Thus, the solutions are 1 and 2. Check the solutions in the original equation.

If you square each side of an equation, the resulting equation is still balanced. However, the solutions may not make the original equation true.

For example, suppose $x = 4$. Then $x^2 = 16$. The solutions to $x^2 = 16$ are 4 and -4. However, -4 is not a solution to the original equation $x = 4$. In this case, -4 is an **extraneous solution.** The extraneous solution resulted from squaring both sides of the original equation.

WORKPLACE COMMUNICATION

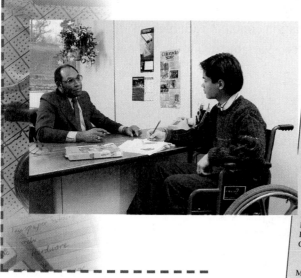

There are many opportunities waiting for you in the modern workplace. Make sure you are prepared!

VACANCY ANNOUNCEMENT

Modern Workplace has immediate openings for entry-level personnel in the following specialty areas. Experience is not required; however, all positions require a high school diploma or GED. Three years of high school mathematics is highly desired for all positions. Some positions require a two-year degree and some require a four-year degree. For additional information and application forms, contact Modern Workplace's Department of Human Resources.

Personnel
Personnel Assistants
Recruiters
Training Associates
Security Specialists

Administration
Administrative Assistants
Word Processors
Office Systems Coordinators
Clerks
Receptionists
Travel Coordinators

Finance
Controllers
Accountants
Accounting Technicians
Budget Analysts
Buyers
Cost Analysts

Data Processing
Computer Systems Analysts
Software Configuration Specialists
Database Specialists
Systems Analysts
Programmers
Data Entry Operators
Computer Operators

Sales and Marketing
Marketing Trainees
Customer Service Representatives
Advertising and Sales Specialists
Telemarketing Trainees

Engineering
Engineers—All Disciplines
Engineering Technicians—All Disciplines
Test Engineers
Research Engineers
Research and Testing Technicians
Product Development Technicians
Atmospheric Space Scientists
Telecommunications Analysts
Telecommunications Technicians
Facilities Engineers and Technicians
Operations Analysts
Chemists
Surveyors
Drafters
Statisticians
Technical Writers
Graphic Illustrators

Production and Maintenance
Material Handlers
Quality Assurance Analysts
Quality Assurance Technicians
Equipment Operators
Maintenance Technicians
Custodians
Machinists
Sheet Metal Workers
Painters
Welders
Electricians
Assemblers
Mechanic's Helpers
Laborers

Continued on Next Page

Now you are ready to complete the Math Lab Activities and Math Applications for this chapter.

MATH LAB

Equipment   Directional compass
Line level
Tape measure
Masking tape
Plumb line
Calculator
Cord or heavy string

Problem Statement

Construction contractors must grade the surfaces of parking lots to ensure adequate drainage of water. The amount of grade is related to the slope of the parking lot surface.

You will lay out a north-south line and an east-west line on a parking lot surface and measure the slopes of the lines. You will also calculate the angles corresponding to these slopes.

Procedure

a Stretch the cord taut and put two marks on the cord, 25 feet apart. These marks will be referred to as A' and B'.

b Choose a location in the parking lot at least 25 feet from any edge of the lot. Mark this location "point A" with a piece of masking tape.

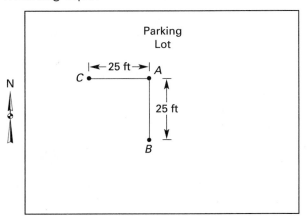

c Use the compass to sight a line directly south of point A. Stretch the cord along this north-south line.

d Have two members of the group hold the ends of the cord at A' and B'. Make sure the cord is taut. One member should stand at point A and hold the cord at a comfortable height. Use the plumb line as shown to locate A' directly above point A. Measure the vertical distance along the plumb line from point A to A'. Label this distance H_A and record its value. The group member holding the cord at point A should keep the vertical distance H_A constant and keep the A' mark directly above point A throughout the remainder of this Activity.

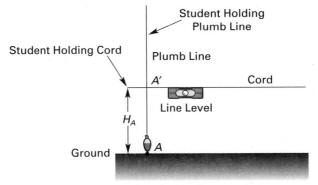

e Attach the line level to the cord near the B' mark. Adjust the height of the cord until the line level indicates the cord is level.

f After leveling the cord, use the plumb line to locate the point on the ground directly beneath the B' mark. Mark this location "point B" with a piece of masking tape. Measure the vertical distance from point B to B'. Label this distance H_B and record its value.

g Return to point A. Use the directional compass to sight a line directly west of point A. Stretch the cord along this east-west line.

h Repeat Steps **d** through **f** to locate point C and to measure the height of the level cord above point C.

i Use the formula

$$\text{Slope} = \frac{\text{Rise}}{\text{Run}}$$

where run is the length of the cord between A' and B', to calculate the slope of the north-south line. In this case

$$Run = 25\ ft$$

$$Rise = H_A - H_B$$

so that

$$Slope = \frac{H_A - H_B}{25}$$

The drawing below shows the geometry used to calculate the slope.

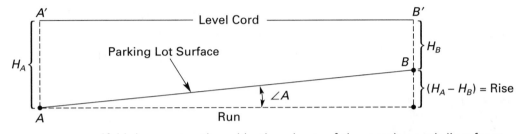

If H_A is greater than H_B, the slope of the north-south line from A to B is uphill, and the slope is positive. If H_A is less than H_B, the slope of the north-south line from A to B is downhill, and the slope is negative. Record the slope of the north-south line.

j Use the same formula to calculate the slope of the east-west line. In this case,

$$Run = 25\ ft$$

$$Rise = H_A - H_C$$

and

$$Slope = \frac{H_A - H_C}{25}$$

If H_A is greater than H_C, the slope of the east-west line from A to C is uphill, and the slope is positive. If H_A is less than H_C, the slope of the east-west line from A to C is downhill, and the slope is negative. Record the slope of the east-west line.

k Now calculate the angles corresponding to the slopes you just measured. Look at the drawing again. Find m∠A from the tangent function. Suppose you measure the rise $(H_A - H_B)$ to be 0.5 feet.

$$\tan A = \frac{opp}{adj} = \frac{0.5}{25} = 0.02$$

To find m∠A, do the following:

- Place your calculator in the degree mode.
- Enter the value 0.02.
- Press the (INV) inverse or 2nd function key.
- Press the (TAN) key.
- Read ∠A (given in degrees) in the window. For tan A = 0.02, ∠A (rounded) is 1.1°.

Calculate the measures of the two angles—one for the right triangle whose hypotenuse is segment AB and the other whose hypotenuse is AC. A positive angle means the parking surface runs uphill from A to C or from A to B. A negative angle means the surface runs downhill. Do you expect good drainage from the parking lot? Explain your answer.

Activity 2: Calculating Diagonal Lengths of Rectangles

Equipment   Tape measure
                Calculator

Problem Statement

Drawing a diagonal across a rectangle results in two identical right triangles.

You will measure the length and the width of several rectangular objects. Then, you will calculate the lengths of the diagonals using the Pythagorean Theorem and compare the calculated values to the measured values. In addition, you will calculate the length of a room diagonal from a corner at the ceiling to an opposite corner at the floor.

Procedure

Complete Steps **a**, **b**, and **c** for each of the following rectangular objects:

- classroom door
- teacher's desk
- chalkboard or dry erase board
- classroom floor

a Measure and record the width (w), length (l), and diagonal (D).

b Use the Pythagorean Theorem to calculate the length of the diagonal.

c Compare the calculated diagonal lengths to the measured diagonal lengths.

d Measure the height (h) of the classroom. Use the length, width, and height to find the diagonal of the room from one corner at the ceiling to the opposite corner at the floor. Find the diagonal using the formula

$$D^2 = l^2 + w^2 + h^2$$

e How does the length of the diagonal of the room compare with the length, width, and height of the room?

Activity 3: Calculating the Length of One Side of a School Building

Equipment   Directional compass
Flat square
String (100 feet)
Calculator

Problem Statement

You will use the tangent ratio to estimate the length of one side of your school building.

Procedure

a Choose a corner of the school building (see drawing). Extend the 100-foot string from the corner of the school building at a right angle to the side being measured. Use the flat square to determine a 90° angle.

b From the end of the string farthest from the building, use the directional compass to find the heading to the chosen corner. Find a second compass heading from the end of the string to the opposite corner of the side of the building being measured (see drawing on next page). Record the two compass headings.

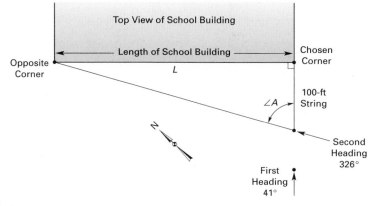

c Subtract the two compass headings. If the difference between the compass headings is less than 90°, this difference is the measure of ∠A in the drawing. If the difference is greater than 90°, subtract the difference from 360° to get m∠A.

For example, suppose the compass headings were 41° and 326°, as shown in the drawing. This gives 326° − 41° = 285°. Since 285° is greater than 90°, subtract 285° from 360° to get 75°. The measure of ∠A is 75°.

d Use the formula below to calculate the distance between the two corners.

$$\tan A = \frac{\text{opp}}{\text{adj}} = \frac{\text{length of side of school building}}{\text{Length of string}}$$

$$\tan A = \frac{L}{100 \text{ feet}}$$

$$L = 100 \tan A$$

e Describe how you could use this method to measure the distance between two points on opposite sides of the Grand Canyon.

MATH APPLICATIONS

The applications that follow are like the ones you will encounter in many workplaces. Use the mathematics you have learned in this chapter to solve the problems.

Wherever possible, use your calculator to solve the problems that require numerical answers.

Two streets intersect at an angle of 60°. The perpendicular distance across each street is 35 feet. At the intersection, the distance (*AB*) from corner to corner is greater than 35 feet.

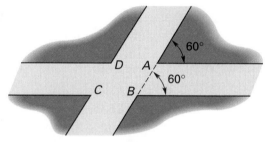

1 Trace the drawing of the intersection onto your own paper. Draw a right triangle using the distance between two corners as the hypotenuse and the perpendicular distance across one of the streets as one of the legs. Label the measures of the angles within your right triangle.

2 Determine the distance between the two street corners *A* and *B*.

3 What is the distance between corners *C* and *D*?

4 You are building a square picture frame from a kit. Each of the four frame pieces is 12 inches long. To check for "squareness," you measure the diagonals to see if they are equal. Find the length of the diagonals connecting the outside corners of the square frame.

5 A trail map shows that you are 20 miles away from a mountain peak. You can see the peak in the distance, and you estimate that it is about 7° above the horizon. Draw a sketch to illustrate this information. Use the tangent ratio to estimate the height (to the nearest 1000 feet) of the mountain peak.

6 A single-engine airplane loses power at an altitude of 6000 feet. The plane can maintain a glide path that makes an angle of depression of at least 18°. What is the maximum horizontal distance the plane can glide before it makes an emergency landing?

7 A television antenna is mounted on a mast at the peak of a roof. When standing 40 feet away from the end of the building, you measure the angle of elevation to the top of the antenna as 50° and the angle to the peak of the roof as 39°. Determine the height of the peak of the roof, the height of the antenna above the ground, and the length of the mast.

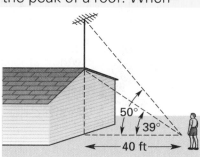

50° 39° ←— 40 ft —→

A research team reflects a laser beam from a mirror placed on the moon by the Apollo astronauts. The laser beam has a very small angle of spread, about 0.001°. The beam travels 239,000 miles to reach the moon.

8 Draw two right triangles representing the beam as it spreads from its source toward the moon. Label the angle of spread (known as the *divergence angle*), each triangle, and the length of the leg that corresponds to the distance to the moon. (*Hint:* Half the beam is used for each right triangle.)

9 Determine the total width of the beam when it reaches the moon.

You measure the angle that the sun makes with the vertical by measuring the shadow cast by a perpendicular object of known height. Suppose a yardstick that is perpendicular to the ground casts a shadow that extends 9 inches from the base of the yardstick.

10 Make a sketch showing the right triangle's dimensions and the angle that the sunlight makes with the yardstick.

11 What is the value of the tangent of the angle formed by the sunlight with the yardstick?

12 When the tangent of angle *A* is known, the inverse tangent function will give the value of angle *A*. Enter the tangent value into your calculator and press the (INV) key followed by the (TAN) key. The calculator will then display the angle. What is the angle of the sun in the measurement above?

Astronomers measure large distances using an angular measurement called *parallax*. The position of a distant object is precisely charted. Six months later, when the Earth is on the opposite side of the sun, the object is charted again. The position of the object in the sky will be at a slightly different angle. The drawing below illustrates this for a certain star. (*Note: the drawing is not to scale.*)

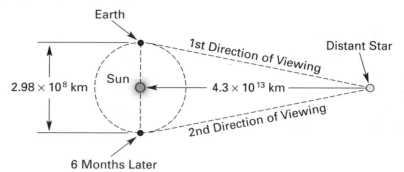

13 Identify the right triangles in the drawing. Note that the small angle in each right triangle is $\frac{1}{2}$ the angle determined from the two chartings. Identify the adjacent and opposite legs with respect to this small angle.

14 Compute the tangent of the small angle in one of the right triangles.

15 Use the INV key followed by the TAN key (see Exercise 12) to find the measure of the small angle in the right triangle.

16 Double the measure of the small angle. This angle is referred to as *stellar parallax*. The distance used in this Exercise is for the closest star to the sun and involves a relatively large amount of stellar parallax. Astronomers use much smaller angles in actual practice.

Pendulum clocks can maintain accurate time because they have a finely adjustable period of swing. The period of swing of a pendulum is given by the formula

$$T = 2\pi \cdot \sqrt{\frac{L}{g}}$$

T is the period of the pendulum in seconds (the period is defined as the time required for one complete swing, back and forth).
L is the length of the pendulum in centimeters.
g is the acceleration due to gravity, 980 centimeters per second.

17 To what length should you adjust the pendulum for a 1.000 second period? Use your calculator's value for π. Write your answer to the nearest thousandth.

18 If the length is shortened to 24.800 centimeters, would the period be longer or shorter than 1 second?

AGRICULTURE & AGRIBUSINESS

While repairing a fence, you need to determine whether the fence's corner forms a 90° angle. Both sections of fence are straight and have poles spaced 8 feet apart. Your only measuring device is a 50-foot tape measure.

19 Describe a procedure incorporating a Pythagorean triple and a 3-4-5 right triangle that will determine whether the fence lines are perpendicular.

20 You are using the Doyle log rule to estimate the amount of wood to haul on a log truck. To use the rule, you measure the diameter of the small end of a log (in inches) and its length (in feet), then use the following formula:

$$BF = L \cdot \left(\frac{d - 4}{4}\right)^2$$

BF is the approximate board feet of usable lumber.
d is the log's small-end diameter in inches.
L is the length of the log in feet.

For cost effectiveness, the truck must haul about 5000 board feet of usable lumber. If the truck can carry 20 logs, each about 27 feet long, estimate the small-end diameter of the logs for a 5000 board-foot load. Write your answer to the nearest inch.

BUSINESS & MARKETING

When a business sends out a mass-mail advertisement, it must consider the expected response rate and its confidence interval. The expected response rate is the expected increase in sales (expressed as a percent) resulting from the advertisement. The confidence interval is specified as the upper and lower limit of the sales increase. For example, a 95% confidence interval means that 95 mass mailings out of 100 will result in at least the lower limit specified and at most the upper limit specified.

The limits for a 95% confidence interval can be calculated from the formulas below:

$$L_U = p + 1.96 \cdot \sqrt{\left(\frac{p \cdot (1 - p)}{n}\right)}$$

$$L_L = p - 1.96 \cdot \sqrt{\left(\frac{p \cdot (1 - p)}{n}\right)}$$

L_U is the upper limit of the response (95% of the time).
L_L is the lower limit of the response (95% of the time).
p is the response rate (a decimal).
n is the number of pieces mailed.

21 A business wants to use an advertisement to achieve a 3% response rate and a 95% confidence interval of 1.34%. Assume half the confidence interval is above the expected response rate and half is below. This means that

$$L_L = 0.03 - \frac{0.0134}{2} = 0.0233$$

How many pieces must be mailed to achieve this lower limit?

22 Substitute this value into the formula for L_U and solve for L_U. Subtract L_L from L_U. Does your answer agree with the desired 95% confidence interval?

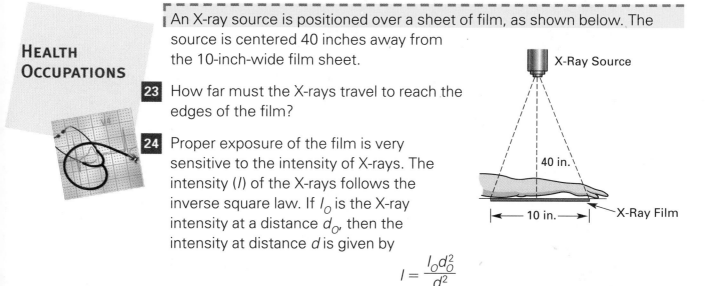

An X-ray source is positioned over a sheet of film, as shown below. The source is centered 40 inches away from the 10-inch-wide film sheet.

X-Ray Source

40 in.

10 in.

X-Ray Film

23 How far must the X-rays travel to reach the edges of the film?

24 Proper exposure of the film is very sensitive to the intensity of X-rays. The intensity (*I*) of the X-rays follows the inverse square law. If I_O is the X-ray intensity at a distance d_O, then the intensity at distance *d* is given by

$$I = \frac{I_O d_O^2}{d^2}$$

What is *I* at the center of the film? What is *I* at the edge of the film? Write each intensity as a function of $I_O d_O^2$.

25 How much more intense are the X-rays at the film's center than at the film's edges? (Write the ratio of the X-ray intensity at the edge of the film to the intensity at the center of the film.)

26 When taking a series of X-rays of the lumbar spine, one view requires the patient to sit and lean against a foam support at a 45° angle. If the distance from the base of the spine to the neck is 66 centimeters, to what height must the neck be raised to create a 45° angle? (Assume the spine is straight.)

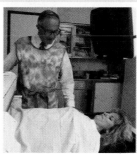

45°

Another X-ray view requires the X-ray source to be positioned at an angle of 20° from the centerline and to be located 40 inches vertically from the table.

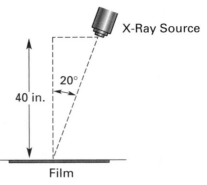

X-Ray Source

20°

40 in.

Film

27 What is the horizontal distance of the X-ray source from the centerline?

28 Will the X-ray source be more than or less than 40 inches from the center of the film? How much more or less will it be than 40 inches?

Nuclear medicine technicians caring for patients with radioactive implants must minimize their own exposure to radiation. One of the factors controlling the amount of radiation health-care workers receive is distance from the patient.

The radiation level follows the *inverse-square law*; a certain radioactive implant produces a radiation intensity I at a distance z given by

$$I = 100 \text{ mrems per hour} \cdot \frac{z_0^2}{z^2}$$

where $z_O = 1$ meter.

29 At what distance is the radiation intensity 50 mrems per hour? 10 mrems per hour? 1 mrem per hour?

An art display includes two sound effects speakers. You want to position the speakers on the wall so that when you stand midway between them, you are 7 feet from each speaker. Both speakers should be the same distance from the corner.

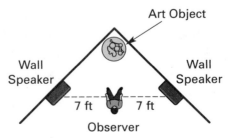

Art Object

Wall Speaker

Wall Speaker

7 ft   7 ft

Observer

30 How far from the corner should you position each speaker?

INDUSTRIAL TECHNOLOGY

You are constructing a form for a concrete sidewalk that makes a 90° angle with the curb. Before you pour the concrete, you must verify that the forms are *square*; that is, whether they form 90° angles at the corners. The forms are 12 feet long and 4 feet wide.

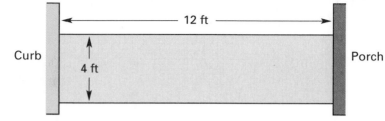

Curb   12 ft   Porch

4 ft

31 Describe a procedure, using the Pythagorean Theorem, your tape measure, and a calculator, which verifies whether your forms are *square*.

A guy wire is used to stabilize a utility pole. The wire is fastened to the pole 25 feet above the ground and staked to the ground 4 feet from the base of the pole. Someone recommends that the pole is more stable if the ground stake is moved out to 6 feet from the pole.

32 How long is the guy wire when it is staked 4 feet from the base of the pole?

33 How long must the guy wire be if it is staked 6 feet from the pole?

In electrical circuits with varying currents and voltages, the combined effect of resistance and reactance is called impedance. The impedance is related to the resistance and the reactance by the following formula:

$$Z^2 = R^2 + X^2$$

Z is the impedance in ohms.
R is the resistance in ohms.
X is the reactance in ohms.

34 Draw and label a right triangle to represent the relation between the impedance, the resistance, and the reactance.

35 A stereo speaker is labeled as having an impedance of 8 ohms. Your measurement of the speaker circuit with an ohmmeter shows a resistance of 1.5 ohms. What is the reactance of the speaker?

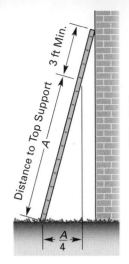

A safety manual illustrates guidelines for the safe use of an extension ladder. Suppose your ladder has a maximum extended length of 24 feet.

36 What is the maximum distance *A* to the top support for your ladder?

37 What is the maximum landing height that you can safely reach with your ladder?

A plane is forced to travel the two-legged path shown below to avoid flying over a restricted area. Point *B* is 3.4 kilometers from a direct route through the restricted area.

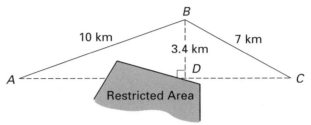

38 How much shorter is the direct route than the two-legged path?

You must load material onto a platform that is 26 inches above the ground. You want to use a ramp, but the ramp angle must not be greater than 30°.

39 Draw a triangle that represents this situation. Determine the shortest ramp you can use.

40 Using a ramp of this length, what is the horizontal distance from the platform to the point where the ramp touches the ground?

41 A surveyor made an angular error of 0.05° when surveying a rectangular lot that is 350 feet wide. What is the error in the width of the lot?

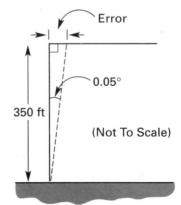

You are designing a gardening shed with a sloped roof.

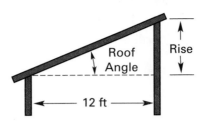

42 Write a formula for the tangent of the roof angle using the rise and the run for the roof.

43 Substitute a run of 12 feet into your formula and isolate the variable for the rise. Use this formula to make a table of the rise for roof angles of 5°, 10°, 15°, 20°, 25°, and 30°.

44 For each angle, compute the length of roof decking needed to span the distance between the two walls. This will be the hypotenuse of the right triangle. Add these values to your table.

45 If the run remains constant at 12 feet, what are some of the consequences of using too large an angle for the roof?

A certain design of a screw has 10 thread grooves in 1.5 inches. Thus, the width of the groove is 0.15 inches. The spread of the groove is 60°. The groove width and spread are shown in the drawing below.

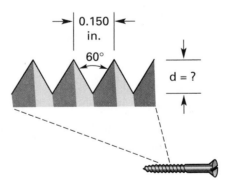

46 Draw and label a right triangle that shows the known angles and dimensions for a thread groove.

47 What is the depth *d* of this thread?

You are asked to help design a cylindrical tank with a diameter of 24 feet. The tank will have a cone-shaped roof rising 5 feet above the lip of the tank. It will be made of steel weighing 40 pounds per square foot.

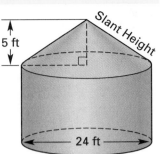

48 What is the slant height of the roof's surface; that is, the distance from the peak to the rim of the cylindrical tank?

49 What is the surface area of the roof? (*Hint:* The surface area of the cone is given by the formula $A = \frac{1}{2}\pi sd$, where s is the slant height and d is the diameter.)

50 Multiply the surface area (in square feet) by the weight of the steel per square foot to find the estimated weight of the tank's steel roof.

To estimate the height of a tall broadcasting antenna tower, you drive exactly $\frac{1}{2}$ mile from the tower's base. Using your protractor, you estimate the top of the tower to be 20° above the horizon.

20°

$h = ?$

$\frac{1}{2}$ mile

51 Write the trigonometric ratio that relates the angle to the dimensions of a right triangle.

52 Find the height of the tower.

53 What are some possible sources of error in this procedure?

A jet maintains a speed of 600 kilometers per hour while climbing at an angle of 30°.

54 You can represent the velocity of the jet using two vectors. One vector shows the horizontal velocity, and the other vector shows the vertical velocity. Draw a right triangle that shows the horizontal and vertical velocity components, as well as the resultant velocity of the jet at a 30° angle. The resultant velocity vector is the hypotenuse.

55 Determine the value of the horizontal and vertical components of the jet's velocity. (*Hint:* Find the length of the legs of the triangle you drew in Exercise 48.)

56 Use the vertical component of the velocity to find how long it will take the jet to climb 10 kilometers.

Mechanical drawing sets have triangles of various sizes. Each dimension in the chart below is the length of the longest leg. For each triangle, determine the length of the other leg and the length of the hypotenuse.

30° - 60° - 90° Drawing Triangles

57 4 in.

58 6 in.

59 10 in.

60 18 in.

61 24 in.

45° - 45° - 90° Drawing Triangles

62 4 in.

63 6 in.

64 10 in.

65 18 in.

66 24 in.

You are installing a television antenna on a mast on the roof of a house. You must determine the lengths of the guy wires that will extend from the mast to the eaves on the corners of the house. The wires will connect 15 feet above the base of the mast. The mast will rest atop the house peak, which is 7 feet above the level of the eaves. The eaves form a square that is 40 feet on a side, and the peak is centered in the square.

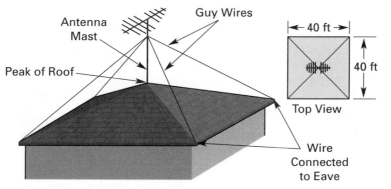

67 Determine the length of guy wire needed between the mast and each corner.

An automotive windshield repair technician is checking the specifications of a replacement for a car's rear window. He must determine the angle that the car's rear window makes with the window ledge. The depth of the ledge is 12 inches, and its corresponding height is $10\frac{1}{2}$ inches.

68 Make a sketch of the right triangle formed by the rear window, the window ledge, and the height. Label the known dimensions. Identify the slope of the angle that the window surface makes with the window ledge.

69 Compute the tangent of this angle.

70 To find the angle, enter the value of the tangent (that is, the ratio) into your calculator. Press the $\boxed{\text{INV}}$ key followed by the $\boxed{\text{TAN}}$ key. What angle should appear in the rear window specifications?

A home buyer has asked the builder to change the design of a house roof. The current plans call for the roof to have a rise-to-run ratio of 1:3, using 16-foot rafters with an 18-inch overhang. The buyer has asked to change the slope to a ratio of 2:5.

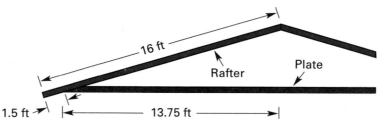

71 What is the tangent of the angle formed by the rafter and the plate for the roof slope of the current design?

72 To find the angle, enter the value of tan *A* into your calculator, and press the $\boxed{\text{INV}}$ key followed by the $\boxed{\text{TAN}}$ key. What angle do the rafters make with the plate for the current design and for the new design?

73 Is the new slope steeper or flatter than the current design slope?

74 The builder has already bought materials. He must still use 16-foot rafters and keep the same run of 13.75 feet. With the new roof slope, will the amount of overhang be more or less than the original overhang?

75 What is the length of the new overhang?

Fire fighters compute the rate of flow of water through a fire hose with the following formula:

$$\text{Flow Rate} = 30 \bullet d^2 \bullet \sqrt{p}$$

Flow rate is in gallons per minute.
d is the nozzle diameter in inches.
p is the nozzle pressure in pounds per square inch (psi).

76 A fire chief is designing a training exercise for his station. He wants the crew to use two hoses with a combined flow rate of 1000 gallons per minute. The nozzle diameter of each hose is 2.5 inches, but the nozzle pressure of one hose is four times the nozzle pressure of the other hose. What nozzle pressure should the chief use for the hoses?

CHAPTER 12 ASSESSMENT

Skills

Solve each equation.

1. $m^2 + 144 = 169$

2. $\sqrt{6} \cdot \sqrt{12} = t$

3. $q = \dfrac{\sqrt{75}}{\sqrt{15}}$

4. $\sqrt{2d - 3} = 5$

Applications

5. Triangle *ABC* and triangle *XYZ* are similar right triangles. Suppose the length of *AC* is 15 meters, *BC* is 9 meters, and *XZ* is 35 meters. What is the length of *XY*?

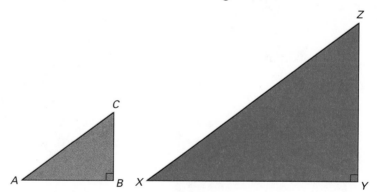

6. One leg of a right triangle is 8 inches. The other leg is 15 inches. How long is the hypotenuse?

7. The hypotenuse of a right triangle is 13 feet. One leg is 5 feet. How long is the other leg?

8. How long is the diagonal of a square that is 3 inches on each side?

9. The legs of a 45°- 45° right triangle are 12 millimeters each. What is the length of the hypotenuse?

10. The hypotenuse of a 30°-60° right triangle is 9 centimeters. How long is each leg?

Write each ratio as a fraction in lowest terms.

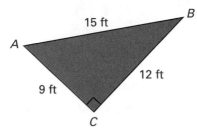

11. Sine of angle *A*.

12. Cosine of angle *A*.

13. Tangent of angle *B*.

14. The angle of elevation to the top of a 125-foot building is 50°. How far from the base of the building was the sighting taken?

15. A ramp is 150 meters long. It rises vertically for a distance of 30 meters. What is the angle of elevation for the ramp?

Math Lab

16. In the drawing below, the cord is held level between *A'* and *B'* 25 feet apart. The height measurement from *A* to *A'* is 3 feet 5 inches. The height measurement from *B* to *B'* is 2 feet 11 inches. Determine the slope of the parking lot.

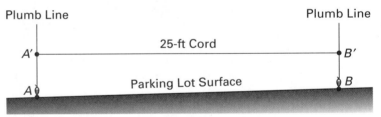

17. The dimensions of a classroom are 10 meters by 6.1 meters by 3 meters. What is the length of a diagonal that runs from the right front ceiling corner of the classroom to the left rear corner of the classroom?

18. Angle *A* is a right angle. The distance from point *A* to point *C* is 50 meters. The two compass headings at point *C* are 213° and 155°. How far is it from point *A* to point *B*?

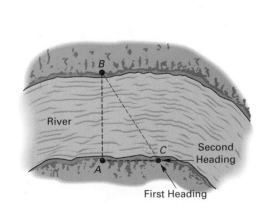

Tables of Conversion Factors

To convert from meters to inches, for example, find the row labeled "1 meter" and the column labeled "in." The conversion factor is 39.37. Thus, 1 meter = 39.37 in.

LENGTH

	cm	m	km	in.	ft	yd	mi
1 centimeter	1	0.01	10^{-5}	0.3937	3.281×10^{-2}	1.094×10^{-2}	6.214×10^{-6}
1 meter	100	1	10^{-3}	39.37	3.281	1.094	6.214×10^{-4}
1 kilometer	10^5	1000	1	3.937×10^4	3281	1094	0.6214
1 inch	2.54	0.0254	2.54×10^{-5}	1	0.0833	0.0278	1.578×10^{-5}
1 foot	30.48	0.3.48	3.048×10^{-4}	12	1	0.3333	1.894×10^{-4}
1 yard	91.44	0.9144	9.144×10^{-4}	36	3	1	5.682×10^{-4}
1 mile	1.6093×10^5	1609.3	1.6093	6.336×10^4	5280	1760	1

AREA

	cm²	m²	in.²	ft²	acre	mi²	ha
1 square centimeter	1	10^{-4}	0.1550	1.076×10^{-3}	2.471×10^{-8}	3.861×10^{-11}	10^{-8}
1 square meter	10^4	1	1550	10.76	2.471×10^4	3.861×10^{-7}	10^{-4}
1 square inch	6.452	6.452×10^{-4}	1	6.944×10^{-3}	1.594×10^{-7}	2.491×10^{-10}	6.452×10^{-8}
1 square foot	929.0	0.09290	144	1	2.296×10^{-5}	3.587×10^{-8}	9.29×10^{-6}
1 acre	4.047×10^7	4047	6.273×10^6	43,560	640	1.563×10^{-3}	0.4047
1 square mile	2.590×10^{10}	2.590×10^6	4.007×10^9	2.788×10^7	640	1	259
1 hectare	10^8	10^4	1.55×10^7	1.076×10^5	2.471	3.861×10^{-3}	1

VOLUME (CAPACITY)

	cm³	m³	in.³	ft³	L	oz	gal
1 cubic centimeter	1	10^{-6}	0.06102	3.531×10^{-5}	1.000×10^{-3}	0.03381	2.642×10^{-4}
1 cubic meter	10^6	1	6.102×10^4	35.31	1000	3.381×10^4	264.2
1 cubic inch	16.39	1.639×10^{-5}	1	5.787×10^{-4}	0.01639	0.5541	4.329×10^{-3}
1 cubic foot	2.832×10^4	0.02832	1728	1	28.32	957.5	7.480
1 liter	1000	1.000×10^{-3}	61.03	0.03532	1	33.81	0.2642
1 ounce	29.57	2.957×10^{-5}	1.805	1.044×10^{-3}	0.02957	1	7.813×10^{-3}
1 gallon	3785	3.785×10^{-3}	231	0.1337	3.785	128	1

1 gallon = 4 quarts (qt) = 8 pints (pt) = 16 cups (c)
1 cup (c) = 8 ounces (oz) = 16 tablespoons (tbsp) = 48 teaspoons (tsp)

MASS/WEIGHT

	g	kg	oz	lb	ton (short)	ton (metric)
1 gram	1	10^{-3}	0.03527	2.205×10^{-3}	1.102×10^{-6}	10^{-6}
1 kilogram	10^{3}	1	35.27	2.205	1.102×10^{-3}	10^{-3}
1 ounce	28.35	0.02835	1	0.0625	3.125×10^{-5}	2.835×10^{-5}
1 pound	453.6	0.4536	16	1	0.0005	4.536×10^{-4}
1 ton (short)	9.072×10^{5}	907.2	3.2×10^{4}	2000	1	0.9072
1 ton (metric)	10^{6}	10^{3}	3.527×10^{4}	2.205×10^{3}	1.102	1

ANGLE

	min (')	deg(°)	rad	rev
1 minute	1	0.01667	2.909×10^{-4}	4.630×10^{-5}
1 degree	60	1	0.01745	2.778×10^{-3}
1 radian	3438	57.30	1	0.1592
1 revolution	2.16×10^{4}	360	6.283	1

1 gallon = 4 quarts (qt) = 8 pints (pt) = 16 cups (c)

1 cup (c) = 8 ounces (oz) = 16 tablespoons (tbsp) = 48 teaspoons (tsp)

TIME

	s	min	hr	d	y
1 second	1	0.01667	2.788×10^{-4}	1.157×10^{-5}	3.169×10^{-8}
1 minute	60	1	0.01667	6.944×10^{-4}	1.901×10^{-6}
1 hour	3600	60	1	0.04167	1.141×10^{-4}
1 day	8.640×10^{4}	1440	24	1	2.738×10^{-3}
1 year	3.156×10^{7}	5.259×10^{5}	8766	365.3	1

Glossary

A

Absolute value The distance of a number from zero on a number line.

Absolute value of a vector The length of a vector, its numerical value. Also called the magnitude.

Addition property of equality When you add the same quantity to both sides of an equation, the equation stays equal and balanced.

Addition property of inequality When you add the same quantity to both sides of an inequality, the inequality stays equal and balanced.

Additive identity Zero is called the additive identity because when you add zero the value of the original quantity remains unchanged. Example: $5 + 0 = 5$

Algebraic expression A mathematical sentence involving at least one variable.

And An operator that connects two mathematical sentences and indicates that all solutions must satisfy *both* sentences. The system would be a conjunction.

Angle of elevation The angle between a horizontal line (parallel to the ground) and a line drawn from a given point to some point above the ground (usually the top of a tall object).

Area The amount of space inside a flat shape, such as a triangle or a circle.

Array A rectangular arrangement of numbers.

Associative property of addition When you add multiple terms, you can change the way they are grouped without changing the answer.
Example: $5 + (3 + 17) = (5 + 3) + 17$

Associative property of multiplication When you multiply several terms, you can change the way they are grouped without changing the answer.
Example: $5 \times (3 \times 17) = (5 \times 3) \times 17$

Axis A number line used to locate points on a graph. Locating points on a plane requires two axes.

Axis of symmetry A line drawn through the vertex of a parabola such that the part of the parabola on one side of the line is the exact mirror image of the part on the other side.

B

Balancing equations Whenever you perform an operation on one side of an equation, you must perform the same operation to the other side in order for the equation to remain true.

Base Whenever a number is raised to a power, that number is called the base.
Example: in $2^3 = 8$, 2 is the base

Basic Property An algebraic expression that is true for every possible value of the variables it contains.

Binomial A polynomial with two terms.

Binomial square The product of two identical binomials. A binomial square produces a perfect square trinomial.

Break-even point The point at which two equations have the same value.

C

Cartesian coordinate system A method of locating points on a plane. The system uses two axes that cross at right angles to each other. The axes are marked off in units starting at zero where the axes cross. One axis runs vertically, and its numbers increase as you move up the axis—this is called the y-axis. The other axis—the x-axis—runs horizontally, and its numbers increase as you move to the right.

Circumference The distance around a circle, starting at one point, going around the circle once and ending up at the same point you started.

Closed interval An interval that includes its endpoints.

Coefficient When a variable is multiplied by a constant, that constant is called a coefficient of the variable. Example: in the expression $4x - 2y = 12$, 4 and 2 are coefficients

Combining vectors Given two vectors A and B, the process of placing the tail of vector A at the head of vector B and then drawing a new vector starting at the tail of vector B and ending at the head of vector A. This new vector is called the resultant vector.

Common denominator The least common multiple of all the denominators in a problem. A common denominator allows you to combine fractions easily.

Common factors When you list all the factors of two or more quantities, the common factors are those items that appear in all of the lists.

Commutative property of addition When you add several terms, you can change the order of addition without affecting the final answer. Example: $23 + 6 = 6 + 23$

Commutative property of multiplication When you multiply several terms, you can change the order of multiplication without affecting the final answer.
Example: $23 \times 6 = 6 \times 23$

Completing the square A process of factoring a polynomial by adding or subtracting a constant in order to create a perfect square trinomial.

Composite number A number that has factors other than one and the number itself; that is, it has more than one factor pair.

Compound inequality A system of more than one inequality joined by AND or OR operators.

Compound interest Interest computed using both the principal and any previously accrued interest.

Conditional inequality An inequality whose truth depends on the value of a variable, such as $x > 5$.

Congruent Two figures are congruent if they are the same point for point.

Conjunction A compound inequality involving the AND operator.

Consistent system A system of equations that has at least one solution.

Constant A quantity whose value does not change, such as 5 or 12. Pi is also a constant since it always equals 3.14, rounded to the nearest hundredth.

Constant term The term in a polynomial that has no variable part.

Constraint Limits on the possible solutions to a problem. In a linear programming problem, each inequality is called a constraint.

Coordinate A number that is used to tell where a point is located on a graph. The ordered pair (7, 2) gives two coordinates that locate a unique point in the Cartesian coordinate system.

Correlation coefficient A number between -1 and 1 that tells how two sets of data are related. A value near 1 or -1 shows high correlation (positive or negative, respectively), while a value near zero shows a lack of correlation.

Corresponding angles In similar triangles, the angles that are equal between the two triangles are called corresponding angles.

Corresponding sides In similar triangles, the corresponding sides are the sides opposite the corresponding angles.

Cosine The ratio of the adjacent side of an angle to the hypotenuse in a right triangle.

Counterexample A set of values that makes an equation false. One counterexample is all that is needed to prove an equation is not a basic property.

Cross-multiplying Multiplying the means and extremes of a proportion. If $\frac{3x}{4} = \frac{5}{12}$, then the result of cross-multiplying is $4(5) = 3x(12)$.

Cube of number The third power of a number. Example: 5^3 is five cubed, or 125.

Cube root If $x^3 = y$, then x is a cube root of y.

Density The property that between any two real numbers there is another real number; the rational numbers are also dense, the integers are not.

Dependent events When the outcome of one event affects the possible outcomes of another, the events are dependent.

Dependent system A system of equations that has an infinite number of solutions. The graphs of the equations are all the same line.

Dependent variable A variable whose value is determined by the value of another variable in an equation.

Determinant The portion of the quadratic formula under the radical. Example: $b_2 - 4ac$

Diameter A line segment passing through the center of a circle and connecting two points on the circle.

Difference The answer to a subtraction operation.

Difference of two squares The binomial $a_2 - b_2$; it is a special product that factors into $(a + b)(a - b)$.

Direct variation A situation in which one variable is proportional to a constant multiple of another. In the equation $y = 5x$, y varies directly with x.

Discriminant A number obtained from a matrix by calculating the difference of the products along the diagonals.

Disjunction A compound inequality involving the OR operator.

Distance formula For two points (x,y) and (a,b), the distance between them can be calculated by using the formula $(a - x)^2 + (b - y)^2$.

Distributive property The distributive property states that $a(b + c) = ab + ac$ for all values of a, b, and c. Example: $5(3 + 2) = 5 \cdot 3 + 5 \cdot 2$

Division property of equality When you divide both sides of an equation by the same nonzero quantity, the equation stays balanced.

Division property of inequality When you divide both sides of an inequality by the same positive number, the inequality stays balanced. When you divide both sides by the same negative number, you must flip over the inequality symbol, then the equation stays balanced. Division by zero is impossible.

Domain The set of all input values for a function; the allowable values of the independent variable.

Elements In a matrix or array, the individual numbers are called elements.

Entries In a table of information, the values within the table are called entries.

Equation A mathematical sentence stating that two quantities are equal.

Expanded notation A way of writing a number so that each digit is written separately and multiplied by a power of ten written in decimal form. All the digits and their multipliers are then written as one sum.

Experimental probability A probability based on the number of times an event has occurred over the number of trials that were performed.

Exponent A quantity representing the power to which a base is raised. Example: in 3^4, the exponent is 4; it means $3 \cdot 3 \cdot 3 \cdot 3$, or 81.

Exponential decay An exponential function in which either the exponent is less than one and the base is greater than one, or the exponent is greater than one and the base is between zero and one.

Exponential function An equation in which at least one variable is used as an exponent. Example: $y = 2^x$

Exponential growth An exponential function in which the exponent is greater than one and the base is greater than one.

Extraneous solution An extra solution obtained in the process of solving an equation. The extraneous solution is not a solution to the original equation.

Extremes In the proportion $\frac{a}{b} = \frac{c}{d}$, a and d are called the extremes.

Factor (noun) A part or piece of something. When parts are multiplied together to produce a quantity, each part is called a factor.

Factor (verb) To resolve a number or a polynomial into its individual factors.

Factor pair Two integers multiplied together to produce a number are considered a factor pair for that number.

Feasible solution Any point that makes all the inequalities in a linear programming problem true.

FOIL technique First, Inside, Outside, Last—a mnemonic for remembering how to multiply binomials.

Formula An equation that expresses a relationship between at least two variables.

Frequency distribution table A table showing the number of individual things in each of certain groups.

Function A relation between two variables such as x and y where there is only one possible value of y for any given value of x.

Function notation The notation used to represent the rule between two variables, usually written $f(x)$ and read "f of x."

Fundamental counting principle To find the number of ways to make a combination of choices, multiply together the number of possible choices in each category.

Greatest integer function A function that rounds input values down to the nearest integer.

Grouping symbols Parentheses or brackets used to set off one part of an equation from another.

Guess and check A method of problem solving in which a guess is made and checked in an equation. Based on the result, a new adjusted guess is made and then checked. This process is repeated until a suitable answer is found.

H

Histogram A bar graph made using the information from a frequency distribution table.

Horizontal line A line with slope zero. All horizontal lines are parallel to the x-axis.

Horizontal translation Moving the graph of a function to the right or left without changing the shape of the graph.

Hyperbola The graph of an inverse variation equation.

Hypotenuse In a right triangle, the side opposite the right angle; it is always the longest side.

Inconsistent system A system of equations that has no solutions.

Independent events Events whose outcomes have no effect on each other.

Independent variable A variable whose value does not depend on the value of any other variable.

Inequality A mathematical sentence stating that two quantities are greater than or less than one another.

Integer A number with no fractional part. The numbers $\ldots-2, -1, 0, 1, 2 \ldots$ represent the set of integers.

Intercept The point at which the graph of a line crosses an axis.

Interest Money charged for the use of money, usually calculated as a percentage of the amount of money borrowed.

Intersection The point at which two lines cross. Also, items that are common to two or more sets are said to be in the intersection of those sets. The AND operator indicates an intersection.

Interval A set of numbers between two given numbers. Example: the set of numbers between 3 and 8 is an interval

Inverse operations Operations that undo each other, such as addition and subtraction.

Inverse variation A situation in which one variable is proportional to the inverse of another. The equation $y = \frac{1}{x}$ shows y varying inversely with respect to x.

Irrational number A number whose decimal representation never terminates or repeats.

Isolating a variable The process of performing operations on an equation in order to leave a chosen variable on one side of the equation and have all other expressions on the other side.

Isosceles triangle A triangle that has two angles exactly the same.

L

Leg Either of the two sides of a right triangle adjacent to the right angle.

Like terms Terms that contain exactly the same variables raised to the same exponents. Example: $3x^2$ and $-2x^2$ are like terms

Line of best fit The line that comes closest to going through all the points on a graph.

Linear equations An equation written in the form $y = mx + b$, where m and b are constants. The graph of a linear equation is a straight line.

Linear function A function that can be represented by a straight line not parallel to the y-axis.

Linear programming A method of finding the best solution to a system of inequalities.

Linear term A polynomial, a term with a variable raised to the first power.

M

Magnitude The numerical value or length of a vector.

Matrix Another name for an array of numbers.

Mean Often called the average, it is the total of a number of quantities divided by the number of items.

Means In the proportion $\frac{a}{b} = \frac{c}{d}$, the means are b and c.

Measure of central tendency A number that somehow represents an entire set of numbers. The mean, median, and mode are all measures of central tendency.

Median The middle number of a set of data.

Metric System A system of weights and measures whose units are all based on powers of ten.

Mode The most common number in a set of data. If no number appears more than once, there is no mode.

Monomial A variable multiplied by a constant.

Multiplication property of equality When you multiply both sides of an equation by the same quantity, the equation remains balanced.

Multiplication property of inequality When you multiply both sides of an inequality by the same positive number, the inequality remains balanced. When you multiply by a negative number, then you must flip the inequality sign for the equation to stay balanced.

Multiplicative identity One is the multiplicative identity because when you multiply any quantity by one, the result is the same quantity. Example: $5 \cdot 1 = 5$

Multiplicative inverse Any two numbers whose product is one are multiplicative inverses; also called reciprocals. Example: 3 and $\frac{1}{3}$ are multiplicative inverses

Mutually exclusive events Two events where the occurrence of one event precludes the occurrence of the other.

Negative correlation A relationship where the increase of one variable implies a decrease in the other.

Negative exponent When a negative number is used as an exponent. Example: $x_{-n} = \frac{1}{x_n}$

Negative integer Any integer whose value is less than zero.

Numerical expression A mathematical expression that contains no variables.

Occurrence The outcome of one event.

Odds The ratio of the probability of an event's occurrence to the probability of its nonoccurrence.

Open interval An interval that does not include its endpoints.

Opposite of a vector The opposite of a vector *A* has the same magnitude as vector *A* but points in precisely the opposite direction.

Opposite of an integer Two integers are opposite if they are the same distance from zero. Example: 5 and –5 are opposites

Optimum solution In linear programming, the best of all the feasible solutions.

Or An operator that joins two mathematical sentences and indicates that solutions will satisfy *either* of the two sentences. The system would be a disjunction.

Order of operations A set of rules determining the way to evaluate a mathematical expression. First, perform all operations within any grouping symbols. Then, perform all calculations involving exponents. Multiply or divide in order from left to right. Finally, add or subtract in order from left to right.

Ordered pair A pair of numbers enclosed by parentheses and separated by a comma. The ordered pair represents a unique point in the Cartesian coordinate plane. Example: (2, 1) is an ordered pair; it is represented by the point 2 units to the right and 1 unit up from the origin

Origin The point in the Cartesian coordinate system where the axes cross and where the values of both *x* and *y* are zero.

Outcome The result of the occurrence of an event.

Outlier A data point that lies far away from all of the other data points in a set.

Parabola The graph of a quadratic equation.

Parallel lines Lines that have the same slope but different *y*-intercepts. Parallel lines never intersect.

Parent function The simplest form of a function that can be changed to generate an entire class of functions with similar graphs. Example: $y = x^2$ is the parent function of $y = x^2 + 2$

Pencil test Another name for the vertical-line test.

Percent Part of 100. Example: 60% is 60 out of 100; $60\% = 0.60 = 0.6$

Perfect square number A number that has one factor pair which is two of the same number.

Perfect square trinomial A quadratic of the form $(ax)_2 + 2abx + b_2$, where *a, b,* and *c* are constants. A perfect square trinomial factors into a binomial square, $(ax + b)_2$.

Perpendicular Two lines are perpendicular if they meet at right angles to each other.

Pi The ratio of a circle's circumference to its diameter.

Polynomial A sum of two or more terms using one or more variables.

Positive correlation A relationship in which the increase of one variable implies an increase in the other.

Positive integer An integer greater than zero.

Power When you multiply the base *b* by itself *n* times, the product is the *n*th power of *b*. Example: the third power of 2 is 2^3, or 8

Power of a power When an exponential expression is raised to a power, the powers are multiplied together.

Power of a product When a product is raised to a power, each factor of the product is raised to the power.

Power of a quotient When a quotient is raised to a power, both the numerator and the denominator are raised to the power.

Power of ten If a number can be written in the form 10^n, where *n* is an integer, the number is a power of ten.

Power of ten notation A form where a quantity is written as a number multiplied by a power of ten.

Prime Anything that cannot be factored is called prime.

Prime factorization Breaking a number down into a list of factors that are all prime numbers.

Prime number A number whose only factor pair is one and the number itself.

Principal The basic amount of money used to calculate interest.

Probability The likelihood of an event's occurrence, often expressed as a ratio or a percent.

Product The result of multiplying two or more quantities.

Product of powers property When two or more exponential expressions involving the same base are multiplied together, the result is the base written with the sum of the exponents as its power.

Property of adding zero When you add zero to any quantity, the result is the same quantity. Example: $5 + 0 = 5$

Property of multiplying by one When you multiply any quantity by one, the result is the same quantity. Example: $5 \cdot 1 = 5$

Proportion A mathematical sentence stating that one ratio is equal to another. Example: $\frac{a}{b} = \frac{c}{d}$, where $b \neq 0$ and $d \neq 0$

Proportion property In a true proportion, the product of the means equals the product of the extremes.

Pythagorean Theorem In a right triangle, the square of the length of the hypotenuse equals the sum of the squares of the lengths of the two legs.

Pythagorean triple A set of three *integers* that satisfies the Pythagorean Theorem.

Quadrants The Cartesian coordinate plane is divided into four regions by the axes; these regions are called quadrants.

Quadratic equation An equation in which the highest power of the independent variable is two.

Quadratic formula A formula used to calculate the roots of a quadratic equation:
$$x = \frac{-b \pm \sqrt{b^2 - 4ac}}{2a}$$

Quadratic function A quadratic equation expressed as a function. The parent function is $y = x^2$.

Quadratic term In a polynomial, the term whose variable is raised to the second power.

Quotient The result of a division operation.

Quotient of powers property When an exponential expression is divided by another expression involving the same base, the result is the base raised to the difference of the exponents.

R

Radical A sign ($\sqrt{\ }$) indicating the square root of a quantity.

Radicand The quantity under the square root sign.

Radius A line segment from the center of a circle to the circle itself.

Range (of data) The interval in which all the data lie, found by calculating the difference between the data points with the largest and the smallest values.

Range (of function) The set of all outputs for a function; the possible values of the dependent variable.

Rate of change The slope of a linear equation.

Rate of interest The ratio of the interest charged to the principal borrowed expressed as a percent.

Ratio The relative sizes of two numbers, usually expressed as $a{:}b$ or $\frac{a}{b}$, where $b \neq 0$.

Rational expression An expression that is a ratio of two quantities.

Rational number Any number that can be written as a ratio of two integers with the second integer not equal to zero. Integers, terminating decimals, and repeating decimals are all rational numbers.

Rationalizing Removing radicals from the denominator of a fraction by multiplying both the numerator and the denominator by the radical.

Real numbers The set of all rational and irrational numbers combined is called the set of real numbers.

Reciprocal If the product of two numbers is one, the numbers are reciprocals. A number's reciprocal is also called its multiplicative inverse.

Reciprocal function The function $y = \frac{1}{x}$, where $x \neq 0$.

Regression line Another name for the line of best fit.

Relation Any set of ordered pairs.

Repeating decimal A decimal that after some point consists of an infinite number of repetitions of the same number sequence. Example: $2\frac{2}{3} = 2.666 \ldots$

Restrictions Limitations imposed on the domain and range of a function so that all the possible inputs and outputs make sense for a real-world situation.

Resultant The one vector that represents the combination of two or more vectors.

Right angle An angle measuring exactly 90 degrees, sometimes called a square angle.

Right triangle Any triangle containing a right angle.

Rise of a line The vertical change in a line between two points, found by calculating the difference between the y values of the points.

Roots The values of x that make a polynomial equal zero, also called the zeros of a polynomial.

Run of a line The horizontal change in a line between two points, found by calculating the difference between the x values of the points.

Sample space The set of all possible outcomes for a given event.

Scale factor The ratio by which the original dimensions of a figure are multiplied to enlarge or reduce the figure.

Scatter plot A graph treating data as individual points, used to check for correlation in data.

Scientific notation A shorthand way of writing very large or small numbers such that they are represented by a number between one and ten multiplied by a power of ten written in exponential form. Example: $3{,}700{,}000 = 3.7 \times 10^6$

Similar triangles Two triangles that have the exact same angles are called similar triangles.

Simple interest Interest calculated only on the original principal.

Sine The ratio of the opposite side of an angle to the hypotenuse in a right triangle.

Slope The ratio of the rise of a line to its run; slope can be thought of as the "steepness" of the line.

Slope-intercept form The graph of a linear equation written in the form $y = mx + b$ has m as its slope and b as its y-intercept.

Special product Perfect square trinomials and differences of squares are called special products.

Special right triangles Right triangles with known relationships between the sides and the angles.

Square of a number A number raised to the second power. Example: 5^2 is five squared

Square root If $y = x^2$, then x is the square root of y. The square root of a is written $\sqrt{a}$. Example: the square root of 25, $\sqrt{25}$, is 5

Square root property The solution for $x^2 = k$, where $k > 0$ is $\sqrt{k}$ or $-\sqrt{k}$.

Standard form For a linear equation, $ax + by = c$. For a quadratic equation, $ax_2 + bx + c = d$.

Statistics The study of interrelationships among data.

Substitute Replacing a variable with any quantity that has the same value.

Subtraction property of equality If the same quantity is subtracted from both sides of an equation, then the equation stays balanced.

Subtraction property of inequality If the same quantity is subtracted from both sides of an inequality, the inequality stays balanced.

Sum (Σ) The result of adding two or more quantities.

System of equations Two or more interrelated equations involving the same variables.

System of inequalities Two or more inequalities involving the same variables and usually related by AND or OR.

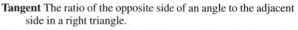

Tangent The ratio of the opposite side of an angle to the adjacent side in a right triangle.

Terminating decimal A decimal that stops after a certain number of digits. Example: $\frac{3}{4} = 0.75$

Test point A value substituted into an inequality to find out on which side of a line the solutions lie.

Theoretical probability The ratio of the number of favorable outcomes to the number of outcomes possible.

Transformation Changing the graph of a parent function by multiplying it by a constant or by adding or subtracting a constant.

Translation Moving a graph or figure up or down or to the left or right without altering its slope.

Tree diagram A diagram showing the successive decomposing of a number into its factors.

Trigonometric ratios The ratios of the lengths of the sides in a right triangle. Example: the tangent is the ratio of the opposite side to the adjacent side

Trigonometry The study of the relationships among the angles and the sides in triangles.

Trinomial A polynomial with three terms.

Union The union of two sets is the set of elements that are contained in either of the sets. A union is indicated by the OR operator.

Variable A quantity, usually represented by a letter, that can take on different values.

Vectors A quantity having both magnitude and direction, often represented by an arrow.

Venn diagram A diagram that uses circles to represent sets and their interrelations.

Vertex The point at which a parabola reverses direction; also the "tip" of the absolute value function's graph.

Vertex (of a triangle) Another name for the corners of a triangle.

Vertical line A line with undefined slope. All vertical lines are parallel to the y-axes.

Vertical-line test If *any* vertical line crosses a graph in more than one place, then the graph is not a function.

Vertical reflection An image in which each point of a graph is replaced with a point an equal distance on the other side of the x-axis. It looks as if the original graph has been turned upside-down.

Vertical translation Moving a graph up or down without changing the shape of the graph.

Volume The amount of space inside a three-dimensional object such as a box or a sphere.

X-axis The horizontal axis of the Cartesian coordinate system.

X-coordinate The first number in an ordered pair. It tells how far to the right or left of the origin a point is located.

X-intercept The x-coordinate of the point at which a line crosses the x-axis.

Y-axis The vertical axis of the Cartesian coordinate system.

Y-coordinate The second number in an ordered pair. It tells how far up or down from the origin a point is located.

Y-intercept The y-coordinate of the point at which a line crosses the y-axis.

Z

Zero exponent Any quantity (except zero) raised to the zero power always equals one.

Zero power The zero power of any number equals one. Example: $3^0 = 1$

Zero product property If two quantities' product is zero, then one or both of the quantities must be zero.

Zeros The values for x that make a polynomial equal zero, also called the roots of the polynomial.

Selected Answers

This section contains selected answers to the Practice and Problem Solving and Mixed Review sections.

Chapter 1 Integers and Vectors

Lesson 1.1, pages 4-7
Practice and Problem Solving
7. +25 minutes   **9.** –2340 feet
11. –8648 meters
13. +$36   **15.** +400 square feet
17. –1000 feet
Mixed Review
19. $3600

Lesson 1.2, pages 8-12
Practice and Problem Solving
7. –8 < –6   **9.** –5 < –4   **11.** –20 < –15
13. –9 < 0   **15.** 7   **17.** 0   **19.** 6   **21.** –25
23. No
Mixed Review
25. $487

Lesson 1.3, pages 13-18
Practice and Problem Solving
7. 0   **9.** –15   **11.** –25   **13.** –34   **15.** –32
17. 0   **19.** –100   **21.** –3 kilowatts   **23.** +2
Mixed Review
25. –86 < –28   **27.** 0 > –86   **29.** +4344 > 0

Lesson 1.4, pages 19-22
Practice and Problem Solving
3. 14   **5.** –1   **7.** 3   **9.** 32   **11.** –20
13. –18   **15.** –18   **17.** $897.66   **19.** $479.53
Mixed Review
21. 6.06 inches   **23.** 2.85 inches
25. 4.91 inches   **27.** –123 feet   **29.** +118 feet

Lesson 1.5, pages 24-28
Practice and Problem Solving

7. 2   **9.** –9   **11.** –7   **13.** –72   **15.** $8\frac{1}{3}$

17. $-4\frac{3}{4}$ **19.** –2; 3; –1; –3; 3
21. 17; –10; –11; 21; –17
Mixed Review
23. 11   **25.** $40
27. Copying 500 sheets costs less.

Lesson 1.6, pages 29-33
Practice and Problem Solving
5. –2   **7.** 11   **9.** 7   **11.** –2   **13.** –53
15. 55   **17.** –85   **19.** –66
Mixed Review
21. $40   **23.** 5217.625   **25.** 5203.5

Lesson 1.7, pages 34-39
Practice and Problem Solving
7.

9. Air speed is greater.
11. Air speed is less.
Mixed Review
13. –$360; +$141; +$317; –$2   **15.** 10; –1; –4;
–10; –16; 8; 0; –3   **17.** 38   **19.** 83

Chapter 2 Scientific Notation

Lesson 2.1, pages 4-8
Practice and Problem Solving
5. 3^2 **7.** $(-5)^2$ **9.** $(-3)^5$ **11.** 27   **13.** –1
15. 128   **17.** 1,000,000 watts   **19.** 50,000
21. 3^4 **23.** 4^5
Mixed Review
25. Cathy, Tim, Jose, Maria, Pat   **27.** –1 pound
29. –505 feet   **31.** –475 feet

Lesson 2.2, pages 9-12
Practice and Problem Solving
7. 0.00412 **9.** 0.1 **11.** 0.04 **13.** 1 **15.** 0.001

17. 3^{-4} **19.** 10^{-2} **21.** $\dfrac{1}{1000000000}$;

0.000000001
Mixed Practice
23. Volga < Mississippi

25. Amazon +2100 miles; Danube –124 miles; Ganges –340 miles; Mississippi +440 miles; Nile +2260 miles; Ohio –590 miles; St. Lawrence –1100 miles; Volga +294 miles

Lesson 2.3, pages 13-15
Practice and Problem Solving
5. 93×10^6 **7.** 55×10^8

9. 25×10^{-6} millimeters **11.** $\$4 \times 10^{12}$

13. $2 \times 10^4 + 3 \times 10^3 + 6 \times 10^2 + 8 \times 10^1 + 1 \times 10^0$

15. $7 \times 10^2 + 2 \times 10^1 + 6 \times 10^0 + 0 \times 10^{-1} + 5 \times 10^{-2}$
Mixed Review
17. –7 **19.** –9 **21.** –19 **23.** $0.35

Lesson 2.4, pages 16-21
Practice and Problem Solving
5. 5.67×10^1 **7.** 2.45×10^5 **9.** 9.3673×10^4

11. 4.5×10^7 **13.** 1.35737×10^2

15. 2.0×10^{19} molecules **17.** 6.5×10^{-6}

19. 0.00001 **21.** 5,870,000,000,000 miles

23. 0.0000003 meters
Mixed Review
25. 6 months **27.** 105 seconds

Lesson 2.5, pages 22-27
Practice and Problem Solving
7. 10^2 **9.** 10^{-3} **11.** 10^{-13} **13.** 10^0

15. 3.5×10^9 **17.** 3.18×10^{-8}

19. 18×10^{-1} meters; 1.8×10^{-2} meters; 1.8 meters

Mixed Review
21. 36°F **23.** 6×10^{3}°C

Lesson 2.6, pages 28-32
Practice and Problem Solving
5. 0.034 kilometers **7.** 58 millimeters

9. 0.350 seconds **11.** 800,000 milligrams

13. 0.083 meters **15.** 0.142 amps **17.** 10^{-12}
Mixed Review
19. –10°F **21.** –15°F

23. 1.392×10^6 kilometers **25.** 5.4×10^{-15}

Chapter 3 Using Formulas

Lesson 3.1, pages 4-8
Practice and Problem Solving
5. 65 **7.** 7 **9.** 2 **11.** –28 **13.** –100

15. 20 **17.** –22 **19.** 48 **21.** 4 **23.** 3

25. $106
Mixed Review
27. 4,700,000,000 years **29.** 3×10^{-2}

Lesson 3.2, pages 9-16
Practice and Problem Solving
5. yes **7.** yes **9.** no **11.** yes **13.** yes

15. 384 square millimeters **17.** 12 miles

19. 81 square feet **21.** 95°F **23.** same

25. $D = RT$ **27.** $T = \dfrac{D}{R}$ **29.** 15 miles
Mixed Review
31. Charlotte, Long Beach, Austin, New Orleans, Cleveland, Seattle

33. 2.76×10^{-8} centimeters **35.** 0.048 meters

Lesson 3.3, pages 17–20
Practice and Problem Solving
5. 12.57 meters; 12.57 square meters **7.** 50.27 centimeters; 201.06 square centimeters

9. 39.27 millimeters; 124.69 square millimeters

11. 38.48 square centimeters

13. 25.13 square meters **15.** 157.08 square feet

Mixed Review

17. Pacific Ocean: –12,925 feet; Atlantic Ocean: –11,730 feet **19.** 4.03×10^2 inches

Lesson 3.4, pages 21-24

Practice and Problem Solving

5. 15.625 cubic feet

7. about 113.1 cubic meters

9. about 2.61×10^{11} cubic miles

11. about 268.1 cubic centimeters

13. 3456 cubic inches

Mixed Review

15. On–Time **17.** –$295; $1655; –$695

19. +$665

Lesson 3.5, pages 25-28

Practice and Problem Solving

3. $3.60 **5.** $90.77 **7.** $1027.57

9. $2382.02

Mixed Review

11. Line B: $105,000 **13.** The order from left to right is B, C, D, A. **15.** 4.167×10^0

17. $135 **19.** 14,400 pounds

Chapter 4 Solving Equations

Lesson 4.1, pages 4-11

Practice and Problem Solving 4-11

7. –8 **9.** 8 **11.** $6\frac{2}{3}$ **13.** $-9\frac{2}{3}$ **15.** $-13\frac{1}{3}$

17. 3 **19.** $\frac{1}{2}$ **21.** –30 **23.** $\frac{V}{r} = i$

25. $\frac{i}{rt} = p$ **27.** 15 feet

Mixed Review

29. 2400 meters **31.** 0.52 liters

Lesson 4.2, pages 12-16

Practice and Problem Solving

5. $\frac{1}{4}$ **7.** $1\frac{1}{9}$ **9.** –84 **11.** 25 **13.** $7\frac{1}{3}$

15. –25 **17.** $A = LW$

Mixed Review

19. 5.6×10^{-5} **21.** 1.4395×10^2 **23.** The change in the temperature is –10°. **25.** 1 gallon of paint costs $8.50.

Lesson 4.3, pages 17-22

Practice and Problem Solving

5. 4 **7.** –40 **9.** 30 **11.** 25% **13.** 66.7%

15. 4.5% **17.** $0.60 **19.** 25% **21.** $54.25

23. $3000

Mixed Review

25. 2.478×10^{-4} **27.** –37

Lesson 4.4, pages 23-29

Practice and Problem Solving

5. 4 **7.** 63 **9.** –1 **11.** –2 **13.** –0.8

15. 3.2 **17.** 4.4 **19.** –12.7 **21.** –3.12

23. 24 workstations

Mixed Review

25. $73 **27.** 25 = yard line

29. 163 miles each day.

Lesson 4.5, pages 30-35

Practice and Problem Solving

3. 7 **5.** –6 **7.** –6 **9.** –12 **11.** –1.78

13. –2 **15.** $x = \frac{c - b}{a}$ **17.** 54 feet

19. 11 feet **21.** $\frac{P - 2W}{2} = L$ **23.** 96 **25.** $125

Mixed Review

27. 1.3312×10^3 **29.** about 9 centimeters

Lesson 4.6. pages 36-40

Practice and Problem Solving

5. $\frac{6}{7}$ **7.** $-3\frac{4}{7}$ **9.** $6 \neq 5$; no solution **11.** –8

13. $-2\frac{6}{7}$ **15.** $\frac{6}{41}$ **17.** $60 + x$ **19** 30 volts;

90 volts. **21.** $1.5(2\pi x) = 2\pi(x + 4)$ **23.** 70 42-inch; 280 38-inch **25.** 10 by 4; 8 by 6.

27. – 126, +111, –7, –40, +32

Chapter 5 Linear Functions

Lesson 5.1, pages 4-10

Practice and Problem Solving

5. $A(-4,0)$; $B(2,-3)$; $C(1,5)$; $D(-2,-2)$
7. Caddo $(-6,7)$; Limestone Gap $(-4,0)$; Ada $(4,1)$; Windom $(-6,-3)$ Hartshorne $(5,-4)$; Bug Tussle $(0,-6)$ **9.** four **11.** two

Mixed Review

13. true **15.** 10^{-5} kilometers **17.** about 132,889.37 centimeters **19.** 120 votes
21. 50% **23.** 1

Lesson 5.2, pages 11-16

Practice and Problem Solving

7. rectangle **17.** straight line
19.

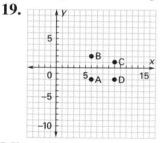

Mixed Review

21. 1.3986×10^7 **23.** 14 **25.** $351
27. -10 **29.** -12 **31.** 166.7

Lesson 5.3, pages 17-23

Practice and Problem Solving

7. $-\dfrac{1}{2}$ **9.** 5 **11.** $\dfrac{1}{5}$ **13.** $-\dfrac{4}{3}$ **15.** 2

17. $-\dfrac{1}{11}$ **19.** 0 **21.** 2.3 feet per mile

23. $\dfrac{1}{10}$

Mixed Review

25. 44.5 gallons per hour **27.** 30 feet **29.** 24

Lesson 5.4, pages 24-32

Practice and Problem Solving

7. slope $= -3$; y–intercept 5 **9.** slope $= -1$;
y-intercept 8 **11.** slope $= \dfrac{5}{2}$; y-intercept $\dfrac{3}{2}$

13. slope $= \dfrac{5}{2}$; y-intercept $\dfrac{3}{2}$ **15.** 20; 10

17. 6 **19.** 0.05; 2

Mixed Review

21. $8333.33 **23.** 7 weeks

Lesson 5.5, pages 33-37

Practice and Problem Solving

5. $\dfrac{5}{3}$; 5 **7.** 8; -4 **9.** $\dfrac{5}{3}$; -2.5 **11.** $-\dfrac{5}{2}$; 10

13. 6; -4 **15.** -8; $\dfrac{8}{3}$ **17.** $d = 4t + 2$

19. $A = 6 + 8h$

Mixed Review

21. 10^{-5} **23.** 10^8 **25.** 306,796.16 square centimeters **27.** 31

Lesson 5.6, pages 38-44

Practice and Problem Solving

5. $y = x + 1$ **7.** $y = \dfrac{3}{2}x + \dfrac{19}{2}$ **9.** $y = x + 5$

11. 65 miles per hour **13.** $5x + 3y = 65$
15. 10 adult tickets and 5 child's tickets

17. $d = \dfrac{1}{500}F$ **19.** 1500 pounds

Mixed Review

21. -12 pounds **23.** $-9°$ **25.** 200 **27.** 120 casings

Lesson 5.7, pages 45-47

Practice and Problem Solving

7. They are equal. **9.** They are perpendicular.

Mixed Review

13. 8 **15.** -33 **17.** 1 **19.** $y = 3x - 2$
21. $2,743.75 **23.** $2,100,000

Chapter 6 Nonlinear Functions

Lesson 6.1, pages 4-9

Practice and Problem Solving

7. $y = 2x - 2$; 8, 10, 12 **9.** $y = 5x + 1$; 26, 31, 36
11. $y = \frac{1}{2}x + \frac{1}{2}$; 3, $\frac{7}{2}$, 4 **13.** -7 **15.** 2 **17.** 0
19. $y = 8x + 15$; slope 8; y-intercept 15

Mixed Review

21. 8.5×10^{-5} **23.** $5\frac{4}{5}$ **25.** -2 **27.** 5 **29.** 61.5%

Lesson 6.2, pages 10-17

Practice and Problem Solving

7. vertical stretch by a factor of 2 **9.** reflection across the x-axis, vertical move up 1 unit, vertical stretch by a factor of 3 **11.** stretch by a factor of 4, vertical move up 6 units **13.** stretch by a factor of 3, vertical move up 2.5 units

Mixed Review

15. $216.08 **17.** 10^{-5} **19.** 45 **21.** 1600 **23.** 20
25. 20°; 60°; 100° **27.** $-\frac{3}{2}$; 6

Lesson 6.3, pages 18-24

Practice and Problem Solving

7. 100 watts **9.** 1600 watts **11.** -4 **13.** -10
15. vertical move 1 unit down **17.** horizontal move 2 units left, vertical move 1 unit up, vertical stretch by a factor of 3 **19.** vertical stretch by a factor of π

Mixed Review

21. 4AM **23.** -7°F **25.** 18,75% **27.** $3\frac{3}{4}$

29.

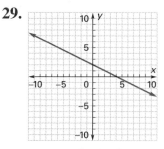

Lesson 6.4, pages 25-32

Practice and Problem Solving

7. -50 **9.** about 24.8 **11.** 4.90 and -4.90
13. 5 meters **15.** 5 feet **17.** 229.04 inches
19. about 10.63 miles **21.** 91 feet

Mixed Review

23. -9 **25.** -4; -9 **27.** about 34.3 gallons
29. $311.52

Lesson 6.5, pages 33-38

Practice and Problem Solving

7.

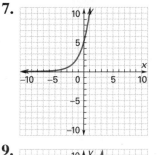

9.

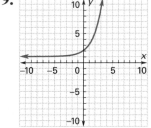

11.

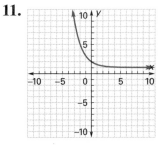

13. Multiply by a negative number.
15. $\frac{1}{2}, \frac{1}{4}, \frac{1}{8}, \frac{1}{16}$
17. $b = p(1.06)^x$ $1338.22 **19.** $14,327.86

Mixed Review

21. $4166.67 **23.** -6 **25.** $y = -\frac{2}{3}x + \frac{11}{3}$

Lesson 6.6, pages 39-44

Practice and Problem Solving
9. The graph moves away from the origin.
13. The graph moves away from the origin and reflects over the *x*-axis. **17.** The graph moves closer to the *x*- and *y*-axes. **19.** all the integers
21. The graph looks like a staircase.
23. 3 stamps; You can only buy a whole number of stamps.

Mixed Practice
25. under **27.** 5 **29.** about 16.97 feet
31. 19 inches

Chapter 7 Statistics and Probability

Lesson 7.1, pages 4-10

7. 10 **9.** 70 **11.** 75.8 **13.** 2.4
15. 262 **17.** 276.3
19. 60 kilograms **21.** 59.4 kilograms
23. It will not affect the mode—60 still occurs the most. It will not affect the median—60 is still the middle term.
25. $5200 **27.** $1127.45

Lesson 7.2, pages 11-17

5.

Class	Tally	Frequency
0 - 9	////	4
10 - 19	//////	6
20 - 29	///////	7
30 - 39	///	3
	$n = 20$	

7. 31 **9.** 20
11.

Classes	Tally	Frequency	(c)(f)
50	//	2	100
60	/	1	60
70	//	2	140
75	///	3	225
80	//////	6	480
85	/////	5	425
90	/	1	90
95	///	3	285
100	//	2	200
		25	2005

13.

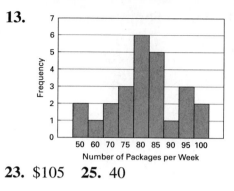

Number of Packages per Week

23. $105 **25.** 40

27.

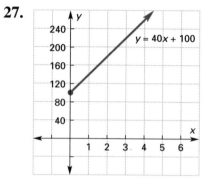

$y = 40x + 100$

Lesson 7.3, pages 18-23

5. The data shows a negative correlation.
7. One example is $y = -\dfrac{3}{5}x + 26$
9. The answer depends on the line drawn.
11. 80 **13.** 2 students; 9 hours and 6 hours

15.

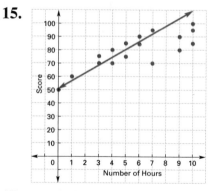

Number of Hours

17. $y = 2.5x + 65$
19.

January	$-$45.00
February	$-$75.00
March	$+$40.00
April	$+$ 5.00
May	$+$20.00

21. about 216.8 square feet **23.** 19.5

Lesson 7.4, pages 24-30

5. 25% **7.** 55% **9.** 88%
11. about 41.8% **13.** 9.1% **15.** The answer depends on the data supplied. **17.** 50%
19. (h, h, h), (h, h, t), (h, t, t), (t, t, t), (t, t, h), (t, h, h), (t, h, t), (h, t, h)
21. 3 to 5, or 60% **23.** 80°
25. $-20, -19$, and -18 **27.** (2, 0)

Lesson 7.5, pages 31-38

3. Answers depend on individual results.
5. about 50 each **7.** (T, H) represents tails for the nickel and heads for the penny.
9. (T, H), (H, T), (H, H), (T, T)
11-19. Answers depend on individual results.
21. 24 miles **23.** 21 defective parts
25. $y = 25,000 - 750x$, where x represents the number of seconds the plane descends and y represents the height at any time x.

Lesson 7.6, pages 39-45

5. $\frac{1}{36}$ **7.** $\frac{1}{4}$ **9.** $\frac{11}{36}$ **11.** $\frac{1}{36}$ **13.** $\frac{15}{36}$
15. $\frac{1}{6}$ **17.** $\frac{5}{6}$ **19.** $\frac{36}{55}$ **21.** $\frac{17}{20}$

23. $r = \frac{i}{pt}$ **25.** In April 800 phones were made, in May 1000 phones were made, and in June 2000 phones were made.

Lesson 7.7, pages 46-51

3. 24 **5.** 12 **7.** 10 **9.** 60 **11.** 45
13. 30 **15.** 210 **17.** $1150 **19.** $50
21.

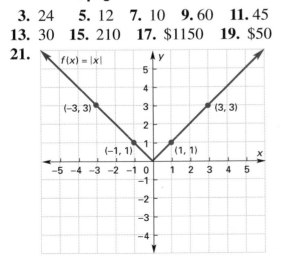

23.

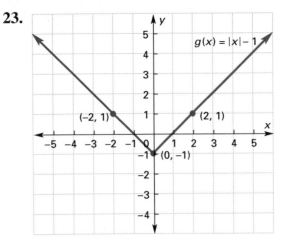

Lesson 7.8, pages 52-57

3. $\frac{1}{8}$ **5.** $\frac{17}{24}$ **7.** $\frac{19}{24}$ **9.** $\frac{3}{8}$ **11.** $\frac{1}{8}$
13. $\frac{5}{33}$ **15.** $\frac{25}{144}$ **17.** $\frac{1}{16}$ **19.** $\frac{9}{16}$
21. 0 **23.** $\frac{1}{22}$ **25.** $\frac{1}{11}$ **27.** $\frac{1}{12}$
29. 11 lamps **31.** $y = -2400x + 26,000$; The car is 4 years old.

Chapter 8 Systems of Equations

Lesson 8.1, pages 4-10

5.

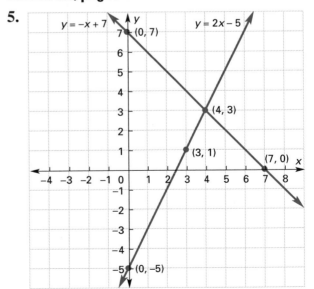

7.

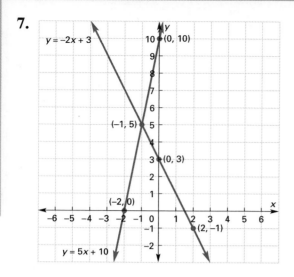

13.

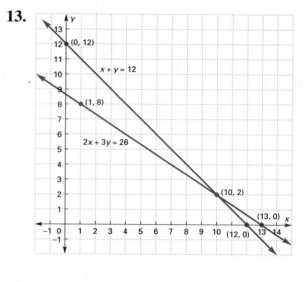

9.

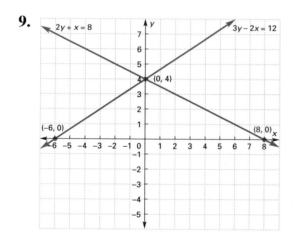

15.

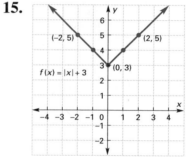

17. $y \geq 3$ **19.** 4, 5, 6, 6, 6, 7, 7, 7, 8, 8
21. 6.5

Lesson 8.2, pages 11-17

7. consistent and independent
9. consistent and dependent
11. inconsistent   **13.** consistent and independent   **15.** inconsistent
17. inconsistent   **19.** independent
21. A. City messenger $+2.0$ miles
    B. Dentist -1.2 miles
    C. Hospital nurse $+3.3$ miles
    D. Hotel employee $+1.2$ miles
    E. Secretary $+0.2$ miles
    F. Teacher -0.3 miles
23. about 4.002389×10^7 kilometers
25. about 5.099×10^{14} square kilometers
27. The circumference is increased by a factor of 2. The surface area is increased by a factor of 4.

11.

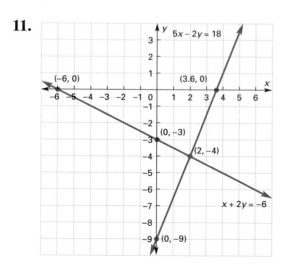

Lesson 8.3, pages 18-24

5. dependent **7.** (11, 1) **9.** (8, 0)
11. (3, −2) **13.** The 3-subject notebook cost is $5.00. The single-subject notebook cost is $2.50. **15.** Joe does 37 haircuts per week, and Nita does 58 haircuts per week.
17. 360° **19.** 45°, 65°, and 70° **21.** $\frac{1}{2}$
23. $\frac{7}{10}$ **25.** $\frac{1}{4}$ **27.** $\frac{9}{100}$

Lesson 8.4, pages 25-31

5. (0, 0) **7.** (3, 3) **9.** (2, 0.5) **11.** (3, 1)
13. One delivery will contain 13 pallets, the other will contain 10 pallets.
15. The flat rate for the first 10 words is $10, each additional word is $0.65.
17. One truck will hold 2.87 tons of sand.
19. $N = \dfrac{x}{(215)(2.87)}$ **21.** 2 **23.** $0.5 \le x \le 5$
25. As x becomes very small, $g(x)$ becomes very large.

Lesson 8.5, pages 32-37

5. (−11, −6) **7.** (−1, −2) **9.** (10, 14)
11. (0, −2) **13.** dependent **15.** The sprayer costs $5 per hour to rent. The generator costs $10 per hour to rent. **17.** One compartment holds 27 pieces of baggage, and the other compartment holds 68 pieces of baggage.
19. 9.5×10^{-8} centimeters **21.** 3.0×10^{10} centimeters per second **23.** 1.5 square feet
25. One machine part weighs 1.8 pounds, the other part weighs 6.8 pounds.

Lesson 8.6, pages 38-43

3. $\left(-\frac{2}{3}, -\frac{5}{3}\right)$ **5.** (0, 4) **7.** $\left(\frac{5}{3}, \frac{5}{3}\right)$ **9.** (−3, 2)
11. $\left(\frac{2}{3}, -3\right)$ **13.** The speed of the truck is 50 miles per hour. The speed of the car is 65 miles per hour. **15.** 3 **17.** 7.6 **19.** 130.5 rentals

21. B = Boy; G = Girl sample consists of (B, B, B, B), (B, B, B, G), (B, B, G, G), (B, G, G, G), (G, G, G, G), (G, G, G, B), (G, G, B, B), (G, B, B, B), (G, B, B, G), (G, B, G, B), (B, G, B, G), (B, G, G, B), (G, B, G, G), (G, B, G, G), (B, G, B, B), (B, B, G, B)
23. $\frac{3}{8}$ **25.** $\frac{7}{8}$

Chapter 9 Inequalities

Lesson 9.1, pages 4-8

5. $x < 5$ **7.** $x \le 0$ **9.** $2 < x \le 7$
11.
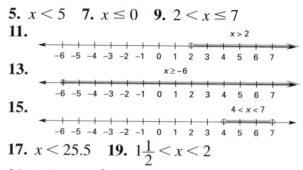
13.

15.

17. $x < 25.5$ **19.** $1\frac{1}{2} < x < 2$
21. 1.24×10^6 Newtons per square meter
23.
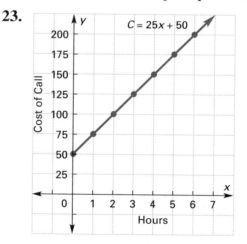

Lesson 9.2, pages 9-13

5.
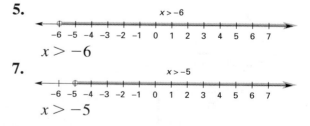
$x > -6$
7.

$x > -5$

9.

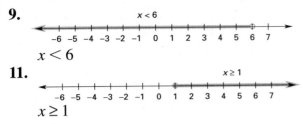

$x < 6$

11.

$x \geq 1$

13. $x + 35 \geq 150$; Tai needs to save at least $115 more to purchase the stereo.

15. $x - 15 < 50$; The larger number is any number less than 65.

17. $x + 210 \leq 350$; The gift shop can spend $140 or less in advertising.

19. 44 **21.** 41 **23.** $\dfrac{1}{23} \approx 4.3\%$

25. about 0.01%

Lesson 9.3, pages 14-19

3.

$x < 2$

5.

$x \geq -10$

7.

$x < -16$

9.

$x \geq -2$

11.

$x > -2$

13. $x > 30\%\ (75)$; at least $22.50

15. $8x \leq 1000$; $125 **17.** $86 + 88 + x \geq 270$; 96 points **19.** $20 - 1.39x \leq 5$; less than or equal to 10.8 gallons **21.** 3

23. $y = 3x + 11$ **25.** 0

Lesson 9.4, pages 20-26

5.

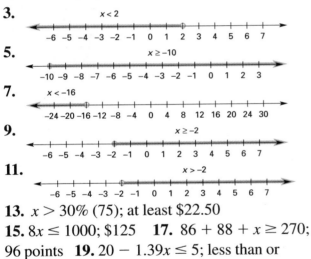

$-1.5 \leq x < 0.8$

7.

$x < -3$ or $x > 1$

9.

$x < \dfrac{3}{2}$

11.

$-1 < x < 2$

13. $x > 72$; above $72°$
15. $145 < 2x + 5 < 155$; between $70 and $75
17. 152.4 pounds **19.** positive correlation
21. $(\dfrac{1}{2}, 2)$ **23.** $x = y + 20,\ x + y = 100$;
Mario saved $40 and Nancy saved $60.
25. $15x + 12y = 34.20,\ 10x + 7y = 21.45$;
One gallon of 87 octane gas costs $1.20, and one gallon of 92 octane gas costs $1.35.

Lesson 9.5, pages 27-30

5.

$x < -4$ or $x > 4$

7.

$x \leq 3$ or $x \geq 12$

9.

$x < -5$ or $x > \dfrac{5}{2}$

11.

$x \leq -\dfrac{5}{3}$ or $x \geq 5$

13. $|x - 120| > 5$; over 125 minutes or under 115 minutes **15.** $|x - 30| > 3$; less than 27 psi and greater than 33 psi **17.** 8.75×10^{6} cubic feet **19.** 875 fans

21. $9 - 0.25(9) = x$; $6.75 **23.** $x + (x + 1) + (x + 2) = 63$; 20, 21, and 22

Lesson 9.6, pages 31-39

5.

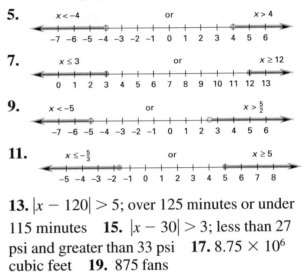

$y > -x - 2$

7.

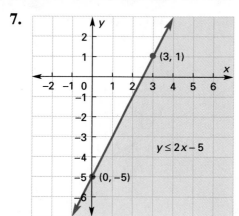

$y \leq 2x - 5$

9.

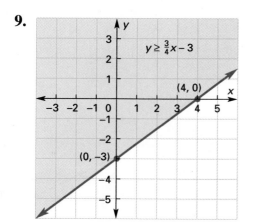

$y \geq \frac{3}{4}x - 3$

11.

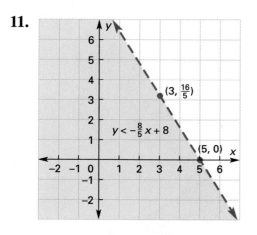

$y < -\frac{8}{5}x + 8$

13. $2L + 2W > 24$ **15.** 10 feet, 8 feet; 11 feet, 9 feet; 12 feet, 10 feet

17.

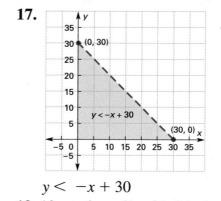

$y < -x + 30$

$$y < -x + 30$$

19. $10x + 6y \leq 60$ **21.** 0 jackets on the faster machine, 10 jackets on the slower machine. Two more solutions that satisfy the constraints are (3, 2) and (4, 3). **23.** The slope of line AB is -1. The slope of line MN is 1.

25. $y = x + 3$

27.

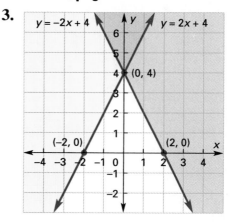

29. $y = -x$

Lesson 9.7, pages 40-44

3.

5.

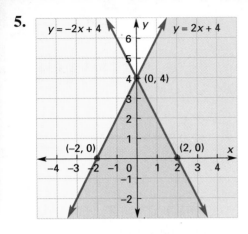

7.

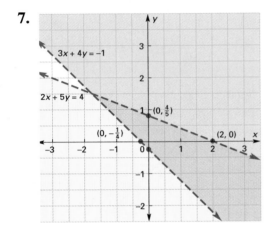

9.

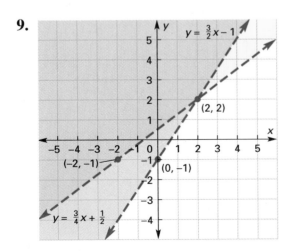

11.

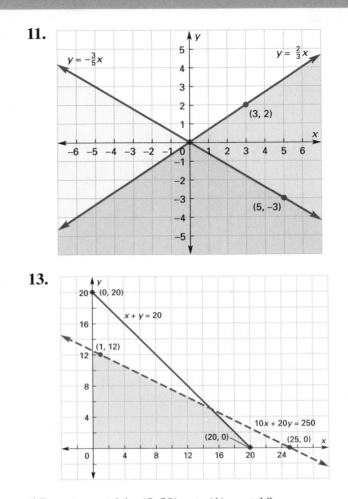

13.

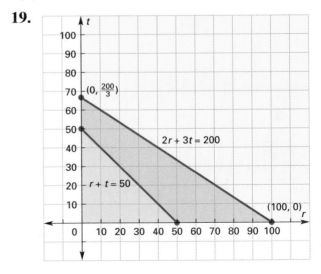

15. $x + y \le 14; \ (0.50)x + (1)y < 10$

17. To maximize profits, the corner point $(8, 6)$ should be used. The profit at this point will be $3(8) + 4(6) = \$48$.

19.

Between 0 and 66 tennis shoes can be produced, and between 0 and 100 running shoes can be produced. Any combination in the solution will work.

21. $\frac{3}{10} = 30\%$ **23.** $\frac{9}{245} \approx$ about 3.67%
25. $\frac{19}{50} = 38\%$ **27.** Tim has invested $42.
Shaneka has invested $54.

Chapter 10 Polynomials and Factors
Lesson 10.1, pages 4-9

5. binomial **7.** binomial **9.** trinomial **11.** $5x + 8$ **13.** $3x^2 - x - 3$
15. $x^2 + 5x - 12$ **17.** $4x^2$ **19.** $x + 3$
21. $x^2 + 7x + 4$ **23.** $4x^2 - 8x$
25. $6x^2 - 2x + 7$ **27.** $r = \frac{d}{t}$
29. 9 **31.** -2 **33.** 3 CDs

Lesson 10.2, pages 10-15

7. $1 \cdot 45; 3 \cdot 15; 5 \cdot 9$ **9.** $1 \cdot 32; 2 \cdot 16; 4 \cdot 8$
11. $1 \cdot 26; 2 \cdot 13$ **13.** $2 \cdot 3 \cdot 7$ **15.** $2 \cdot 3 \cdot 3$
17. $2 \cdot 2 \cdot 7$ **19.** 6 **21.** 12 **23.** 13 **25.** 6
27. $(-1, -3)$ **29.** The pieces are 32.5 inches and 35.5 inches.

Lesson 10.3, pages 16-20

3. $ab + 3a$ **5.** $7x^2 + 35xy$
7. $8x^4 + 14x^2 - 10x$ **9.** $x^2 + 2y^2$
11. $x^3y - 2xy^3$ **13.** $V = 3\pi r^3$ **15.** $y(3x - 5y)$
17. $cx^2(x + 2x^2)$ **19.** $\frac{13}{20}$
21. all whole numbers less than 7
23. all whole numbers less than 4
25. $\frac{3}{500} = \frac{12}{n}$; $2000

Lesson 10.4, pages 21-26

5. a^7 **7.** $10x^7$ **9.** $\frac{8x^3}{27}$ **11.** $6x^5 + 8x^3$
13. $4s$ **15.** $C = \frac{Dx}{48}$
17. $2w + 2(w + 4) = 36$; 77 square inches

19.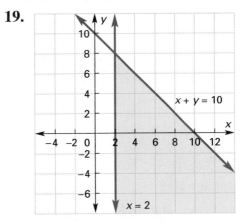

Lesson 10.5, pages 27-31

5. $\frac{5x}{4}$ **7.** $\frac{15 - 8x}{10x}$ **9.** $\frac{x^2 + 2}{2x}$ **11.** $\frac{-x + 8}{x^2}$
13. $\frac{3x^2 - x + 4}{2x^2}$ **15.** 10 hours
17. 75 hours **19.** 24, 38, and 29 students
21. The shirts are $35, and the shorts are $22.

Lesson 10.6, pages 32-38

5. $x^2 + 7x + 12$ **7.** $2x^2 + 5x + 2$
9. $x^2 + 7x + 10$ **11.** $3x^2 + 10x + 3$
13. $10x^2 - 13x - 3$ **15.** $15x^2 - 11x + 2$
17. -9 **19.** $50 **21.** $2L + 2W = 19$ and $4W - 3 = L$; The width is 2.5 centimeters. The length is 7 centimeters.

Lesson 10.7, pages 39-41

3. $(a + 1)(a + 7)$ **5.** $(x - 1)(x - 1)$
7. $(a + 3)(a - 2)$ **9.** $(w - 2)(w - 4)$
11. $(3x + 5)(2x - 3)$ **13.** 18 square feet
15. ± 5 **17.** -12 **19.** $\frac{1}{273}$

Lesson 10.8, pages 42-46

5. $(x + 2)(x + 3)$ **7.** $(x - 2)(x - 3)$
9. $(x - 1)(x + 6)$ **11.** $(x + 1)(5x + 2)$
13. $(3x + 5)(x + 3)$ **15.** $(2x + 5)(3x - 4)$
17. $x + 15 = 18$; $3 **19.** $25x + 10x = 280$; 8 feet **21.** $(-4, 1)$

Lesson 10.9, pages 47-52

7. $x^2 - 9$ **9.** $x^2 - y^2$ **11.** $x^2 - 9$
13. $4y^2 + 12xy + 9$ **15.** $(x + 5)(x - 5)$
17. $(3 - 4y)(3 + 4y)$ **19.** $(y - 9)^2$
21. $(y - 1)^2$ **23.** $(x - a)^2$ **25.** $b = \dfrac{P - 2a}{2}$
27. $y = \dfrac{1}{5}x + \dfrac{3}{5}$ **29.** $y = -3$ **31.** $(-\dfrac{1}{2}, -2)$

Chapter 11 Quadratic Functions
Lesson 11.1, pages 4-9

7. $y = 3x^2 + 5$, translation 5 units up, shrink by a factor of 3, vertex $(0, 5)$

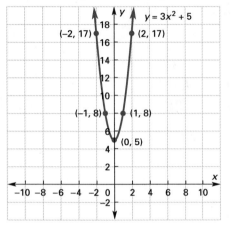

9. $y = (x - 2)^2 + 1$, translation 2 units right and 1 unit up, vertex $(2, 1)$

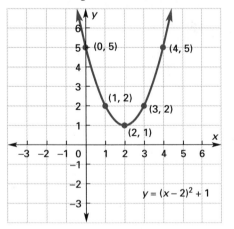

11. $y = -\dfrac{1}{2}(x + 1)^2 + 2$, translation 1 unit left and 2 units up, reflection across the x-axis, shrink by a factor of $\dfrac{1}{2}$, vertex $(-1, 2)$

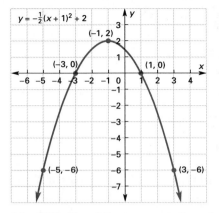

13. $78.41 **15.** $x = y + 2$ and $4x + 4y = 40$; 4 inches and 6 inches
17. $(85 + 90 + 93 + x) \div 4 \geq 90$; 92
19. $y = \dfrac{3}{2}x + \dfrac{5}{2}$

Lesson 11.2, pages 10-16

7. $-2; -4; V(-3, -1)$ **9.** $-1; V(-1,1)$
11. $2; 5; V(-\dfrac{3}{2}, -\dfrac{13}{4})$ **13.** $5; -1; V(2, -9)$
15. $1; 3; V(2, -1)$ **17.** $4.98; 2.56;$
$V(1.7, -4.37)$ **19.** $900 **21.** 420 cubic centimeters **23.** 2 **25.** $-1; 6$ **27.** $(0, 6)$

Lesson 11.3, pages 17-22

3. $-2; -3$ **5.** $-3; 1$ **7.** $-2; -1$ **9.** $\dfrac{3}{2}; -4$
11. $-8; 2$ **13.** $-\dfrac{3}{2}; \dfrac{4}{3}$
15. $5^2 = 4s + 12$; 6 meters **17.** 6×10^{25}
19. $2(W + 1) + 2W = 180$
21. $2L + 2W = 180; L - W = 1$
23. the same **25.** 0

Lesson 11.4, pages 23-29

5. $y = -(x - 4)^2 + 16$; maximum $= 16$; $V(4, 16); x = 4$
7. $y = -(x - \dfrac{3}{2})^2 + \dfrac{17}{4}$; maximum $= \dfrac{17}{4}$; $V(\dfrac{3}{2}, \dfrac{17}{4}); x = \dfrac{3}{2}$
9. $y = -(x + \dfrac{7}{2})^2 + \dfrac{89}{4}$; maximum $= \dfrac{89}{4}$; $V(\dfrac{7}{2}, \dfrac{89}{4}); x = -\dfrac{7}{2}$
11. $y = -(x + 2)^2 - 8$; maximum $= -8$; $V(-2, -8); x = -2$ **13.** 8%

15. $4x + 2y = 2880$ and $x + y = 960$; 480 hammers

Lesson 11.5, pages 30-33

5. $-2; 6$ **7.** $-1.32; 5.32$ **9.** $0.8; -8.8$
11. $-1.3; 2.3$ **13.** $-0.54; 5.54$
15. $-4.24; 0.24$ **17.** $0; 8$ **19.** 10 meters by 12 meters **21.** $1,050,000 **23.** $\frac{1}{8}$ **25.** $\frac{1}{2}$

Lesson 11.6, pages 34-41

3. 13 **5.** 0 **7.** $-3; 7$ **9.** none **11.** $-1.1; 0.22$ **13.** none **15.** 7 meters by 12 meters
17. 24 inches and 36 inches **19.** 328 sensors
21. 7.75 seconds **23.** 11 **25.** 18
27. 19

Chapter 12 Right Triangle Relationships
Lesson 12.1, pages 4-11

3. $1; 12$ **5.** $17\frac{1}{3}; 39$ **7.** 144 yards
9. 12.5 inches; 18.75 inches **11.** 10 feet
13. undefined **15.** $y = -2$
17.

$x \leq 3$

‹———+———+———+———+———+———+———•———+———+———+———+———+———+———+———›
       -6 -5 -4 -3 -2 -1   0   1   2   3   4   5   6   7

19. $10y - y > 16$; at least 17.7 feet

Lesson 12.2, pages 12-17

3. no **5.** yes **7.** no **9.** 21.9 yards
11. 50.3 feet **13.** 5 **15.** 17
17. $(AB)^2 + (AC)^2 = (BC)^2$; $160 = 160$
19. $216 **21.** 2 gallons; $9

Lesson 12.3, pages 18-22

3. $2\sqrt{3}$ **5.** $10\sqrt{2}$ **7.** $\frac{\sqrt{6}}{3}$ **9.** $3\sqrt{2}$

11. $r = \sqrt{\dfrac{A}{\pi}}$ **13.** $20\sqrt{15}$ feet per second

15.+17.

$y=\sqrt{x}$		$y=\sqrt{2x}$		$y=\sqrt{\frac{1}{2}x}$	
x	y	x	y	x	y
0	0	0	0	0	0
1	1	1	$\sqrt{2}$	1	$\frac{\sqrt{2}}{2}$
4	2	4	$2\sqrt{2}$	4	$\sqrt{2}$
9	3	9	$3\sqrt{2}$	9	$\frac{3}{2}\sqrt{2}$
16	4	16	$4\sqrt{2}$	16	$2\sqrt{2}$

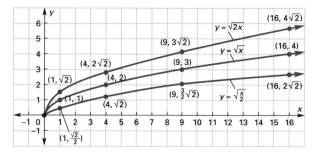

19. $2L + 2(L - 25) = 250$; 75 by 50 feet
21. $x + y = 24$ and $y = 2x + 6$; 18 feet; 6 feet

Lesson 12.4, pages 23-32

5. a, c, and d **7.** $10\sqrt{2}$ feet **9.** $8\sqrt{2}$ inches
11. 12 feet; $6\sqrt{3}$ feet **13.** 10 inches; $10\sqrt{3}$ inches **15.** 180 inches **17.** about 127 feet
19. 4 trucks **21.** 1200 **23.** $x + y = 30$ and $y = x + 6$; 18 bolts
25. $n^2 + (n + 2)^2 = (n + 1) + 155$; 8 and 10

Lesson 12.5, pages 33-44

5. $\sin A = \frac{4}{5}$, $\sin B = \frac{3}{5}$; $\cos A = \frac{3}{5}$, $\cos B \frac{4}{5}$; $\tan A = \frac{4}{3}$, $\tan B = \frac{3}{4}$ **7.** $\sin A = \frac{\sqrt{2}}{2}$, $\sin B = \frac{\sqrt{2}}{2}$; $\cos A = \frac{\sqrt{2}}{2}$, $\cos B = \frac{\sqrt{2}}{2}$; $\tan A = 1$, $\tan B = 1$ **9.** 0.707
11. $\sin 35° = \frac{40}{x}$; about 69.74 feet
13. about 47.67 feet
15. $\tan 87° = \frac{x}{55}$; about 1049 feet
17. $-\frac{11}{2}$
19.

$x + y = 1$; $2x - y = 2$; (0, 1); (1, 0); (2, 2); Solution set

21. -4

Index

Credits

Chapter 7

2: Archive Photos. 3: Rhoda Sidney/PhotoEdit. 4: Lawrence Migdale/Stock Boston. 5: Alan Brown/Photonics Graphics. 9: Hewlett-Packard. 11: Aaron Chang/Stock Market. 15: David Parker/Photo Researchers, Inc. 17: Coco McCoy/Rainbow. 18: Matthew McVay/Stock Boston. 20: Michael Newman/PhotoEdit. 23: David M. Grossman/Photo Researchers, Inc. 24: George Goodwin/Picture Cube. 28: Loren Santow/Tony Stone Images. 31: Hill/The Image Works. 38: © 1995 PhotoDisc, Inc. 41: Daemmrich/The Image Works. 42: © 1995 PhotoDisc, Inc. 44: COREL. 46: Nancy Sheehan/PhotoEdit. 51: Mary Kate Denny/PhotoEdit. 57: Ford Aerospace and Communications Corporation. 61: CORD. 64: David R. Frazier/Photo Researchers, Inc. 67: Nathan Benn/Stock Boston. 69: Jeff Greenberg/Photo Researchers, Inc. 72: © 1995 PhotoDisc, Inc. 73: Cathy Ferris. 75: Charlie Westerman/Liaison International. 76: Bachman/Stock Boston.

Chapter 8

2: COREL. 3: © 1995 PhotoDisc, Inc. 4: Jon Riley/Tony Stone Images. 7: © 1995 PhotoDisc, Inc. 9: Richard Clintsman/Tony Stone Images. 9: © 1995 PhotoDisc, Inc. 12: David Young Wolff/PhotoEdit. 16: Courtesy of Delta Airlines. 18: John Coletti/The Picture Cube. 22: Tommaso Guicciardini/Photo Researchers, Inc. 23: Miro Vintoniv/The Picture Cube. 25: Jim Pickerell/Tony Stone Images. 28: Lien/Nibauer/Liaison International. 30: © 1995 PhotoDisc, Inc. 31: COREL. 33: Bob Daemmrich/Stock Boston. 36: Jeff Greenberg/Photo Researchers, Inc. 38: Andrew Errington/Tony Stone Images. 41: Tom Pantages. 42: © 1995 PhotoDisc. 44: CORD. 45: CORD. 49: © 1995 PhotoDisc, Inc. 51: Mark Burnett/Photo Researchers, Inc. 53: © 1995 PhotoDisc Inc. 55: top, David Shopper/Stock Boston; bottom, © 1995 PhotoDisc, Inc. 58: Tony Stone Images. 60: Stanley Rowin/The Picture Cube.

Chapter 9

2: Eurotunnel/QA Photos 3: Eurotunnel/QA Photos. 6: William Taufic/The Stock Market. 8: top, © 1995 PhotoDisc, Inc.; bottom, Stephen McBrady/PhotoEdit. 12: center, Nik Kleinberg/Stock Boston. 14: John Henley/The Stock Market. 16: Frank Siteman/Stock Boston. 17: Jeff Greenberg/PhotoEdit. 18: Richard Pasley/Stock Boston. 20: Jeff Persons/Stock Boston. 23: Dennis MacDonald/PhotoEdit. 25: Michele Bridwell/PhotoEdit. 27: © 1995 PhotoDisc, Inc. 29: © 1995 PhotoDisc, Inc. 31: © 1995 PhotoDisc, Inc. 35: Mark Gibson. 39: The Image Works. 40: Tom Pantages. 41: D. Chidester/The Image Works. 43: Henry Horenstein/Stock Boston. 52: David Carriere/Tony Stone Images. 54: Ed Lallo/Liaison International. 57: top, Seth Resnick/Liaison International; bottom, Mark Segal/Tony Stone Images. 59: Stuart McCall/Tony Stone Images. 60: Tony Stone Images. 62: Scott Slobodian/Tony Stone Images.

Chapter 10

2: © 1995 PhotoDisc, Inc. 3: © 1995 PhotoDisc, Inc. 4: Tom Carroll. 7: Phyllis Picardi/Stock Boston. 9: Granitsas/The Image Works. 15: Phil Degginger/Tony Stone Images. 17: Courtesy of Drexel University. 20: Eric Brissaud/Liaison International. 23: Tom Pantages. 25: Charles Gupton/Stock Boston. 27: Robert Reichert/Liaison International. 31: bottom, Pedrick/The Image Works. 32: David Young-Wolff/PhotoEdit. 38: Lee Snider/The Image Works. 39: CORD. 45: Robert Ginn/The Picture Cube. 46: Mark Gibson. 52: Matthew Borkoski/Stock Boston. 55: CORD. 58: Hewlett-Packard. 59: Jane Lotta/Photo Researchers, Inc. 61: Robert Fried/Stock Boston. 62: Miro Vintoniv/The Picture Cube.

Chapter 11

2: Tom Carroll. 3: Liaison International. 5: Bob Daemmrich/Stock Boston. 9: Reinstein/The Image Works. 10: Tom Carroll. 16: William Johnson/Stock Boston. 19: Esbin-Anderson/The Image Works. 21: Cary Wolinsky/Stock Boston. 23: Gregg Grosse. 27: Bob Daemmrich/The Image Works. 29: Sheila Nardulli/Liaison International. 31: Addison Geary/Stock Boston. 32: Courtesy of Tennessee Photo Services. 33: © 1995 PhotoDisc, Inc. 36: © 1995 PhotoDisc, Inc. 39: © 1995 PhotoDisc, Inc. 40: © TRW Inc. 42: CORD. 48: CORD. 50: © 1995 PhotoDisc, Inc. 52: top, Lynn Stone/Natural Selection; bottom, Giboux/Liaison International. 53: Oil Spill Public Information Center. 55: William Taufic/The Stock Market. 58: James Mejuto. 63: © 1995 PhotoDisc, Inc.

Chapter 12

2: COREL. 3: COREL. 6: COREL. 8: Blair Seitz/Photo Researchers, Inc. 11: Navy photo by Gerry Winey. 15: COREL. 16: Dick Davis/Photo Researchers, Inc. 17: Thomas Braise/The Stock Market. 18: Dennis MacDonald/PhotoEdit. 21: © 1995 PhotoDisc, Inc. 25: Tony Arruza/The Image Works. 29: Robert Rathe/Stock Boston. 31: © 1995 PhotoDisc, Inc. 33: © 1995 PhotoDisc, Inc. 38: © 1995 PhotoDisc, Inc. 40: © 1995 PhotoDisc, Inc. 42: Collins/The Image Works. 48: CORD. 49: CORD. 52: © 1995 PhotoDisc, Inc. 54: © 1995 PhotoDisc, Inc. 55: © 1995 PhotoDisc, Inc. 60: COREL. 62: Mark Gibson. 64: Tim Lynch/Stock Boston.